Synthetic Biology

Synthetic Biology
Tools and Applications

Edited by

Huimin Zhao

University of Illinois at Urbana-Champaign
Urbana, IL, USA

AMSTERDAM • BOSTON • HEIDELBERG • LONDON
NEW YORK • OXFORD • PARIS • SAN DIEGO
SAN FRANCISCO • SINGAPORE • SYDNEY • TOKYO
Academic Press is an imprint of Elsevier

Academic Press is an imprint of Elsevier
32 Jamestown Road, London NW1 7BY, UK
225 Wyman Street, Waltham, MA 02451, USA
525 B Street, Suite 1800, San Diego, CA 92101-4495, USA

Notice
No responsibility is assumed by the publisher for any injury and/or damage to persons
or property as a matter of products liability, negligence or otherwise, or from any use or
operation of any methods, products, instructions or ideas contained in the material herein.
Because of rapid advances in the medical sciences, in particular, independent verification of diagnoses
and drug dosages should be made

British Library Cataloguing-in-Publication Data
A catalogue record for this book is available from the British Library

Library of Congress Cataloging-in-Publication Data
A catalog record for this book is available from the Library of Congress

ISBN: 978-0-12-394430-6

For information on all Academic Press publications
visit our website at www.store.elsevier.com

Typeset by MPS Limited, Chennai, India
www.adi-mps.com

Printed and bound in China
13 14 15   10 9 8 7 6 5 4 3 2 1

Working together
to grow libraries in
developing countries

www.elsevier.com • www.bookaid.org

CONTENTS

v

SECTION IV • Future Prospects

Adam P. Arkin
Department of Bioengineering, University of California, Berkeley, California, USA; Physical Biosciences Division, Lawrence Berkeley National Laboratory, Berkeley, California, USA

David Baker
Department of Biochemistry, University of Washington, Seattle, Washington, USA

Konstantinos Biliouris
Department of Chemical Engineering and Materials Science, University of Minnesota, Minneapolis, Minnesota, USA

Gregory Bokinsky
Joint BioEnergy Institute, Emeryville, California, USA

Bradley C. Bundy
Department of Chemical Engineering, Brigham Young University, Provo, Utah, USA

Paolo Carrara
Biology Department, University of Roma Tre, Rome, Italy

George M. Church
Wyss Institute for Biologically Inspired Engineering, Harvard University, Boston, Massachusetts, USA; Department of Genetics, Harvard Medical School, Boston, Massachusetts, USA

Patrick C. Cirino
Department of Chemical and Biomolecular Engineering, University of Houston, Houston, Texas, USA

Ryan E. Cobb
Department of Chemical and Biomolecular Engineering and Institute for Genomic Biology, University of Illinois at Urbana-Champaign, Urbana, Illinois, USA

Dawn T. Eriksen
Department of Chemical and Biomolecular Engineering, Energy Biosciences Institute, University of Illinois at Urbana-Champaign, Urbana, Illinois, USA

Todd Freestone
Department of Chemical and Biomolecular Engineering and Institute for Genomic Biology, University of Illinois at Urbana-Champaign, Urbana, Illinois, USA

Martin Fussenegger
Department of Biosystems Science and Engineering, ETH-Zurich, Basel, Switzerland

vii

Dan Groff
Joint BioEnergy Institute, Emeryville, California, USA

Joao C. Guimaraes
Department of Bioengineering, University of California, Berkeley, California, USA; Department of Informatics/CCTC, University of Minho, Braga, Portugal

Boon Chin Heng
Department of Biosystems Science and Engineering, ETH-Zurich, Basel, Switzerland

Sui Huang
Institute for Systems Biology, Seattle, Washington, USA

Mitsuhiro Itaya
Institute for Advanced Biosciences, Keio University, Tsuruoka, Japan

Michael C. Jewett
Department of Chemical and Biological Engineering, Northwestern University, Evanston, Illinois, USA; Chemistry of Life Processes Institute, Northwestern University, Evanston, Illinois, USA; Member, Robert H. Lurie Comprehensive Cancer Center, Northwestern University, Feinberg School of Medicine, Northwestern University, Chicago, Illinois, USA

Yiannis Kaznessis
Department of Chemical Engineering and Materials Science, University of Minnesota, Minneapolis, Minnesota, USA

Jay Keasling
Joint BioEnergy Institute, Emeryville, California, USA; QB3 Institute, University of California; Department of Bioengineering; Physical Biosciences Division, Lawrence Berkeley National Laboratory, Berkeley, California; Department of Chemical and Biomolecular Engineering, University of California, Berkeley, California, USA

Jin Eyun Kim
Metabolic and Biomolecular Engineering National Research Laboratory, Department of Chemical and Biomolecular Engineering (BK21 program), Center for Systems and Synthetic Biotechnology, Institute for the BioCentury, KAIST, Daejeon, Republic of Korea

Tae Yong Kim
Bioinformatics Research Center, KAIST, Daejeon, Republic of Korea

Yu Bin Kim
Metabolic and Biomolecular Engineering National Research Laboratory, Department of Chemical and Biomolecular Engineering (BK21 program), Center for Systems and Synthetic Biotechnology, Institute for the BioCentury, KAIST, Daejeon, Republic of Korea; Bioinformatics Research Center, KAIST, Daejeon, Republic of Korea

Yutetsu Kuruma
Department of Medical Genome Sciences, The Tokyo University, Tokyo, Japan

Sang Yup Lee
Metabolic and Biomolecular Engineering National Research Laboratory, Department of Chemical and Biomolecular Engineering (BK21 program), Center for Systems and Synthetic Biotechnology, Institute for the BioCentury, KAIST, Daejeon, Republic of Korea; Bioinformatics Research Center, KAIST, Daejeon, Republic of Korea; Department of Bio and Brain Engineering and Bioinformatics Research Center, KAIST, Daejeon, Republic of Korea

Sijin Li
Department of Chemical and Biomolecular Engineering, Energy Biosciences Institute, University of Illinois at Urbana-Champaign, Urbana, Illinois, USA

Chang C. Liu
Department of Bioengineering, University of California, Berkeley, California, USA; Miller Institute for Basic Research in Science, Berkeley, California, USA

Pier Luigi Luisi
Biology Department, University of Roma Tre, Rome, Italy

Yunzi Luo
Department of Chemical and Biomolecular Engineering, University of Illinois at Urbana-Champaign, Urbana, Illinois, USA

Siying Ma
Department of Biomedical Engineering and the Institute for Genome Sciences and Policy, Duke University, Durham, North Carolina, USA

Michael T. Mee
Department of Biomedical Engineering, Boston University, Massachusetts, USA; Department of Genetics, Harvard Medical School, Boston, Massachusetts, USA

Tereza Pereira de Souza
Instituto de Biociências Letras e Ciências Exatas de São José do Rio Preto, Universidade Estadual Paulista Júlio de Mesquita Filho; Sao Jose do Rio Preto, Brazil

Shuai Qian
Department of Chemical and Biomolecular Engineering, University of Houston, Houston, Texas, USA

Arnaz Ranji
Department of Chemical and Biological Engineering, Northwestern University, Evanston, Illinois, USA; Chemistry of Life Processes Institute, Northwestern University, Evanston, Illinois, USA

Florian Richter
Institut für Biologie, Humboldt-Universität zu Berlin, Berlin, Germany

Claudia Schmidt-Dannert
University of Minnesota, Department of Biochemistry, Molecular Biology and Biophysics, St. Paul, Minnesota, USA

Jae Ho Shin
Metabolic and Biomolecular Engineering National Research Laboratory, Department of Chemical and Biomolecular Engineering (BK21 program), Center for Systems and Synthetic Biotechnology, Institute for the BioCentury, KAIST, Daejeon, Republic of Korea

Patrick Smadbeck
Department of Chemical Engineering and Materials Science, University of Minnesota, Minneapolis, Minnesota, USA

Robert P. Smith
Department of Biomedical Engineering, Duke University, Durham, North Carolina, USA

Seung Bum Sohn
Metabolic and Biomolecular Engineering National Research Laboratory, Department of Chemical and Biomolecular Engineering (BK21 program), Center for Systems and Synthetic Biotechnology, Institute for the BioCentury, KAIST, Daejeon, Republic of Korea; Bioinformatics Research Center, KAIST, Daejeon, Republic of Korea

Pasquale Stano
Biology Department, University of Roma Tre, Rome, Italy

Nicholas Tang
Department of Biomedical Engineering and the Institute for Genome Sciences and Policy, Duke University, Durham, North Carolina, USA

Yu Tanouchi
Department of Biomedical Engineering, Duke University, Durham, North Carolina, USA

Jingdong Tian
Department of Biomedical Engineering and the Institute for Genome Sciences and Policy, Duke University, Durham, North Carolina, USA; Tianjin Institute of Industrial Biotechnology, Chinese Academy of Sciences, Tianjin, China

Ilya Tikh
University of Minnesota, Department of Biochemistry, Molecular Biology and Biophysics, St. Paul, Minnesota, USA

Katherine Volzing
Department of Chemical Engineering and Materials Science, University of Minnesota, Minneapolis, Minnesota, USA

Harris H. Wang
Columbia University, New York, USA

Jeffrey C. Wu
Department of Microbiology and Molecular Biology, Brigham Young University, Provo, Utah, USA

Lingchong You
Department of Biomedical Engineering, Institute for Genomic Biology, Institute for Genome Sciences and Policy and Center for Systems Biology, Duke University, Durham, North Carolina, USA

Huimin Zhao
Department of Chemical and Biomolecular Engineering and Energy Biosciences Institute, Departments of Chemistry, Biochemistry, and Bioengineering, University of Illinois at Urbana-Champaign, Urbana, Illinois, USA

Joseph Xu Zhou
Institute for Systems Biology, Washington, USA

SYNTHETIC BIOLOGY: WHAT IS IN A NAME?

The era of synthetic biology has arrived. A Google search using 'synthetic biology' as a query yielded close to three million hits. A quick search of the scientific literature using 'synthetic biology' as a query on the Web of Sciences indicated that the number of research publications increased exponentially from less than three per year in early 2000 to more than 400 in 2011. A growing number of major research centers and institutes were set up in academia to explore exciting research opportunities in synthetic biology in the past few years. Examples include, but are not limited to, the Synthetic Biology Engineering Research Center (SynBERC) funded by the National Science Foundation in the United States, and the Synthetic Biology Institute at the University of California at Berkeley, the Wyss Institute for Biologically Inspired Engineering at Harvard University, the Centers for Synthetic Biology at MIT and University of California at San Francisco, the J. Craig Venter Institute, the recently formed Synthetic Biology Center in the United Kingdom, and the Chinese Academy of Sciences' Key Laboratory of Synthetic Biology in Shanghai, China. In parallel, there are rapidly growing efforts in industry to explore the medical and biotechnological applications of synthetic biology. Exemplary companies include Synthetic Genomics, Amyris, Gevo, Life Technologies, and Ginkgo Bioworks. So what is synthetic biology? Why has synthetic biology suddenly become a buzzword everywhere?

According to Wikipedia (http://en.wikipedia.org/wiki/Synthetic_biology), it was the Polish geneticist Wacøaw Szybalski who first clearly laid out a vision for synthetic biology in 1974: 'Let me now comment on the question "what next?" Up to now we are working on the descriptive phase of molecular biology. . .. But the real challenge will start when we enter the synthetic biology phase of research in our field. We will then devise new control elements and add these new modules to the existing genomes or build up wholly new genomes. This would be a field with the unlimited expansion potential and hardly any limitations to building "new better control circuits" and. . . finally other "synthetic" organisms, like a "new better mouse". . .. I am not concerned that we will run out of exciting and novel ideas,. . . in the synthetic biology, in general.' Amazingly, although these comments were made almost four decades ago, they are still accurate descriptions of the synthetic biology field today.

Due to the interdisciplinary nature of the field, it is not surprising that synthetic biology means different things to different people, even to leading practitioners in the field. To avoid possible confusion for the reader, the term synthetic biology is defined in this book as 'design, construction, and characterization of improved or novel biological systems using engineering design principles.' Synthetic biology not only has broad applications in medical, chemical, food, and agricultural industries, but may also revolutionize our understanding of basic life sciences. The main aim of synthetic biology is to develop the foundational technologies and knowledge that make the engineering of biology safer, more reliable, and more predictable. The key engineering design principles that synthetic biologists attempt to incorporate into their experiments are modularization and standardization of biological parts, which is analogous to modularization and standardization of electronic parts such as inverters, switches, counters, and amplifiers. By doing so, the biological parts may become reusable, and can be readily assembled to build large and complex biological systems.

Based on the above-mentioned definition of synthetic biology, it is clear that synthetic biology is still rooted in genetic engineering. However, synthetic biology has become a field of its own, largely thanks to the advances in systems biology and the development of new powerful tools for DNA synthesis and sequencing. For many experiments, researchers are now not limited to the physical DNA isolated from naturally occurring organisms any more. They can synthesize entire genes and pathways from scratch, and may also incorporate some desirable genetic modifications with ultimate flexibility. In addition, the development of new tools for construction of large DNA molecules such as pathways, plasmids, and whole genomes enables researchers to build ever-increasingly complex biological systems with desired features. Furthermore, computational modeling tools allow researchers to better predict the properties of a biological system prior to physical construction. Therefore, synthetic biology represents a quantum leap in genetic engineering, which can greatly facilitate the development of new biological products and processes.

It should be noted that synthetic biology is sometimes confused with systems biology or metabolic engineering. Simply put, systems biology focuses on characterization of complex interactions within biological systems using a more holistic perspective, which is complementary to synthetic biology. Systems biology and synthetic biology are essentially two sides of the same coin. Many systems biology tools are indispensable in synthetic biology. Metabolic engineering focuses on the engineering of cellular metabolisms to produce a certain substance using recombinant DNA technologies. Both synthetic biology and metabolic engineering emphasize systems-level understanding and engineering. However, metabolic engineering is generally limited to single cells, and the main goal of metabolic engineering is to maximize the formation of a specific product. The key enabling research tool in metabolic engineering is recombinant DNA technology. In comparison, synthetic biology deals with a biological system (living and nonliving), which could be a single protein or protein complex, a pathway, a biological machinery (e.g. photosynthesis, transcription, translation), a genome, a cell, a cell consortia, a tissue or organ, and an ecosystem. The scientific questions synthetic biologists attempt to address are also broader, ranging, from studying the stochastic nature of gene expressions to studying the cell–cell communications in microbial consortium, syntrophony, and symbiosis to studying the origin of life by constructing artificial cells and minimal genomes. Moreover, the tools in synthetic biology are not limited to recombinant DNA technology. Key enabling tools for synthetic biology include DNA synthesis, construction of large-size DNA molecules, and bio-orthogonal systems such as the introduction of unnatural amino acids into proteins, artificial gene regulation systems (circuits) controlled by light, small molecules, pH, etc. and the interface of nanotechnology/microtechnology and the living biological systems.
In addition, the capabilities to synthesize DNA and proteins in vitro enable the construction of functional pathways for synthesis of value-added compounds and artificial cells. Therefore, synthetic biology and metabolic engineering are two related but distinct fields.

SYNTHETIC BIOLOGY: WHAT'S NEW?

The purpose of the *Synthetic Biology: Tools and Application* book is to provide a systematic and integrated framework to examine key enabling components in the emerging field of synthetic biology. This book will uniquely address tools and methodologies developed for engineering biological systems at a wide range of levels including molecular, pathway, network, whole cell, and multicell levels. It will highlight many exciting examples of practical applications of synthetic biology such as microbial production of biofuels and drugs, new therapeutic regimens for treatment of human diseases, artificial cells, and artificial photosynthesis. In addition, it will discuss challenges and future prospects in synthetic biology. The overall scope of this book was largely based on a review article we wrote in 2010 (Synthetic biology: putting synthesis into biology. Jing Liang,

Yunzi Luo, and Huimin Zhao. *Wiley Interdisciplinary Reviews: Systems Biology and Medicine*, 3, 7−20 (2011)).

This book consists of 17 chapters contributed by thought leaders and leading practitioners in the field of synthetic biology. Chapters 1−8 describe various state-of-the-art experimental and computational tools for synthetic biology, whereas Chapters 9−13 highlight a number of important applications of synthetic biology in both basic and applied biological research. To conclude, Chapters 14−17 discuss a few ambitious synthetic biology efforts in which tremendous challenges and opportunities coexist.

More specifically, Chapter 1 provides an overview of the new tools for cost-effective DNA synthesis including microfluidic systems, ink-jet printing technology, digital photolithography, and electrochemical arrays. DNA synthesis is one of the foundational technologies for synthetic biology. However, it remains the main cost-limiting step in synthetic biology experiments. This chapter not only describes the technologies for DNA synthesis and their limitations, but also the detection and separation methods used to eliminate the errors generated during chemical oligonucleotide synthesis and in subsequent gene assembly steps. In addition, it highlights a few applications such as codon optimization, construction of genetic circuits, and genome synthesis. It is expected that cheaper and faster DNA synthesis may result in transformative advances in basic and applied research.

Chapter 2 provides an overview of the key methods for protein engineering, with a particular focus on directed evolution approaches and their applications in synthetic biology. Directed evolution mimics the Darwinian evolution process in the test-tube, and involves repeated cycles of genetic diversification and selection or screening. One key advantage of directed evolution is that it does not require any knowledge of the structure and function of the target protein. Proteins are essential components in any natural or synthetic biological system. However, their functions are often not optimized for a synthetic biological system. Therefore, the ability to rapidly tailor protein functions or create novel protein functions is critical to synthetic biology. For example, engineered protein-based biosensors can be used for construction of synthetic genetic circuits, integration of novel regulatory signals, and optimization of biosynthetic pathways for production of fuels and chemicals.

Chapter 3 provides an overview of the key computational and experimental tools that have been developed to design, construct, and optimize biochemical pathways, and the use of these tools to develop new biological processes for synthesis of industrial chemicals from renewable biomass. Pathway engineering is indispensable in the design of microorganisms (or cellular factories) for production of value-added products such as drugs, biofuels, and specialty chemicals. Similar to protein engineering, the key methods can be grouped into rational design and directed evolution or combinatorial approaches. It is relatively straightforward to introduce a biochemical pathway in a heterologous host. However, there are many issues in the engineering of a pathway, and even more challenges in optimizing it for the highest product yield.

Chapter 4 discusses the challenges and opportunities in the standardization and modularization of biological parts, which is one of the key distinguishing features of synthetic biology. Biological parts are genetically encoded modular units with defined biological function that can be used and reused in different contexts. In principle, the standardized biological parts can be assembled to more complex biological systems in a predicative fashion, which should greatly simplify the engineering of biology. However, due to the heterogeneous nature of the biological parts and inherent complexity of the biological systems, it remains an overwhelming challenge to create standard, reliable, and reusable biological parts. This chapter uses the design of genetic circuits as an example to highlight the strategies for construction of complex biological functions to order.

Chapter 5 describes some theoretical considerations for reprogramming multicellular systems. Unlike nonliving systems, multicellular living systems exhibit multistability, i.e. the coexistence of multiple stable attractors which arise from gene regulatory networks, therefore resulting in a rugged landscape. Consequently, cellular reprogramming must not only consider the robustness of an attractor, but also the dynamics operating in the regime of frequent transitions among the attractors. Such theoretical considerations offer new guidelines in designing synthetic biology experiments for reprogramming cells for biomedical and biotechnological applications.

Chapter 6 provides an overview of the computational protein design that is complementary to the directed evolution approaches described in Chapter 2. Computational protein design refers to the design of amino acid sequences based on computational structural modeling of the to-be-designed protein, and is one of the most important rational design strategies for protein engineering. Computational protein design can search a very large protein sequence space and create novel protein functions that are extremely difficult, if not impossible, to achieve by directed evolution. However, due to our limited understanding of protein structure, function, and dynamics, the ability of computational protein design is still limited. For example, computationally designed enzymes are often sluggish compared to their naturally occurring equivalents. Therefore, the combination of computational protein design and directed evolution will remain as the preferred method for engineering novel protein functions in most applications in the foreseeable future.

Chapter 7 highlights the importance of mathematical modeling in synthetic biology and introduces a new computational synthetic biology tool, the Synthetic Biology Software Suite (*SynBioSS*). Two examples were used to illustrate the power of this tool, including the design of a set of AND gates and a set of inducible activators. The combination of these biological parts will lead to a wide variety of regulating devices such as a repressor and an activator. This tool may be extended to incorporate the design of more complex biological systems.

Chapter 8 provides an overview of the computational tools for the design and engineering of biological systems at different levels, including proteins, genetic elements, genetic circuits, biochemical pathways, and whole microorganisms. Some of these tools have also been discussed in other chapters such as Chapters 2–4 and Chapters 6 and 7. A wide variety of successful examples are discussed to highlight the effectiveness of these computational tools.

Chapter 9 focuses on the potential therapeutic applications of synthetic biology in various aspects of human health and disease. It not only provides a comprehensive overview of the molecular toolkits developed for synthetic biology applications in mammalian cells such as genetic switches, memory elements, oscillators, and communication modules, but also discusses the challenges and safety issues associated with the application of synthetic biology to clinical practice.

Chapter 10 also discusses the potential applications of synthetic biology in human health, but with a particular focus on the discovery, characterization, and engineering of natural product biosynthetic pathways. Natural products are an important source of therapeutic drugs. For example, the majority of existing antibacterial and anticancer drugs are natural products or their derivatives. However, the rate of discovering novel natural products is diminishing. This chapter highlights a wide variety of synthetic biology tools developed for discovery and production of novel natural products. As an example, microbial production of the key precursors of the antimalarial drug artemisinin and the anticancer drug taxol are discussed.

Chapter 11 highlights the potential industrial applications of synthetic biology in the energy and chemical industries, with a particular focus on the engineering of microorganisms for the cost-effective production of biofuels. This chapter describes a selection of the methods available for engineering gene expression, enzyme function, and host cell physiology, many of which have been mentioned in other chapters such as Chapters 2, 3, and 8. Notably, this chapter also discusses the challenges associated with biofuels production on an industrial scale, such as the scale-up issues and efficient utilization of the actual plant biomass.

Chapter 12 provides an overview of the available technologies and strategies for designing and constructing bacterial genomes, with a particular focus on the use of *Bacillus subtilis* as a host to build large-size DNAs. Although it has been shown that a microbial genome of up to 0.5 Mb can be synthesized from oligonucleotides, more effective and efficient genome synthesis tools are still needed. This chapter discusses the challenges in building bacterial genomes, and the knowledge gap on the genome structure and function that has become a bottleneck in genome synthesis.

Chapter 13 provides an overview of the recent efforts on the construction and analysis of microbial multipopulation systems, or microbial consortia. Compared to single population systems such as those described in Chapters 8 and 11, microbial consortia may be advantageous due to their ability to generate more complex dynamics and being more robust to environmental variations. This chapter also discusses key considerations when designing a consortium, and how such consortia may be engineered to implement cooperation or competition between populations as well as the applications of engineered microbial consortia in human health and industrial biotechnology.

Chapter 14 describes the current status of designing and constructing artificial or synthetic cells, which is one of the most ambitious goals in synthetic biology. The engineered synthetic cells will not only open up new doors for biotechnological applications, but will also provide new insights on the origin of life and the function and dynamics of biological systems. This chapter focuses on the 'semi-synthetic minimal cells' (SSMCs), namely, cell-like structures containing the minimal number of biological compounds that are required to reconstruct a function of interest. It discusses the key considerations in the construction of SSMCs, and provides an overview of the biotechnological applications of SSMCs when SSMCs are combined with cell-free, liposome, and microfluidic technologies.

Chapter 15 provides a comprehensive overview of recent advances in cell-free synthetic biology. While most synthetic biology efforts have focused on the genetic manipulation and biochemical analysis of living organisms, cell-free synthetic biology aims to reconstitute the biological systems or processes in vitro. Compared to cell-based synthetic biology, cell-free synthetic biology offers an unprecedented level of control and freedom of design for understanding, harnessing, and expanding the capabilities of natural biological systems. This chapter describes not only the tools for cell-free synthetic biology, but also some important applications such as cell-free protein synthesis, cell-free metabolic engineering, and minimal cells.

Chapter 16 provides an overview on the use of synthetic biology tools to engineer light-energy conversion and carbon fixation in various nonphotosynthetic hosts. The resulting microorganisms should be energetically more efficient than the organisms without the engineered artificial photosynthesis machinery, which is highly desirable when they are used to produce chemicals and biofuels from renewable biomass. In addition, these microorganisms may be able to directly use inexpensive substrate carbon dioxide and sunlight to produce a wide variety of chemicals and fuels. Since engineered artificial photosynthesis is still in its embryonic stage, this chapter mostly highlights the challenges of implementing light-energy capture and conversion into microbial hosts, and provides some preliminary results.

Chapter 17 briefly describes the recent developments in engineering synthetic microbial ecosystems and future prospects in this emerging area of synthetic biology. The engineered synthetic ecosystems can sense, degrade, and produce various value-added compounds in both natural and artificial environments. Clearly, engineering synthetic ecology is more ambitious than the current efforts in engineering microbial consortia that typically deal with two or three types of cells as described in Chapter 13.

SYNTHETIC BIOLOGY: WHAT'S NEXT?

Synthetic biology has already made a resounding impact in basic and applied research after it became a field of its own around 2005. Milestone accomplishments include, but are not limited to, the engineering of a yeast strain for cost-effective production of artemisinic acid, a key precursor of the commonly used antimalarial drug artemisinin, the reconstruction of a complete microbial genome by the J. Craig Venter Institute, and the synthesis of a yeast chromosome. Partly stimulated by these early successes, the interest and investment in synthetic biology continue to grow rapidly around the world. As such, the future prospects of synthetic biology are extremely exciting.

As elaborated in this book, although challenges abound across all scales from the modularization and standardization of the biological parts to the integration of all parts together at different levels, there are numerous opportunities in synthetic biology. It has become clear that both new and improved foundational tools for synthetic biology and more exciting applications in basic and applied research are urgently needed to transform synthetic biology into an independent engineering discipline. In particular, new design rules are needed in synthetic biology that will truly make engineering of biology easier, faster, safer, more predictable, and more reliable. Given the high potential of synthetic biology in addressing those most daunting societal challenges in human health, energy, and sustainability, the field of synthetic biology will continue to grow rapidly in the foreseeable future. The era of synthetic biology should last for a long time.

SECTION I

Synthesis and Engineering Tools in Synthetic Biology

New Tools for Cost-Effective DNA Synthesis

Nicholas Tang[1], Siying Ma[1] and Jingdong Tian[1,2]
[1]Duke University, Durham, NC, USA
[2]Chinese Academy of Sciences, Tianjin, China

INTRODUCTION

DNA synthesis is a powerful enabling technology and yet a limiting step in synthetic biology. Decreasing costs in DNA synthesis will open new frontiers and project concepts that would not be feasible to most scientists at current levels. The cost of column synthesized oligonucleotides has dropped 10-fold over the past 15 years,[1] and is currently around USD0.08−0.2 per nucleotide.[2] Gene synthesis is a more expensive process, and involves generating longer DNA constructs from overlapping oligonucleotides. In 2000, the market cost of gene synthesis was approximately USD10 per base.[3] Since then, prices have dropped nearly 50-fold in 10 years, to as low as USD0.2 per base pair, with average error rates of about 1 in 300−600 bases.[1,2] Prices continue to drop by a factor of 1.5 per year, in a manner akin to Moore's law.[1] By 2005, there existed 39 gene synthesis vendors worldwide, and the number has increased since then. In the past couple of years, more companies have begun to emerge to take on the challenge of adapting new synthesis technologies for the market. It would be interesting to see dramatic changes to the gene synthesis market in the near future.

OLIGONUCLEOTIDE SYNTHESIS
Column Oligonucleotide Synthesis

Chemical DNA synthesis can be used for applications as common as primer, linker, or probe synthesis. Here we discuss oligonucleotide synthesis for the application of gene and genome synthesis. Chemical assembly of DNA using programmable synthesizers is now a routine procedure. The most common and reliable system for chemical synthesis involves synthesizing individual oligonucleotides in small columns. A series of valves and pumps introduce the correct nucleotide monomers and reagents required for growing oligomers of DNA in a stepwise manner. Chemical oligodeoxynucleotide synthesis is different from enzymatic DNA synthesis in living cells in that it is a cyclical process that elongates nucleotides from the 3′ to the 5′-end. The starting complex for chemical synthesis of DNA consists of an initial acid-activated nucleoside phosphoramidite tethered with a spacer to a solid-support controlled pore glass (CPG) or polystyrene (PS) bead. The advent of solid-phase DNA synthesis made automation possible by eliminating purification steps to remove intermediates or unreacted reagents. The column is simply rinsed with

3

Synthetic Biology. DOI: http://dx.doi.org/10.1016/B978-0-12-394430-6.00001-7

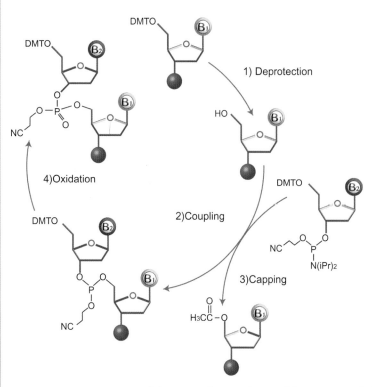

FIGURE 1.1

The solid-phase, four-step oligodeoxynucleotide synthesis cycle.

anhydrous acetonitrile to remove these reagents, and then purged with argon to remove the remaining acetonitrile.

The cyclic addition of additional monomers to the existing oligonucleotide chain occurs in four steps: deprotection, activation/coupling, capping, and oxidation (Fig. 1.1). Each additional nucleoside which is added to the growing chain has a 5′DMT protection group. This assembly is called a phosphoramidite. The four-step phosphoramidite chemistry is the method of choice for most commercial DNA synthesizers because the yields are more accurate and homogeneous than other methods.[4] First, a strong acid is used to de-block the 5′-O-4,4′-dimethoxytrityl (DMT) group, removing the protecting group from the nucleotide chain and exposing a reactive OH group. In the next step, 1H-tetrazole and the dissolved phosphoramidite are simultaneously added to the column. Tetrazole, a weak acid, protonates the trivalent phosphorus on the 3′-end of the monomer. This results in a slow displacement of the secondary amine and formation of a highly reactive tetrazolide that then immediately couples with the OH group. At this point, the added phosphoramidite is coupled to the existing chain. Uncoupled 5′-OH groups are blocked by an acylating capping reagent, usually acetic anhydride, to minimize deletion products. Finally, the unstable phosphite triester internucleotide linkage between nucleotides is oxidized to a more stable pentavalent phosphotriester. The end results of this process are oligonucleotide strands that are bound to beads. Each phosphate bond contains a methyl group, which can be removed by chemical treatment in the reaction column. The 5′ terminus of the last nucleotide can be deprotected through detritylation of the DMT group, and phosphorylated by T4 kinase. DNA strands can also then be cleaved from the spacer linker off the solid support.

The described four-step synthesis procedure has been the basis of fully automated DNA synthesizers with up to 1536 sequence throughputs.[5] Throughput evolved from 2−4 individual sequences in initial synthesizers manufactured by Applied Biosystems, to 96 well plates in 1995.[6] Lashkari et al. used computer-controlled solenoid valves to deliver bulk reagents through Teflon tubes into a microwell plate. Since then, parallel synthesis using multiplexed reagent delivery lines has allowed for synthesis in other microwell plate formats.

Optimizations in reaction chemistry include a two-step cycle synthesis, which reduces costs by eliminating several reagents.[7] A peroxy anion is used as a nucleophile to remove a 5′-carbonate and oxidize the internucleotide phosphite triester. Deprotection with peroxy anion under mildly basic conditions can eliminate depurination, a side reaction that leads to mutations in synthetic DNA. If further developed, the two-step synthesis process can make oligonucleotide synthesis simpler, and consequently more robust.

Microarray Oligonucleotide Pool Synthesis

The major costs for gene synthesis are attributed to oligonucleotide synthesis, sequence verification, and labor for processing steps. Microarray-enabled oligonucleotide pool synthesis effectively tackles oligonucleotide costs. Since their inception in 1995, microarrays have dramatically revolutionized genomics with massive parallelism and automation.

Microarrays are 2D solid-phase arrays used to assay or screen biological materials like nucleic acids, proteins, or cells. Oligonucleotide arrays can be used for a variety of designs, including gene expression screens, SNP genotyping, comparative genomic hybridization (CGH), tiling, ChIP-on-chip, microRNA, resequencing, and aptamer screening.[8] More relevantly, oligonucleotide microarrays can offer significant reductions in DNA synthesis cost due to their dense and massively parallel feature designs. For example, reducing costs by scaling down reagent volumes in resin-based synthesis is restricted to decreasing the diameter of capillaries. Simply eliminating the rinsing of lines in instrumentation makes at least a 10-fold reduction in cost.[6] This is because acetonitrile is often used in high volumes for rinsing and dissolving reagents, making it one of the most expensive bulk reagents.[9] Cost for reagents in a custom inkjet microarray slide is less than USD50 per slide, due to the low volumes of phosphoramidite and tetrazole necessary for the miniaturized platform.[10] Microarray synthesis not only lowers costs, but also makes synthesis more environmentally friendly. High-density arrays can offer 10^4-10^6 unique oligonucleotides, and can reduce costs by at least an order of magnitude.[11,12] The price for 3912 90-mers from LC Sciences is about USD 1000; the Agilent 55 k chip, ∼USD 7000, which translates to USD 0.0025 per base.[13] These costs do not include downstream costs for gene assembly.

The variety in technologies and techniques for microarray-based oligonucleotide synthesis is expected to offer reductions in cost and improvements in throughput in the coming years. In contrast, the cost of column-synthesized oligonucleotides has remained constant over the past six years and is unlikely to decrease significantly.[14] The microarray technologies that exist in the DNA synthesis market include ink-jet printing (Agilent, Protogene), photosensitive 5′ deprotection (Nimblegen, Affymetrix, Flexgen), photo-generated acid deprotection (Atactic/Xeotron/Invitrogen, LC Sciences), and electrolytic acid/base arrays (Oxamer, Combimatrix/Customarray).

Optimizations that have been made to microarray synthesis include synthesis on PDMS (poly(dimethylsiloxilane))[15] and COC (cyclic olefin copolymer)[16] substrates as low-cost and flexible alternatives to glass. Ma and coworkers showed that oxCOC, a hybrid substrate composed of COC and RF sputtered SiO_2, takes benefits from both constituents.[17] COC offers low density, resistance to organic solvents, high stiffness, and UV transparency, while SiO_2 offers useful surface linkers for phosphoramidite chemistry. Furthermore, oxCOC can be manipulated for large-scale production. With soft lithography, a PDMS stamp can be made to imprint channels or wells on COC before thin-film deposition of SiO_2.[18] The stamps are disposable and can be created in a non-cleanroom setting with silicon molds, although the silicon molds must be fabricated via photolithography. A study involving inkjet microchip synthesis demonstrated that using a COC chip with patterned silica features reduced the error rate of synthesized oligonucleotides from one in 200 bases to one in 600 bases,[16] which is equivalent to high-quality column synthesis.

The dominant costs are now enzymatic processing, cloning, and sequencing.[12] A recent study by Tian's group addresses the limitations of processing steps using an integrated combination of isothermal nicking, strand displacement amplification (nSDA), and polymerase cycling assembly (PCA), reducing the cost even further to USD 0.005 per base with an error frequency of <0.2 errors/kb.[19] Such reduced cost could make gene and gene library synthesis more widely accessible.

Microfluidic and Fluidic Systems

Microfluidics allows for the control and manipulation of small volumes of liquids. These features make it particularly useful for a variety of applications like PCR and cell screening. More specifically, microfluidics can decrease the space and cost requirements of DNA synthesis. A recent study reports a programmable microfluidic synthesis platform that can synthesize ∼100 pmol of each unique oligonucleotide, which is substantial enough not to require PCR amplification steps before gene assembly.[20] The output levels from this study better match the amounts necessary for gene assembly. Using microfluidics for DNA synthesis reduces reagent consumption by 100-fold compared to conventional column synthesis which generates far more oligonucleotides than necessary for gene assembly. On the other hand, although microarrays can synthesize a large number of oligonucleotides, they produce only 10^6–10^8 molecules per spot and yield less than 2 fmol. This yield is six orders of magnitude lower than those of column-synthesized oligonucleotides,[20,21] which are too low for gene assembly without amplification. Although amplification methods have been refined, PCR amplification can introduce errors and increase overall labor and processing costs. Additionally, there is a 100-fold reduction of reagent consumption compared to other solid-phase synthesis technologies that have less efficient deblocking steps.[20] DNA synthesis with microfluidics as performed in this study can further advance lab-on-a-chip (LOC) or Micro Total Analysis Systems (μTAS). However, the 17–24 mer length and 1 in 153 bp error rates could still be refined to be suitable for longer gene constructs.

Although microfluidics allows for massively parallel microscale DNA synthesis, there are a few key technical challenges. One challenge is organization of addressable synthesis units. Xiao et al.[22] used soft lithography to synthesize oligonucleotide arrays on glass surfaces. During the coupling step of synthesis, pre-cast PDMS microstamps transferred a mixture of nucleoside monomer and tetrazole onto the glass slides surface. Like photomasks in photolithography, different stamps dictate predefined areas for the coupling of transferred monomer. 20 mers were synthesized with high coupling efficiency, with a stepwise yield of 97%.

Blair et al.[23] used 365 nm wavelength ultraviolet-light-emitting diodes (UV-LEDs) as a cost-effective alternative light source for addressing and directing oligonucleotide synthesis inside glass capillaries. A string of UV-LEDs were positioned along the length of a glass capillary. The inside walls were functionalized for oligonucleotide synthesis using photolabile 2-nitrophenyl propoxycarbonyl (NPPOC) chemistry.[24] The glass capillaries were designated capillary synthesis cells (CSC) in their fluidic system. Because the spectrum of the UV-LED contained no emission below 360 nm, no filters were needed to prevent DNA damage from short wavelength radiation. 70 mers were successfully synthesized and used for gene assembly.

Another technical challenge is the chemical resistance of microfluidic substrates. For example, elastomers are a popular material for microfluidic devices because of their fabrication cost and labor benefits over silicon. However, elastomers must be chosen or modified to be chemically resistant to the organic solvents used in oligonucleotide synthesis. Popular elastomers like poly(dimethylsiloxane) (PDMS) degrade, swell and clog in contact with DNA synthesis reagents, which results in as high as a 90% flow rate drop.[25] Moorcroft et al.[15] circumvented this issue by substituting both the oxidation and deprotection solvents in conventional DNA synthesis. PDMS microchannels were first molded by a soft lithography process. Then, the surface was functionalized by silanizing with 3-glycidoxypropyltrimethoxysilane (GPTMS) and adding a PEG spacer. 21 mers were synthesized on the PEG-Silane-PDMS chip. The Quake group used a chemically resistant photocurable perfluoropolyether (PFPE) for their microfluidic synthesis chambers.[26] To avoid functionalization, the synthesis chambers were made of PFPE, while oligonucleotide synthesis was performed on porous silica beads. The device was used to synthesize 60 pmol of 20 mers oligonucleotides with 60-fold less reagent consumption than conventional synthesis.

Another chemically resistant substrate is carbon. Carbon offers superior heat and chemical stabilities compared to glass, allowing for robust linkage of oligonucleotide probes.[27]

Phillips et al. functionalized carbon surfaces for light-directed oligonucleotide array synthesis using RF plasma treatment and 9-decene-1-ol. The authors used two types of carbon as substrates for DNA synthesis: glassy carbon and CVD diamond. Oligonucleotide synthesis was then performed on these surfaces with photolabile NPPOC chemistry. Despite robust oligonucleotide linkage, carbon-based DNA probe arrays[28,29] are not yet commonly used, due to their low-density probe coverage.

Hua and Gulari[25] reported a platform that integrated a pneumatic microvalve array in a microfluidic system to precisely control the flow of fluidics for parallel oligonucleotide synthesis. They coated the PDMS valve membranes with parylene to make them chemically resistant and compatible with aggressive chemical reagents. These valve membranes were sandwiched between a PDMS and a silicon layer. The PDMS layer provided air channels, while the fluid reactions occurred on the silicon layer. The pressurized air channels closed the valves and restricted fluids from leaving. Sixteen multiplexed air channels were sufficient to control 12 870 reactors. 30 mers were successfully synthesized with a stepwise synthesis yield of ~99.5%.

Photolithography

Some of the earliest efforts to synthesize oligonucleotides on microarrays involved using physical photolithographic masks to direct a light pattern on nucleoside monomers with photolabile protecting groups.[30,31] Light patterns are directed to areas on the array to remove photolabile protecting groups from the growing oligomers. After deprotection, a selected phosphoramidite monomer is spread over the entire surface and only the exposed areas are activated for coupling of that particular phosphoramidite. Masks must be prefabricated for each cycle so that the next bases on the synthesized oligonucleotides are mapped to the areas of exposure. Affymetrix Inc. applied this technology to large-scale fabrication of high-density GeneChip probe arrays.

The costs for optics, light sources, and pre-fabricated photolithographic masks can be prohibitive. This is partly because, in principle, the number of necessary masks is four times the number of bases in the oligonucleotides. Texas Instrument's DMD (Digital Micromirror Device) eliminates the necessity for physical masks, greatly reducing processing time and costs for light-directed oligonucleotide synthesis. A DMD device consists of an electromechanically controlled array of micromirrors.[32] It is typically used for DLP (digital light processing) projection display systems, with resolutions as high as 2073 600 pixels (SVGA). With flexible programmable light patterns, the generation of arrays can be automated.

(R,S)-1-(3,4-(methylenedioxy)-6-nitrophenyl)ethyl chloroformate (MeNPOC) or 2-(2-nitrophenyl)propoxycarbonyl (NPPOC) can be used in photo-deprotection to block function groups on the linker or monomers.[32,33] In one study, a 600 × 800 DMD array of 16 μm wide micromirrors was used to synthesize 41 mer oligonucleotides.[32,33] In this work, sub-populations of oligonucleotides were eluted, amplified, and assembled. The authors extrapolated that a chip with 786 432 unique 40 mer oligonucleotides could potentially be used to assemble a >15 Mb construct. FlexGen's in-house synthesizer, the Flexarrayer, generates oligonucleotides in a similar manner. Nucleotide deprotection is activated by a laser before the nucleotides are washed and bound to activated spots to produce 60 mers.

Another study used a photoacid generator (PAG), triarylsulfonium hexafluoroantimonate, to perform the deblocking step.[34] As with conventional acid-deblocking nucleotide phosphoramidite chemistry, acids are generated in the ditritylation step, but with a UV photolytic process. The resulting PGA solution is at a high enough mM concentration to effectively remove the DMT group on a nucleotide, and is more efficient than photo-cleavage methods. The main advantage of this method is that it relies on conventional phosphoramidite chemistry, without the cost or availability limitations of photolabile protection groups. Photoacid generator (PAG) may liberate free radicals to produce errors in synthesis. Serafinowski and Garland[35] developed and used two photosensitive esters, R-phenyl-4,5-dimethoxy-2-nitrobenzyltrichloroacetate and R-phenyl-4,5-dimethoxy-2,6-dinitrobenzyltrichloroacetate, as PAG agents for synthesis to avoid these problems.

Zhou et al.[11,36] coupled the PAG technique with a parallelized microfluidic platform. An array chip was installed in a flow-through cartridge connected to a commercial oligonucleotide synthesizer (Expedite 8909). The chips used differential surface tension to isolate reaction sites, and allowed for the miniaturization of synthesis. Cleaved oligonucleotides were of high enough quantities and qualities to be successfully assembled in 10 kb constructs with PCR and ligation.

Electrochemical Arrays

Another method of DNA synthesis involves using localized electrochemical reactions for the deblocking step. An array of platinum microelectrodes is covered with an electrolyte solution. Current is simultaneously applied to individually addressable microelectrodes with a semiconductor circuit, so that the electrolytes near activated anodes are oxidized and release acid for deblocking. In one study, the electrolyte used was 25 mM hydroquinone and 25 mM benzoquinone with 25 mM tetrabutylammoniumhexafluorophosphate in anhydrous acetonitrile.[37] By pairing each anode with adjacent cathodes which can reduce acid, the produced acid is confined. The acid then diffuses to the layer of substrate which contains the oligonucleotides being synthesized. The electrodes were made by thin-film photolithography of 50-nm thick iridium metal and were durable enough to use over 500 cycles without deteriorations. The authors have demonstrated the synthesis of short 17 bp oligonucleotides with complete electrochemical deblocking in as little as 9 seconds.

Customarray's oligonucleotides arrays are also synthesized using a semiconductor-based electrochemical synthesis process. Rather than using cathodes to localize generated acids, synthesis is performed on a polymer surface over the surface of the semiconductor. The porous polymer slows down acid diffusion to prevent cross-contamination between local electrodes. The surface also increases the oligonucleotide density during synthesis. Customarray's 90 K microarrays contain 94 000 unique oligonucleotides and 25 μm features, and are capable of synthesizing oligonucleotides of up to 50 bp lengths.

Besides feature density, another advantage of electrochemical synthesis is flexibility. Because photolithographic masks are not involved, it is less expensive and requires less labor to change the array design. Additionally, electrochemical detrytilation is a relatively efficient chemistry; the deblocking step can be complete in seconds and any side reactions from the electrolyte chemicals are minimal. Efficient chemistry and a lack of moving parts contribute to a total synthesis time of less than 24 hours.

Inkjet Printing

In 1981, Hood and Caruther modified liquid phase phosphite-triester chemistry for solid-phase DNA synthesis on polymer supports.[38] Solid-phase synthesis not only became the method of choice for conventional DNA synthesis, but also opened DNA synthesis to automated and miniaturized strategies on microarrays by eliminating the need to purify synthetic intermediates or unreacted reagents. The first inkjet DNA synthesizers used in-house microfabricated piezoelectric actuators.[8] Piezoelectric inkjet DNA synthesizers operate similarly to commercial inkjet printers. Instead of ink, the inkjet head contains four phosphoramidite fluid channels, one activator channel, and one optional modified base

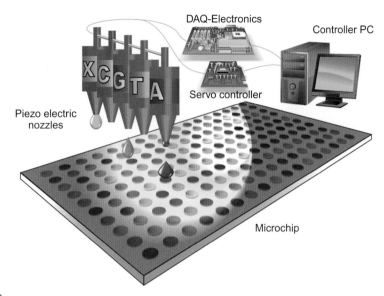

FIGURE 1.2

DNA microchip synthesis with piezoelectric inkjet technology. The microarray synthesis platform consists of a controller PC, DAQ-electronics, and a servo controller, which together control the movement and the printing of the inkjet print head onto slide surfaces. Droplets are dispensed onto as many as 10 000 spot features that are each less than 100 μm in diameter.

channel. In this work, the discrete features on the substrate could hold and segregate 100 pL droplets using pre-patterned surface tension (Fig. 1.2).

9

Piezoelectric print head technology is primarily commercialized by Agilent. In 2001, Hughes et al. used Agilent's second-generation system for gene-expression profiling of biological samples from various organisms.[39] The flexibility of the inkjet system allowed them to create new arrays and update sequences rapidly. Since then, gene expression profiling has been conventionally performed on Agilent microarrays.[39–42] Through refinement of surfaces, firing parameters, and solvent mixtures, Agilent has evolved its 22 575 feature throughput to a current 244 000 feature throughput on a single glass slide. Their current systems now also include computer visualization systems, print head maintenance modules, and automated slide handling.

GENE ASSEMBLY

Chemical oligonucleotide synthesis accumulates errors due to side reactions and inefficiencies in the stepwise reactions.[43] Although oligonucleotides of up to 600 bp in length can be synthesized, yields are extremely low.[44] Gene assembly becomes necessary for synthesis of longer constructs. Almost all current assembly techniques use a combination of PCR or ligation-based assembly. The advantage of these methods over restriction digestion/ligation methods is that they can perform scarless and sequence-independent assembly.

Ligation-based assembly uses thermo-stable DNA ligase to join pre-synthesized oligonucleotides. Successful use of ligation-based assembly has been demonstrated in research and commercial synthesis platforms.[45–47] Thermo-stable ligase is advantageous compared to T4 DNA ligase, because fewer DNA secondary structures will form at elevated ligation temperatures.[48–50] Ligation-based assembly involves ligation and PCR amplification. Overlapping oligonucleotides are designed to completely cover both strands of the sequence. They are phosphorylated at the 5′-ends for the ligation reaction. The oligonucleotides are first heat denatured, and then cooled for annealing and ligation at 50−65°C. Because the ligation reaction is linear, the full construct is designed with flanking primer sequences so that PCR can amplify the construct.

PCR-based assembly, otherwise known as polymerase cycling assembly (PCA), remains one of the most cost-effective gene assembly methods.[51,52] In a two-step procedure, partially overlapping oligonucleotides are designed to span the whole sequence.[53] Because gaps between oligonucleotides on the same strand are allowed, the number of starting oligonucleotides is fewer than that required for ligation-based assembly. A PCR reaction is carried out so that overlapping oligonucleotides anneal and extend. The resulting double-stranded construct can serve as the template for the PCR amplification reaction. In a single-step procedure, the amplification primers are included with the oligonucleotides for a combined assembly and amplification reaction.[54,55] Although extra cycles may be needed, this procedure is more easily multiplexed than assembly with multiple steps.[53] Although PCR-based methods are efficient, constructs involving repetitive sequences or secondary structures may have difficulties during PCR, and thus can better be assembled with ligase-based assembly.[56]

There are a number of related overlapping extension techniques for sequence-independent assembly. For the sequence-independent assembly of circular double-stranded constructs or plasmids, circular polymerase extension cloning (CPEC)[57] can successfully assemble not only multiple-gene constructs, but also combinatorial sequence libraries.[19] In a single-step reaction, overlapping oligonucleotides are assembled and circularized. Other approaches suitable for plasmid construction include the In-Fusion commercial kit from Clontech,[58] Uracil-specific excision reagent (USER)[59] and Sequence- and Ligation-Independent Cloning (SLIC).[60] Errors may arise during the assembly of large constructs. Gibson isothermal or 'chewback and anneal' assembly avoids length-dependent errors with the T5 DNA polymerase, allowing for assembly of genome length constructs,[61] such as a 16.3-kb mitochondrial genome.[62]

Rather than cleaving oligonucleotides from microchips, Quan et al. uses an approach involving isothermal nicking and strand displacement amplification (nSDA) of immobilized microarray oligonucleotides.[19] This simultaneously amplifies and releases oligonucleotides, which are then PCA assembled into 1-kb constructs. A microfluidic system serves to integrate synthesis, amplification and assembly (Fig. 1.3).[19,63] By performing all steps on chip, high-throughput assembly can be easily coupled with downstream reactions.

Although high-fidelity DNA microchips can hold up to a million unique oligonucleotides, they are difficult to scale. Microarray oligonucleotide pools are highly complex, and become problematic in complications like potential cross-hybridization between assembled fragments. More successful scaling will lower the cost of high-quality gene synthesis. To address this issue, Kosuri et al. combined selective oligonucleotide pool amplification, optimized gene assembly, and enzymatic error correction.[12] The authors cleaved microchip-synthesized oligonucleotides. These oligonucleotides were synthesized with flanking amplification primer annealing sites corresponding to several subpools. A quarter-million specific amplification primers were then used to selectively PCR amplify specific subpools of oligonucleotides from the original complex background of oligonucleotides. The flanking amplification primer sequences were then cleaved with restriction enzymes, which allows for seamless subsequent PCR assembly into gene constructs. The authors tested the system on 47 genes encoding for proteins and antibodies, including 40 error-free single-chain antibody genes that had previously been shown to be difficult to synthesize due to high GC content and repetitive sequences. The assembly optimization and enzymatic treatment allowed for accurate and low-cost synthesis, bringing costs down to an estimated USD 0.01/bp.

On-chip amplification and assembly

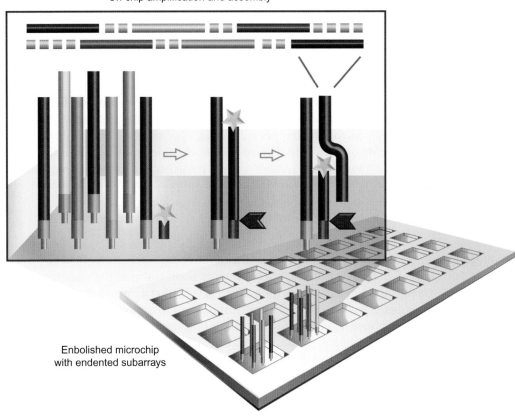

Enbolished microchip
with endented subarrays

FIGURE 1.3
Integrated on-chip DNA amplification and gene assembly. The microchips are divided into physically isolated sub-arrays where oligonucleotides are amplified by isothermo nicking and strand displacement amplification. The released strands are the assembled into 0.5–1 kb gene fragments within the wells.

11

QUALITY CONTROL

Eliminating errors is critical for most gene synthesis projects, and is probably the most costly and time-consuming step in gene synthesis. Errors are generated mainly during chemical oligonucleotide synthesis and also in subsequent gene assembly steps.

In order to reduce the number of faulty oligonucleotides, chromatographic purification by high-performance liquid chromatography (HPLC) or polyacrylamide gel electrophoresis (PAGE) purification is typically used by conventional oligonucleotide synthesis vendors. However, chromatographic purification is expensive and time-consuming, and PAGE purification is even more costly. In this section, we will detail alternative purification methods that couple with high-throughput synthesis technologies.

Fluorescence Selection

Reporter genes can be used to select in-frame gene fusions. Kim et al. used a fluorescence selection method to improve the fidelity of gene synthesis by focusing on the elimination of deletions and insertions in chemically synthesized genes. The authors constructed vectors where GFP is expressed if the insertion of an in-frame DNA construct into appropriate cloning sites shifts the GFP gene in frame.[64] The absence of insertions or deletions results in fluorescent cell colonies when transformed into *E. coli* cells, which can be selected using a transilluminator. The authors synthesized six similar-length genes from overlapping oligonucleotides using ligase-based assembly and subsequent PCR, ligated the genes into pBK reporter vectors, and transformed for overnight growth of colonies on agar plates. On average, error rates improved from 1 in 629 bp (blind selection) to 1 in 6552 bp (green fluorescence selection). This error rate allows for the straightforward construction of error-free genes larger than 1000 bp by one-cycle synthesis.

Hybridization Selection

Bringing down error rates of microarray oligonucleotide synthesis would involve substantial efforts. Microarrays typically produce oligonucleotides with lower product quality, lower product yields, and higher complexity than other sources. Tian et al. used parallel amplification and purification of oligonucleotides from complex oligonucleotide mixtures using microfluidics to allow for a 2~5-fold increase in quality over PAGE purified microarray derived oligonucleotides and a 10-fold increase in quality over unpurified microarray synthesized oligonucleotides.[53] This was accomplished using generic amplification primer sequences for the amplification of all oligonucleotides in a single PCR reaction. The low yields of microchip-derived oligonucleotides are simply insufficient for gene assembly without amplification. Additionally, distinct subsets of primer sequences allow for the fractioning of the oligonucleotide pool into subpools. With a concentrated, highly complex oligonucleotide starting pool, the potential for undesired cross-hybridization becomes an issue. Selective amplification of subsets of oligonucleotides reduces this cross-hybridization.

The difficulty with restriction digested double-stranded microchip oligonucleotides is that after denaturation, matched antisense oligonucleotides will compete with designated overlapping oligonucleotides, reducing assembly efficiency. Tian et al. avoided this problem by enriching single-stranded oligonucleotides by hybridization to antisense selection oligonucleotides and then eluting them. This was also a purification process, which allowed for selection against oligonucleotides with errors. These oligonucleotides produce mismatches during hybridization and elute at a lower temperature. Using selective amplification and purification via hybridization enabled the synthesis of 21 genes of the *E. coli* 30 S ribosomal subunit cluster from a single pool.

In a similar approach, Borovkov et al. combined hybridization-based oligo selection with parallel amplification.[14] Erroneous oligos were eluted and excluded from assembly. By assembling from unpurified pools of microarray oligos, the authors eliminated purification steps, reducing assembly costs 100-fold. This approach allows for the robust use of inexpensive, but imperfect microarray oligonucleotides for gene synthesis.

Mismatch Recognition

Besides using hybridization selection or improved chemistries with reduced depurination, a general approach for eliminating errors in gene synthesis is to use DNA mismatch recognition proteins or enzymes.[65] Mismatches are small single-strand loops that are generated by hybridization between correct and incorrect sequences. There are two categories of enzymes for mismatch removal: the mismatch binding proteins and the mismatch cleaving enzymes.[66−70]

Mismatch binding proteins like MutS selectively bind to mismatch sites. Incorrect sequences can be excluded by the removal of protein−DNA complexes with gel-shift assay or affinity columns. Smith and Modrich first reported the use of MutHLS to remove mutant sequences.[71] MutL couples MutH endonuclease to the MutS bound site, leading to MutH mediated cleavage. The cleaved heteroduplex was then removed by gel electrophoresis. Using MutS from *Thermus aquaticus*, Carr et al. established a method that reduced error rate to 1 per 10 kb in two cycles.[67] In order to tolerate more errors in longer starting DNA fragments, MutS treatment can be applied to smaller constituent fragments. In a process called consensus shuffling, the longer fragment is cleaved into overlapping fragments using restriction endonucleases. Using column filtration, immobilized MutS can bind to and filter mismatch fragments. Eluted fragments are then reassembled with PCR. Two iterations improved the error rate 3.5-fold to 1 per 3.5 kb.[68] Because erroneous gene fragments are eliminated in mismatch filtration, enough initial fragments must exist in the pool to survive MutS binding. This may prevent mismatch binding from being applied to low-quality and low-yield microarray-derived oligonucleotides.Mismatch cleaving enzymes will bind and cleave at the mismatch sites. These include phage resolvases like T4 endonuclease VII, T7 endonuclease I, or *E. coli* endonuclease V, mismatch repair endonucleases like MutH, and single-strand specific nucleases like the S1, P1, or CEL − 1 nucleases.[65] Cleaved sequences can be removed by degradation or size selection with gel electrophoresis.[56,72] Cleaved sequences can also be repaired by trimming off the mismatched base pairs with exonuclease and then extending with polymerase. Treatment with T4 endo VII and *E. coli* endo V resulted in a four-fold reduction of error rate to 1 in 606 bp.[69] Bang and Church integrated circular assembly amplification with mismatch cleavage error correction.[56] Circular DNA-containing errors were linearized with *E. coli* endonuclease V. Exonuclease degraded linearized and uncircularized polynucleotides, reducing the error rate seven-fold to 1 in 4 kb.

Mismatch recognition can be readily integrated into automated gene synthesis procedures to significantly reduce error rates. Two recent microarray studies by Kosuri et al. and Quan et al. used CEL-based endonucleases to cleave and eliminate mismatch complexes from the hybridization of erroneous sequences in a population of microarray oligonucleotides.[12,19] This selectively removed erroneous sequences from the population, allowing for enrichment of error-free populations. Two iterations of enzymatic error correction reduced the error rate 16-fold to 1 in 8701 bp.[63]

High-Throughput Sequencing

To obtain error-free starting oligos, Church's group employed a novel next-generation sequencing (NGS) approach to address error correction costs in conventional gene synthesis.[73] They started with a pool of microarray-synthesized oligonucleotides and screened them using high-throughput pyrosequencing with the Roche 454 GS FLX platform. Input oligonucleotides were initially amplified via emulsion PCR, where unique oligonucleotides were attached to streptavidin-coated beads and amplified to millions of copies in water−oil aqueous solutions. Each bead was then loaded into a well in a PicoTiter plate and subjected to pyrosequencing. Unlike conventional Sanger sequencing, pyrophosphates get released by synthesis and the resulting chemiluminescence yields intensity peaks indicative of each base's location in the sequence. The beads containing amplicons with the desired sequences were then extracted from the flowcell to a multiwall plate with a camera-guided micropipette tool. These oligonucleotides were used as starting points, where they were enriched via PCR. The sequence-confirmed enriched pool has high-fidelity sequences that can be used to construct synthetic genes.

This approach holds promise for high-quality gene construction. The current setup relies on human handling, which contributes to an approximate 10% bead extraction failure rate. Further optimization, like the automation of emulsion PCR and bead retrieval, would improve efficiency and accuracy, and reduce labor and cost. Furthermore, making this technology compatible with other NGS platforms and other more robust microarray synthesis platforms has the potential to further scale-up the megacloning process. Using these methods, an applicable platform can be designed for large-scale and high-quality gene synthesis.

APPLICATIONS OF DNA SYNTHESIS
Gene Circuits

Gene circuits are synthetic DNA constructs which contain regulatory and functional elements to control biological responses.[74] Genetically engineered machines can serve as effective gene delivery, drug production, and metabolic platforms while shedding fundamental insight on natural biological systems. Design, construction, and implementation of synthetic gene networks provide a natural framework for designing larger artificial biological systems.

Partially due to a lack of facile gene writing tools, groups in synthetic biology took on a standardized assembly approach to construct gene circuits based on the collection and standardization of biological parts termed BioBricks. The BioBrick library is an open source standardized collection of biological parts, which is growing in size and documentation. BioBrick[75] or BglBrick[76] methods offer standardization in assembly. Standardized flanking sequences containing restriction sites allow for pairwise assembly of modular constructs into larger constructs. This makes assembly more parallel than traditional restriction digest and ligation methods. Efforts have been made to refine and automate the process.[77] However, larger constructs are often incompatible with BioBrick assembly due to additional restriction sites within the construct. Also, BioBrick digest scars cannot be avoided. The sequential BioBrick assembly procedure also makes it laborious.

While we now have the ability to sequence entire genomes and are able to computationally simulate dynamic systems to a certain extent, the ability to rapidly implement designed circuits that perform as expected still eludes us. Gene synthesis can serve to replace laborious recombinant DNA methods used in traditional genetic engineering. It can serve to replace BioBrick assembly, or at least circumvent the need to assemble smaller constructs. It would enable the transition from ad hoc restriction enzyme methods to systematic gene circuit and pathway designs.

Gene synthesis would ease the transition of synthetic biology from modules to systems.[78] This intellectual investment in DNA synthesis may enable scientists and engineers to quickly design, construct, and validate gene circuits that may do anything from curing diseases to generating alternative energy in a synthetic fashion.[79]

Codon Optimization

The demand for efficient recombinant protein production has grown in recent years. Commercially valuable proteins can be expressed in heterologous hosts like *E. coli* for large-scale production. These hosts can be used to produce a wide array of useful proteins for applications like pharmaceutics, cell therapies, and nano-material production. Even though the chemical and industrial setups have been engineered and optimized, biological protein expression can remain a bottleneck in production systems.

It is acknowledged that modifying a synthetic gene design for optimal synonymous codon usage using DNA synthesis can help achieve increased protein expression levels.[80–83] Unfortunately, robust rules for codon optimization are not available beyond the notion of mimicry of naturally occurring highly expressed genes. There is a lack of experimentally supported design principles based on codon usage, because of the limitation of small sample size for study imposed by existing synthesis technologies. Current methods of codon optimization are slow and expensive, due to the lack of fast and economical gene synthesis technology and incomplete understanding of protein translation.[82] Translation in bacteria involves initiation, elongation, termination, and ribosome turnover. At the same time, interactions with the expression environment vary protein solubility, mRNA stability, protein localization, and cell viability. Most factors are not fully understood, making it difficult to postulate causality. A number of theories have been proposed in the past for codon optimization. Some of the traditional schemes include CAI (codon adaptation index) maximization, codon sampling, dicodon optimization, and codon frequency matching.[82] Other theories for codon optimization include the importance of number of rare codons, pairs of consecutive rare codons, frequency of rare-pairs, number of potential RNaseE cleavage sites, Shine-Dalgarno occlusion by secondary structure, number of predicted rho-independent transcription termination signals, and many others.[84]

Kudla et al.[84] performed a systematic study where they generated a library of 154 GFP codon variants in which the codon usage was varied randomly at synonymous sites.

They expressed these proteins from a standard expression vector in *E. coli* for a 250-fold variation in protein levels as measured by GFP intensity with a microplate reader. The stability of mRNA folding near the start codon explained more than half the variation in protein levels. Secondary structures can inhibit rates of translation initiation and were the dominant factors in influencing gene expression levels. Codon variation can influence secondary structures by varying mRNA-free energies. The authors concluded that the key factor is avoidance of mRNA secondary structures near the ribosomal binding site, and that codon bias is important in fewer cases than previously thought.

Welch et al.[80] presented a novel systematic analysis of the relationship between sequence and expression. They first demonstrated the utility of synonymous codon variation by designing and synthesizing 40 codon variants of genes encoding two structurally and functionally different proteins: a DNA polymerase and a scFv; and realized a 40-fold variation in expression from bacterial lysates in polyacrylamide gels. The variants were designed using a Monte Carlo repeated random sampling algorithm. Experimental expression data poorly correlated with the Codon Adaptation Index (CAI), a previously widespread gene design principle that measures and mimics usage of preferred codons in a particular host genome. Welch et al. made the claim that favorable codons are not codons that are most abundant in highly expressed *E. coli* proteins. This was significant because CAI and related rules have been used to predict gene expression levels in the past by various gene synthesis vendors. Furthermore, they examined whether sequence characteristics affecting expression were local (mRNA structures, codon clusters) or global (codon usage, GC content). They divided the best and worst expressers into thirds and made chimeras. Some chimeras showed highly distributed effects, while some showed strong dependence on the parental origin of particular segments. Their findings showed that expression levels were not region-specific. Variation among the chimeras is largely explained by their predictive partial least squares model used to generate gene variants. Finally, when their model was cross-referenced with various design parameters, results suggested a biochemical basis of preferred codons in the sensitivity of amino-acylated tRNA during starvation conditions, based on predictive modeling work by Elf et al.[85]

The authors avoided overselling their claims by not offering a simple solution for codon optimization, but focusing on their systematic analysis as a design tool. A larger sample size for gene variants is still needed to make more specific conclusions about codon optimization.

Advances in throughputs of de novo DNA synthesis have allowed combinatorial library screens to become more pervasive.[86–88] With low-cost DNA synthesis, a large library of gene variants can be synthesized and analyzed with high-throughput expression screening platforms. Tian's group[19] applied this strategy by using their on-chip gene synthesis technology to synthesize oligonucleotide libraries, and assembled them into libraries of codon variants for LacZα and 74 *Drosophila* transcription factors. In one round of synthesis and screening, clones were selected with gene expression levels which varied from 0% to almost 60% of the total cell protein mass.

Genome Synthesis

While assembly of regular-sized genes in less than two weeks is now routine, assembly of longer sequences is still costly and unpredictable. On the genome scale, costs become prohibitive. A 10^6 bp genome would cost on the order of USD 100 000 for oligonucleotides alone, an unreasonable cost for an average laboratory. If microarrays are utilized, however, oligo costs could potentially be reduced to <USD 100,[21] and it would take less than a single chip to make all of the oligonucleotides. However, high gene synthesis costs have not deterred researchers from de novo synthesis of longer sequences. Synthesis length records have historically increased at a logarithmic rate. In 2002, the 7.5 kb poliovirus genome was

chemically synthesized.[2] In 2004, a 14 kb ribosomal gene cluster and a 32 kb polyketide synthase gene cluster were chemically synthesized. In 2008, the 583 kb *Mycoplasma genitalium* genome was synthesized at an estimated USD 10 million. And in 2010, a 1.1 Mb *Mycoplasma mycoides* genome was synthesized at an estimated USD 40 million. While the costs have deterred smaller laboratories from large genome synthesis, gene and genome synthesis will eventually be integrated into standard bioengineering routines.

Viral genomes are more feasible engineering targets because of their smaller sizes. In 2002, Cello et al. successfully synthesized a 7.5 kb long poliovirus genome.[89] The genome cDNA was assembled from 70 mer oligonucleotides to overlapping 400−600 bp fragments and to three large fragments. These fragments were finally ligated together to form a full genome. Working on intermediate cassettes allowed for sequence verification and parallelization. The cDNA was then transcribed into viral RNA using RNA polymerase in cell-free extract. The resulting genome allowed for translation of viral proteins and replication. Tissue culture experiments confirmed the pathogenic characteristics of the wild-type poliovirus. The only modification was the addition of genetic markers.

Smith et al. improved on previous methods of genome synthesis by demonstrating a method that would only require two weeks.[52] The well-studied 5386 bp PhiX174 bacteriophage genome, which encodes for 11 proteins, was assembled from synthetic oligonucleotides. This was one of the earliest projects conducted by the Venter Institute in genome synthesis. Three key modifications to the synthesis protocol were made to improve quality and decrease time for genome synthesis. One improvement was the addition of a gel purification step for purchased oligonucleotides in order to eliminate malformed or truncated oligos. In order to prepare oligonucleotides for PCA, top and bottom strands were purified separately so that annealing did not occur. The second improvement was the ligation of oligonucleotides under stringent annealing conditions, which reduced errors. The final improvement was the assembly of 5−6 kb ligation products into a full-length genome by PCR assembly, which is much faster than traditional restriction digest and ligation. After PCR amplification and another round of gel purification, functional synthetic phages were electroporated into *E. coli* cells and sequenced. Their method contained 9−10 inactivating mutations per synthetic genome, and one sequenced bacteriophage exactly matched the original sequence. Although most bacteriophages did not match the original sequence, the team created a reliable and quick method of creating a genome with expected physical properties.

Viral genome synthesis already has useful applications. There is a demand for effective influenza vaccines that can be rapidly produced. In recent methods, live viruses can be attenuated for vaccine development using global codon deoptimization.[90,91] By replacing codon pairs with suboptimal synonymous pairs in the genome, viral gene product synthesis is significantly diminished in otherwise functional and replicating viruses. Coleman and coworkers applied large-scale computer-aided viral genome redesign to perturb genome codon-pair bias in a process called synthetic attenuated virus engineering (SAVE).[90] This allows for vaccines to be rapidly designed and synthesized. At the same time, there is a large safety margin. Reversion to virulence is unlikely because codon-pair deoptimization results from the accumulated effects of thousands of synonymous mutations. SAVE-attenuated influenza viruses were created and used for effective vaccination by targeting genes critical for the assembly and replication of the influenza virus.[91] Further developments in the predictability of SAVE attenuation and additional assessment of the genetic stability of the attenuation phenotype may eventually lead to the use of SAVE-attenuated viruses in human trials. Improvements in DNA synthesis technology can accelerate vaccine development with the rapid synthesis of custom viral genomes.

In 2008, the synthesis of a 582 970 bp *M. genitalium* genome called JCVI − 1.0 was completed.[92] JCVI − 1.0 only differed from the wild-type genome in the exclusion of

MG408 which blocks pathogenicity, and identification watermarks. The synthesis of 101 starting 5−7 kb cassettes was outsourced to commercial vendors. Cassettes were assembled via PCR amplification with a primed bacterial artificial chromosome (BAC) vector, which allowed for recombination in *E. coli*. The process of assembly and recombination was repeated until quarter-length genome sequences were generated. These were amplified with a primed yeast artificial chromosome (YAC) vector for final recombination in *Saccharomyces cerevisiae*. The extracted genome was sequence confirmed.

Many improvements have been made to genome synthesis. For example, the Venter Institute has been able to utilize yeast cells to facilitate genome synthesis.[93] Lartigue et al. first reported the transformation of modified *M. mycoides* genome, YCpMmyc1.1, into yeast spheroplasts.[94] The authors then used tools exclusive to yeast to engineer the genome, before transplanting it into *Mycoplasma capricolum* cells. These tools allow for combinations of insertions, deletions, and rearrangements that were not previously possible with bacterial culture techniques. For example, the deletion of the ORF of a nonessential gene, the type III restriction enzyme, from the genome was performed by introducing a knockout cassette with a selection marker. Endonuclease-induced double-stranded breaks then promoted homologous recombination to remove the cassette, enabling seamless gene deletion. They finally devised a method to avoid restriction endonuclease cleavage of donor DNA in *M. capricolum* through in vitro methylation treatment and inactivation of the recipient restriction enzyme coding region. Methylation had to be performed in vitro due to incompatible methylases. The authors avoided problems with gene toxicity because a difference in stop codon results in truncated gene expression in yeast. Also, transplantation was limited to bacterial systems of the same genus to ensure compatible cell compositions between donor and recipient cell types. Despite these notable limitations, the achievement was groundbreaking. Being able to isolate and transform bacterial genomes into yeast, engineer them, and retransplant them creates opportunities to manipulate genomes in a way that was not previously possible, which ultimately results in the assembly of larger genomes.[95]

In 2010, a 1.1 Mb bacterial genome was synthesized using an accumulation of assembly techniques.[95] This was a technical feat, achieving the synthesis of the largest piece of DNA thus far with the accuracy to drive a functional synthetic organism. Although synthesis and assembly presented formidable labor and supply costs, the group was able to establish reliable methods for the assembly of 1 Mb constructs. A five-stage assembly scheme was used for complete assembly. The authors detail the progression from the assembly of oligonucleotides to 1078 1-kb cassettes, to 109 10-kb cassettes, to 11 100-kb cassettes, and to the full 1-Mb genome. Above the 1-kb stage, the authors used the previously mentioned yeast homologous recombination for assembly. The study was driven by several modern technologies including fast sequencing and outsourced oligonucleotide synthesis to commercial companies. The accomplishment was not without roadblocks. One notable problem was a single base pair deletion that caused a frameshift in *dnaA*, an essential gene for chromosomal replication. The problem was traced back to the 811−900 assembly, which was corrected through reconstruction and genome engineering in yeast. Similar problems will continue to present themselves in future genome synthesis projects and will become technical challenges for researchers in the field. Although the parallel assembly was efficient, each transition step can introduce random synthesis errors. Additionally, while homologous recombination in yeast can enable synthetic genome recombination,[95] it has a 20% efficiency and the underlying mechanisms are still unknown.[96] Problematic genome fragments with GC-rich noncoding sequences may also be necessary to discourage yeast repair mechanisms.[97] Finally, fully automating the steps of genome synthesis and transplantation would ease the labor and time involvement, enabling synthetic biologists to test hypotheses efficiently.

This accomplishment not only demonstrates the power of current synthesis methods, but also suggests the possibility of breaking new ground in synthetic life. However, we perhaps cannot build a truly synthetic cell until we can build a genome bottom-up from its functional parts. At this point, researchers simply do not understand the biological processes enough to produce an economically useful bacterium. However, the Venter Institute already has plans for building synthetic bacteria for carbon capture and biofuels. A novel synthetic minimal genome may offer a simplified and well-understood chassis for synthetic biologists. Perhaps, one may envision a hybrid genome or even a second orthogonal genome. For example, the successful fusion of *Bacillus subtilis* and *Synechocystis* genomes resulted in proliferating hybrids, which suggests the addition of new functions like photosynthesis to a species through genome synthesis.[98]

CONCLUSION

In four decades, the length record of synthetic DNA has grown from less than 100 bp to over 1 Mb.[95,99] Increases in productivity have allowed for DNA synthesis to be applied to a wide range of applications, including DNA nanotechnology, metabolic engineering, gene circuits, and genome synthesis.[79,100,101] However, the engineering infrastructures of DNA synthesis and assembly have not caught up to the growing demands of synthetic biology. For the engineering of large gene circuits and pathways, de novo DNA synthesis is responsible for only a small fraction of the entire constructs. Although the technologies discussed show promise for low-cost automated gene synthesis, further refinements are needed to improve manufacturing costs and turnover times before these new technologies can replace current column-based oligonucleotide synthesizers. High-throughput strategies are still compromised by high error rates. With further improvements and implementation of the instruments and strategies discussed here, DNA synthesis could become a routine method in bioengineering in the near future. The ability to rapidly create novel biological machines could transform biological and medical research.

Acknowledgments

The authors are thankful for financial support from NIH (R01HG005862) and the National Basic Research Program of China (2011CBA00805, 2012CB721102).

References

1. Mueller S, Coleman JR, Wimmer E. Putting synthesis into biology: a viral view of genetic engineering through de novo gene and genome synthesis. *Chem Biol.* 2009;16:337−347.

2. Carr P, Church GM. Genome engineering. *Nat Biotechnol.* 2009;27:1151−1162.

3. Bügl H, Danner JP, Molinari RJ, et al. DNA synthesis and biological security. *Nat Biotechnol.* 2007;25:627−629.

4. Caruthers MH. Gene synthesis machines: DNA chemistry and its uses. *Science (New York, N.Y.).* 1985;230:281−285.

5. Cheng J-Y, Chen H-H, Kao Y-S, Kao W-C, Peck K. High throughput parallel synthesis of oligonucleotides with 1536 channel synthesizer. *Nucleic Acids Res.* 2002;30:e93.

6. Lashkari DA, Hunicke-Smith SP, Norgren RM, Davis RW, Brennan T. An automated multiplex oligonucleotide synthesizer: development of high-throughput, low-cost DNA synthesis. *Proc Nat Acad Sci.* 1995;92:7912.

7. Sierzchala AB, Dellinger DJ, Betley JR, Wyrzykiewicz TK, Yamada CM, Caruthers MH. Solid-phase oligodeoxynucleotide synthesis: a two-step cycle using peroxy anion deprotection. *J Am Chem Soc.* 2003;125:13427−13441.

8. Blanchard A, Kaiser R, Hood L. High-density oligonucleotide arrays. *Biosens Bioelectron.* 1996;11:687−690.

9. Ellington A, Pollard JD. Introduction to the synthesis and purification of oligonucleotides. Curr Protoc Nucleic Acid Chem. Edited by Serge L. Beaucage et al. 2001; Appendix 3: Appendix 3C.

10. Lausted C, Dahl T, Warren C, et al. POSaM: a fast, flexible, open-source, inkjet oligonucleotide synthesizer and microarrayer. *Genome Biol.* 2004;5:R58.

11. Zhou X, Cai S, Hong A, et al. Microfluidic PicoArray synthesis of oligodeoxynucleotides and simultaneous assembling of multiple DNA sequences. *Nucleic Acids Res.* 2004;32:5409−5417.

12. Kosuri S, Eroshenko N, Leproust EM, et al. Scalable gene synthesis by selective amplification of DNA pools from high-fidelity microchips. *Nat Biotechnol.* 2010;28:1295−1299.

13. Kim H, Jeong J, Bang D. Hierarchical gene synthesis using DNA microchip oligonucleotides. *J Biotechnol.* 2011;151:319−324.

14. Borovkov AY, Loskutov AV, Robida MD, et al. High-quality gene assembly directly from unpurified mixtures of microarray-synthesized oligonucleotides. *Nucleic Acids Res.* 2010;38:e180.

15. Moorcroft MJ, Meuleman WRA, Latham SG, Nicholls TJ, Egeland RD, Southern EM. In situ oligonucleotide synthesis on poly(dimethylsiloxane): a flexible substrate for microarray fabrication. *Nucleic Acids Res.* 2005;33:e75.

16. Saaem I, Ma K-S, Marchi AN, LaBean TH, Tian J. In situ synthesis of DNA microarray on functionalized cyclic olefin copolymer substrate. *ACS Appl Mater Interfaces.* 2010;2:491−497.

17. Ma K-S, Reza F, Saaem I, Tian J. Versatile surface functionalization of cyclic olefin copolymer (COC) with sputtered SiO_2 thin film for potential BioMEMS applications. *J Mater Chem.* 2009;19:7914.

18. Ashley JF, Cramer NB, Davis RH, Bowman CN. Soft-lithography fabrication of microfluidic features using thiol-ene formulations. *Lab Chip.* 2011;11:2772−2778.

19. Quan J, Saaem I, Tang N, et al. Parallel on-chip gene synthesis and application to optimization of protein expression. *Nat Biotechnol.* 2011;29:449−452.

20. Lee C-C, Snyder TM, Quake SR. A microfluidic oligonucleotide synthesizer. *Nucleic Acids Res.* 2010;38:2514−2521.

21. Kong DS, Carr P a, Chen L, Zhang S. Parallel gene synthesis in a microfluidic device. *Nucleic Acids Res.* 2007;35:e61.

22. Xiao PF, He NY, Liu ZC, He QG, Sun X, Lu ZH. In situ synthesis of oligonucleotide arrays by using soft lithography. *Nanotechnology.* 2002;13:756−762.

23. Blair S, Richmond K, Rodesch M, Bassetti M, Cerrina F. A scalable method for multiplex LED-controlled synthesis of DNA in capillaries. *Nucleic Acids Res.* 2006;34:e110.

24. Beier M, Hoheisel JD. Versatile derivatisation of solid support media for covalent bonding on DNA-microchips. *Nucleic Acids Res.* 1999;27:1970−1977.

25. Hua Z, Xia Y, Srivannavit O, et al. A versatile microreactor platform featuring a chemical-resistant microvalve array for addressable multiplex syntheses and assays. *J Micromec Microeng.* 2006;16:1433.

26. Huang Y, Castrataro P, Lee C-C, Quake SR. Solvent resistant microfluidic DNA synthesizer. *Lab Chip.* 2007;7:24−26.

27. Phillips MF, Lockett MR, Rodesch MJ, Shortreed MR, Cerrina F, Smith LM. In situ oligonucleotide synthesis on carbon materials: stable substrates for microarray fabrication. *Nucleic Acids Res.* 2008;36:e7.

28. Strother T, Knickerbocker T, Russell Jr. JN, Butler JE, Smith LM, Hamers RJ. Photochemical functionalization of diamond films. *Langmuir.* 2002;18:968−971.

29. Sun B, Colavita PE, Kim H, et al. Covalent photochemical functionalization of amorphous carbon thin films for integrated real-time biosensing. *Langmuir.* 2006;22:9598−9605.

30. Fodor SP, Read JL, Pirrung MC, Stryer L, Lu AT, Solas D. Light-directed, spatially addressable parallel chemical synthesis. *Science.* 1991;251:767.

31. Barone AD, Beecher JE, Bury PA, et al. Photolithographic synthesis of high-density oligonucleotide probe arrays. *Nucleosides Nucleotides Nucleic Acids.* 2001;20:525−531.

32. Singh-Gasson S, Green RD, Yue Y, et al. Maskless fabrication of light-directed oligonucleotide microarrays using a digital micromirror array. *Nat Biotechnol.* 1999;17:974−978.

33. Richmond KE, Li M-H, Rodesch MJ, et al. Amplification and assembly of chip-eluted DNA (AACED): a method for high-throughput gene synthesis. *Nucleic Acids Res.* 2004;32:5011−5018.

34. Gao X, LeProust E, Zhang H, et al. A flexible light-directed DNA chip synthesis gated by deprotection using solution photogenerated acids. *Nucleic Acids Res.* 2001;29:4744−4750.

35. Serafinowski PJ, Garland PB. Novel photoacid generators for photodirected oligonucleotide synthesis. *J Am Chem Soc.* 2003;125:962−965.

36. Srivannavit O, Gulari M, Gulari E, et al. Design and fabrication of microwell array chips for a solution-based, photogenerated acid-catalyzed parallel oligonucleotide DNA synthesis. *Sens Actuators A Phys.* 2004;116:150−160.

37. Egeland RD, Southern EM. Electrochemically directed synthesis of oligonucleotides for DNA microarray fabrication. *Nucleic Acids Res.* 2005;33:e125.

38. Matteucci M. Synthesis of deoxyoligonucleotides on a polymer support. *J Am Chem.* 1981;544:3185−3191.

39. Hughes TR, Mao M, Jones AR, et al. Expression profiling using microarrays fabricated by an ink-jet oligonucleotide synthesizer. *Nat Biotechnol.* 2001;19:342−347.

40. Fisher W, Zhang M. A biochip microarray fabrication system using inkjet technology. Automation Science and Engineering, IEEE Transactions On 2007;4:488−500.

41. Fisher W. An automated biological fluid dispensing system for microarray fabrication using inkjet technology. Proceedings 2006 IEEE International Conference on Robotics and Automation, 2006. ICRA 2006. 2006:1786−1793.

42. Wolber PK, Collins PJ, Lucas AB, De Witte A, Shannon KW. The Agilent in situ-synthesized microarray platform. *Methods Enzymol.* 2006;410:28−57.

43. LeProust EM, Peck BJ, Spirin K, et al. Synthesis of high-quality libraries of long (150 mer) oligonucleotides by a novel depurination controlled process. *Nucleic Acids Res.* 2010;38:2522−2540.

44. Ciccarelli RB, Gunyuzlu P, Huang J, Scott C, Oakes FT. Construction of synthetic genes using PCR after automated DNA synthesis of their entire top and bottom strands. *Nucleic Acids Res.* 1991;19:6007−6013.

45. Itakura K, Hirose T, Crea R, et al. Expression in *Escherichia coli* of a chemically synthesized gene for the hormone somatostatin. *Science.* 1977;198:1056.

46. Goeddel DV, Kleid DG, Bolivar F, et al. Expression in *Escherichia coli* of chemically synthesized genes for human insulin. *Proc Natl Acad Sci USA.* 1979;76:106−110.

47. Au LC, Yang FY, Yang WJ, Lo SH, Kao CF. Gene synthesis by a LCR-based approach: high-level production of leptin-L54 using synthetic gene in *Escherichia coli. Biochem Biophys Res Commun.* 1998;248:200−203.

48. Barany F. Genetic disease detection and DNA amplification using cloned thermostable ligase. *Proc Natl Acad Sci USA.* 1991;88:189−193.

49. Pfeffer M, Wiedmann M, Batt CA. Ligase chain reaction (LCR)−overview and applications. *Genome Res.* 1995;19:375−407.

50. Barany F. The ligase chain reaction in a PCR world. *Genome Res.* 1991;1:5−16.

51. Jayaraman K, Shah J, Fyles J. PCR mediated gene synthesis. *Nucleic Acids Res.* 1989;17:4403.

52. Smith HO, Hutchison CA, Pfannkoch C, Venter JC. Generating a synthetic genome by whole genome assembly: X174 bacteriophage from synthetic oligonucleotides. *Proc Natl Acad Sci USA.* 2003;100:15440.

53. Tian J, Gong H, Sheng N, et al. Accurate multiplex gene synthesis from programmable DNA microchips. *Nature.* 2004;432:1050−1054.

54. Stemmer W, Crameri A, Ha K, Brennan T. Single-step assembly of a gene and entire plasmid from large numbers of oligodeoxyribonucleotides. *Gene.* 1995.

55. Prodromou C, Pearl LH. Recursive PCR: a novel technique for total gene synthesis. *Protein Eng.* 1992;5:827−829.

56. Bang D, Church GM. Gene synthesis by circular assembly amplification. *Nat Methods.* 2007;5:37−39.

57. Quan J, Tian J. Circular polymerase extension cloning of complex gene libraries and pathways. *PloS One.* 2009;4:e6441.

58. Sleight SC, Bartley BA, Lieviant JA, Sauro HM. In-fusion BioBrick assembly and re-engineering. *Nucleic Acids Res.* 2010;38:2624−2636.

59. Nour-eldin HH, Geu-flores F, Halkier BA. USER cloning and USER fusion: the ideal cloning techniques for small and big laboratories. *Methods Mol Biol.* 2010;643:185−200.

60. Li MZ, Elledge SJ. Harnessing homologous recombination in vitro to generate recombinant DNA via SLIC. *Nat Methods.* 2007;4:251−256.

61. Gibson DG, Young L, Chuang RY, Venter JC, Hutchison CA, Smith HO. Enzymatic assembly of DNA molecules up to several hundred kilobases. *Nat Methods.* 2009;6:343−345.

62. Gibson DG, Smith HO, Hutchison III CA, Venter JC, Merryman C. Chemical synthesis of the mouse mitochondrial genome. *Nat Methods.* 2010;7:901−903.

63. Saaem I, Ma S, Quan J, Tian J. Error correction of microchip synthesized genes using surveyor nuclease. *Nucleic Acids Res.* 2011:1−8.

64. Kim H, Han H, Shin D, Bang D. A fluorescence selection method for accurate large-gene synthesis. *Chembiochem.* 2010;11:2448−2452.

65. Ma S, Saaem I, Tian J. Error correction in gene synthesis technology. *Trends Biotechnol.* 2012;30:147−154.

66. Modrich P. Mechanisms and biological effects of mismatch repair. *Annu Rev Genet.* 1991;25:229−253.

67. Carr PA, Park JS, Lee Y-J, Yu T, Zhang S, Jacobson JM. Protein-mediated error correction for de novo DNA synthesis. *Nucleic Acids Res.* 2004;32:e162.

68. Binkowski BF, Richmond KE, Kaysen J, Sussman MR, Belshaw PJ. Correcting errors in synthetic DNA through consensus shuffling. *Nucleic Acids Res.* 2005;33:e55.

69. Fuhrmann M, Oertel W, Berthold P, Hegemann P. Removal of mismatched bases from synthetic genes by enzymatic mismatch cleavage. *Nucleic Acids Res.* 2005;33:e58.

70. Yang W. Structure and function of mismatch repair proteins. *Mutat Res.* 2000;460:245−256.

71. Smith J, Modrich P. Removal of polymerase-produced mutant sequences from PCR products. *Proc Natl Acad Sci USA.* 1997;94:6847−6850.

72. Young L, Dong Q. Two-step total gene synthesis method. *Nucleic Acids Res.* 2004;32:e59.

73. Matzas M, Stähler PF, Kefer N, et al. High-fidelity gene synthesis by retrieval of sequence-verified DNA identified using high-throughput pyrosequencing. *Nat Biotechnol.* 2010;28:1291−1294.

74. Voigt CA. Genetic parts to program bacteria. *Curr Opin Biotechnol.* 2006;17:548−557.

75. Knight T. Idempotent vector design for standard assembly of biobricks standard biobrick sequence interface. *DSpace.* 2003:1−11.

76. Anderson JC, Dueber JE, Leguia M, et al. BglBricks: a flexible standard for biological part assembly. *J Biol Eng.* 2010;4:1.

77. Canton B, Labno A, Endy D. Refinement and standardization of synthetic biological parts and devices. *Nat Biotechnol.* 2008;26:787−793.

78. Purnick PEM, Weiss R. The second wave of synthetic biology: from modules to systems. *Nat Rev Mol Cell Biol.* 2009;10:410−422.

79. Reza F, Chandran K, Feltz M, et al. Engineering novel synthetic biological systems. *Synth Biol, IET.* 2007;1:48−52.

80. Welch M, Govindarajan S, Ness JE, et al. Design parameters to control synthetic gene expression in *Escherichia coli*. *PloS One.* 2009;4:e7002.

81. Saunders R, Deane CM. Synonymous codon usage influences the local protein structure observed. *Nucleic Acids Res.* 2010;44:1−10.

82. Welch M, Villalobos A, Gustafsson C, Minshull J. You're one in a googol: optimizing genes for protein expression. *J R Soc Interface.* 2009;6(suppl 4):S467−S476.

83. Rosano GL, Ceccarelli EA. Rare codon content affects the solubility of recombinant proteins in a codon bias-adjusted *Escherichia coli* strain. *Microb Cell Fact.* 2009;8:41.

84. Kudla G, Murray AW, Tollervey D, Plotkin JB. Coding-sequence determinants of gene expression in *Escherichia coli*. *Science (New York, N.Y.).* 2009;324:255−258.

85. Elf J, Nilsson D, Tenson T, Ehrenberg M. Selective charging of tRNA isoacceptors explains patterns of codon usage. *Science (New York, N.Y.).* 2003;300:1718−1722.

86. Salis HM, Mirsky EA, Voigt CA. Automated design of synthetic ribosome binding sites to control protein expression. *Nat Biotechnol.* 2009;27:946−950.

87. Yokobayashi Y, Weiss R, Arnold FH. Directed evolution of a genetic circuit. *Proc Nat Acad Sci USA.* 2002;99:16587−16591.

88. Ellis T, Wang X, Collins JJ. Diversity-based, model-guided construction of synthetic gene networks with predicted functions. *Nat Biotechnol.* 2009;27:465−471.

89. Cello J, Paul AV, Wimmer E. Chemical synthesis of poliovirus cDNA: generation of infectious virus in the absence of natural template. *Science (New York, N.Y.).* 2002;297:1016−1018.

90. Coleman JR, Papamichail D, Skiena S, Futcher B, Wimmer E, Mueller S. Virus attenuation by genome-scale changes in codon pair bias. *Science (New York, N.Y.).* 2008;320:1784−1787.

91. Mueller S, Coleman JR, Papamichail D, et al. Live attenuated influenza virus vaccines by computer-aided rational design. *Nat Biotechnol.* 2010;28:723−726.

92. Gibson DG, Benders GA, Andrews-Pfannkoch C, et al. Complete chemical synthesis, assembly, and cloning of a *Mycoplasma genitalium* genome. *Science (New York, N.Y.).* 2008;319:1215−1220.

93. Gibson DG, Benders GA, Axelrod KC, et al. One-step assembly in yeast of 25 overlapping DNA fragments to form a complete synthetic *Mycoplasma genitalium* genome. *Proc Natl Acad Sci USA.* 2008;105:20404−20409.

94. Lartigue C, Vashee S, Algire MA, et al. Creating bacterial strains from genomes that have been cloned and engineered in yeast. *Science (New York, N.Y.).* 2009;325:1693−1696.

95. Gibson DG, Glass JI, Lartigue C, et al. Creation of a bacterial cell controlled by a chemically synthesized genome. *Science (New York, N.Y.).* 2010;329:52−56.

96. Benders GA, Noskov VN, Denisova EA, et al. Cloning whole bacterial genomes in yeast. *Nucleic Acids Res.* 2010;38:2558−2569.

97. Stinchcomb DT, Thomas M, Kelly J, Selker E, Davis RW. Eukaryotic DNA segments capable of autonomous replication in yeast. *Proc Natl Acad Sci.* 1980;77:4559.

98. Itaya M. Stable positional cloning of long continuous DNA in the bacillus subtilis genome vector. *J Biochem.* 2003;134:513−519.

99. Agarwal K, Caruthers M, Gupta N. Total synthesis of the gene for an alanine transfer ribonucleic acid from yeast. *Nature.* 1970.

100. McDaniel R, Weiss R. Advances in synthetic biology: on the path from prototypes to applications. *Curr Opin Biotechnol.* 2005;16:476−483.

101. Benner SA, Sismour AM. Synthetic biology. *Nat Rev Genet.* 2005;6:533−543.

Protein Engineering as an Enabling Tool for Synthetic Biology

Patrick C. Cirino and Shuai Qian
University of Houston, Houston, TX, USA

INTRODUCTION

Proteins are essential ingredients in any natural or synthetic biological system, and the extent to which properties of one or more proteins can be modified in direct and desired ways often dictates success in synthetic biology and metabolically engineered systems. Common examples of protein properties that may need modification include stability (e.g. to solvents, low pH, or thermal denaturation), binding affinity, catalytic activity, and substrate specificity. Enzymes are often engineered to carry out new functions in the context of synthetic metabolic pathways. As synthetic biologists are often concerned with achieving novel control over gene expression, regulatory proteins and transcription factors engineered to have new or improved regulatory properties can also play a powerful role in this discipline. This chapter describes a variety of the most common and effective methods in which existing proteins are engineered for improved or even novel properties. Note that a 'property' may also be referred to as a 'fitness' parameter, or 'activity' (even for the case of noncatalytic proteins). We first introduce the general concept of protein engineering and the two broadest categories of this practice – rational design versus directed evolution. Following a closer look into rational protein design, we then describe a variety of techniques used by protein engineers to construct protein libraries and to screen these libraries using high-throughput assays or cell growth-based selections. Finally, we present several recent and/or pioneering examples in which protein engineering served as an enabling tool in synthetic biology applications.

The protein engineering methods described here all involve introducing changes at the DNA level (i.e. altering the corresponding gene sequence). Specified changes to a DNA sequence originate from DNA synthesis (refer to Chapter 1 for detailed discussion of DNA synthesis technologies). Short oligonucleotides may be constructed (e.g. PCR primers) and used to introduce desired mutations using any number of gene amplification, assembly, or recombination techniques, or alternately the entire gene may be synthesized and directly cloned into an expression vector. While this chapter does not describe details behind the molecular biology used to engineer proteins, it is important to note that in all cases the changes made are at the genetic level and the engineered protein is expressed in an appropriate host organism. In some cases a single protein scaffold serves as the 'wild-type'

23

Synthetic Biology. DOI: http://dx.doi.org/10.1016/B978-0-12-394430-6.00002-9

or 'parent' template for engineering. In other cases more than one parent serves as the starting point for engineering chimeric proteins (i.e. proteins whose gene sequences are hybrids of more than one gene, resulting from gene recombination). In addition to the citations provided in context throughout this chapter, we refer readers interested in more methods and/or details of experimental protocols in protein engineering to several volumes dedicated to this field.[1–4]

PROTEIN ENGINEERING METHODS

As depicted in Figure 2.1, methods of protein engineering can be broadly categorized as rational design and combinatorial design or directed evolution, although a combination of these is often used. Rational design involves the use of protein sequence, structure, and function information, and any known relationships between these, to devise specific mutagenesis

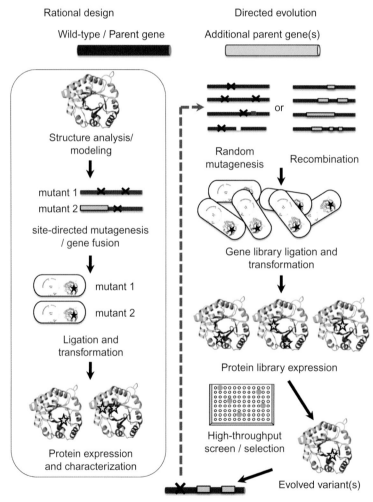

FIGURE 2.1

Approaches to engineer proteins can be broadly categorized as rational design or directed evolution. Rational design involves the use of sequence/structure/function information to predict site-directed mutagenesis strategies (e.g. point mutations or gene fusions) that will improve properties of interest. The protein's gene sequence is modified accordingly, and the protein variant is expressed and characterized. Directed evolution involves the generation of gene libraries based on one or more parent gene sequence. The library of mutants are expressed and assayed in a high-throughput screen. Improved variants are selected and the process can be repeated, thus mimicking natural evolution. As described in the text, there are many variations on these methods, including semirational design and in vitro expression.

strategies to alter protein properties in a defined and predicted manner. Computational tools that assist in predictive protein design strategy are continuously improving.

Owing to the great difficulties in understanding protein sequence–structure–function relationships, protein engineers must often accept some level of naivety and turn to a less predictive design approach. In general, the alternative strategy is to clone and express many different variants (create a combinatorial library), and then use a functional assay to screen each library member to select variants with improvements in the property of interest. Often this strategy of mutagenesis and selection is implemented iteratively, essentially mimicking Darwinian evolution. Parameters such as the positioning of mutations, extent of sequence diversity, and number or percentage of library members assayed can vary considerably and depend primarily on the extent to which sequence–function relationships are understood and the nature of the function assay used to screen or select variants (e.g. throughput and accuracy).

Rational design by site-directed mutagenesis has the potential to effect drastic improvements in protein function, for example when coordinated mutations in a binding pocket successfully improve binding or activity toward a nonnative substrate. Unfortunately, this approach is not always as fruitful as one may hope, and predicted, targeted mutations often do not furnish any improvements. Directed evolution by random mutagenesis, on the other hand, has a high rate of success. However, since this approach involves successive accumulation of low numbers of mutations, and since in any given round of screening only a small number of possible mutations are tested, only moderate improvements in function are typical in each round of screening (e.g. <two-fold). Recombination of multiple parental sequences offers opportunities for more drastic improvements, given that a much larger number of functional sequence changes are accessible (i.e. a large number of 'fitness-conferring' point mutations relative to a single parent are presented in the recombinant library). As understanding of individual protein sequence–structure relationships improve, strategies of site-directed and combinatorial mutagenesis can be combined and often result in the most drastic improvements in protein engineering.

Rational Design

Accurate prediction of mutagenesis strategies that improve protein properties of interest are becoming increasingly possible due to: continued progress in protein structure determination, increasing numbers of sequences and sequence–structure relationships populating research databases, growing insights into structure–function relationships, and advanced computational and molecular modeling tools to predict protein structure and dynamics. Visual inspection of 'static' crystal structures of proteins in complex with ligands has been a common approach to rational design altered ligand binding. Structure analysis might identify locations within a binding pocket to substitute hydrogen bond donors or acceptors to improve ligand affinity, or replace a bulky residue with a smaller one to allow access of a larger substrate. Once one or more desired mutations are determined, the corresponding gene is synthesized and cloned. While sometimes successful, there are obvious limitations with this technique alone, given the dynamic nature of proteins and amino acid side-chains.

Improvements upon this 'inspection' approach involve levels of increasing computational complexity. This topic is covered in detail in Chapter 7, and is therefore only mentioned briefly here. As an example, a ligand of interest may be computationally docked into the protein binding pocket and its lowest energy position can be predicted. Side-chain replacements can then be iteratively tested, each followed by energy minimization calculations in which ligand and/or neighboring side-chain conformations are sampled to reach a lowest energy complex. This may be performed with or without fixing the protein's backbone coordinates.[5,6] An alternate approach to using molecular modeling to modify an existing protein scaffold is the more challenging de novo design of a protein, whereby amino acids and/or protein secondary structure elements are assembled (computationally) to achieve the desired structure/function.[6]

In the absence of structural information, a number of other experimental and molecular modeling tools can be used to predict protein structure or structure–function relationships. Homology modeling is commonly used to guide rational design of a protein, based on existing structural information for a homologue.[7] Often experimental results from studies of a protein's homologues or mutants lead to successful rational protein design (perhaps in combination with structural insights). For example, one may compare homologous proteins having nearly identical sequences (or variants of a single wild-type protein) and identify specific amino acid substitutions that are largely responsible for differences in specific properties between the homologues. Techniques such as alanine scanning mutagenesis,[8] in which individual amino acid positions are substituted with alanine (chosen since it eliminates the side chain beyond the β carbon and is least likely to change the protein conformation or impose electrostatic or steric effects) are also commonly used to search for residues critical to a protein's function and can furnish insights for rational design.

Rational design often involves a single amino acid position targeted for site-directed mutagenesis to another amino acid (e.g. to relax an enzyme's substrate specificity by increasing the volume of the binding pocket[9]). The more difficult task of identifying multiple positions to mutate simultaneously to achieve improvements in function due to coordinating point mutations is most commonly accomplished with the aid of molecular modeling, but has also been demonstrated via visual structure analysis.[10]

Another form of rational protein design involves the construction of site-specific protein chimeras, fusions, and truncations. Common, simple examples are the addition of in-frame tags (e.g. 6-histidines) and fluorescent protein fusions. More complex examples involve the rearrangement, removal, or addition of specific protein domains or modules. Here, a domain is defined as a structurally separable unit within a protein, while a module is a functionally minimal unit that is transferable from one protein context to another.[11] With sufficient structural and functional information about a domain or module, one may be able to rationally combine their respective functions by careful, site-directed insertions or fusions. However, even a seemingly straightforward protein construction may not properly fold or retain function for unpredicted reasons. As described below, the combinatorial approach of constructing chimeric libraries followed by screening for active proteins can be more effective.

A growing area of rational protein design involves the incorporation of noncanonical amino acids at specific sites in a protein sequence.[12,13] This allows the introduction of novel chemical functionalities. While the techniques and efficiency of this technology continue to evolve, the general methodology involves nonsense codon suppression with suppressor tRNAs and their corresponding aminoacyl tRNA synthetases that have been engineered (using directed evolution methods) to accept a desired noncanonical amino acid. While proteins engineered to contain noncanonical amino acids have largely been developed for in vitro protein applications and as molecular tools to probe biological functions, their use holds promise for creating new biological functions and chemistries in synthetic biology.

Collectively, rational design has furnished many successfully engineered proteins. Examples include proteins with improved thermal stability, altered substrate specificity, proteins engineered to bind metals, and even enzymes with novel catalytic activity.[3,14,15] Despite such successes, in many cases it is overly difficult or impossible to predict a mutagenesis strategy that will yield a desired protein property.

DIRECTED EVOLUTION

Even if a protein structure is known, 'rational' site-directed mutations predicted by structural and/or computational analyses to improve a function of interest are often deleterious or do not provide the expected change in function due to unknown complexities of protein structure and biochemistry. It is important to remember that properly folded and functional proteins are typically only marginally stable relative to their unfolded or non-functional state(s).[16] Even seemingly minor perturbations in sequence can lead to significantly reduced stability or function. A simple demonstration of this is the observation that a large majority of protein variants containing a single amino acid substitution relative to the wild-type or parent sequence result in reduced or eliminated parent fitness. One can therefore imagine a number of ways in which a mutation predicted to improve binding toward a specific ligand results in a less stable protein, even if the mutation does indeed improve binding (which may or may not be detected, depending on the binding assay conditions).

For these reasons an evolutionary approach to protein engineering is often preferred. In general this entails generating genetic diversity in the form of a mutant library, followed by screening the resulting protein variants for the improved properties of interest. The more closely the screening conditions reflect the actual conditions in which the protein is to have the desired functions, the more likely the resulting variant isolated will be useful. Hence the directed evolution adage 'you get what you screen for' (see below for examples of screening assays used in directed evolution). For example, if stability is not included as a fitness parameter in a screening assay, one might end up with a less stable but more active variant. The required stability of course depends on the specific application.

Below we outline commonly used methods to generate libraries of protein variants. The parent or wild-type protein may be a single sequence, and amino acid changes are simply substitutions at various positions throughout that sequence. Alternately, for the case of recombination, more than one gene sequence (corresponding to more than one protein sequence) serves as the template for generating genetic (and hence protein) diversity, and diversity is assessed by the number of crossovers between templates, as well as the relative differences in amino acid sequence between variants of interest and their parents.

GENERATING SEQUENCE DIVERSITY

Random Mutagenesis

Random mutagenesis can be an attractive approach under a variety of circumstances in which a protein function is to be improved by directed evolution. When little or nothing is known about a protein's structure or sequence—function relationships, this is the obvious choice both to gain functional insights and to isolate improved variants. But even when a great deal is known about the target protein, screening random mutations throughout the sequence is still likely to yield improved variants when more targeted mutagenesis techniques do not. A common strategy is to first generate and screen random mutation libraries to identify potential 'hot spot' amino acid positions and regions of a protein, which are then targeted for further mutagenesis. Random mutagenesis can also be applied to specific gene segments, such as a gene region corresponding to one domain of a multidomain protein.

A variety of techniques have been developed to randomly introduce nucleotide substitutions into gene segments. Early random mutagenesis methods involved exposing whole cells to mutagenic conditions, such as UV exposure,[17] X-ray radiation,[18] or chemical mutagens.[19,20] 'Mutator' strains have been constructed, for example *E. coli* XL1-red which is deficient in three primary DNA repair pathways (*mutS*, *mutD*, and *mutT*).[21] Amplifying a plasmid via growth in a mutator strain produces randomly mutagenized plasmid libraries. Since mutations are not targeted to only the gene of interest with these methods, subsequent cloning steps are necessary before screening. Mutations accumulating in host DNA cause growth deficiencies, so several rounds of growth, plasmid purification, transformation, and regrowth are often required.

The most widely used method for creating random mutation gene libraries is error-prone polymerase chain reaction (error-prone PCR).[22] This technique involves PCR-based gene amplification using a low-fidelity DNA polymerase (e.g. *Taq* polymerase, which lacks 3′ to 5′ exonuclease proofreading activity), and any number of PCR conditions that promote a relatively controlled frequency of deoxynucleotide misinsertion during DNA replication. The primer sequences selected dictate the gene region to be amplified and therefore mutated. Elevated concentrations of $MgCl_2$ (e.g. ~7 mM instead of ~1.5 mM helps to stabilize non-complementary pairs), addition of Mn^{2+} (which replaces Mg^{2+} as a mutagenic cofactor during DNA replication), or a deoxynucleotide analogue such as 8-oxo-dGTP or dITP, and using unequal concentrations of deoxynucleotides all increase frequency of mutation during template amplification.[22] Other parameters such as the template concentration and number of PCR cycles can also be used to adjust mutation frequency. Depending on the conditions used, mutation rates can be varied from one to ~20 mutations per kb. It is important to remember that many nucleotide substitutions result in silent mutations (no change in protein amino acid sequence), and amino acid changes requiring more than one nucleotide substitution within a single codon occur with low probability.

While a powerful and essentially universally effective protein engineering technique, there are several limitations associated with random mutagenesis via error-prone PCR. For example, often the distribution of mutations appearing during PCR amplification is not even, and in some cases mutation 'hot spots' arise. *Taq* polymerase also preferentially introduces A→T, T→A, A→G, and T→C mutations, and rarely incorporates G→C and C→G transversions. Many reports describe optimized protocols and reagents to help alleviate such limitations, and commercial error-prone PCR kits featuring these improvements are available. For example, DNA polymerases have been engineered to exhibit altered mutation biases. Fuji and coworkers developed error-prone rolling circle amplification, allowing for direct transformation of the product.[23] This method requires no ligation step, but mutations are distributed throughout the entire expression vector.

Recombination

Gene recombination involves the exchange or sharing of genetic information between multiple parental sequences to create new 'progeny' genes. A key benefit of using recombination to generate genetic diversity is the fact that the parental sequence fragments being combined correspond to pieces of related and functional proteins (the homologues provide 'functional diversity'). Compared to random mutagenesis, where increasing the number of mutations drastically reduces the probability of obtaining a folded or functional protein, recombination allows for a higher probability of a greater number of amino acid changes (relative to a given parent sequence) to retain the protein's ability to fold and carry traits from both parents, as well as potentially new properties. In vitro recombination for the sake of creating gene libraries is classified as either homologous, when the parent genes share a high level of sequence similarity, or non-homologous, when the parents are not homologues or share low sequence similarity.

In Vitro Homologous Recombination DNA shuffling is the first reported method of in vitro recombination.[24] Here, DNAse I is first used to randomly fragment the parental genes. Then, provided there is sufficient sequence similarity between overlapping regions of DNA fragments from different parents, the fragments can anneal to one another and reassemble into a full-length gene using PCR. Depending on the total gene size, the degree of sequence similarity between parents, and desired number of crossovers (regions where different parental fragments are assembled), fragment sizes isolated from DNAse I treatment for reassembly can vary from as low as 10–50 bp to greater than 1 kb. Random point mutations tend to occur at low rates during recombination even with a high-fidelity polymerase, and researchers will often intentionally employ error-prone PCR during PCR-based gene recombination to further diversify their library. In a noteworthy, early example of DNA shuffling, a family of 20 human interferon-alpha genes was shuffled followed by selection of antiviral and antiproliferation activities in murine cells, resulting in variants having 285 000-fold increased activity.[25] The best chimeras were composed of up to five parental genes and contained no random point mutations. In another example, shuffling of 26 homologous protease genes generated many chimeric proteases that were significantly improved over any of the parental enzymes for four different properties assayed.[26]

A number of other PCR-based and homology-dependent protocols have since been developed which accomplish essentially the same as DNA shuffling, and many of these are summarized in a methods volume.[1] Examples include the staggered extension process,[27] random-priming in vitro recombination,[28] and random chimeragenesis on transient templates.[29] In general these methods work well for DNA sequences sharing at least 65% identity, although higher sequence identity more readily yields a greater number of crossovers.

In vitro recombination methods are also often used in directed evolution, even when the only genetic diversity is introduced by random mutagenesis of a single parent gene. Here, one or more rounds of mutagenesis and screening to isolate improved variants results in a handful of mutant genes, each carrying a different set of point mutations. By shuffling these highly identical mutant DNA sequences, one can readily obtain a library containing all combinations of point mutations. Beneficial mutations can be combined and may show additive effects, while any potentially deleterious mutations that have accumulated will be eliminated by 'back-crossing' with the wild-type sequence.

In Vivo Homologous Recombination Whereas gene libraries generated using PCR are most commonly ligated into an expression vector and transformed into *E. coli* (not necessarily for expression and screening, but at least for plasmid library recovery), a more recently developed technique that does not require ligation takes advantage of yeast homologous recombination.[30–33] The general approach involves cotransforming in yeast a linear vector expressing the target gene with linear, homologous DNA fragments. In vivo homologous recombination between these genes results in a library of mutants cloned into the vector. The homologous DNA fragments providing sequence diversity can be PCR products of in vitro recombination, a family of homologues, or even a library of synthetic oligonucleotides.

Nonhomologous Recombination Often it is desirable to generate protein fusions or higher-order chimeras by combining two or more genes that are either not homologues or do not share sufficient sequence similarity for the above methods. One approach is rational design by assembling protein domains or modules, as described above (fusion or crossover positions are selected and the corresponding chimera is cloned and expressed). Typically, however, the choice of positions to insert protein domains, modules, or nonhomologous crossover points is not obvious, or individually constructed chimeras do not work. A variety of nonhomologous recombination methods have therefore been developed for the generation of chimeric libraries which can subsequently be screened for function. The methods of 'incremental truncation for the creation of hybrid enzymes' (ITCHY),[34] ITCHY combined with DNA shuffling ('SCRATCHY'),[35] and 'sequence homology-independent protein recombination' (SHIPREC)[36] involve in vitro construction of gene fusions by ligating libraries of nonhomologous genes which have been randomly truncated. More recently, alternate techniques have been developed which utilize more advanced molecular biology tools, in efforts to enhance crossovers.[37,38]

Whereas these previous methods involve randomizing the locations of nonhomologous sequence crossovers, parent protein structural information can also be used to guide the design

29

of combinatorial crossover libraries.[39–41] Here, efforts are made to swap distinct structural elements and to minimize disruptions to the overall protein architecture. PCR primers corresponding to the selected crossover regions are then designed and used to direct the desired recombination events. This method is particularly useful for chimeragenesis from distantly related family members, where structure is largely conserved though sequence similarity is low.

Site-Directed Diversification

Above, site-directed mutagenesis in which amino acid positions are intentionally substituted with alternate, specified residues is considered 'rational' protein design. The classification becomes a hybrid between rational and combinatorial design when targeted positions are instead randomized. With this approach, information from structural and/or biochemical analyses is used to predict which positions in a protein should be mutated in order to improve a desired property. However, rather than constructing a single new variant, a focused gene library is instead constructed, in which the codon positions to be mutated are diversified to a desired extent. Traditionally this involves synthesizing a series of oligonucleotides corresponding to the regions to be mutated, but where the bases at the codons of interest are replaced by mixtures of nucleotides. The diversified oligonucleotides are used as primers to amplify the gene, and the fragments are assembled into a gene library using PCR. The mixture of nucleotides at each base position can be specified, such as using an equimolar combination of A, T, G, and C (referred to as 'N'), or using a mixture of only G and C (referred to as 'S'). The choice of codon diversification depends on how many and which amino acids one wants to be represented in the protein library.

Site-saturation mutagenesis refers to a codon set that includes all 20 amino acids. There are 64 possible codons and three possible stop codons in an NNN set, while the set NNS still encodes all 20 amino acids and only contains one stop codon (UAG). Still, 32^n possible combinations from n simultaneously randomized positions quickly becomes an intractable number of library members to screen as n grows, depending on the screening method used. One can instead iteratively perform saturation mutagenesis at a small number of positions (e.g. residues lining an enzyme's substrate binding pocket randomized in pairs or in triplicate), again depending on the number of variants that can be screened.[42,43] Alternately, individual positions throughout the protein can be subjected to saturation mutagenesis and screened. After screening each library, mutations showing improvements can be combined and tested for additivity, further mutagenesis can be performed, etc. Other approaches involve PCR-based gene assembly using primers or gene fragments which have been doped with randomized codons.[44] In this manner, any one position targeted for mutagenesis still has a high probability of being coded for the wild-type amino acid, depending on the level of mutant doping. This technique is therefore similar to random mutagenesis, except only selected positions are mutated, and the mutations are specified (on oligonucleotides) and can therefore be entire codons or even multiple codons on a single oligonucleotide. Many variations on these types of codon mutagenesis techniques have been reported.

In order to examine a greater number of simultaneously diversified residue positions without creating intractable library sizes, one may consider ways of restricting the codon alphabet to limit the set of amino acids encoded at each position. A variety of such methods have been developed and applied, with the chosen codon usage resulting from experimental observations, chemistry of the corresponding amino acids, and/or predictions from molecular modeling.[45] Researchers have explored protein libraries where n positions are diversified to as few as two possible amino acids.[46] In a noteworthy example, Reetz and coworkers compared the use of fully randomized codons (NNK, where 'K' represents G or T) to NDT codons ('D' represents A, G, or T), which corresponds to 12 codons that code for 12 possible amino acids.[47] Given the same number of codon positions diversified, the NDT library provided a significantly higher frequency of positive variants (the protein function was epoxide hydrolysis by a hydrolase). This study demonstrates that given limited library screening capabilities, functional redundancy in the genetic code can be a limiting factor.

SCREENING AND SELECTION

The speed and accuracy of a screening platform typically dictate both the mutagenesis strategy chosen, as well as the number of variants screened. Maximum attainable protein library sizes are considered limited by cloning efficiency (ligation and transformation) at $\sim 10^9 - 10^{10}$ members. For cell-free expression or in vitro display systems (see below), since no transformation is required, protein libraries of up to 10^{14} members can be constructed.[48] As outlined below, the throughput of screening systems requiring transformation can vary from $\sim 10^3$ to $\sim 10^9$, and perhaps 10^{10} for genetic selections and phage display. The universal theme in any protein screening assay is the retention of a direct link between the expressed protein's function and the protein's genetic code (i.e. the mutations conferring function different from the parent). Thus, if a protein variant is not to be assayed in its host, one must either keep track of the corresponding cell/plasmid, or physically link the protein to its genetic information (display technologies).

In general, the accuracy and quantitative nature of a protein assay vary inversely with the throughput of the assay. This is particularly true for enzyme screening. For example, very low-throughput analyses such as chromatography, NMR, or various spectroscopic methods can provide quantitative information (concentration, structure, kinetics), while many high-throughput methods that monitor cell growth or fluorescence usually provide primarily qualitative information. Protein engineers are often willing to sacrifice data accuracy and confidence in order to observe much larger numbers of clones, at the risk of selecting false positives as well as missing positive clones.

An example scenario for engineering an enzyme's activity toward a nonnative substrate follows: a chromatography (HPLC) assay provides with high accuracy the amount of desired compound produced by each enzyme variant (throughput of $\sim 10^2$ clones per day per instrument), a 384-well-microtiter plate spectrophotometric assay provides kinetic information about each variant acting on a surrogate substrate (similar to but not exactly the intended substrate) (throughput of $\sim 10^4$ clones per day), and finally a growth selection has been developed, in which a minimal level of the enzyme's desired activity is required for cell viability, but cell growth rate does not otherwise correlate well with activity ($\sim 10^8$ clones per day). Note that other screening technologies could easily substitute for these three options in similar scenarios — e.g. FACS and phage display are similar to growth selection in terms of throughput and information garnered.

If a good deal is known about the enzyme, perhaps a few attempts at site-directed mutagenesis would be sufficient. Alternately, n active-site residues may be individually mutated to all possible amino acids, resulting in $19 \times n$ variants that could easily be tested using HPLC for small values of n. The plate reader assay becomes preferable when saturation mutagenesis is instead performed at many or all residues (individually) in the enzyme, or other mutagenesis strategies result in library sizes greater than ~ 1000. Even for very large or intractable library sizes (from random mutagenesis, doped mutation oligonucleotide amplification, or family shuffling), as long as the average mutation frequency remains relatively low (1–3 amino acid substitutions per clone), or if the library is a product of homologous DNA shuffling, the frequency of positive hits should remain high enough to enable use of the plate reader assay, with the benefit of having quantitative information about the relative activities of variants. Positive clones can later be verified using the desired substrate and the HPLC assay. However, since the cell growth selection is capable of quickly eliminating a large number of nonfunctional mutants, it may be desirable to use the selection as a preliminary screen, even for the case of random mutation or DNA shuffling libraries. For larger libraries containing >3 amino substitutions per variant, or when it is desired to have a high probability of observing all members of a large library, the selection is an essential first step, though likely to unintentionally eliminate positive hits, furnish false positives, and provide no kinetic data.

Acceptable screening accuracy depends on the functional quality of the library. For example, random point mutations are not likely to result in drastic improvements in function. The screen must be sensitive enough to detect low levels of activity/function (particularly in initial rounds of screening), and also accurate enough to reliably identify variants with relatively small improvements in function (e.g. 30% improved). There are no steadfast rules for choosing a mutagenesis strategy or screening system, and a desired level of library coverage or oversampling is case-dependent. While not discussed here, a variety of research papers and reviews address statistics and probability relating to protein library construction and evaluation.[49–51] These can help guide experimental design and interpretation of screening results and library quality.

High-Throughput Screening in Microtiter Plates

The use of microtiter plate assays is a conventional and straightforward high-throughput approach to protein library screening. The requirement is that the protein property of interest can be directly or indirectly measured in the microtiter plate, most commonly via spectrophotometry or fluorometry. Enzymes have most commonly been engineered using this approach, since spectroscopic methods are excellent for monitoring rates of substrate consumption or product formation. Often the assay will employ a surrogate, colorimetric/fluorometric substrate that is similar in structure/chemistry to the intended substrate (but includes a chromophore/fluorophore). Linking protein function to expression of a fluorescent protein is also common. A typical procedure is as follows: a gene library is constructed and transformed into the expression host of interest. Individual colonies (representing individual clones) are picked and used to inoculate cultures in microtiter plate wells. The cultures are replicated for storage and retrieval. Appropriate growth conditions are used for protein expression and ultimately each culture, supernatant, or cell lysate, is transferred to a microtiter assay plate to measure optical properties linked to protein function. Cultures and assays may be carried out in 96-, 384-, or even 1536-well plates. Colony picking and pipetting robotics are commonly used. Often a great deal of work goes into assay development, and while this approach is relatively time-consuming and low-throughput, it can provide much functional detail for each variant in the library.

Flow Cytometry (Fluorescence-Activated Cell Sorting)

Flow cytometry is a technique that allows for the rapid measurement of multiple optical properties of individual cells, particles, or compartmentalized droplets. Fluorescence-activated cell sorting (FACS) involves the use of flow cytometry to separate members of a cell population having specified fluorescent properties (and is not limited to only cells). Individual FACS instruments are now capable of sorting more than 20 000 cells per second. The use of FACS for protein engineering requires the generation of a fluorescence signal that reflects the protein's property of interest. In the case of engineering proteins for controlling gene expression, this is as simple as incorporating expression of a fluorescent protein into the assay. For the case of engineering a protein's ligand-binding properties, a variety of fluorescent conjugates can be used. The application becomes more complicated for enzyme engineering, but clever methods have been developed to accomplish this, as summarized in many reviews.[52,53] An example is the use of a fluorescence sensor for intracellular pH shifts ('pHluorin'), which could be triggered by hydrolytic enzyme reactions.[54] The high-throughput nature of FACS has led to the use of flow cytometry as one of the most enabling tools in protein engineering and synthetic biology. Below we describe some key applications of FACS for protein engineering. Furthermore, the 'Applications' section of this chapter describes the use of FACS for engineering effector specificity of a regulatory protein.

Cell Surface Display If a protein's fluorescence-linked property can be measured intracellularly, then FACS isolation of the corresponding cell and genetic information is straightforward. However, if the protein does not fold or function properly intracellularly, or if the protein's substrate, ligand, or function-linked fluorophore do not readily cross the cell membrane, then the assay must take place extracellularly, presenting the challenge of linking genotype to phenotype. For FACS applications, a variety of cell surface display methods have been developed and implemented for bacteria, yeast, insect, and mammalian cells.[55] The general approach is to recruit the host's native machinery for protein translocation and membrane anchoring by fusing the protein of interest to a 'carrier protein' system.

In one example, Georgiou and coworkers used *E. coli* surface display and multiparameter flow cytometry to engineer a native outer membrane protease with altered substrate specificity.[56] In this system, the peptide substrate containing a fluorophore and FRET quenching partner separated by the target scissile bond coat the *E. coli* surface displaying the protease variants. Cleavage at the target peptide disrupts the FRET interaction, resulting in fluorescence. To include specificity in the screen, an alternate fluorescent counterselection substrate containing an undesired protease cleavage site was also used. This approach resulted in a protease variant having high specificity and activity on a nonnative substrate.

A wide variety of proteins can be displayed on the yeast surface.[55] While cloning and transformation efficiencies are often lower in yeast compared to bacteria, yeast surface display is often preferred due to its native eukaryotic secretory pathway. Yeast surface display is a proven system for selecting antibodies with improved affinity, specificity, and stability. Wittrup and coworkers developed the Aga1-Aga2 display system in *Saccharomyces cerevisiae*.[57] The Aga1 protein is naturally covalently attached to the cell wall and forms disulfide bonds with Aga2. Fusing the protein of interest to Aga2 results in surface display. Epitope tags can be included in the protein fusions to allow immunofluorescent quantification of displayed protein and normalization of activity or binding. This system has been widely used for affinity maturation of antibody—antigen interactions, as well as other protein—protein interactions.[55,58]

In Vitro Compartmentalization An alternate approach to using cells as expression hosts is to express protein libraries in vitro.[59] In order to maintain a link between genotype and phenotype, however, compartmentalization is needed. In this approach, microscopic emulsion compartments are generated as artificial cells. The gene library is transcribed and translated within the emulsion droplets, so as long as an assay can be developed in which droplet fluorescence correlates with protein function, FACS can be used for sorting while maintaining the genotype—phenotype linkage. Reaction conditions or solubilities that may be incompatible with in vivo systems may be more compatible with emulsions, and limitations associated with transport across cell membranes are eliminated. This technique was first reported for the selection of a phosphotriesterase from 3.4×10^7 library members with 63-fold improved activity relative to wild-type.[60] Water-in-oil (w/o) emulsions with 10^{10} droplets per ml of emulsion can be created, with the majority containing a single variant, enabling screening of 10^8-10^{11} variants.[61]

Similarly, emulsions can also be used for compartmentalizing whole cells, e.g. if the fluorescence signal would otherwise diffuse away from the cell. In the direction evolution of serum paraoxonase (PON1), a library of variants was expressed in *E. coli*, and single cells were encapsulated in w/o/w emulsion droplets.[59] Droplets with improved thiolactonase activity were isolated by FACS using a fluorogenic thiol-detecting dye that is retained in the droplets.

Phage Display

Phage display is the longest-standing platform for protein library display, and remains a powerful and widely used technique for engineering binding properties of proteins (i.e. affinity maturation).[62] With this technique, phage DNA is modified such that the gene encoding the protein of interest is fused to a gene encoding a protein comprising the phage coat. Phage DNA is packaged into phage particles upon infecting *E. coli*. For the case of gene libraries inserted into the coat gene fusion (and introduced into *E. coli*), each assembled phage particle displays a single variant, and the corresponding mutant gene is packaged inside that particle, thus retaining a genotype–phenotype linkage. The phage display selection process involves binding of the displayed protein to an immobilized target. Unbound phage can be washed away, followed by elution and recovery of the bound phage particles. Several rounds of this selection process can be performed with increasing stringency until the desired enrichment has been achieved and individual clones are recovered, sequenced, and further characterized. Libraries of 10^9–10^{11} members are readily screened with this method. While this strategy selects for improved binding properties, catalytic activities have been engineered using phage display, e.g. by selecting for variants with strong affinity for transition state analogues.[63]

Cell-Free Ribosome and mRNA Display

To maintain a genotype–phenotype linkage and avoid the library size limitations associated with transformation and in vivo expression, researchers have developed cell-free display systems for library screening. Two popular methods are ribosome display,[64] in which the protein and its mRNA sequence are noncovalently attached to a stalled ribosome, and mRNA display, in which the protein becomes covalently linked to its mRNA sequence.[65] These systems potentially enable searching libraries with up to 10^{10}–10^{14} members, and with fewer constraints than in vivo methods. Other similar methods include covalent and noncovalent DNA display.[66,67]

In the ribosome display system, the gene variants are transcribed and translated in vitro. The mRNA does not contain a stop codon, but rather a C-terminal tether that allows the expressed protein to fully exit the ribosome and hopefully maintain fold and function. Techniques are used to stabilize the stalled ribosome complex, and screening is performed using immobilized targets similar to in phage display. The bound mRNA sequences are eluted, reverse transcribed to cDNA, and amplified.[68] Iterative rounds of in vitro expression, ribosome display and screening can be performed as desired/needed. Ribosome display has been successfully applied to improve protein stability and affinity, and even select for enzymatic activities.[69,70] Jermutus and coworkers evolved the affinity and stability of single-chain Fv antibody fragments (scFvs) through DNA shuffling and ribosome display screening.[71] A scFv with 30-fold improved affinity for fluorescein compared to wild-type was isolated from the library.

In mRNA display, transcription and translation again are carried out in vitro, but the mRNA template is ligated to an ssDNA linker bonded to puromycin at the 3′ end. The puromycin acts as a translational inhibitor in the ribosome by forming an amide linkage with the polypeptide chain and causing the mRNA–protein fusion to be released. The mRNA–polypeptide fusion is then purified and at this point the mRNA can be reverse transcribed to form a mRNA/cDNA duplex (prior to selection). Selection is performed as in ribosome display, and the cDNA is then isolated and amplified and subjected to further rounds of in vitro display and selection. Compared to ribosome display, the genotype–phenotype linkage in mRNA display can withstand harsher biochemical treatments (e.g. during the selection stage). Fukuda and coworkers used mRNA display to evolve single-chain Fv antibody fragments for fluorescein binding.[72] After four rounds of selection, they isolated six affinity matured mutants whose off-rates were decreased more than one order of magnitude and dissociation constants improved ~30-fold.

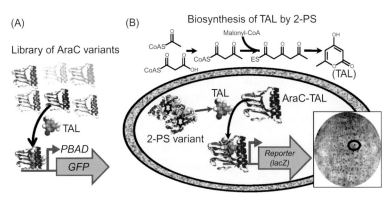

FIGURE 2.2

Design and application of a TAL-responsive reporter based on the AraC regulatory protein. (A) An *araC* gene library was constructed and expressed in *E. coli*, and variants induced by TAL were isolated using FACS screening. (B) TAL is produced by the enzyme 2-PS from *G. hybrida*. TAL-responsive AraC variant 'AraC-TAL' was used as an endogenous reporter to screen for improved TAL production by 2-PS variants expressed in *E. coli*.

Auxotrophic Complementation (Selections)

Though not always possible, an alternative to screening each library member is to develop a cellular selection system, where cell growth or survival depends on some threshold level of function by the protein variant of interest (auxotrophic complementation). This typically involves genetically modifying the host and/or formulating the growth medium so that the protein function to be evolved provides an essential metabolite for its host. For very large libraries consisting of many mutations per variant, a vast majority of library members will have relatively little or no fitness (compared to what is being sought). A suitably developed selection system allows for large libraries to be screened without having to physically observe 'unfit' library members. For example, whereas a plasmid library of 10^8 members may be transformed into a bacterial host, subsequent plating of all transformants onto appropriately formulated agar plates may result in formation of only 10^4 colonies, corresponding to variants with sufficient function to provide auxotrophic complementation. In some cases the rate of colony growth may also correlate with activity of the protein of interest. A liquid culture-based growth selection may alternately be developed, where the fittest members out-grow others.

In one example, Boersma and coworkers developed a selection to improve the enantioselectivity of *Bacillus subtilis* Lipase A (LipA).[73] Here, *E. coli*'s native pathways for aspartate synthesis were deleted, and the auxotrophic strain was used to select LipA variants that could hydrolyze the aspartate ester of the desired enantiomer (*S*)-(+)-1,2-*O*-isopropylidene-sn-glycerol. A phosphonate ester of the opposite enantiomer (*R*)-(−)-1,2-*O*-isopropylidene-sn-glycerol was also added to inhibit growth in the presence of less enantioselective variants. After three rounds of selection with increasing stringency (by increasing the phosphonate ester concentration) a mutant was isolated with greatly improved enantioselectivity. Additional examples of using selections for protein engineering are provided below and depicted in Figure 2.2.

APPLICATIONS OF PROTEIN ENGINEERING IN SYNTHETIC BIOLOGY

Engineering Protein-Based Biosensors

Construction of synthetic genetic circuits and integration of novel regulatory signals are common goals in synthetic biology. Bacterial regulatory proteins which naturally regulate gene expression in response to small molecule inputs ('effectors') serve as useful tools in synthetic biology applications. Furthermore, the ability to engineer these proteins allows synthetic biologists to develop customized genetic circuits. Regulatory proteins engineered to respond to novel small molecule inputs are also finding use as molecular reporters for high-throughput screening of enzyme and metabolic pathway libraries (production of the effector/input molecule activates expression of an easily detectable reporter gene/protein such as GFP). Examples in which regulatory proteins were engineered are provided below.

THE XYLR REGULATOR

The transcriptional activator XylR from the soil bacterium *Pseudomonas putida* is natively induced by benzene derivatives such as xylene. de Lorenzo and coworkers sought to develop biosensors that detect toxic benzene derivatives in soil.[74,75] A mutant library of XylR was constructed by randomly mutating (via error-prone PCR) the XylR effector-binding domain and the 'B connector' to the DNA-binding domain (DBD). A selection/counterselection system was developed in a *P. putida* uracil auxotroph, analogous to the yeast URA3 selection system. Here, XylR variants that respond to the presence of 2,4-dinitrotoluene (DNT), a xenobiotic used in the polymer industry, activate expression of *pyrF* encoding orotidine-5'-phosphate decarboxylase, thereby complementing uracil auxotrophy. Meanwhile, in the presence of uracil and fluoroorotic acid (FOA), leaky XylR variants or variants that respond to the decoy compound xylene express *pyrF* in the absence of DNT, resulting in production of a toxic compound and eliminating nonspecific clones. Using this system, a XylR variant with four amino acid substitutions showed a strong response to 1 mM DNT. A DNT-sensing *P. putida* strain was engineered with luciferase as the reporter, and thus emits light in the presence of DNT within a soil sample.[75,76] Additional work has focused on developing stable and environmentally safe strains and formulations for delivery of aromatic biosensors to soil using water-soluble gelatin capsules.[77]

THE ARAC REGULATOR

The well-characterized AraC regulatory protein represses transcription from promoter PBAD in the absence of effector, and activates transcription in the presence of its native effector L-arabinose.[78] Cirino and coworkers sought to alter the effector specificity of AraC, to create novel molecular reporters. They initially engineered AraC to specifically respond to D-arabinose, the rarer isomer of the native inducer.[79] Based on analysis of the AraC L-arabinose binding pocket, four amino acid positions were chosen for simultaneous saturation mutagenesis. Dual screening using FACS isolated AraC variants that induce expression of GFP (at PBAD) in the presence of D-arabinose. Variants that cause leaky GFP expression or show responsiveness to L-arabinose (in the absence of D-arabinose) were selected against. AraC variants were recovered that showed strong induced gene expression in the presence of 0.1 mM D-arabinose, with essentially no response to L-arabinose. It is noteworthy that similar attempts to isolate AraC variants from a random mutagenesis library were unsuccessful.

Isoprenoids receive increasing attention as a valuable class of secondary metabolites, and have been the target in numerous synthetic biology studies. Mevalonate is the product of HMG-CoA reductase and a key intermediate in isoprenoid biosynthesis.[80] To facilitate combinatorial library screening of pathways involving HMG-CoA reductase, Tang and Cirino next sought a mevalonate-responsive AraC variant.[81] A similar strategy as described for D-arabinose was used, only this time a five-site saturation mutagenesis library was created, and dual screening via FACS was performed in the presence and absence of 30 mM mevalonate. After several rounds of screening, a mevalonate-responsive variant was selected that showed a relatively linear increase in expression of reporter protein (GFP or LacZ) from PBAD as the mevalonate concentration was increased from 10 mM to 100 mM. The authors then demonstrated the utility of this mevalonate reporter system by using it to screen a library of mutations in the ribosomal binding site (RBS) region upstream of

HMG-CoA reductase. Clones with increased LacZ activity (expressed from PBAD) also produced up to four-fold increased titers of mevalonate in liquid culture.[81]

A third example from this group is depicted in Figure 2.2.[82] Here, an AraC library was screened (using FACS) to isolate a variant responding to 4-hydroxy-6-methyl-2-pyrone (triacetic acid lactone; TAL), which is a precursor to industrially useful chemicals and bioactive compounds.[83–85] The TAL reporter was then used for screening in the directed evolution of TAL production (Fig. 2.2B). Random mutation libraries of the TAL-synthesizing enzyme 2-pyrone synthase (2-PS) from *Gerbera hybrida* were expressed in *E. coli* containing the reporter system. As in the case for mevalonate, clones with increased reporter activity also produced increased titers of TAL.

New Biosynthetic Pathways

Synthetic biology now plays an important role in the metabolic engineering of microorganisms for novel or enhanced production of biochemicals. Here we provide examples in which protein engineering was critical to the success of designing new biosynthetic pathways.

LONG-CHAIN ALIPHATIC ALCOHOLS

Liao and coworkers sought a nonnatural metabolic pathway in *E. coli* to produce aliphatic alcohols with carbon chain lengths greater than 5 (attractive biofuel targets) during growth on glucose. As depicted in Figure 2.3, they developed a strategy to produce 7- (C7) to 9-carbon (C9) 2-keto-acids, which could then be decarboxylated (by 2-keto acid decarboxylase KIVD from *Lactococcus lactis*) and reduced to the alcohol by ADH6 from *Saccharomyces cerevisiae*.[86] *E. coli* naturally produces 2-keto-3-methylvalerate from L-threonine, as a precursor to L-isoleucine. The L-valine analogue of 2-keto-3-methylvalerate is 2-ketoisovalerate (containing one less methyl group), which is converted to

37

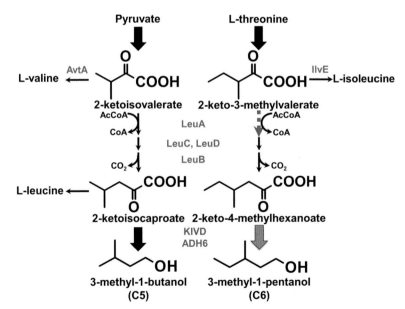

FIGURE 2.3

Synthetic metabolic pathway engineered for the production of the C6 alcohol 3-methyl-1-pentanol by *E. coli*. The native L-leucine biosynthetic pathway from pyruvate uses the enzymes LeuA, LeuC, LeuD, and LeuB. Heterologous expression of the nonnative decarboxylase KIVD and dehydrogenase ADH6 results in production of 3-methyl-1-butanol (C5). As described in the text, C6 production from 2-keto-3-methylvalerate was achieved by enhancing the selectivity of KIVD and enlarging the substrate binding pocket of LeuA using protein engineering.

2-ketoisocaproate by the LeuACDB pathway (Fig. 2.3). The authors started from an *E. coli* strain previously designed to overproduce the precursor threonine. KIVD and ADH6 are naturally promiscuous, so to reduce byproduct formation and drive flux to the C6 alcohol, KIVD was rationally engineered to have higher selectivity toward 2-keto-4-methylhexanoate. Based on homology models of the KIVD substrate-binding region built from structures of homologues, the authors identified residue positions that could influence selectivity. After testing a handful of site-directed point mutations, a variant with two amino acid substitutions that produced the highest titer of 3-methyl-1-pentanol was selected, and this variant was shown to have greatly enhanced specificity towards 2-keto-4-methylhexanoate versus 2-ketoisovalerate.

Also using homology modeling and site-directed mutagenesis, the enzyme LeuA, which naturally catalyzes the condensation of acetyl-CoA with 2-ketoisovalerate, was engineered to have a larger binding pocket, thus better accommodating the nonnative substrate 2-keto-3-methylvalerate and further increasing production of the C6 alcohol. It is also worth noting that the LeuA enzyme used in this study additionally carried a single amino acid substitution previously identified to reduce feedback inhibition by leucine.[87]

A result of increasing tolerance to larger substrates by LeuA is the ability of these variants to carry out additional condensations of acetyl-CoA with the carboxylate product of the LeuCDB pathway, effectively adding an additional carbon per cycle. Thus, in a follow-up study, a combination of quantum mechanical modeling, protein–substrate complex modeling, and structure-based protein engineering were used to further expand the substrate range of LeuA from branched-chain ketoacids to linear-chain and even aromatic-chain 2-ketoacids, allowing for the production of their respective decarboxylated alcohols.[88] A variant with the largest active site volume has six amino acid substitutions and produced 80 mg L^{-1} 1-heptanol and 2 mg L^{-1} 1-octanol.

NONNATURAL AMINO ACID: L-HOMOALANINE

L-homoalanine is a nonnatural amino acid that can serve as a chiral precursor to a variety of important pharmaceutical products. Glutamate dehydrogenase (GDH) catalyzes the reductive amination of 2-ketoglutarate with ammonia, producing glutamate. Zhang and coworkers sought a GDH variant capable of catalyzing reductive amination of 2-ketobutyrate for the production of L-homoalanine.[89] While L-valine is synthesized in *E. coli* via transamination between glutamate and 2-ketoisovalerate catalyzed by AvtA and IlvE (see Fig. 2.3), an alternate route could involve reductive amination of 2-ketoisovalerate. Since 2-ketobutyrate is similar in structure to 2-ketoisovalerate, the authors reasoned that a GDH variant capable of acting on 2-ketoisovalerate to make L-valine would also synthesize L-homoalanine from 2-ketobutyrate. Auxotrophic complementation was therefore used to isolate GDH variants with altered substrate specificity. Based on analysis of the GDH crystal structure from *Clostridium symbiosum* and sequence comparison with the enzyme from *E. coli*, four *E. coli* GDH substrate binding pocket resides were targeted for simultaneous saturation mutagenesis. A valine auxotroph *E. coli* strain was constructed (by deleting genes *avtA* and *ilvE*) and used to isolate GDH variants conferring valine synthesis and growth on selective media. A resulting GDH variant was characterized and shown to have greatly enhanced activity on 2-ketobutyrate, allowing for 5.4 g L^{-1} of L-homoalanine to be produced from *E. coli* growing on glucose.

ENGINEERING PRODUCTS OF THE AROMATIC AMINO ACID PATHWAY

In a final example of using protein engineering to create synthetic pathways, Frost and coworkers sought increased production of shikimic acid by *E. coli*, which naturally produces shikimic acid as an intermediate toward aromatic amino acids and vitamins.[90] Shikimic acid can be used to synthesize Tamiflu®, an orally effective antiinfluenza agent.[91,92] The shikimate pathway starts with the condensation of phosphoenolpyruvate (PEP) and erythrose-4-phosphate (E4P) to 3-deoxy-D-arabino-heptulosonate-7-phosphate (DAHP), catalyzed by DAHP synthase. To avoid competition for PEP by the *E. coli* phosphotransferase system (PTS) and therefore improve flux to DAHP during growth on glucose, the authors sought to replace PEP with pyruvate in a novel enzyme-catalyzed condensation with E4P. 2-Keto-3-deoxy-phosphogalactonate (KDPGal) aldolase catalyzes the reversible condensation of pyruvate and glyceraldehyde-3-phosphate (G3P) to KDPGal, and the authors reasoned this enzyme could be engineered to use E4P instead of G3P, thus producing DAHP. Directed evolution of KDPGal aldolase activity toward E4P was accomplished by screening mutants based on their ability to improve growth of an *E. coli* strain engineered to be deficient in DAHP synthase. Starting from three different parent gene sequences encoding KDPGal aldolase (from *E. coli*, *K. pneumoniae*, and *S. typhimurium*), a combination of random mutagenesis and DNA shuffling, and later site-directed mutagenesis based on homology models, were used to create mutant libraries and evolve activity toward E4P utilization and DAHP production. Ultimately, a variant with >60-fold improved catalytic activity on E4P compared to wild-type *E. coli* KDPGal aldolase was created, resulting in an effective bypass route to DAHP and its downstream products.

CONCLUSIONS

This chapter summarizes key methods in protein engineering, and describes noteworthy and thought-provoking examples in which they were effectively applied to achieve improved or novel protein properties, leading to improved or novel cellular properties. Molecular modeling and other computational tools are improving, advancing rational and semirational protein design efforts. Owing to their relative ease of genetic manipulation and growth, microbial systems have served as the workhorses for protein engineers. More recently, advances in the in vitro production and screening of proteins have provided a new experimental framework. Continuous development of genetic and protein expression systems in more complex cell lines offer protein engineers opportunities to design more complex protein functions within the context of eukaryotic hosts. As increasingly complex synthetic biological systems are sought, the diversity and complexity of required protein properties also increase. Protein engineering is thus taking on a central role in synthetic biology.

References

1. Arnold FH, Georgiou G. *Directed Evolution Library Creation: Methods and Protocols*. Totowa: Humana Press; 2003.

2. Arnold FH, Georgiou G. *Directed Enzyme Evolution: Screening and Selection Methods*. Totowa: Humana Press; 2003.

3. Park SJ, Cochran JR. *Protein Engineering and Design*. Boca Raton: CRC Press; 2009.

4. Robertson DE, Noel JP. *Protein Engineering*. San Diego: Elsevier Academic Press; 2004.

5. Saven JG. Computational protein design: engineering molecular diversity, nonnatural enzymes, nonbiological cofactor complexes, and membrane proteins. *Curr Opin Chem Biol*. 2011;15:452−457.

6. Pantazes RJ, Grisewood MJ, Maranas CD. Recent advances in computational protein design. *Curr Opin Struct Biol*. 2011;21:467−472.

7. Green DF. Computer graphics, homology modeling, and bioinformatics. In: Park SJ, Cochran JR, eds. *Protein Engineering and Design*. Boca Raton: CRC Press; 2009:223−238.

8. Cunningham BC, Wells JA. High-resolution epitope mapping of hGH-receptor interactions by alanine-scanning mutagenesis. *Science*. 1989;244:1081−1085.

9. Schwaneberg U, Schmidt-Dannert C, Schmitt J, Schmid RD. A continuous spectrophotometric assay for P450 BM-3, a fatty acid hydroxylating enzyme, and its mutant F87A. *Anal Biochem*. 1999;269:359−366.

10. Bornscheuer UT, Pohl M. Improved biocatalysts by directed evolution and rational protein design. *Curr Opin Chem Biol*. 2001;5:137−143.

11. Koide S. Generation of new protein functions by nonhomologous combinations and rearrangements of domains and modules. *Curr Opin Biotechnol*. 2009;20:398−404.

12. Yüksel D, Pamuk D, Ivanova Y, Kumar K. Protein engineering using noncanonical amino acids. In: Park SJ, Cochran JR, eds. *Protein Engineering and Design*. Boca Raton: CRC Press; 2009:206−218.

13. Link AJ, Mock ML, Tirrell DA. Non-canonical amino acids in protein engineering. *Curr Opin Biotechnol.* 2003;14:603−609.

14. Cedrone F, Ménez A, Quéméneur E. Tailoring new enzyme functions by rational redesign. *Curr Opin Struct Biol.* 2000;10:405−410.

15. Yang W, Jones LM, Isley L, et al. Rational design of a calcium-binding protein. *J Am Chem Soc.* 2003;125:6165−6171.

16. Fersht A. *Protein stability. Structure and Mechanism in Protein Science: A Guide to Enzyme Catalysis and Protein Folding.* New York: W.H. Freeman; 1999:508−539.

17. Doudney C, Haas F. Mutation induction and macromolecular synthesis in bacteria. *Proc Natl Acad Sci USA.* 1959;45:709−772.

18. Ong T, De Serres F. Mutagenicity of chemical carcinogens in *Neurospora crassa. Cancer Res.* 1972;32:1890−1893.

19. Myers RM, Lerman LS, Maniatis T. A general method for saturation mutagenesis of cloned DNA fragments. *Science.* 1985;229:242−247.

20. Lai YP, Huang J, Wang LF, Li J, Wu ZR. A new approach to random mutagenesis in vitro. *Biotechnol Bioeng.* 2004;86:622−627.

21. Muteeb G, Sen R. Random mutagenesis using a mutator strain. *Methods Mol Biol.* 2010;634:411−419.

22. Cirino PC, Mayer KM, Umeno D. Generating mutant libraries using error-prone PCR. *Methods Mol Biol.* 2003;231:3−10.

23. Fujii R, Kitaoka M, Hayashi K. One-step random mutagenesis by error-prone rolling circle amplification. *Nucleic Acids Res.* 2004;32:e145.

24. Stemmer WPC. Rapid evolution of a protein in vitro by DNA shuffling. *Nature.* 1994;370:389−391.

25. Chang CCJ, Chen TT, Cox BW, et al. Evolution of a cytokine using DNA family shuffling. *Nat Biotechnol.* 1999;17:793−797.

26. Ness JE, Welch M, Giver L, et al. DNA shuffling of subgenomic sequences of subtilisin. *Nat Biotechnol.* 1999;17:893−896.

27. Aguinaldo AM, Arnold FH. Staggered extension process (StEP) in vitro recombination. *Methods Mol Biol.* 2003;231:105−110.

28. Esteban O, Woodyer RD, Zhao H. In vitro DNA recombination by random priming. *Methods Mol Biol.* 2003;231:99−104.

29. Coco WM. RACHITT: gene family shuffling by random chimeragenesis on transient templates. *Methods Mol Biol.* 2003;231:111−127.

30. Abécassis V, Pompon D, Truan G. Producing chimeric genes by CLERY: in vitro and in vivo recombination. *Methods Mol Biol.* 2003;231:165−173.

31. Wingler LM, Cornish VW. Reiterative recombination for the in vivo assembly of libraries of multigene pathways. *Proc Natl Acad Sci.* 2011;108:15135−15140.

32. Swers JS, Kellogg BA, Wittrup KD. Shuffled antibody libraries created by in vivo homologous recombination and yeast surface display. *Nucleic Acids Res.* 2004;32:e36.

33. Mézard C, Pompon D, Nicolas A. Recombination between similar but not identical DNA sequences during yeast transformation occurs within short stretches of identity. *Cell.* 1992;70:659−670.

34. Ostermeier M, Lutz S. The creation of ITCHY hybrid protein libraries. *Methods Mol Biol.* 2003;231:129−142.

35. Lutz S, Ostermeier M. Preparation of SCRATCHY hybrid protein libraries. *Methods Mol Biol.* 2003;231:143−151.

36. Udit AK, Silberg JJ, Sieber V. Sequence homology-independent protein recombination (SHIPREC). *Methods Mol Biol.* 2003;231:153−164.

37. Villiers B, Stein V, Hollfelder F. USER friendly DNA recombination (USERec): a simple and flexible near homology-independent method for gene library construction. *Protein Eng Des Sel.* 2010;23:1−8.

38. Bittker JA, Le BV, Liu JM, Liu DR. Directed evolution of protein enzymes using nonhomologous random recombination. *Proc Natl Acad Sci USA.* 2004;101:7011−7016.

39. Hiraga K, Arnold FH. General method for sequence-independent site-directed chimeragenesis. *J Mol Biol.* 2003;330:287−296.

40. Meyer MM, Hochrein L, Arnold FH. Structure-guided SCHEMA recombination of distantly related β-lactamases. *Protein Eng Des Sel.* 2006;19:563−570.

41. O'Maille PE, Bakhtina M, Tsai MD. Structure-based combinatorial protein engineering (SCOPE). *J Mol Biol.* 2002;321:677−691.

42. Reetz MT, Carballeira JD. Iterative saturation mutagenesis (ISM) for rapid directed evolution of functional enzymes. *Nat Protoc.* 2007;2:891−903.

43. Chockalingam K, Chen Z, Katzenellenbogen JA, Zhao H. Directed evolution of specific receptor–ligand pairs for use in the creation of gene switches. *Proc Natl Acad Sci USA.* 2005;102:5691–5696.

44. Herman A, Tawfik DS. Incorporating synthetic oligonucleotides via gene reassembly (ISOR): a versatile tool for generating targeted libraries. *Protein Eng Des Sel.* 2007;20:219–226.

45. Chen MM, Snow CD, Vizcarra CL, Mayo SL, Arnold FH. Comparison of random mutagenesis and semi-rational designed libraries for improved cytochrome P450 BM3-catalyzed hydroxylation of small alkanes. *Protein Eng Des Sel.* 2012;25:171–178.

46. Fellouse FA, Li B, Compaan DM, Peden AA, Hymowitz SG, Sidhu SS. Molecular recognition by a binary code. *J Mol Biol.* 2005;348:1153–1162.

47. Reetz MT, Kahakeaw D, Lohmer R. Addressing the numbers problem in directed evolution. *Chembiochem.* 2008;9:1797–1804.

48. Leemhuis H, Stein V, Griffiths AD, Hollfelder F. New genotype–phenotype linkages for directed evolution of functional proteins. *Curr Opin Struct Biol.* 2005;15:472–478.

49. Firth AE, Patrick WM. Statistics of protein library construction. *Bioinformatics.* 2005;21:3314–3315.

50. Patrick WM, Firth AE, Blackburn JM. User-friendly algorithms for estimating completeness and diversity in randomized protein-encoding libraries. *Protein Eng.* 2003;16:451–457.

51. Bosley AD, Ostermeier M. Mathematical expressions useful in the construction, description and evaluation of protein libraries. *Biomol Eng.* 2005;22:57–61.

52. Georgiou G. Analysis of large libraries of protein mutants using flow cytometry. *Adv Protein Chem.* 2001;55:293–315.

53. Cirino PC, Frei CS. Combinatorial enzyme engineering. In: Park SJ, Cochran JR, eds. *Protein Engineering and Design.* Boca Raton: CRC Press; 2009:131–152.

54. Schuster S, Enzelberger M, Trauthwein H, Schmid RD, Urlacher VB. pHluorin-based in vivo assay for hydrolase screening. *Anal Chem.* 2005;77:2727–2732.

55. Moore SJ, Olsen MJ, Cochran JR, Cochran FV. Cell surface display systems for protein engineering. In: Park SJ, Cochran JR, eds. *Protein Engineering and Design.* Boca Raton: CRC Press; 2009:23–50.

56. Varadarajan N, Gam J, Olsen MJ, Georgiou G, Iverson BL. Engineering of protease variants exhibiting high catalytic activity and exquisite substrate selectivity. *Proc Natl Acad Sci USA.* 2005;102:6850–6855.

57. Boder ET, Wittrup KD. Yeast surface display for screening combinatorial polypeptide libraries. *Nat Biotechnol.* 1997;15:553–557.

58. Miller KD, Pefaur NB, Baird CL. Construction and screening of antigen targeted immune yeast surface display antibody libraries. *Curr Protoc Cytom.* 2008 [chapter 4: unit 4.7]. doi: 10.1002/0471142956.cy0407s45.

59. Aharoni A, Griffiths AD, Tawfik DS. High-throughput screens and selections of enzyme-encoding genes. *Curr Opin Struct Biol.* 2005;9:210–216.

60. Griffiths AD, Tawfik DS. Directed evolution of an extremely fast phosphotriesterase by in vitro compartmentalization. *EMBO J.* 2003;22:24–35.

61. Miller OJ, Bernath K, Agresti JJ, et al. Directed evolution by in vitro compartmentalization. *Nat Methods.* 2006;3:561–570.

62. Ernst A, Sidhu SS. Phage Display Systems for Protein Engineering. In: Park SJ, Cochran JR, eds. *Protein Engineering and Design.* Boca Raton: CRC Press; 2009:1–22.

63. Baca M, Scanlan TS, Stephenson RC, Wells JA. Phage display of a catalytic antibody to optimize affinity for transition-state analog binding. *Proc Natl Acad Sci.* 1997;94:10063–10068.

64. Schaffitzel C, Hanes J, Jermutus L, Plückthun A. Ribosome display: an in vitro method for selection and evolution of antibodies from libraries. *J Immunol Methods.* 1999;231:119–135.

65. Takahashi TT, Austin RJ, Roberts RW. mRNA display: ligand discovery, interaction analysis and beyond. *Trends Biochem Sci.* 2003;28:159–165.

66. Bertschinger J, Neri D. Covalent DNA display as a novel tool for directed evolution of proteins in vitro. *Protein Eng Des Sel.* 2004;17:699–707.

67. Odegrip R, Coomber D, Eldridge B, et al. CIS display: in vitro selection of peptides from libraries of protein–DNA complexes. *Proc Natl Acad Sci USA.* 2004;101:2806–2810.

68. Barendt PA, Sarkar CA. Cell-free display systems for protein engineering. In: Park SJ, Cochran JR, eds. *Protein Engineering and Design.* Boca Raton: CRC Press; 2009:51–82.

69. Yan X, Xu Z. Ribosome-display technology: applications for directed evolution of functional proteins. *Drug Discov Today.* 2006;11:911–916.

70. Amstutz P, Pelletier JN, Guggisberg A, et al. In vitro selection for catalytic activity with ribosome display. *J Am Chem Soc.* 2002;124:9396–9403.

71. Jermutus L, Honegger A, Schwesinger F, Hanes J, Plückthun A. Tailoring in vitro evolution for protein affinity or stability. *Proc Natl Acad Sci.* 2001;98:75–80.

72. Fukuda I, Kojoh K, Tabata N, et al. In vitro evolution of single-chain antibodies using mRNA display. *Nucleic Acids Res.* 2006;34:e127.

73. Boersma YL, Dröge MJ, Van Der Sloot AM, et al. A novel genetic selection system for improved enantioselectivity of *Bacillus subtilis* lipase A. *Chembiochem.* 2008;9:1110−1115.

74. Galvão TC, Mencía M, de Lorenzo V. Emergence of novel functions in transcriptional regulators by regression to stem protein types. *Mol Microbiol.* 2007;65:907−919.

75. de las Heras A, de Lorenzo V. Cooperative amino acid changes shift the response of the σ54-dependent regulator XylR from natural m-xylene towards xenobiotic 2, 4-dinitrotoluene. *Mol Microbiol.* 2011;79:1248−1259.

76. de Las Heras A, Carreño CA, de Lorenzo V. Stable implantation of orthogonal sensor circuits in gram-negative bacteria for environmental release. *Environ Microbiol.* 2008;10:3305−3316.

77. de las Heras A, de Lorenzo V. In situ detection of aromatic compounds with biosensor *Pseudomonas putida* cells preserved and delivered to soil in water-soluble gelatin capsules. *Anal Bioanal Chem.* 2011;400:1093−1104.

78. Schleif R. AraC protein, regulation of the l-arabinose operon in *Escherichia coli*, and the light switch mechanism of AraC action. *FEMS Microbiol Rev.* 2010;34:779−796.

79. Tang SY, Fazelinia H, Cirino PC. AraC regulatory protein mutants with altered effector specificity. *J Am Chem Soc.* 2008;130:5267−5271.

80. Kirby J, Keasling JD. Biosynthesis of plant isoprenoids: perspectives for microbial engineering. *Annu Rev Plant Biol.* 2009;60:335−355.

81. Tang SY, Cirino PC. Design and application of a mevalonate-responsive regulatory protein. *Angew Chem Int Ed.* 2011;50:1084−1086.

82. Gredell JA, Frei CS, Cirino PC. Protein and RNA engineering to customize microbial molecular reporting. *Biotechnol J.* 2011.

83. Hansen CA, Frost J. Deoxygenation of polyhydroxybenzenes: an alternative strategy for the benzene-free synthesis of aromatic chemicals. *J Am Chem Soc.* 2002;124:5926−5927.

84. Agrawal J. Recent trends in high-energy materials. *Prog Energy Combust Sci.* 1998;24:1−30.

85. Eckermann S, Schröder G, Schmidt J, et al. New pathway to polyketides in plants. *Nature.* 1998;396:387−390.

86. Zhang K, Sawaya MR, Eisenberg DS, Liao JC. Expanding metabolism for biosynthesis of nonnatural alcohols. *Proc Natl Acad Sci.* 2008;105:20653−20658.

87. Gusyatiner MM, Lunts MG, Kozlov YI, Ivanovskaya LV, Voroshilova EB. DNA coding for mutant isopropylmalate synthase L-leucine-producing microorganism and method for producing L-leucine. In United States Patent, 2002.

88. Marcheschi RJ, Li H, Zhang K, et al. A synthetic recursive "+1" pathway for carbon chain elongation. *ACS Chem Biol.* 2012;7:689−697.

89. Zhang K, Li H, Cho KM, Liao JC. Expanding metabolism for total biosynthesis of the nonnatural amino acid L-homoalanine. *Proc Natl Acad Sci.* 2010;107:6234−6239.

90. Ran N, Draths K, Frost J. Creation of a shikimate pathway variant. *J Am Chem Soc.* 2004;126:6856−6857.

91. Ran N, Frost JW. Directed evolution of 2-keto-3-deoxy-6-phosphogalactonate aldolase to replace 3-deoxy-D-arabino-heptulosonic acid 7-phosphate synthase. *J Am Chem Soc.* 2007;129:6130−6139.

92. Krämer M, Bongaerts J, Bovenberg R, et al. Metabolic engineering for microbial production of shikimic acid. *Metab Eng.* 2003;5:277−283.

Pathway Engineering as an Enabling Synthetic Biology Tool

Dawn T. Eriksen, Sijin Li and Huimin Zhao
University of Illinois at Urbana-Champaign, Urbana, IL, USA

INTRODUCTION

Pathway engineering has proven indispensable in the design of microbes for production of value-added products such as drugs, biofuels, and specialty chemicals.[1-4] Typically, a heterologous metabolic pathway producing the value-added compound is introduced into a host microorganism. Ideally, the pathway must be efficient, producing the target compound with high titer, yield, and productivity while balancing the metabolic burden of the microorganism.[5] To achieve this goal, a wide variety of strategies have been developed for pathway engineering. The following chapter will focus on the computational and experimental tools for the design, construction, and optimization of a heterologous metabolic pathway. Engineering of a recombinant microorganism capable of producing a target compound with high titer, yield, and productivity is certainly not limited to the optimization of the heterologous metabolic pathway. However, methods for engineering the host microorganism are beyond the scope of this chapter. For further information on metabolic engineering of the host microorganism itself, see Chapter 12. For additional examples of the applications of pathway engineering tools for biofuels production, see Chapter 11, and for drug discovery, see Chapter 10.

There are many challenges in pathway engineering. For example, one major challenge is in the design of the pathway, which may require sorting through thousands of possible enzymes and reactions wherein all the parameters have different substrate preferences and kinetic features. Computational algorithms have proven vital in pathway design and have the ability to discover novel pathways by combining enzymes from various sources to produce nonnatural products. Another major challenge is the development of efficient and reliable tools for pathway construction. The pathway of interest may contain many structural and regulatory genes with sizes ranging from tens to hundreds of kilobases. Thus it is imperative that the pathway can be readily constructed with high efficiency and fidelity.

However, pathway engineering is more than just recruiting various enzymes and stringing them together.[6] In early recombinant gene expression studies, it was discovered that there was significant growth inhibition in cells overexpressing superfluous genes. This inhibition was attributed to the competition of protein synthesis machinery for essential proteins required for cell growth versus the production of the overexpressed proteins, and the

43

Synthetic Biology. DOI: http://dx.doi.org/10.1016/B978-0-12-394430-6.00003-0

demand for nucleotides in the replication of the DNA itself. This became known as the metabolic burden, and reduction of the metabolic burden has become essential in the pursuit to identify an optimal metabolic pathway.[7-10] Minimizing the metabolic burden is challenging enough, but more problems arise when often one or two enzymes in the pathway will not express well, or will have low activity. This causes a bottleneck or a build-up of intermediates in the pathway, thereby reducing the overall titer. All these challenges combined have prompted the development of powerful computational and experimental tools for pathway engineering (Table 3.1).

TABLE 3.1 A List of Available Tools for Pathway Engineering

Tool	Description	Advantages	References
Pathway design tools			
BNICE	The Biochemical Network Integrated Computational Explorer	Design pathways based on specific functional groups via matrix operation and evaluate by their thermodynamic properties	[12,17,18]
ReBiT	Retro-Biosynthesis Tool	Pathway design and evaluation, step-by-step suggestion on cofactors and enzymes involved	http://www.retro-biosynthesis.com/
PPS	The Pathway Prediction System	Predict biodegradation pathways based on chemical structures	http://umbbd.msi.umn.edu/predict/
Pathway construction tools			
DNA Assembler	Uses overlapping sequences	In vivo one-step pathway assembly tool with high efficiency, especially feasible in *S. cerevisiae*-based research	[23,99]
Gibson Isothermal Assembly	Uses overlapping sequences	In vitro one-step assembly tool	[24,25]
Ordered Gene Assembly in *B. subtilis* (OGAB)	Type II restriction enzyme digestion	One-step pathway assembly in *B. subtilis*	[26,27]
BioBrick™/BglBrick	Modular restriction digestions	Utilize restriction digestion, BioBrick™ assembly kit commercially available	[29]
Golden Gate	Type II restriction enzyme digestion	One-step, one-pot assembly in short time	[33]
OE-PCR/ CPEC	PCR with overlapping sequences	PCR-based tool, easy to manipulate	[34,35]
USER	Uses overlapping sequences	USER-friendly cloning kit commercially available	[36–38]
SLIC	Uses overlapping sequences	Exonuclease involved, overhang length is determined by chewing back time	[39]
Pathway optimization tools			
Optimizing gene expression			
Plasmid Copy-Number	Modify copy number of plasmid to reduce metabolic burden and still produce high titer	Gene expression can be easily modified	[39,41–43,86]
CiChE	Integrate the pathway into the chromosome	Avoid allele segregation and allow multiple copies of gene integration	[44]

(Continued)

TABLE 3.1 (Continued)

Tool	Description	Advantages	References
Codon Optimization	Express a heterologous enzyme, change the DNA sequence to complement host codon bias	Can be done by DNA synthesis	[50−53]
Tuning Intergenic Regions	Improve gene expression far from promoter, improve stability of mRNA by introducing secondary structures to reduce mRNA degradation	Control of gene expression in an operon structure	[54]
TIGR Library	Avoid rationally designing the intergenic regions, creation of a library	Library approach allows for nonintuitive mutations	[55]
Promoter Engineering	Engineer different promoters with different strengths and apply them to a pathway to overexpress rate-limiting enzymes	Allows fine-tuning of gene expression for a single gene	[57]
Combinatorial Promoter Library	Library approach to promoter engineering	Fine tuning of multiple gene expression simultaneously	[58]
Dynamic Transcriptional Regulation	Gene circuit designed to express certain genes to redirect flux towards desired product	Express superfluous genes only when needed, reducing burden	[59,60]
RBS Engineering	Improve translation efficiency by optimizing the RBS site by mathematical models	Reduce the library size required to screen	[63]

Optimizing protein activity

Tool	Description	Advantages	References
Increasing Protein Activity	In vitro enzyme engineering to improve activity increases product titer	Proteins with higher activity do not need high expression	[64]
Improving Substrate Specificity	In vitro enzyme engineering to alter specificity to desired substrate	Reduce side reactions and redirect the flux towards the product	[65]
Alter Cofactor Specificity	In vitro enzyme engineering to alter specificity to desired cofactor, can create internal regeneration mechanisms for cofactor	Reduce the competition for cofactors between required metabolism reactions and the desired pathway	[72−74]
Library of Homologous Enzymes	In vivo library combination of different enzyme homologues with different activities	Can select proteins for pathway with best activity, most substrate specificity, and cofactor usage simultaneously	[75]
Protein Scaffolds	Anchor the protein in a special manner to stimulate active site channeling	Shuttle intermediates directly to the next reaction, reducing diffusion	[79,80]

45

DESIGN AND CONSTRUCTION OF PATHWAYS

Pathway Design Tools

In the design of metabolic pathways for synthesis of target compounds, traditional methods simply modify existing endogenous pathways or recruit partial/entire pathways into a heterologous host. Recent advances in bioinformatics have allowed a new approach, wherein the computational tools can be used to design novel and specific pathways for the desired compound. This strategy relieves the inherent regulatory restrictions of endogenous pathways, and exploits computational and theoretical design tools to maximize the efficiency of the designed pathways.

Several databases containing enzymes and enzymatic reactions have been used as the platform for the computational design of novel pathways. These databases include but are not limited to: Kyoto Encyclopedia of Genes and Genomes (KEGG);[11] BRaunschwig ENzyme DAtabase (BRENDA);[12] MetaCyc;[13] the University of Minnesota Biocatalysis/ Biodegradation Database (UM-BBD);[14,15] Retro-Biosynthesis Tool (ReBiT);[16] and the Universal Protein Resource (Uniprot).[17] With these databases, it is possible to identify a staggering number of different pathways for synthesis of the same target compound. Thus, a major challenge in computational design is to rank candidate pathways to predict which pathway will have the highest yield of the desired compound.

The Biochemical Network Integrated Computational Explorer (BNICE) designs pathways via matrix operation and evaluates candidate pathways by their thermodynamic properties.[12,17,18] The KEGG LIGAND database serves as the enzyme library, and enzymes involved in pathway design are selected by the first three levels of their EC numbers. By this method, the range of enzymes involved is broad because they are not substrate specific but functional group specific instead. Then the Gibbs free energy of each individual reaction is calculated and unfavorable pathways are eliminated. By employing the BNICE framework, over 400 000 theoretical novel biochemical pathways from the substrate chorismate to an aromatic amino acid were discovered.[18] Thousands of novel linear polyketide structures have also been explored using similar approaches.[19]

The database ReBiT provides enzyme and enzymatic reaction information, but it is also used as a tool for pathway design. Similar to BNICE, ReBiT also utilizes functional groups of enzymes for enzyme assignment. The user inputs a target compound and then the program identifies the functional groups of enzymes, based on the EC number, that could theoretically produce that compound. Suggestions on which specific enzymes could be involved and the related cofactors are also displayed in the selection step of each reaction.[16]

The UMBBD-Pathway Prediction System (UM-PPS) applies natural precedents to eliminate infeasible steps in predicting plausible biodegradation pathways.[15] This prediction system incorporates the UM-BBD as its database.[21] It allows the user to input the chemical structure of a substrate, and the functional groups are analyzed. Then atom-to-atom mapping is implemented until all possible resulting compounds are displayed. Different from ReBiT, users select compound intermediates instead of enzymes and the operation iteratively repeats until a complete pathway concludes on the desired product. A series of other restrictions and rules for pathway design have been implemented as well. One restriction is to rank UM-PPS results into five aerobic likelihood groups from very likely to very unlikely, thus reducing the number of predictions.[20] Another rule focuses on specific functional groups in a molecule that almost invariably have the potential for the desired reaction, and thus extracts a set of priority rules.[21] These restrictions reduce the number of predicted products based on existing known pathways and avoid false-positive compounds which do not exist in nature. The UM-PPS tool has been used to identify novel routes for biodegradation of multiple substances, and some of the predicted pathways matched the ones identified by human expert predictions.[14]

A recently developed framework by Cho et al. employs a retrosynthesis model to generate pathways and evaluates them by a prioritization scoring algorithm.[22] In this framework, two databases are established: one is a reaction rule database and the other is a binding site rule database. Fifty reaction rules were developed, and binding site rules were generated based on the functional groups. It was claimed that 81% of the enzymes present in the KEGG database are covered in these new databases, enabling a near-comprehensive analysis. Five factors including binding site covalence, chemical similarity, thermodynamic favorability, pathway distance, and organism specificity are integrated in the prioritization scoring algorithm, and each pathway deduced is assigned a specific score based on those factors. As proof of concept, novel biofuel synthetic pathways for the production of isobutanol, 3-hydroxypropionate, and butyryl-CoA were designed, and the predictions matched experimentally proven synthetic pathways.[22]

Pathway Construction Tools

ONE-STEP CONSTRUCTION TOOLS

The DNA assembler method developed by Shao et al. is an efficient one-step in vivo pathway construction tool applied to a large number of DNA fragments that have overlapping homologous regions (Fig. 3.1A).[23] Using in vivo homologous recombination in

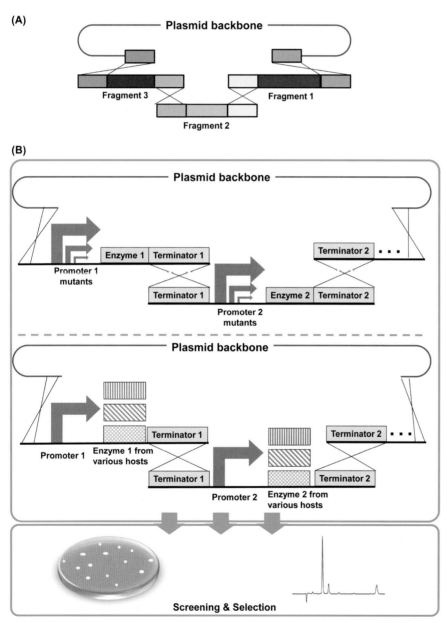

FIGURE 3.1

Scheme of one representative pathway engineering tool. (A) DNA assembler for pathway construction; (B) DNA assembler-based combinatorial pathway engineering

Saccharomyces cerevisiae, an entire biochemical pathway can be constructed in a single step. This tool has been directly applied to pathways up to 45.3 kb in length with high efficiency. As proof of concept, the biosynthetic pathway of a rare nitroaryl-substituted polyketide, aureothin, was assembled into *S. cerevisiae*. Seven fragments with sizes ranging from 4 to 5 kb were amplified and cotransformed with the three helper fragments to make a single circular DNA molecule of 35.7 kb. Six out of ten randomly picked colonies were confirmed as correct by restriction digestion, which indicates 60% efficiency.

The isothermal assembly method developed by Gibson et al. uses an in vitro recombination system to assemble overlapping DNA fragments.[24,25] An enzyme cocktail containing T5 exonuclease, Phusion DNA polymerase, and *Taq* DNA ligase executes overhang chewback, gap-filling, and ligation all in one step. Gibson et al. reported the construction of a 16.3 kb mouse mitochondrial genome from 600 overlapping 60-mers in vitro, demonstrating that this method is a successful approach for multiple-fragment, multiple-size assembly.[25] Employing the same method, Smith et al. reported a one-step combinatorial assembly of promoter and gene cassettes in *E. coli*. An acetate utilization pathway containing one acetate kinase gene and one phosphotransacetylase gene was constructed, and a synthetic library containing three promoters and four gene variants was constructed with 81% assembly efficiency.[24]

Ordered Gene Assembly in *Bacillus subtilis* (OGAB) is a one-step assembly method to combine multiple DNA fragments into a *B. subtilis* plasmid.[26,27] DNA fragments with overhanging sequences generated by Type IIS restriction enzymes are amplified and ligated by T4 DNA ligase in a specified order. In 2003, Itaya et al. successfully assembled five antibiotic resistance genes in a designed order and orientation.[27] Later, the zeaxanthin operon consisting of five carotenoid biosynthetic genes *crtE*, *crtY*, *crtI*, *crtB*, and *crtZ* was assembled into a plasmid by the OGAB method, proving the capability of OGAB to construct pathways in one step.[26]

MULTISTEP CONSTRUCTION TOOLS

The BioBrick™ method is used to construct a metabolic pathway in a step-by-step manner. Briefly, restriction digestions and ligations are implemented with compatible sticky ends generated by a set of specific restriction enzymes *EcoRI*, *XbaI*, *SpeI*, and *PstI*. The standard prefix and suffix sequences containing corresponding restriction sites are cleaved by *EcoRI* and *SpeI*, or *EcoRI* and *XbaI*. During the construction step, since the *XbaI* recognition site TCTAGA shares the overhang sequence CTAG with the *SpeI* recognition site ACTAGT, the two digested sticky ends fuse together to generate a 6-bp scar sequence ACTAGA between the original sequence and the inserted sequence.[28] Based on this method, the BglBrick method was developed in which *XbaI*, *SpeI*, and *PstI* are replaced with *BglII*, *BamHI*, and *XhoI*, respectively, thus alternating the scar sequence, along with the removal of a possible in-frame stop codon.[29] These two approaches allow operon construction, serving as building blocks for pathway assembly. However, the time-consuming manipulation and the existence of a scar in key locations prevent its wide application in pathway construction.

Based on Type IIS restriction enzymes, a series of pathway construction tools have been developed such as pairwise selection assembly method,[30] Golden Gate assembly method,[31] and Golden Braid method.[32] Among the methods utilizing Type IIS restriction enzymes, Golden Gate assembly method attracts much attention because it can rapidly assemble multiple DNA fragments in a one-pot, one-step manner. Type IIS enzymes recognize asymmetric sequences and cleave outside of these recognition sites, resulting in overhangs that can be ligated by a DNA ligase. Using the Golden Gate method, a model protein trypsinogen was efficiently integrated into a vector.[33]

Overlap extension polymerase chain reaction (OE-PCR) is a scarless method for DNA assembly.[34] Primers are designed to contain sequences identical to the upstream and downstream DNA sequences of the target DNA, so that the complementary overhangs will anneal to each other and be extended during the PCR process, which allows ordered assembly. Circular Polymerase Extension Cloning (CPEC) was further developed and enables the assembly of a target gene into a linearized plasmid in PCR implementation.[35] These PCR-based technologies avoid restriction digestion and ligation, though their application is limited by high GC content sequences. Scale-up is problematic due to the error rate of PCR, especially in construction of DNA molecules with sizes over 10 kb.

Similar to the overlap extension strategy, but not PCR-based, the uracil specific excision reagent (USER) method can also be used for large pathway construction.[36–38] One uracil nucleotide is recruited in DNA fragments, close to the 5′-end of the linear fragment, and the fragments are excised by uracil DNA glycosylase along with cleavage by apurinic/apyrimidinic (AP) endonuclease. This produces complementary overhangs which can be fused together by T4 DNA ligase which makes the desired product ready for PCR amplification. Sequential implementation enables the construction of an entire pathway in plasmid, or even in the chromosome.

Sequence and Ligation Independent Cloning (SLIC) utilizes in vitro homologous recombination and single-strand annealing to assemble multiple fragments in a single reaction.[39] In both the inserted fragments and vector, single-stranded DNA overhangs are produced by the chew-back mechanism of exonucleases. T4 DNA polymerase bears some exonuclease activity, but this reaction can be then stopped by the addition of dNTP. Thus, this enzyme was chosen to chew-back the DNA. Assembly of 10 fragments with sizes ranging from 275 bp to 980 bp into one plasmid has been proven with the SLIC method.

PATHWAY OPTIMIZATION

Strategies for Optimizing a Metabolic Pathway Based on Gene Expression

The methods which allow facile assembly of a full pathway have opened many doors for biosynthetic processes. However, it is not as simple as stringing multiple genes together because the pathway must be optimized to overcome the metabolic burden. Initial efforts to balance the gene expression in a pathway to minimize the metabolic burden were focused on varying inducer concentrations. It has been shown that different inducer concentrations can greatly affect the titer of the desired product.[40] However, more advanced tools have since been developed to diversify the available toolkit.

COPY NUMBER OF PLASMIDS EXPRESSING A TARGET PATHWAY

During initial work in this field, a high-copy-number plasmid, which replicates 30–40 times within the cell, was thought to be the vector of choice when biosynthetically producing value-added compounds because it produced a larger amount of proteins compared to low-copy-number plasmids. However, Jones et al.[41] proved that low-copy-number plasmids, which replicate only 1–2 times within the cell, could actually produce a higher yield of the product. Two independent pathways were analyzed: the production of polyphosphate (polyP) and lycopene. These pathways were chosen for the study because the substrates required to make the metabolites were critical for general cell metabolism, and each metabolite could be produced/manipulated by the overexpression of a single gene. The metabolic burden was quantified based on the final optical density (OD) reading of the cultures containing each of the two independent metabolic pathways. It was shown that the single copy plasmids overexpressing the key enzyme for both pathways had a higher OD than the culture with the analogous high-copy plasmids. This higher OD increased the overall yield of the desired products in the low-copy plasmid system. Thus, more cells producing a smaller amount of compound can be more productive than fewer cells producing a lot of the compound. This method of lowering the copy number of the plasmid has been met with great success, especially with multigene pathways. Other successful examples include the mevalonate pathway and the biosynthesis of antibiotics.[39,42,43]

CHROMOSOMAL INTEGRATION

As shown by Jones et al. comparing high- and low-copy plasmids,[41] the plasmid expression of gratuitous proteins can be successful. However, there are disadvantages to the plasmid-based system. Plasmid instability has been shown to be a major problem. Due to the metabolic burden on the plasmid-bearing organism, it has a selective disadvantage compared to a plasmid-free cell. Without selection pressure, an entire culture could be dominated by the plasmid-free cell. The loss of the productive pathway (allele segregation) has been described as the fundamental flaw in plasmid-based gene expression. Tyo et al.[44] developed a technique called chemical-inducible chromosomal evolution (CiChE) to integrate the heterologous genes into the chromosome at multiple sites, with up to 40 consecutive copies. Integration into the chromosome leads to long-term stability and circumvents the loss of productivity by allele segregation due to use of plasmid. More than offering stability over a long period like other integration methods, this particular method is a facile way to finely tune the gene expression by multiple insertions and having exact copy-numbers, allowing genes with low activity to have higher expression and thereby reducing bottlenecks.

CODON OPTIMIZATION

The degeneracy of the genetic code is exploited systematically in different ways by different organisms.[45,46] In the early days of recombinant DNA expression, it became abundantly clear that codon optimization would be crucial for efficient protein expression. Studies have shown that the preferred codons correlate with the abundance of cognate tRNAs available within the cell.[47–49] Thus, if a gene from an organism with a certain codon bias is recombinantly expressed in a host with a different codon bias, the translation of the recombinant protein will be suboptimal. This can be overcome by codon optimization, wherein the codon sequence of the target gene is altered to match that of the codon bias of the host. Codon optimization is commercially available from DNA synthesis companies and has proven to increase protein expression in multiple projects.[50–53]

OPERONS AND TUNING INTERGENIC REGIONS

Grouping the genes of interest for the pathway into a single polycistronic mRNA can be beneficial to optimize the enzyme expression. This one-step gene expression allows for simplified induction processes and improved regulatory mechanisms. However, the design of the basic operon can greatly affect expression. For example, the genes in closest proximity to the promoter will be expressed the greatest. To improve the expression of genes further downstream from the promoter, the efficiency of the transcription can be controlled by optimizing transcriptional processes. The transcription termination, mRNA stability, and translation initiation can all affect how well the proteins are expressed. The DNA regions which control these processes are the intergenic regions.

Smolke et al.[54] showed that altering the mRNA stability by engineering the intergenic regions would increase protein production. By introducing RNA hairpin turns into the 3′ and 5′ intergenic region of the gene of interest, the mRNA was less susceptible to mRNA degradation by the RNaseE nuclease. Activity analysis of the enzyme of interest suggested that with increasing hairpin turns, the enzyme activity increased. This was the first example of differential control of gene expression established in an operon. This idea of independently controlling gene expression within an operon was further explored by Pfleger et al.[55] By creating a library of tunable intergenic regions (TIGRs), the expression of several genes within an operon was simultaneously tuned for optimal expression. The TIGRs contained control elements that include mRNA secondary structures, RNase cleavage sites, and ribosomal binding site (RBS) sequestering sequences. The library of mutated intergenic regions was constructed and applied to the mevalonate pathway. To select, the library was transformed into a melavonate auxotroph strain and then screened by a mevalonate biosensor. The product concentration of the final mutant pathway was increased seven-fold compared to that of the original wild-type pathway. It was interesting to note that the reason for improved production was actually a down-regulation of two genes within the operon, which was counterintuitive. Thus, the ability to tune multiple regulatory mechanisms simultaneously without specific rational design is a major strength of the TIGR approach.

PROMOTER STRENGTH

More than increasing mRNA stability, the promoter strength can be tuned to increase the gene expression by producing more quantities of the mRNA. The strength is based on efficient promoter recognition and rapid binding of the DNA polymerase.[56] However, multiple genes in a pathway do not all need to be expressed at the same level.

Promoters can be mutagenized to achieve precise strength and regulation. These promoters of varying strengths have been applied to exhibit a broad range of genetic control, which can be selected to construct an optimal metabolic pathway. Alper et al.[57] used a library of mutant promoters with varying strengths to regulate the expression of the *dxs* gene, which led to the improvement of the volumetric productivity of lycopene. It was found that optimal gene expression did not involve the strongest promoter. This library method allows for the precise quantitative control of gene expression in vivo. A more broadly applicable method was recently established by Du et al.,[58] which relies on the DNA assembler method to automatically combine multiple libraries of mutant promoters of different strengths in a multigene pathway. Through a rigorous screening and selection process, the optimal combination of mutant promoters in the pathway can be identified (Fig. 3.1B). This technique is especially powerful because it involves multiple promoters in a multigene pathway, optimizing the expression of all genes simultaneously. By balancing the flux through the pathway, no bottlenecks were observed. Additionally, it was shown that the technique allowed for identification of customized pathways for specific *Saccharomyces cerevisiae* strains.

DYNAMIC TRANSCRIPTIONAL REGULATION

Genetic circuits have great potential for dynamic transcriptional regulation through inducible transcription factors and *cis*-regulatory elements. They can be used to reduce metabolic burden and toxic byproduct formation. For example, when protein translation was coupled to glucose availability and other metabolic byproducts of the system, a dynamic control was introduced to regulate flux through the desired pathway.[59,60] The circuit was designed to detect the concentration of acetyl phosphate (ACP), a byproduct of the lycopene pathway. If ACP was detected, two enzymes were turned 'on' by the genetic circuit, and high expression of those enzymes rerouted the carbon flux from the byproduct back to the product. Thus, these significant enzymes would only be expressed when needed, and did not overburden the cell too significantly.

COMPUTATIONAL METHOD FOR RBS ENGINEERING

Similar to transcriptional engineering, such as promoter engineering and combinatorial engineering of the tunable intergenic regions of operons, translational engineering can also be used to effectively regulate gene expression. RBS and other regulatory RNA sequences are the major control elements for translation initiation. It is possible to mutate the RBS sequence and create a library with varying RBS strengths,[55,61,62] and it was shown that an optimal RBS can improve product titer. However, the library size that must be screened increases combinatorially with the number of RBS sites to engineer. Thus, attempts to optimize gene expression within a system, without specific rational design, can be prohibitively inefficient and time-consuming. To simplify the process, a computational method for predicting the translation rate of mRNA was developed by Salis et al.[63] The translation initiation of mRNA is comprised of several different molecular interactions. The Gibbs free energy of initiation can be measured and quantified, and consequently the total Gibbs free energy of the reaction can be determined. An expression of the translation efficiency was modeled on the basis of the ΔG. This computational tool was successfully applied to predict the reliable operation of a synthetic AND gate by balancing the expression of the proteins in the circuit. In principle, this method can also be used to predict the optimal RBS sequences of bottleneck enzymes to balance the metabolic flux.

Strategies for Optimizing a Metabolic Pathway on the Protein Level

Changing gene expression can greatly optimize the pathway; however, in many circumstances the inherent limitations associated with activities and specificities of the desired enzymes in the pathway cannot be overcome by optimizing the gene expression. Thus tools for engineering the pathway on the protein level are also needed.

INCREASING ENZYME ACTIVITY OR IMPROVING SUBSTRATE SPECIFICITY

Leonard et al.[64] engineered a diterpenoid biosynthetic pathway for levopimaradiene production. Two rate-limiting enzymes, geranylgeranyl disphosphate synthase (GGPPS) and levopimaradiene synthase (LPS), were identified as bottlenecks. These enzymes were engineered independently for higher activity. The mutant enzymes with the highest in vitro activity were selected and subsequently cloned into the full pathway. The pathways with the engineered GGPPS and LPS enzymes increased production and also increased the selectivity toward the desired product. This resulted in a 2600-fold increase in levopimaradiene production.

Another interesting example of protein engineering in pathway optimization is altering substrate specificity. Zhang et al.[65] constructed a pathway to produce 3-methyl-1-pentanol, a nonnatural alcohol, by engineering a promiscuous enzyme to have altered specificity towards the desired substrate, (S)-2-keto-methylhexanoate. KivD, the promiscuous key enzyme, needed to be engineered to reduce the formation of byproducts, thus driving the carbon flux toward the desired alcohol. The mutant exhibited a specificity towards (S)-2-keto-methylhexanoate 40-fold higher than its natural substrate. The wild-type enzyme specificity towards (S)-2-keto-methylhexanoate was only four-fold higher than the natural substrate. Another enzyme in the system also needed engineering: LeuA required a higher activity toward (S)-2-keto-3-methylvalerate to reduce the bottleneck. The combination of the mutated KivD and LeuA produced 685.7 mg/L of (S)-3-methyl-1-pentanol, whereas the control consisting of the two wild-type enzymes produced no detectable amounts of the desired product. By introduction of these mutants, it was possible to expand the E. coli metabolism to produce C5 and C8 alcohols.

ALTERING COFACTOR SPECIFICITY

Many pathways involved in the synthesis of value-added compounds employ oxidation-reduction reactions catalyzed by enzymes using cofactors $NAD(P)^+$ and $NAD(P)H$. These cofactors are just as critical in natural metabolism as in the heterologous pathways. Thus, the competition for the cofactors between the enzymes requiring them can be a major limiting factor.[66] Engineering of cofactor recycling systems has proven to be beneficial: by altering the cofactor specificity of the enzymes within the pathway, an internal regeneration mechanism can be incorporated. A particular project which has received much attention in this area is the conversion of xylose to ethanol. Two of the major enzymes in the pathway, the xylose reductase (XR) and xylitol dehydrogenase (XDH), have an imbalance in their cofactor usage, as XR prefers NADPH and XDH prefers NAD^+. A number of studies analyzing this pathway have shown that the redox imbalance generated during the xylose assimilation limits cell growth, producing unwanted byproducts and reducing ethanol production.[67–71] Multiple studies have focused on engineering the XDH enzyme to have altered specificity towards $NADP^+$, which would match with the XR cofactor preference of NADPH.[72,73] Similarly, the XR was altered to have a preference for NADH, which led to a 40-fold increase in ethanol productivity over the wild-type enzyme.[74]

COMBINATORIAL ASSEMBLY OF HOMOLOGOUS ENZYMES IN A PATHWAY

In the design of a pathway, there are often many different homologues of the desired enzymes that conduct the specific chemistry required. Identifying the proper enzymes to be used in the pathway can be challenging. Often, enzymes are chosen based on the highest activity, but this is not always advantageous. Enzymes could have unknown properties, such as side reactions and unknown regulation, which would reduce the overall productivity. Also, balancing the enzyme activities in the pathway is difficult for rational design. Thus, a library approach encompassing the automatic assembly of hundreds of genes was developed (Fig. 3.1B).[75] In this method, the fungal xylose utilization pathway, which consists of three heterologously expressed enzymes (XR, XDH, and XKS) was constructed. A total of 20 homologues of XR, 22 XDHs, and 19 XKSs were cloned into the library. Combinatorial assembly of these genes using the DNA assembler method resulted in a library of over 8000 xylose-utilizing pathways. After colony-size-based high-throughput screening, colonies with the best utilization of xylose as a sole-carbon source were selected and subsequently screened for ethanol production. This particular method is powerful, because it was able to identify different combinations of the three types of enzymes in the pathway for different media conditions and strains.

PROTEIN SCAFFOLDS AND SUBSTRATE CHANNELING

Industrial bioprocesses have taken advantage of immobilized enzymes to facilitate enzyme reutilization and substrate–enzyme diffusion for years. It is well known that by positioning enzymes in a specific organized manner, more efficient mass transfer of the substrate will occur. For example, cellulosomes are enzyme complexes that can degrade cellulose from plant cell walls to fermentable sugars.[76,77] This concept has also been developed into Consolidated Bioprocessing or CBP,[78] wherein cellulosomal degradation of biomass and ethanol production can be simultaneously achieved. To this end, Wen et al. constructed a cellulosome on the yeast surface.[79] By docking three key cellullases including endoglucanase, cellobiohydrolase, and a β-glucosidase onto a scaffolding displayed on the yeast surface, these enzymes could efficiently break down the cellulose and convert the glucose to ethanol.[79]

Another interesting example is the enzyme tryptophan synthase, which contains a largely hydrophobic tunnel connecting two active sites about 25 Å apart. In addition to protecting the reactive intermediate, the tunnel allows for the chemical to be immediately channeled from one active site to another. The same concept was applied to a pathway by Dueber et al.[80] The enzymes within the pathway were colocalized using a synthetic scaffold built from protein–protein interaction domains, which specifically bind the corresponding ligands of the metabolic enzymes. This created an easy access for the substrate to diffuse from one enzyme active site to another. This technique prevents substrates from diffusing away from the active site, and could make up for enzymes with low activity. Low-activity enzymes can be expressed and coupled within the scaffold in multiple numbers. This technique was applied to the mevalonate biosynthetic pathway and improved the titer by 22-fold compared to the nonscaffold wild-type control. The method was also shown to lower the metabolic burden, thus improving cell growth.

APPLICATIONS OF PATHWAY ENGINEERING TOOLS

The tools and methods for pathway construction and optimization are varied. This is significant because not every tool will work in each system, thus it is important to apply multiple tools to the pathway to get the best results. The following examples show how different tools were applied.

Production of Glucaric Acid

A culmination of some of these pathway construction and optimization methods is exemplified in the production of glucaric acid. This compound has been identified as a 'Top Value-Added Chemical from Biomass' by the US Department of Energy,[81] and is used in cholesterol-reduction studies[82] and cancer therapies.[83,84] Currently, its primary use is as a starting material for hydroxylated nylons.[40] Moon et al. used the ReBiT program to identify three enzymes from different sources that can form a novel metabolic pathway to produce glucaric acid in *E. coli*.[16,40,80] The pathway design process was based on an organic chemist's methodology-retrosynthesis, which considers the product first and then designs a synthetic route backwards towards the substrate. It was hypothesized that glucaric acid can be produced from D-glucuronic acid by uronate dehydrogenase (UDH). The D-glucuronic acid can be produced from *myo*-inositol by the *myo*-inositol oxygenase (MIOX), whereas the *myo*-inositol can be produced from glucose via a combination of endogenous enzymes and a heterologous enzyme: *myo*-inositol-1-phosphate synthase (INO1) (Fig. 3.2A).

The MIOX gene was identified from a mouse genome and was synthesized with codon optimization. The INO1 gene was cloned from the cDNA of *S. cerevisiae*, while the UDH gene was cloned from the cDNA of *Pseudomonas syringae*. Traditional cloning techniques incorporated these enzymes into both high- and low-copy plasmids, with a T7 promoter for each enzyme. The resulting *E. coli* strains produced 0.73 g/L of glucaric acid. Initial optimization by varying the isopropyl- β-D-thiogalactopyranoside (IPTG) concentrations increased the production to 1.13 g/L.[40] Experiments determined the bottleneck to be the MIOX-catalyzed step.[80] The high MIOX activity in *E. coli* was found to be strongly influenced by exposure to high concentrations of its substrate *myo*-inositol. Thus it was hypothesized that by reducing diffusion distance and transit time, the recruitment of the pathway enzymes to a protein synthetic scaffold could increase the concentration of *myo*-inositol concentrated around the active site of the bottleneck enzyme. The scaffolds were constructed by assembling three protein–protein interaction domains linked together by glycine-serine linkers. The strains were constructed with varying numbers of proteins linked and ratios of INO1 and MIOX in the scaffold; UDH was determined to be the most active and thus did not need to be overexpressed in the scaffold. It was found that a 1:1 of INO1 to MIOX ratio was the best, with four of each protein linked together. The glucaric acid production increased five-fold compared to the nonscaffold system. The results suggested that the improved production was not only due to increased stability of the MIOX enzyme, but also because the substrate of the enzyme was in close proximity to the enzyme (shown by high titers with increasing the number of INO1 linked to the scaffold).[80,85]

Production of 1,4-Butanediol

Another example of a microbe producing a nonnatural compound is 1,4-butanediol (BDO). BDO is a major commodity chemical used to annually make over 2.5 million tons of valuable polymers such as polyesters, plastics, and spandex fibers.[86] It is currently produced from petroleum-based substrates such as acetylene, butane, propylene, and butadiene. Due to the instability of petroleum-based substrates, it is desirable to identify a metabolic pathway for BDO production from biomass. The design of this novel metabolic pathway was carried out by Genomatica using an in-house pathway prediction software called SimPheny Biopathway Predictor, which elucidated all the potential pathways from the *E. coli* central metabolism that could possibly produce BDO. The Biopathway Predictor algorithm is based on transformations of functional groups by known chemistry, termed 'reaction operators.' The algorithm identified over 10 000 pathways within four to six steps for the synthesis of BDO from substrates acetyl-CoA, α-ketoglutarate, succinyl-CoA, and glutamate. The pathways were ranked based on maximum theoretical yield, pathway length, number of nonnative steps, number of novel steps, and thermodynamic feasibility. The highest-ranking pathways were found to proceed through the 4-hydroxybutyrate intermediate. The optimal pathway determined is shown in Figure 3.2B. In initial results, the expression of the six genes under a single strong promoter in a multicopy plasmid resulted in a substantial metabolic burden on the cell. The subsequent experiments

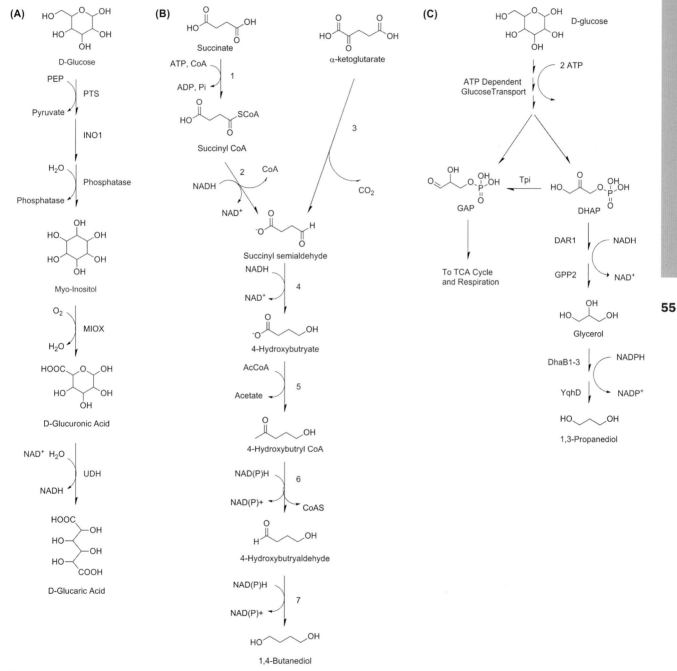

55

FIGURE 3.2

(A) Novel heterologous pathway for the production of glucaric acid. (B) BDO biosynthetic pathways introduced into *E. coli*. Enzymes for each numbered step are: (1) succinyl-CoA synthetase; (2) CoA-dependent succinate semialdehyde dehydrogenase; (3) 2-oxoglutarate decarboxylase; (4) 4-hydroxybutyrate dehydrogenase; (5) 4-hydroxybutyryl-CoA transferase; (6) 4-hydroxybutyryl-CoA reductase; (7) alcohol dehydrogenase. Steps 1 and 7 occur naturally in *E. coli*, but the rest were heterologously expressed. (C) The engineered metabolic pathway from D-glucose to 1,3-propanediol rearranged the gene order onto two plasmids with medium and low copy numbers, which helped to optimize the expression balance. Additionally, the genes were codon optimized, which further improved the gene expression and reduced the metabolic burden. This led to the production of 18 g/L BDO. To optimize the pathway even further, a systems-wide approach was utilized. This was based on constraint-based modeling and the OptKnock framework to design a strain with enhanced anaerobic operation of the oxidative tricarboxylic acid cycle. This made the pathway to produce BDO the only way to balance redox in aerobic growth. This optimized pathway required microaerobic growth, but still produced BDO at about 20 g/L. Additionally, it was shown that the BDO could be produced at high titers grown on biomass hydrolysate, composed of various oligosaccharides and other impurities, which is significant if this will be an industrial process using renewable sources.[86]

Production of 1,3-Propanediol

1,3-Propanediol (1,3-PD) is a valuable compound used in polymer production. Since 2008, the annual market for 1,3-propanediol is over 100 million pounds and grows increasingly.[87] It is naturally produced only from glycerol by special microorganisms such as *Klebsiella*,[88] *Citrobacter*,[89] *Enterobacter*,[90] and *Clostridium*[91] under anaerobic conditions. In order to establish an efficient and economically viable pathway to produce 1,3-PD from glucose, three approaches have been developed. The first approach is to introduce the metabolic pathway from glucose to glycerol into native producers;[92,93] the second is to introduce the pathway from glycerol to 1,3-PD into native glycerol producers such as *S. cerevisiae*.[94] The last, and the most interesting, is to introduce both pathways into a platform host such as *E. coli* which has no capacity to produce either glycerol or 1,3-PD.[95]

In the pathway converting glucose to glycerol, glucose is converted to glycerol-3-phosphate after a series of reactions, and glycerol-3-phosphate is converted to glycerol by the enzyme glycerol-3-phosphatase. *S. cerevisiae* possesses two genes named GPP1/2 which encode two glycerol-3-phosphatases, enabling the production of glycerol.[96] To produce 1,3-PD from glycerol, the glycerol is converted by glycerol dehydratase (GDHt) to 3-hydroxypropion aldehyde (3-HPA), which is an intermediate that can be converted to 1,3-PD by 1,3-propanediol oxidoreductase (PDOR). GDHt is a vitamin-B12-dependent enzyme. During catalysis, the inactive coenzyme B12 binds to GDHt and catalysis stops. When an auxiliary enzyme, glycerol dehydratase reactivase releases the coenzyme, the resultant free GDHt is able to catalyze the conversion from glycerol to 3-HPA.[97]

After the *S. cerevisiae dhaB* gene encoding GDHt, the *dhaT* gene encoding PDOR, and the GPP1/2 genes were introduced, the resultant *E. coli* strain was able to produce 1,3-PD from glucose. Nevertheless, a lot more improvements were needed to optimize the final titer and yield. Nakamura and Whited discovered that glucose transportation by the phosphotransferase system (PTS) and glyceraldehyde-3-phosphate dehydrogenase (*gap*) down-regulation are two key limitations hampering efficient 1,3-PD production.[93] To relieve the excessive glucose transportation, an ATP-dependent glucose phosphorylation system was observed to be energetically more efficient than the original phosphoenolpyruvate (PEP)-dependent PTS in *E. coli*. As a result, two endogenous genes *galP* and *glk* encoding galactosepermease and glucokinase were introduced, replacing the endogenous PTS system. To prevent the waste of glycerol in the central carbon metabolism, the glycerol kinase gene *glpK* and the glycerol dehydrogenase gene *gldA* were deleted. Additionally, a triosephosphate isomerase gene *tpi* was also deleted, which eliminates the conversion between dihydroxyacetone phosphate (DHAP) and glyceraldehyde 3-phosphate (GAP) that imposes an artificial ceiling on 1,3-PD yield. To overcome this limitation, *gap* down-regulation concomitant with reinstatement of *tpi* was achieved, which led to a titer of 135 g/L and a yield of 0.6 mol 1,3-propanediol/mol glucose.[93]

CONCLUSIONS AND FUTURE PROSPECTS

The ability to design, construct, and optimize a metabolic pathway for synthesis of value-added chemicals with high efficiency and fidelity has opened the doors for many possibilities. Biological catalysis via engineered microorganisms can offer a competitive advantage over traditional chemical synthesis, and will have not only economic benefits, but also environmental benefits. Many chemicals, fuels, and materials can now be produced by recombinant microorganisms from renewable feedstock such as plant biomass. However, there are still many future challenges. With the thrust to consider and design cellular systems like computer 'parts,' modularity across systems has become significant, as demonstrated by the BioBrick™ methodology. With modularity will come a drive for customizable strains that can be engineered rapidly for a single specific task at an industrial level.[98] This has already been shown by entrepreneurial companies like Ginkgo BioWorks. The rapid construction and optimization of a pathway will become paramount to be competitive with industrial standards of traditional chemical synthesis. Perhaps more relevant than rapid construction and optimization is a full understanding of heterologous gene expression. Pathway design tools like BNICE can identify a significant number of genes to be used for a pathway, but there is no guarantee that the gene will be expressed in the host organism. This has already proven to be a significant setback in *S. cerevisae*, wherein many pathways which have been successful in *E. coli* will not be processed similarly in yeast, based mainly on gene expression. Beyond pathway engineering, another future direction will be controlling pathways of whole microbial consortia which communicate via molecular signaling. Mixed populations within a culture will be able to independently coordinate their different gene expression to achieve more complex tasks than just a single pathway.

Acknowledgments

We thank the National Institutes of Health (GM077596), the National Academies Keck *Futures Initiative* on Synthetic Biology, the Energy Biosciences Institute, and the National Science Foundation as part of the Center for Enabling New Technologies through Catalysis (CENTC), CHE-0650456 for financial support in our synthetic biology projects.

References

1. Ro DK, Paradise EM, Quellet M, et al. Production of the antimalarial drug precursor artemisinic acid in engineered yeast. *Nature*. 2006;440:940−943.

2. Atsumi S, Hanai T, Liao JC. Non-fermentative pathways for synthesis of branched-chain higher alcohols as biofuels. *Nature*. 2008;451:86−90.

3. Chang MCY, Eachus RA, Trieu W, Ro DK, Keasling JD. Engineering *Escherichia coli* for production of functionalized terpenoids using plant P450s. *Nat Chem Biol*. 2007;3:274−277.

4. Du J, Shao ZY, Zhao HM. Engineering microbial factories for synthesis of value-added products. *J Ind Microbiol Biotechnol*. 2011;38:873−890.

5. Chotani G, Dodge T, Hsu A, et al. The commercial production of chemicals using pathway engineering. *Biochim Biophys Acta-Protein Struct Molec Enzym*. 2000;1543:434−455.

6. Khosla C, Keasling JD. Timeline−metabolic engineering for drug discovery and development. *NRD*. 2003;2:1019−1025.

7. Bentley WE, Mirjalili N, Andersen DC, Davis RH, Kompala DS. Plasmid-encoded protein − the principle factor in the metabolic burden associated with recombinant bacteria. *Biotechnol Bioeng*. 1990;35:668−681.

8. Vind J, Sorensen MA, Rasmussen MD, Pedersen S. Synthesis of proteins in *Escherichia coli* is limited by the concentration of free ribosomes − expression from reporter genes does not always reflect functional messenger-RNA levels. *J Mol Biol*. 1993;231:678−688.

9. Birnbaum S, Bailey JE. Plasmid presence changes the relative levels of many host-cell proteins and ribosome components in recombinant *Escherichia coli*. *Biotechnol Bioeng*. 1991;37:736−745.

10. Glick BR. Metabolic load and heterologous gene-expression. *Biotechnol Adv*. 1995;13:247−261.

11. Kanehisa M, Goto S. KEGG: kyoto encyclopedia of genes and genomes. *Nucleic Acids Res*. 2000;28:27−30.

12. Schomburg I, Chang A, Ebeling C, et al. BRENDA, the enzyme database: updates and major new developments. *Nucleic Acids Res.* 2004;32:D431–D433.

13. Caspi R, Foerster H, Fulcher CA, et al. MetaCyc: a multiorganism database of metabolic pathways and enzymes. *Nucleic Acids Res.* 2006;34:D511–D516.

14. Hou BK, Ellis LBM, Wackett LP. Encoding microbial metabolic logic: predicting biodegradation. *J Ind Microbiol Biotechnol.* 2004;31:261–272.

15. Ellis LBM, Roe D, Wackett LP. The University of Minnesota biocatalysis/biodegradation database: the first decade. *Nucleic Acids Res.* 2006;34:D517–D521.

16. Prather KLJ, Martin CH. De novo biosynthetic pathways: rational design of microbial chemical factories. *Curr Opin Biotechnol.* 2008;19:468–474.

17. Wu CH, Apweiler R, Bairoch A, et al. The Universal Protein Resource (UniProt): an expanding universe of protein information. *Nucleic Acids Res.* 2006;34:D187–D191.

18. Hatzimanikatis V, Li CH, Ionita JA, Henry CS, Jankowski MD, Broadbelt LJ. Exploring the diversity of complex metabolic networks. *Bioinformatics.* 2005;21:1603–1609.

19. Gonzalez-Lergier J, Broadbelt LJ, Hatzimanikatis V. Theoretical considerations and computational analysis of the complexity in polyketide synthesis pathways. *J Am Chem Soc.* 2005;127:9930–9938.

20. Cho A, Yun H, Park J, Lee S, Park S. Prediction of novel synthetic pathways for the production of desired chemicals. *BMC Systems Biology.* 2010;4:35.

21. Fenner K, Gao JF, Kramer S, Ellis L, Wackett L. Data-driven extraction of relative reasoning rules to limit combinatorial explosion in biodegradation pathway prediction. *Bioinformatics.* 2008;24:2079–2085.

22. Cho A, Yun H, Park JH, Lee SY, Park S. Prediction of novel synthetic pathways for the production of desired chemicals. *BMC Systems Biology.* 2010:4.

23. Shao ZY, Zhao H, Zhao HM. DNA assembler, an in vivo genetic method for rapid construction of biochemical pathways. *Nucleic Acids Res.* 2009:37.

24. Ramon A, Smith HO. Single-step linker-based combinatorial assembly of promoter and gene cassettes for pathway engineering. *Biotechnol Lett.* 2011;33:549–555.

25. Gibson D, Smith H, Hutchison C, Venter JC, Merryman C. Chemical synthesis of the mouse mitochondrial genome. *Nat Methods.* 2010;7:901–U905.

26. Nishizaki T, Tsuge K, Itaya M, Doi N, Yanagawa H. Metabolic engineering of carotenoid biosynthesis in *Escherichia coli* by ordered gene assembly in *Bacillus subtilis. Appl Environ Microbiol.* 2007;73:1355–1361.

27. Tsuge K, Matsui K, Itaya M. One step assembly of multiple DNA fragments with a designed order and orientation in *Bacillus subtilis* plasmid. *Nucleic Acids Res.* 2003;31:e133.

28. Knight Jr TF. Idempotent vector design for standard assembly of biobricks. *DSpace.* 2003:<http://hdl.handle.net/1721.1/21168>

29. Anderson JC, Dueber JE, Leguia M, et al. BglBricks: a flexible standard for biological part assembly. *J Biol Eng.* 2010;4:1.

30. Blake WJ, Chapman BA, Zindal A, Lee ME, Lippow SM, Baynes BM. Pairwise selection assembly for sequence-independent construction of long-length DNA. *Nucleic Acids Res.* 2010;38:2594–2602.

31. Engler C, Kandzia R, Marillonnet S. A one pot, one step, precision cloning method with high throughput capability. *PLoS ONE.* 2008;3:e3647.

32. Sarrion-Perdigones A, Falconi EE, Zandalinas SI, et al. GoldenBraid: an iterative cloning system for standardized assembly of reusable genetic modules. *PLoS ONE.* 2011;6:e21622.

33. Engler C, Gruetzner R, Kandzia R, Marillonnet S. Golden gate shuffling: a one-pot DNA shuffling method based on type IIs restriction enzymes. *PLoS ONE.* 2009;4:e5553.

34. Horton RM, Hunt HD, Ho SN, Pullen JK, Pease LR. Engineering hybrid genes without the use of restriction enzymes: gene splicing by overlap extension. *Gene.* 1989;77:61–68.

35. Quan J, Tian J. Circular polymerase extension cloning of complex gene libraries and pathways. *PLoS ONE.* 2009;4:e6441.

36. Smith C, Day PJ, Walker MR. Generation of cohesive ends on PCR products by UDG-mediated excision of dU, and application for cloning into restriction digest-linearized vectors. *PCR Methods Appl.* 1993;2:328–332.

37. Nour-Eldin HH, Geu-Flores F, Halkier BA. In: FettNeto AG, ed. *Plant Secondary Metabolism Engineering: Methods and Applications.* Vol. 643. Totowa: Humana Press Inc; 2010:185–200.

38. Hansen BG, Salomonsen B, Nielsen MT, et al. Versatile enzyme expression and characterization system for *Aspergillus nidulans,* with the *penicillium brevicompactum* polyketide synthase gene from the mycophenolic acid gene cluster as a test case. *Appl Environ Microbiol.* 2011;77:3044–3051.

39. Pitera DJ, Paddon CJ, Newman JD, Keasling JD. Balancing a heterologous mevalonate pathway for improved isoprenoid production in *Escherichia coli. Metab Eng.* 2007;9:193–207.

40. Moon TS, Yoon SH, Lanza AM, Roy-Mayhew JD, Prather KLJ. Production of glucaric acid from a synthetic pathway in recombinant *Escherichia coli*. *Appl Environ Microbiol*. 2009;75:4660.

41. Jones KL, Kim SW, Keasling JD. Low-copy plasmids can perform as well as or better than high-copy plasmids for metabolic engineering of bacteria. *Metab Eng*. 2000;2:328−338.

42. La Clair JJ, Foley TL, Schegg TR, Regan CM, Burkart MD. Manipulation of carrier proteins in antibiotic biosynthesis. *Chem Biol*. 2004;11:195−201.

43. Alper H, Miyaoku K, Stephanopoulos G. Construction of lycopene-overproducing *E. coli* strains by combining systematic and combinatorial gene knockout targets. *Nat Biotechnol*. 2005;23:612−616.

44. Tyo KEJ, Ajikumar PK, Stephanopoulos G. Stabilized gene duplication enables long-term selection-free heterologous pathway expression. *Nat Biotechnol*. 2009;27:760−765.

45. Gustafsson C, Govindarajan S, Minshull J. Codon bias and heterologous protein expression. *Trends Biotechnol*. 2004;22:346−353.

46. Ernst JF. Codon usage and gene expression. *Trends Biotechnol*. 1988;6:196−199.

47. Ikemura T. Codon usage and transfer-RNA content in unicellular and multicellular organisms. *Mol Biol Evol*. 1985;2:13−34.

48. Grosjean H, Fiers W. Preferential codon usage in prokaryotic genes − the optimal codon anticodon interaction energy and the selective codon usage in efficiently expressed genes. *Gene*. 1982;18:199−209.

49. Ikemura T. Correlation between the abundance of *Escherichia coli* transfer-RNAs and the occurrence of the respective codons in its protein genes − a proposal for a synonymous codon choice that is optimal for the *Escherichia coli* translational system. *J Mol Biol*. 1981;151:389−409.

50. Kink JA, Maley ME, Ling KY, Kanabrocki JA, Kung C. Efficient expression of the paramecium calmodulin gene in *Escherichia coli* after 4 TAA-to-CAA changes through a series of polymerase chain-reactions. *J Protozool*. 1991;38:441−447.

51. Nambiar KP, Stackhouse J, Stauffer DM, Kennedy WP, Eldredge JK, Benner SA. Total synthesis and cloning of a gene coding for the ribonuclease-s protein. *Science*. 1984;223:1299−1301.

52. Tokuoka M, Tanaka M, Ono K, Takagi S, Shintani T, Gomi K. Codon optimization increases steady-state mRNA levels in *Aspergillus oryzae* heterologous gene expression. *Appl Environ Microbiol*. 2008;74:6538−6546.

53. Martin VJJ, Pitera DJ, Withers ST, Newman JD, Keasling JD. Engineering a mevalonate pathway in *Escherichia coli* for production of terpenoids. *Nat Biotechnol*. 2003;21:796−802.

54. Smolke CD, Carrier TA, Keasling JD. Coordinated, differential expression of two genes through directed mRNA cleavage and stabilization by secondary structures. *Appl Environ Microbiol*. 2000;66:5399−5405.

55. Pfleger BF, Pitera DJ, Smolke CD, Keasling JD. Combinatorial engineering of intergenic regions in operons tunes expression of multiple genes. *Nat Biotechnol*. 2006;24:1027−1032.

56. Paul BJ, Ross W, Gaal T, Gourse RL. rRNA transcription in *Escherichia coli*. *Annu Rev Genet*. 2004;38:749−770.

57. Alper H, Fischer C, Nevoigt E, Stephanopoulos G. Tuning genetic control through promoter engineering. *PNAS*. 2005;102:12678−12683.

58. Du J, Yuan Y, Si T, et al. Customized optimization of metabolic pathways by combinatorial transcriptional engineering (COMPACTER). *Nucleic Acids* Res. 2012;40:e142.

59. Farmer WR, Liao JC. Improving lycopene production in *Escherichia coli* by engineering metabolic control. *Nat Biotechnol*. 2000;18:533−537.

60. Fung E, Wong WW, Suen JK, Bulter T, Lee SG, Liao JC. A synthetic gene-metabolic oscillator. *Nature*. 2005;435:118−122.

61. Basu S, Gerchman Y, Collins CH, Arnold FH, Weiss R. A synthetic multicellular system for programmed pattern formation. *Nature*. 2005;434:1130−1134.

62. Anderson JC, Voigt CA, Arkin AP. Environmental signal integration by a modular AND gate. *Mol Syst Biol*. 2007;3:133.

63. Salis HM, Mirsky EA, Voigt CA. Automated design of synthetic ribosome binding sites to control protein expression. *Nat Biotechnol*. 2009;27:946−950.

64. Leonard E, Ajikumar PK, Thayer K, et al. Combining metabolic and protein engineering of a terpenoid biosynthetic pathway for overproduction and selectivity control. *PNAS*. 2010;107:13654−13659.

65. Zhang KC, Sawaya MR, Eisenberg DS, Liao JC. Expanding metabolism for biosynthesis of nonnatural alcohols. *PNAS*. 2008;105:20653−20658.

66. Berrios-Rivera SJ, Bennett GN, San KY. Metabolic engineering of *Escherichia coli*: increase of NADH availability by overexpressing an NAD(+)-dependent formate dehydrogenase. *Metab Eng*. 2002;4:217−229.

67. Bro C, Regenberg B, Forster J, Nielsen J. *In silico* aided metabolic engineering of *Saccharomyces cerevisiae* for improved bioethanol production. *Metab Eng*. 2006;8:102−111.

68. Jeppsson M, Johansson B, Hahn-Hagerdal B, Gorwa-Grauslund MF. Reduced oxidative pentose phosphate pathway flux in recombinant xylose-utilizing *Saccharomyces cerevisiae* strains improves the ethanol yield from xylose. *Appl Environ Microbiol*. 2002;68:1604−1609.

69. Roca C, Nielsen J, Olsson L. Metabolic engineering of ammonium assimilation in xylose-fermenting *Saccharomyes cerevisiae* improves ethanol production. *Appl Environ Microbiol*. 2003;69:4732−4736.

70. Sonderegger M, Schumperli M, Sauer U. Metabolic engineering of a phosphoketolase pathway for pentose catabolism in *Saccharomyces cerevisiae*. *Appl Environ Microbiol*. 2004;70:2892−2897.

71. Verho R, Londesborough J, Penttila M, Richard P. Engineering redox cofactor regeneration for improved pentose fermentation in *Saccharomyces cerevisiae*. *Appl Environ Microbiol*. 2003;69:5892−5897.

72. Krahulec S, Klimacek M, Nidetzky B. Engineering of a matched pair of xylose reductase and xylitol dehydrogenase for xylose fermentation by *Saccharomyces cerevisiae*. *Biotechnol J*. 2009;4:684−694.

73. Matsushika A, Watanabe S, Kodaki T, et al. Expression of protein engineered NADP+-dependent xylitol dehydrogenase increases ethanol production from xylose in recombinant *Saccharomyces cerevisiae*. *Appl Microbiol Biotechnol*. 2008;81:243−255.

74. Runquist D, Hahn-Hagerdal B, Bettiga M. Increased ethanol productivity in xylose-utilizing *Saccharomyces cerevisiae* via a randomly mutagenized xylose reductase. *Appl Environ Microbiol*. 2010;76:7796−7802.

75. Kim BD, Du J, Eriksen D. Combinatorial design of a highly efficient xylose utilization pathway for cellulosic biofuels production in *Saccharomyces cerevisiae*. *Appl Environ Microbiol*. 2013;79:931−941.

76. Fierobe HP, Bayer EA, Tardif C, et al. Degradation of cellulose substrates by cellulosome chimeras − substrate targeting versus proximity of enzyme components. *J Biol Chem*. 2002;277:49621−49630.

77. Fierobe HP, Mechaly A, Tardif C, et al. Design and production of active cellulosome chimeras − selective incorporation of dockerin-containing enzymes into defined functional complexes. *J Biol Chem*. 2001;276:21257−21261.

78. Girio FM, Fonseca C, Carvalheiro F, Duarte LC, Marques S, Bogel-Lukasik R. Hemicelluloses for fuel ethanol: a review. *Bioresour Technol*. 2010;101:4775−4800.

79. Wen F, Sun J, Zhao H. Yeast surface display of trifunctional minicellulosomes for simultaneous saccharification and fermentation of cellulose to ethanol. *Appl Environ Microbiol*. 2010;76:1251−1260.

80. Dueber JE, Wu GC, Malmirchegini GR, et al. Synthetic protein scaffolds provide modular control over metabolic flux. *Nat Biotechnol*. 2009;27:753−759.

81. Top value added chemicals from biomass, vol. 1: results of screening for potential candidates from sugars and synthesis gas. 2004. http://www1.eere.energy.gov/biomass/pdfs/35523.pdf.

82. Walaszek Z, et al. D-glucaric acid content of various fruits and vegetables and cholesterol-lowering effects of dietary d-glucarate in the rat. *Nutr Res*. 1996;16:673−681.

83. Singh J, Gupta KP. Calcium glucarate prevents tumor formation in mouse skin. *Biomed Environ Sci*. 2003;16:9−16.

84. Singh J, Gupta KP. Induction of apoptosis by calcium D-glucarate in 7,12- dimethyl benz[a]anthracene-exposed mouse skin. *J Environ Pathol Toxicol Oncol*. 2007;26:63−73.

85. Moon TS, Dueber JE, Shiue E, Prather KLJ. Use of modular, synthetic scaffolds for improved production of glucaric acid in engineered *E. coli*. *Metab Eng*. 2010;12:298−305.

86. Yim H, Haselbeck R, Niu W, et al. Metabolic engineering of *Escherichia coli* for direct production of 1,4-butanediol. *Nat Chem Biol*. 2011;7:445−452.

87. Kraus GA. Synthetic methods for the preparation of 1,3-propanediol. *Clean*. 2008;36:648−651.

88. Forage RG, Foster MA. Glycerol fermentation in *klebsiella-pneumoniae* − functions of the coenzyme-b12-dependent glycerol and diol dehydratases. *J Bacteriol*. 1982;149:413−419.

89. Homann T. Fermentation of glycerol to 1,3-propanediol by *Klebsiella* and *Citrobacter* strains. *Appl Microbiol Biotechnol*. 1990;33:121−126.

90. Barbirato F. Physiologic mechanisms involved in accumulation of 3-hydroxypropionaldehyde during fermentation of glycerol by *Enterobacter agglomerans*. *Appl Environ Microbiol*. 1996;62:4405−4409.

91. Forsberg CW. Production of 1,3-Propanediol from glycerol by *Clostridium acetobutylicum* and other *Clostridium* species. *Appl Environ Microbiol*. 1987;53:639−643.

92. Hartlep MH, Hussmann WH, Prayitno NP, Meynial-Salles IM-S, Zeng APZ. Study of two-stage processes for the microbial production of 1,3-propanediol from glucose. *Appl Microbiol Biotechnol*. 2002;60:60−66.

93. Nakamura CE, Whited GM. Metabolic engineering for the microbial production of 1,3-propanediol. *Curr Opin Biotechnol*. 2003;14:454−459.

94. Ma Z, Rao Z, Xu L, et al. Expression of *dha* operon required for 1,3-PD formation in *Escherichia coli* and *Saccharomyces cerevisiae*. *Curr Microbiol*. 2010;60:191−198.

95. Li R, Zhang H, Qi Q. The production of polyhydroxyalkanoates in recombinant *Escherichia coli*. *Bioresour Technol*. 2007;98:2313−2320.

96. Saxena RK, Anand P, Saran S, Isar J. Microbial production of 1,3-propanediol: recent developments and emerging opportunities. *Biotechnol Adv.* 2009;27:895—913.

97. Celinska E. Debottlenecking the 1,3-propanediol pathway by metabolic engineering. *Biotechnol Adv.* 2010;28:519—530.

98. Keasling JD. Manufacturing molecules through metabolic engineering. *Science.* 2010;330:1355—1358.

99. Shao Z, Luo Y, Zhao H. Rapid characterization and engineering of natural product biosynthetic pathways via DNA assembler. *Mol Biosyst.* 2011;7:1056—1059.

CHAPTER 4

From Biological Parts to Circuit Design

Joao C. Guimaraes[1,2], Chang C. Liu[1,3] and Adam P. Arkin[1,4]
[1]University of California, Berkeley, CA, USA
[2]University of Minho, Braga, Portugal
[3]Miller Institute for Basic Research in Science, Berkeley, CA, USA
[4]Physical Biosciences Division, Lawrence Berkeley National Laboratory, Berkeley, CA, USA

INTRODUCTION

The last 20 years have witnessed incredible progress in both our understanding of molecular biological systems and the technologies available for manipulating them. As a result, biology is quickly expanding from its historical tradition as a discovery-based science into an engineering science, where biological matter (e.g. RNA, proteins, regulatory circuits, and cells) is treated as building material for the construction of custom biological function. This transition has already brought biological solutions to advanced problems in chemical and pharmaceutical production,[1,2] therapeutics and diagnostics,[3] as well as agricultural and environmental engineering.[4] In fact, it has originated the new field of synthetic biology.

The distinguishing feature of synthetic biology involves the discovery and development of biological parts, which are genetically encoded modular units with defined biological function that can be used and reused in different contexts. Once collections of parts are available, the next level involves their predictable hierarchical assembly into composite function of greater and greater complexity.[5] In theory, predictable assembly is feasible as long as parts are well-characterized and behave consistently in a multitude of contexts. However, it has become clear that parts are not always modular — that is, the functions of many parts are not neatly encapsulated in the part itself, but rather are significantly influenced by interactions with different contexts.[6] As a result, many recent efforts in synthetic biology have focused on the development of strategies to overcome the functional variability of parts that prevents predictable hierarchical assembly. These strategies include the establishment of detailed sequence/activity models that provide better predictions of context effects,[7-10] methods for physically insulating parts from their genetic surroundings to remove undesired interactions,[11-15] and directed evolution and combinatorial library screening techniques that use power in numbers to overcome variability.[16-18] Ultimately, the combination of reliable parts sets with these effective strategies for their predictable functional assembly and rationally tunable function will result in an efficient design cycle for the construction of complex biological functions to order.

In this chapter, we review the status of the development of standard, reliable parts and their assembly to create more complex predictable function and propose strategies for

Synthetic Biology. DOI: http://dx.doi.org/10.1016/B978-0-12-394430-6.00004-2
2013 Published by Elsevier Inc.

63

optimization. We focus this review on genetic circuits for three reasons: first, genetic circuits are ubiquitous in the engineering of custom regulatory programs; second, they span across several levels of biological organization, from individual regulatory elements and genes, to operons, through to networks of regulatory nodes; and third, they are a well-developed area in synthetic biology, thus offering many examples to discuss.

THE PARTS

Genetic circuits link groups of genes so as to execute a coordinated function. The principle mechanism by which coordination occurs is through regulation of gene expression. Therefore, the construction of genetic circuits relies on the availability of parts that control gene expression. Though such parts can be taken from the extraordinary variety of components that nature offers, only a few natural examples are well studied. Thus, the practical approach is to take a well-characterized natural regulatory part and use it as the basis for designing collections of synthetic variants to control each aspect of gene expression. Since the three basic layers of gene expression are transcription, translation, and mRNA degradation, we discuss efforts in engineering large collections of synthetic regulators acting on these three processes (Fig. 4.1).

Transcription

For transcriptional regulation, the key regulatory parts are promoters, which determine the frequency by which RNA polymerase initiates transcription of an mRNA; transcription factors, which bind promoters to change their activity often in response to effector signals; and *cis*-regulatory RNAs, which can abort the continuation of RNA polymerase during transcription. We describe engineered variants of these parts here.

The simplest efforts for deriving parts that control transcription aim to make collections of constitutive promoters of different strengths. For example, Alper et al., starting from the bacteriophage PL-l promoter, created a library of constitutive promoters driving the expression of GFP and screened nearly 200 random variants spanning a wide range of GFP fluorescence for ones that displayed consistent behavior across both bulk replicates and single-cell measurements.[19] The 22 well-behaved mutant promoters populated a 196-fold range in promoter strength. These parts can then be used to control the relative level of a gene's expression. For instance, by varying promoter strength through the use of different library members, the authors demonstrated that maximizing lycopene biosynthesis in *E. coli* requires an intermediate level of *dxs* expression, as too much expression results in the toxic

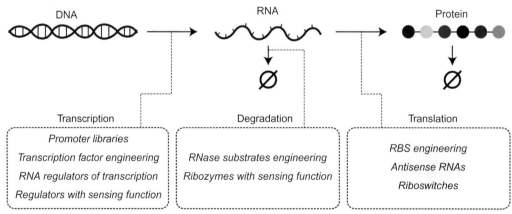

FIGURE 4.1
Biological parts target all processes of central dogma. Construction of genetic circuits relies on the availability of parts that control gene expression. The three basic layers of gene expression are transcription, translation, and mRNA degradation. The figure depicts multiple biological parts that are included in each of these three processes.

build-up of DXP (unless the downstream genes in the lycopene biosynthesis pathway are also overexpressed). This use of a well-characterized library with tunable strength to search for optimal circuit performance — thereby creating a map of quantitative circuit parameters to performance — is one of the key innovations available when rationally designed, well-characterized 'standard' parts are available.

Large constitutive promoter libraries, which are continuously being populated by synthetic biologists with more and more members across a variety of different organisms,[19−21] are important to any application where a gene needs to be expressed at a static level. But genetic circuits are dynamic objects and thus require regulatory parts that can link the expression of one gene to another. To satisfy this demand, several laboratories have aimed to engineer transcriptional 'wires' that can be encoded in the same operon as one gene, and whose production will activate or repress a second gene upon the first operon's transcription.

The most obvious choice for a transcriptional wire is a transcription factor such as those regulating the well-known *lac, ara,* and *tet* promoters of *E. coli*. Indeed, the use of natural transcription factors has already led to the construction of many synthetic circuits, ranging from the repressilator[22] to ones that carry out the difficult XOR and EQUALS functions.[15] These circuits, however, have relied on a very limited number of natural transcription factors.[23] As circuits become more complex, the number of wires required grows. Therefore, there have been several efforts to engineer natural transcriptional regulators into large collections of variants that are mutually orthogonal such that they can be utilized together in the same cell.

For example, Lucks et al. recently demonstrated that orthogonal transcriptional wires could be created from RNA by exploiting the specificity of base-pairing to design orthogonal variants.[24] The authors showed that the pT181 transcriptional regulation system, which uses an antisense RNA to trigger the formation of a transcriptional terminator structure, could be rationally mutated to create variant antisense/attenuator loop pairs that have similar functional properties but that do not exhibit crosstalk, thus allowing up to three genes to be independently repressed in the same cell. To demonstrate that these transcriptional regulators can indeed act as wires in a regulatory circuit, the authors created a synthetic regulatory cascade where an antisense variant was used to repress the expression of a second antisense variant, thus resulting in the overall activation of a controlled gene. The development of 'orthogonal' members of a parts family, derived through minimal changes to a parent molecule, allows increasingly complex circuits to be constructed from more physically homogeneous elements.[25] Ideally, this simplifies circuit analysis since parts will have been characterized through similar methods and will likely have similar biophysical behaviors and failure modes. As more circuits employ these standard parts, the resultant data on behavior are a resource for all who design with these elements.

Though the availability of basepairing rules makes the design of transcriptional wires based on RNA attractive, protein−promoter interactions have also been engineered to produce mutually orthogonal sets of transcriptional control parts. For example, Zhan et al. recently engineered the *lac* promoter system to produce a set of five mutually orthogonal lac repressor (LacI)/lac operator (OLac) pairs.[26] When an operator variant is attached to a desired gene, its transcription is repressed in the presence of the cognate LacI variant. In this study, the authors further showed that combinations of OLac sites could be placed together in front of a regulated gene to achieve multi-input logics. Moreover, one may place LacI variants in the same operon as desired genes, thus linking that gene's expression to the repression of a second desired gene placed under the control of the cognate OLac.

The pT181 and LacI engineering examples have yielded only a handful of mutually orthogonal transcriptional regulators (three and five, respectively). This is largely due to the fact that the design rules behind engineering pT181 and LacI are somewhat unreliable as they rely on intricate RNA−RNA and protein−DNA interactions, respectively. However, there are two promising platforms for designing mutually orthogonal sets of transcription factors where the rules are clearer. The first is based on engineering zinc finger protein transcription factors (ZFP-TFs). ZFP-TFs use a zinc finger protein domain to recognize specific DNA sequences and an effector domain to recruit transcriptional initiation components, thus activating transcription from a cognate DNA sequence. The advantage of ZFP-TFs from an engineering perspective is that the ZFPs are composed of an array of finger domains, each of which recognizes an overlapping 4-bp DNA sequence.[27] Therefore, modular engineering schemes have been developed where libraries of fingers are designed to recognize specific three or four base motifs; the fingers can then be fused together into polydactyl ZFPs and artificial ZFP-TFs that recognize custom promoter sequences.[28,29] In theory, this design scheme facilitates the construction of an arbitrarily large number of mutually orthogonal ZFP-TF/promoter pair libraries that can be used for transcriptional circuit engineering. In combination with standard promoter sequences these should lead to orthogonal and homogeneously acting families of transcriptional regulators.

A related and more recent approach is the use of transcription activator-like effector proteins (TALEs) to engineer synthetic transcription factors. TALEs are DNA-binding proteins that contain tandem repeats of 33−34 amino acid segments that are responsible for DNA recognition.[30] Each repeat utilizes two amino acids at position 12 and 13 of the repeat domain to recognize a single nucleotide in the corresponding DNA binding site. Since there is a code-like consistency in which amino acid pairs recognize single DNA bases, custom TALEs that recognize custom DNA sequences can be engineered. As the natural function of TALEs is to activate transcription, they become an ideal platform for engineering custom transcription factors. Indeed, recent efforts have demonstrated the feasibility of this approach. For example, a study by Zhang et al. generated a collection of 13 custom TALEs, rationally designed to activate transcription at 13 cognate sequences in mammalian cells.[31] Over 77% (10 out of 13) of these TALEs activated expression of their target gene with >10-fold induction, suggesting that this platform can indeed be a reliable method for generating multiple specific transcription factor/promoter pairs. Engineering of synthetic regulatory circuits, especially in mammalian cells, where well-characterized parts for transcriptional regulation are few, will benefit from the continued establishment of TALEs as a platform for custom transcription factor design.

As methods for the construction of transcriptional 'wires' expand, the ability to sense custom extracellular signals must expand alongside so that regulatory circuits can be interfaced with external signals. Towards this end, transcriptional regulators that themselves sense external signals such as small molecules have been engineered. For example, Liu et al. recently reported the construction of transcriptional regulators that attenuate or activate the expression of controlled genes in response to specific unnatural amino acid inputs.[32] Though the engineering of molecular recognition between a transcriptional regulator and a small molecule is often a time-consuming task, in this study, the authors took advantage of the fact that the expanded genetic code field, whose goal is to add new amino acids to the genetic code, already provides nearly 100 engineered synthetases that recognize custom unnatural amino acids and charge them onto engineered tRNAs that suppress blank codons.[33] Therefore, by placing blank codons into leader peptide elements that control the transcription of bacterial operons, the authors automatically gained a large number of transcriptional regulators that are triggered by specific unnatural amino acids.

Another instance of sensing in transcriptional regulators was reported recently by Qi et al., who showed that the rational fusion of aptamer domains with various RNA control systems could add a sensing function to their control.[34] In one example, the authors fused a theophylline-specific aptamer to the pT181 RNA-based transcriptional attenuation system (described above), and showed that the resulting fusion would be active only in the presence of the aptamer's cognate ligand. This was rationally designed by creating a linker between the aptamer and the pT181 regulator that would interfere with pT181 if the aptamer was not folded properly around theophylline. Since in vitro aptamer selection through SELEX is a general method for engineering aptamers that bind specific ligands, this strategy is a versatile method for creating sensors that attenuate transcription of desired genes. If the use of standard platforms for design of allosteric regulators of orthogonal expression controllers is successful, then expression circuits may be composed of a set of homogeneously acting, physically similar elements with simplified models of function. This is a critical step to permitting effective abstraction of circuit design from the underlying biophysical descriptions of the network. Such abstraction has been argued to be a significant driver in the success of other engineering disciplines.[5]

Translation

Once mRNA is produced during transcription, it then proceeds to template the synthesis of protein. Here, another level of regulation, translational control, may occur. In bacteria, translational control involves the ribosome binding site (RBS), which if weak or blocked will prevent protein expression but if strong and unblocked will allow protein expression. Therefore, RBS-based engineering of translational regulators has become a highly developed field in prokaryotic synthetic biology.[35]

At the simplest level, translation initiation is modulated by RBSs of different strength. Like constitutive promoters, engineered RBSs of defined strengths offer the ability to test different levels of a gene's expression, which is particularly useful in the optimization of expression levels in multigene pathways. For example, Salis et al. recently developed a biophysical model to capture the idiosyncrasies of the interactions between the RBS, the 5′ untranslated regions (5′-UTR) that contain it, and the downstream gene sequence to generate specified initiation strengths.[10,36] Using more than 100 predictions coupled with experimental validation, the authors elegantly show how to design, in silico, RBS sequences that hit the desired level of expression in *E. coli*. Nonetheless, the authors acknowledge that the RBS strength calculation model developed has only a ~47% of predicting an RBS that achieves protein expression within a two-fold range of the target level. Still, this provides a good starting point for RBS tuning and the important step of mapping sequence to activity.

As with constitutive promoters, the ability to hit particular expression values does not by itself provide a way to link the expression of multiple genes to form regulatory circuits. For this, RBSs that are controlled by *trans* elements, acting as 'wires,' are necessary. It has recently become clear that the ability to exploit basepairing rules in RNA design makes the construction of large classes of *trans* regulator RNA parts feasible.

For example, in a seminal study, Isaacs et al. showed how to take advantage of RNA basepairing to engineer artificial riboregulators that activate translation in vivo.[37] To do this, the authors exploit a system where a hairpin structure containing the RBS is used to prevent translational initiation. However, activation of gene expression can be achieved by expressing a *trans*-regulatory RNA that hybridizes with part of the hairpin to reveal the RBS and trigger successful translation of the controlled gene. Empirical rules for RNA hybridization were then used to design two orthogonal regulators, and a more recent study from the same laboratory has shown how to generate up to four mutually orthogonal regulators using an extension of this strategy.[38]

In a related approach, Mutalik et al. used the IS10 system, wherein an antisense RNA molecule hybridizes to the 5′-UTR of the target mRNA in order to sequester the RBS, to create multiple mutually orthogonal variants of the IS10 antisense repression system.[39] To do this, the authors first created a series of 23 mutant IS10 pairs rationally designed through varying base pairs in key regions. They then characterized the repression profile of all 529 interactions and reported translation inhibition ranging from 0 to 90%. These data were used to build a quantitative model that was able to explain 87% of the variability of in vivo performance of the RNA regulators. The authors then demonstrated the utility of this model by forward-engineering new RNA regulators, as well as families of up to six mutually orthogonal regulators. More interestingly, the quantitative model described enables designers to explore regulators that vary across a continuous range of regulation.

To add sensing functions to translational regulators, one may fuse aptamer domains to RNA control parts. For example, Qi et al. utilized their aptamer grafting strategy (described above for pT181) to also create IS10 variants that responded to protein and small molecule inputs.[34] This effectively creates artificial riboswitches. These can also be accessed by taking natural riboswitches and changing their specificity for their cognate ligand. For example, Dixon et al. reported the construction of synthetic variants of the *add* A-riboswitch that respond not to the natural ligand adenine, but to related heterocyclic analogues.[40] The *add* A-riboswitch sequesters an embedded RBS in the OFF state but releases the RBS to create the ON state when its cognate adenine ligand, recognized by a handful of hydrogen-bonding contacts, is present. To make variants of this translational regulatory riboswitch, the authors first synthesized a small library of mutants randomized in the ligand-binding site. These mutants were then attached to control chloramphenicol acetyl transferase (CAT) expression and screened in vivo for translational activation of CAT in the presence and absence of a collection of heterocyclic compounds. This screen resulted in a number of riboswitches specifically activated by nonnative ligands that can now be used as inducers of translational activation. In a more recent study, one riboswitch variant triggered by ammeline was used to construct a mutually orthogonal riboswitch for the inducible independent expression of two genes under the control of the two orthogonal riboswitches.[41]

As in transcription, the emerging theme is to create layers of orthogonal but similarly acting, sensing and regulatory elements derived from common parents such that relatively simple models of quantitative function, derived from sequence, may be obtained. This enables a designer to more rationally search a 'parameter' space rather than a sequence space to achieve a desired target.

Degradation

For an mRNA to be translated, it must be stable inside the cell. Therefore, the stability of an mRNA transcript is also a key determinant of gene expression. Parts that control mRNA stability have thus been constructed.

For example, Babiskin and Smolke sought to extend the regulatory tools to control gene expression in eukaryotic organisms by engineering an endogenous RNA control system that regulates the stability of mRNAs.[11] In eukaryotic cells, endoribonucleases can cleave mRNA transcripts and induce their rapid degradation by exoribonucleases. In *Saccharomyces cerevisiae*, the RNase III enzyme, Rnt1p, recognizes specific RNA hairpins and cleaves its substrates. The authors therefore used a screening strategy to generate Rnt1p substrate variants that exhibited variable cleavage activity and consequently, a wide range of gene expression (between 8% and 85%). They also examined the modularity of their devices by verifying it would perform reliably with two sequence divergent fluorescent reporters. The resulting library is a set of post-transcriptional RNA-based controllers that can be inserted into the 3'-UTR of any transcript to yield predictable levels of gene expression.

Degradation-based regulatory parts that sense inputs have also been created. For example, in a number of systems developed by the Smolke laboratory, the addition of engineered ribozymes in the 3'-UTR of yeast mRNA transcripts allows the regulated cleavage of poly-A tails, which are required for the proper export, translation, and stability of mRNAs.[42] By linking ribozymes to aptamer domains in various configurations, Smolke has demonstrated the ability to activate gene expression in response to custom ligands that inhibit ribozyme cleavage. Various instantiations of this method have created regulatory nodes displaying logic functions, as well as higher-order forms such as sigmoidal transfer functions. The control of mRNA stability, especially in eukaryotes where active processing of mRNA requires a specific 3'-UTR, offers a promising strategy to control gene expression.

Together, common principles concerning the design and use of elements controlling transcription initiation/elongation, translation initiation, and degradation are emerging. Standard molecular platforms are being discovered and developed to create families of parts with similar physics of function and mapping of sequence to activity. Each element class controls different elements of gene expression dynamics, including steady-state levels, timing, and noise, and each allows for modulation of their activity through a sensing layer.[43] Thus, each may be used to sculpt complex expression control functions for a given set of genes.

ASSEMBLING PARTS

However, once parts controlling gene expression are developed, they are meant to be used and reused in different applications. This means that they will be taken out of the context in which they were initially developed and characterized — usually in standard strains with GFP as the reporter gene and small culture volumes — and used to control other genes, possibly in specialized host strains and varied culturing conditions. In addition, a part will rarely be used in isolation; more often, it will be composed with other regulatory parts into genetic switchboards, complex nodes, and regulatory circuits. Therefore, the success of the parts-based approach to genetic circuit engineering depends on the ability of parts to function predictably in novel arrangements. However, it has become clear that most biological parts are not robust across different contexts.[6] As a result of these context effects, engineering of synthetic biological systems from parts is often reduced to an ad hoc process wherein random libraries circuits composed of parts and their variants are screened for desired function.

As an example, consider the simplest and most necessary combination of parts: for a gene to be expressed, its coding sequence must be prefixed by a promoter and a 5'-UTR, which in bacteria contains the RBS. This already leaves room for interactions among the promoter, RBS, and coding sequence, from which unpredictable effects may emerge. To explore these interactions, Mutalik et al. recently assembled and experimentally measured all the combinations between seven promoters, eleven 5'-UTRs, and two genes (fluorescent reporters).[44] They then used the analysis of variance (ANOVA) framework to accurately quantify average part performance across different contexts and part composability, which is the variability in performance that can be attributed to the interaction between functionally different genetic elements. In this case, they found that promoters could be reliably composed with 5'-UTRs and genes without significant emergent behavior. However, 5'-UTRs presented unreliable performance when combined with different genes. Though perhaps unsurprising — it has been known for some time that 5'-untranslated RNA can, for example, form structures with particular coding sequences that sequester ribosome binding sites[8,9] — this example shows that even the simple task of taking constitutive promoters and combining them with unregulated 5'-UTRs and two fluorescent reporter genes is subject to emergent properties that influence expression. Are there strategies for making parts and their assembly more robust?

Models of Function

If one has a complete model of a part's behavior in different contexts, then the issue of variability is moot as behavior in any context can be predicted in silico. Of course, achieving accurate models that consider all relevant contextual variables is an ambitious goal; but for several systems, it has become within reach. For instance, in the RBS calculator example discussed above,[36] the algorithm used to generate RBS parts does not treat the RBS in isolation but explicitly models its interaction with surrounding sequences, using the range of

35 nucleotides before and after the start codon as the input. This is predicated on a basic model of translational initiation where the mRNA unfolds as the ribosome initiation complex is created. In this model, RBS strength depends on the energy required to unfold the sequence upon translational initiation, the energy released from hybridization of the mRNA to the 16S rRNA, the energy released upon initiating tRNA hybridization, the energy of unfolding the standby site, and the energy cost of suboptimal spacing to the start codon. To compute the first four energy terms, standard RNA folding algorithms were used; to compute the last term, an empirically determined relationship was used. The result is therefore an RBS part generator that encompasses a large range of context variations in the several energy terms of its model. As discussed above, the RBS calculator essentially reduces the library that must be screened to hit a desired translational efficiency. However, in this case, UTRs and part of the sequence of the gene of interest must be recoded in each context, and the sequence to activity relationship is complex — something that makes online methods of 'sequence tuning' to achieve a particular activity, such as accomplished in multiplex automated genome engineering (MAGE, see below), more difficult, and reuse of parts more costly.

In a similar vein, Keasling and colleagues used a model-driven approach to create a family of RNA-regulated devices to control gene expression through mRNA degradation in E. coli, in which ribozyme cleavage in a transcript's 5′-UTR leads to a 5′OH-RNA that is more stable than the precursor 5′PPP-RNA.[7] The authors describe an expression control unit composed of a promoter, a variable ribozyme/aptazyme, and an RBS that can be coupled to any gene of interest. Performance of this unit is then modeled in great detail using a set of 25 chemical reactions that capture a variety of physical processes that emerge from the context in which the unit must perform, including transcriptional initiation and elongation, ribozyme cleavage and folding, and differential mRNA degradation. Some of these variables are experimentally determined (e.g. ribozyme cleavage rate), while others are inferred from computational simulations (e.g. ribozyme folding rate). Using this model, they created a number of ribozyme- and aptazyme-controlled expression systems that hit predicted levels of gene expression with high precision and integrated these expression control units into a model metabolic pathway to control the flux of metabolite production. Like the RBS calculator, this example represents the development of physics-based sequence-activity models that capture a part's behavior across different contexts through the consideration of a significant number of variables.

Models of these sorts are key to the success of the field. They capture the relevant biophysics of the parts and, sometimes, their interactions, and suggest places where good design might eliminate complexities (e.g. the UTR–gene interaction discovered in the ANOVA analysis above). The simpler the description of sequence-activity can be made by good design choices, and the more compact the sequence region that encodes a given parameter of the model, the easier it will be for designers to rationally explore a parameter space to optimize the behavior of their circuits given the uncertainty in the parts and their operation.

Compositional Insulation

As physical models for predicting the behavior of parts become more complex, parameterization requires more measurements, which can become excessively time-consuming if these parameters depend heavily on each different context in which a part will be used. Therefore, a number of techniques designed to insulate parts from surrounding contexts have been developed. These techniques mitigate part variability by removing possible interactions with surrounding contexts, thus reducing the number of variables that can affect a part's behavior and simplifying parts-based assembly of new behavior for all approaches.

For example, Davis et al. recently created a set of promoters that contained flanking sequences around the minimal ∼50 bp functional promoter in E. coli.[12] The inclusion of ~100 extra nucleotides surrounding the functional motif resulted in a promoter library that showed less variation than standard promoters across different contexts. For instance, promoters from this set, when placed downstream of UP sequences known to enhance transcriptional activity, were largely insulated from additional stimulation. Therefore, the combination of these insulated promoters with the more standard inclusion of transcriptional terminators upstream of promoters to prevent residual transcription from nearby promoters should allow the part to become more robust to genetic context.

If the mechanism of regulation permits, insulation can be made more definite by the insertion of ribozyme sequences between regulatory parts.[14] For example, Lucks et al. wished to create an RNA regulatory cascade where an antisense RNA (A1) would attenuate the expression of a second antisense RNA (A2) under the control of a *cis*-RNA attenuator (AT1) that sensed A1.[24] This involved the fusion of AT1 to A2 such that A1 would regulate the production of A2. However, once generated, A2's function is significantly diminished if AT1 is attached. Therefore, a self-cleaving hammerhead ribozyme sequence was inserted between AT1 and A2 such that once A2 was produced it would be cleaved from AT1. In short, the part, A2, was made more robust through the insertion of a ribozyme sequence that physically cleaved A2 from a destructive context.

Recently, Qi et al. developed a more general approach to regulatory part cleavage from physical context. This strategy involves the insertion of CRISPR repeats between parts such that once transcribed, a coexpressed Csy4 protein cleaves the parts from each other.[13] This allows the generation of homogenous 5′-UTRs regardless of promoter (which often have ill-defined transcriptional start sites), and the separation of multigene operons into separate mRNAs for each gene. It also allows multiple *cis*-regulatory RNA parts, such as riboswitches that control transcriptional continuation, to be reliably fused in tandem, since CRISPR cleavage will ensure that each subsequent *cis*-regulatory RNA part is cleaved from the previous ones before it exerts its regulatory control, mitigating possible interference from upstream mRNA structure on regulatory function. Using this CRISPR strategy, they demonstrated dramatic decreases in part variation across a number of operon assembly studies. In addition, this strategy was essential to the recent construction of multi-input NOR gates through the fusion of up to four mutually orthogonal RNA transcriptional attenuators in series, as each attenuator required a free 5′-terminus for function.[45]

Finally, if one moves beyond the scale of intracellular circuits to intercellular circuits, one can rely on the physical separation between cells for insulation of part behavior from unpredicted compositional effects. For example, recent work by Tamsir et al. shows that cells containing single NOR logic gates, consisting of inducible promoters, some of which respond to quorum molecules and output quorum molecules, can be spatially patterned such that computation progresses from one manually spotted location of cells to the next (for example, from left to right on solid plates).[15] Since cells sense and output different quorum molecules, and since quorum molecules become dilute as they diffuse away from their source, the arrangement of different spots creates different logics that can be integrated in a final spot which outputs a signal according to desired logic functions. In fact, all 16 possible two-input logic gates were constructed from different arrangements of NOR gates encapsulated within whole cells.

The key point of insulation is to remove unwanted interactions and uncertainty from the operation of standard biological parts. Ideally, the parts above would have characterized function that was largely invariable or at least predictable across different immediate genetic contexts, host cells, and environments. There remains a fair amount of low hanging fruit in this area: methods for minimizing unwanted physical interaction among synthetic and host elements; methods for reducing the energetic load of designs on the host cells; methods that reduce competition for common components of the circuit (fan-out issues/retroactivity).

However, the variability of host physiology and its sensitive dependence on environmental conditions, and the complex biochemistry of interactions among circuit elements and products with each other and host, suggests that all uncertainty cannot be removed even from the basal function of a circuit. Thus, a combination of rational design and screening will likely be necessary.

CIRCUIT DESIGN

A major challenge in circuit engineering is the optimization of every component in the circuit such that desired behavior is attained. So far, due to the lack of reliable genetic tools, engineers often have to tinker the system via fine-tuning of every component, resulting in design cycles that are laborious and cost-ineffective.[46,47] Conversely, the advent of well-characterized parts libraries and the ability to assemble them in a predictable manner, as discussed above, can dramatically shorten the design cycle for genetic circuits by providing the designer with the ability to specifically target the desired circuit parameters (Fig. 4.2).

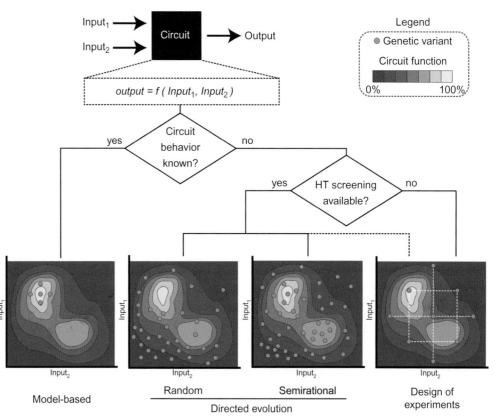

FIGURE 4.2

Comparison between the different methodologies for circuit design. An illustrative circuit composed by two inputs and one output, wherein the circuit function is dependent on concentration of the two input genes, is depicted in the figure. When circuit models are available, functional regions can be readily targeted to yield functional systems (heatmaps depict circuit functionality). Conversely, when such models are unknown, one needs to employ discovery methods to characterize circuit output as a function of the inputs. When high-throughput (HT) screening is available, we can take advantage of the capacity to generate vast numbers of genetic variants provided by directed evolution. However, performing a random mutagenesis strategy does not ensure an effective search of the solution space. Semirational approaches that more precisely target features that efficiently alter inputs can provide a more comprehensive search. Conversely, when HT screening or selection methods are not available, systematic design strategies, such as design of experiments (DoE), can be coupled with reliable genetic parts to provide an effective search strategy.

As an example of such an approach, Salis et al., using the RBS calculator tool described before, aimed to optimize the function of an AND-gate circuit that induces the expression of GFP under the control of a T7 promoter only when two inputs, one activating the expression of an amber suppressor tRNA and the other activating the expression of a T7 polymerase amber mutant, are both present.[36] The accuracy of the AND gate, however, depends on how strongly the T7 polymerase mutant is expressed, since weak expression results in little expression of GFP under all conditions and overexpression results in leakiness that produces GFP even when an input is absent. Using a quantitative model for this specific AND-gate circuit,[46] the authors simulated the ideal level of T7 polymerase expression and then used their predictive design of RBS program to generate an RBS sequence targeting that ideal regime of expression, resulting in optimal AND gate function.

Semirational Directed Evolution

Nonetheless, models that describe circuit dynamics are not trivial to obtain and, when absent, it becomes difficult to exactly estimate the circuit parameters that optimize functionality. Directed evolution techniques offer a very flexible framework to optimize small-scale circuits with a somewhat high probability of success.[47] The challenge with using directed evolution for circuit design, however, is that regulatory circuits contain many parts and thus are much further removed from the level of point mutations than traditional directed evolution targets such as enzyme-binding sites. Therefore, it is necessary to use rational approaches in conjunction with directed evolution techniques to reduce the search space and increase the efficiency of the screening methodology.[48]

This semirational directed evolution approach has been quite valuable to the field of metabolic engineering, where the simple overexpression of genes does not always result in high production of the target compound, as overexpression places metabolic burdens on the cell.[49,50] As a result, expression and activity of enzymes participating in a metabolic pathway need to be properly balanced to maximize titers for the desired final compound, avoiding the accumulation of toxic intermediates.[1] Therefore, methods that rationally target combinatorial diversity to multienzyme expression levels are useful.

For example, Pfleger et al. designed a library of tunable intergenic regions (TIGR) that can be placed in operons to generate differential gene expression for each gene within the operon.[17] These elements are composed of multiple regulatory parts that include mRNA secondary structures, RNase cleavage sites, and RBS sequestering sequences. Upon transcription, mRNA is cleaved at the RNase sites, thereby generating multiple transcripts with various secondary structures that have variable RNA stabilities and translation efficiencies. Using libraries of TIGRs inserted between the three genes of the operon for mevalonate biosynthesis, the authors were able to optimize, through screening, the flux of the mevalonate pathway, which produces a precursor to the antimalarial drug artemisinin. Because TIGRs rationally varied the performance of key parts that control expression variables for key enzymes in the pathway, screening of only ~600 different genetic circuits yielded strains that had up to seven-fold greater production than the original expression system. Further characterization of these strains revealed that balance between multiple enzymes was significantly different from the starting regulatory circuit.

On a larger scale, Church and colleagues have developed a platform for multiplex automated genome engineering (MAGE).[18] This method utilizes oligonucleotide-directed mutagenesis, mediated by an optimized system for Lambda red recombination, to modify multiple loci in the genome of *E. coli* in parallel. The result is a highly scalable method for generating and screening/selecting multiloci phenotypes, especially since the authors have created a setup for automating the MAGE process. In short, MAGE facilitates the ability to generate variation at scales of biological organization much larger than single nucleotides, allowing the application of directed evolution to multicomponent genetic circuits. To demonstrate the

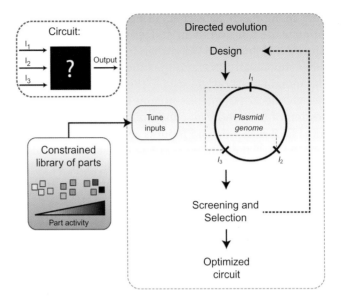

FIGURE 4.3

Constrained libraries of parts improve the development cycle of circuit optimization using directed evolution. An illustrative circuit composed by three inputs is depicted in the figure. Directed evolution permits a rapid and exhaustive search methodology to optimize circuit function when screening and/or selection procedures are available. Its efficiency to investigate the vast space of solutions can be further increased if combined with biological elements that ensure differential and reliable function.

power of MAGE, the authors optimized the 1-deoxy-D-xylulose-5-phosphate (DXP) metabolic pathway to overproduce lycopene. They used MAGE to vary the translational efficiency (through RBS degeneration) of 20 different genes previously reported to affect the pathway, and also to inactivate four other genes that redirect the flux to competing pathways. From as many as 15 billion genetic variants, the authors found strains, by screening for color intensity, with up to a five-fold increase in lycopene production as compared with the wild-type system. Indeed, the ability to rationally target key variables such that sequence variation maps to known parameter ranges, distributed across the entire genome, in a multigene pathways' performance increases search efficiency such that optimization of pathways with large numbers of relevant genes (in this case 20) becomes possible.

It is worthwhile to point out that in these directed evolution experiments, the diversity and number of rounds of selection available is limited (by the inefficiency and time-consuming nature of transformation), but knowledge of where variation should be targeted (rational design) combined with technologies for producing targeted libraries are used to spend the limited diversity wisely. In fact, biological parts with well-defined sequence–activity relationships present an effective starting material to target and mutate circuit components and generate genetic variability that more efficiently cover the search space of circuit functionality (Figs 4.2 and 4.3). Alternatively, one may invest in strategies that dramatically expand the diversity and the number of rounds of selections by developing continuous targeted evolution systems. Such efforts are underway in the directed evolution and synthetic biology communities.[51]

Design of Experiment Approaches

For directed evolution to be applicable, one must have a phenotype that is either selectable or can be screened in high throughput. Otherwise, the amount of time required to characterize the often staggering number of variants considered in a directed evolution experiment becomes prohibitive. Although the semirational directed evolution experiments

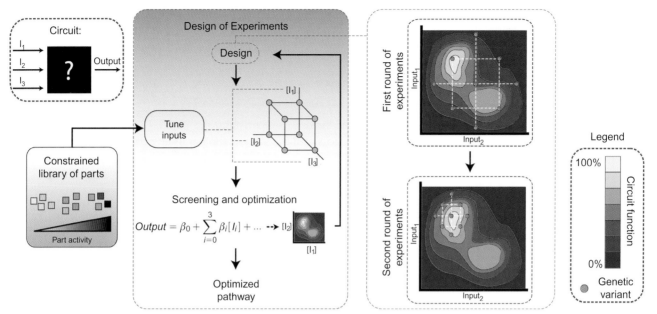

FIGURE 4.4

Design of experiments can be performed if libraries of parts with predictable gene expression are available. An illustrative circuit composed by three genes is depicted in the figure. Design of experiments provides a formal and efficient way to browse solution space. However, the ability to do that requires that input parameters can be reliably manipulated. Constrained libraries of parts that provide reliable function can be used to efficiently search the space of solutions. Design of experiments also enables continuous loop between design and analysis phases, wherein the next round of experiments is driven by the analysis of previous experiments.

described above limit the number of variants that need to be searched by targeting variation only to the key parts that most likely influence the desired function, statistical design of experiments (DoE) can reduce the space of search even further.

DoE is a general framework that integrates the planning and analysis phases. It describes a systematic approach to designing a curated minimal set of experiments that produce the maximum amount of relevant information about the impact of multiple factors in the response of a system. DoE is particularly useful in screening and optimization, where the former identifies the most influential factors and their ranges, and the latter defines the combination of factors that give optimal response. Initial planning of experiments involves three steps: first, define the set of factors that can impact the response variable; second, define the range of levels for each factor; third, define a model, usually polynomial with interaction between factors, to relate factors with response variable.

The use of DoE is widespread across a variety of applications that include pharmaceutical research,[52] chromatography optimization,[53] and recombinant protein production.[54] There are proposals to use it for biological model discrimination and parameterization as well.[55] Though DoE has not been routinely applied to genetic circuit design, it has much potential for this purpose. Indeed, metabolic pathway circuit optimization is perfectly suited for DoE. Take the hypothetical process described in Figure 4.4. As shown, one can use a screening approach to discover factors (or inputs) that have the greatest influence on a response (e.g. the concentration of a desired end-product) in a metabolic pathway composed of three enzymes (inputs). Here, the availability of well-characterized parts libraries that enable precise manipulation of the circuit parameters (i.e. concentration of enzymes) is crucial. DoE can then provide a minimal set of experiments that is feasible to test in the laboratory and will most effectively populate the large space of solutions. There are a number of different ways to choose the initial set of trials, each informative in a different way. For example, in a system composed of three factors each with five putative levels of enzyme concentration, we could use DoE and the central composite design to test experimentally an initial set of only 20 genetic variants out of 125 (5^3) combinations. The central composite model is particularly useful for response surface methods,[56] and essentially fits a second order polynomial model to the observed productivity as a function of variables similar to the one below:

$$Y = \beta_0 + \sum_{i=1}^{n} \beta_i^1 X_i^2 + \sum_{i=1}^{n} \beta_i X_i + \sum_{1 \le i < j}^{n} \beta_{ij} X_i X_j$$

where Y is the response, $X_{1\ldots n}$ the factors, β_0 a constant, $\beta_{1\ldots n}$ and $\beta'_{1\ldots n}$ model coefficients for the first- and second-order factors and $\beta_{12\ldots(n-1)n}$ the parameter for the interaction between variables. This type of design is particularly important when one wants to evaluate the possibility of nonlinear responses and interactions between variables, which is often the case in genetic circuits.

DoE effectively reduces the overall number of experiments without compromising the quality of the data. In addition, it enables a close integration between the design and analysis phases of an experiment such that models derived in the analysis phase can feed back into the next round of experiments to be designed (Fig. 4.4). We believe this strategy will take a central role as our ability to model circuits accurately and the number of parts necessary for perturbing them grows.

For both directed evolution and DoE strategies, the ability to vary circuit parameters in efficient and predictable ways, thus creating rationally targeted and distributed variation, is crucial. Therefore, these methods rely fundamentally on the availability of well-characterized parts. In addition, they become only more powerful as part assembly becomes more predictable, since the more reliably one can target relevant regions of functional space, the faster directed evolution and DoE strategies can hone in on exact behavior.

CONCLUSION

Synthetic biology aims first to develop modular biological parts and then to assemble them into composite function.[5] There is an increasing number of parts that can now be easily designed and tuned to achieve desired functionality using models that describe their behavior. Additionally, there are emerging paradigms that ensure that parts can be composed reliably to yield predictable behavior. Together, these elements provide a more comprehensive representation of a part datasheet,[57,58] and are starting to move synthetic circuit design towards a more computer-aided design (CAD) framework.[59] However, for the foreseeable future there will always be significant uncertainty in part function and interaction with the host and other components. These issues of parasitic interaction and load, including competition for molecular resources and unpredictable chemical toxicity, currently limit an entirely rational approach to circuit design.

Instead, we have argued in favor of an integrated approach for circuit design consisting of parts design, assembly and modeling, and the use of well-characterized parts as sources for systematic genetic variation for two distinct search methodologies: directed evolution and DoE (Fig. 4.2). The former has the exceptional ability to quickly generate millions of variants, but strongly depends on availability of high-throughput screening methods. On the other hand, DoE offers a very flexible strategy to minimize experimental effort and provide learning about system dynamics. This integrated approach should result in efficient design cycles for the creation of custom biological function and a rich knowledge base for interaction and load that can ultimately inform the design of future applications.

Acknowledgments

JCG acknowledges financial support by the Portuguese Fundaçãopara a Ciência e a Tecnologia (SFRH/BD/47819/2008). This work was supported by the Synthetic Biology Engineering Research Center under NSF grant number 0540879 and by the BIOFAB under NSF grant number 0946510.

References

1. Keasling JD. Manufacturing molecules through metabolic engineering. *Science*. 2010;330:1355−1358.

2. Lee SY, Kim HU, Park JH, Park JM, Kim TY. Metabolic engineering of microorganisms: general strategies and drug production. *Drug Discov Today*. 2009;14:78−88.

3. Ruder WC, Lu T, Collins JJ. Synthetic biology moving into the clinic. *Science*. 2011;333:1248−1252.

4. Zhao H, Chen W. Chemical biotechnology: microbial solutions to global change. Editorial overview. *Curr Opin Biotechnol*. 2008;19:541−543.

5. Endy D. Foundations for engineering biology. *Nature*. 2005;438:449−453.

6. Cardinale S, Arkin AP. Contextualizing context for synthetic biology. *Biotechnol J*. 2012;7:856−866. doi:10.1002/biot.201200085.

7. Carothers JM, Goler JA, Juminaga D, Keasling JD. Model-driven engineering of RNA devices to quantitatively program gene expression. *Science*. 2011;334:1716−1719.

8. de Smit MH, van Duin J. Secondary structure of the ribosome binding site determines translational efficiency: a quantitative analysis. *Proc Natl Acad Sci USA*. 1990;87:7668−7672.

9. Kudla G, Murray AW, Tollervey D, Plotkin JB. Coding-sequence determinants of gene expression in *Escherichia coli*. *Science*. 2009;324:255−258.

10. Salis HM. The ribosome binding site calculator. *Methods Enzymol*. 2011;498:19−42.

11. Babiskin AH, Smolke CD. A synthetic library of RNA control modules for predictable tuning of gene expression in yeast. *Mol Syst Biol*. 2011;7:471.

12. Davis JH, Rubin AJ, Sauer RT. Design, construction and characterization of a set of insulated bacterial promoters. *Nucleic Acids Res*. 2011;39:1131−1141.

13. Qi L, Haurwitz RE, Shao W, Doudna JA, Arkin AP. Predictable engineering of prokaryotic gene expression. 2012;doi:10.1038/nbt.2355.

14. Smolke CD, Carrier TA, Keasling JD. Coordinated, differential expression of two genes through directed mRNA cleavage and stabilization by secondary structures. *Appl Environ Microbiol*. 2000;66:5399−5405.

15. Tamsir A, Tabor JJ, Voigt CA. Robust multicellular computing using genetically encoded NOR gates and chemical 'wires.' *Nature*. 2011;469:212−215.

16. Cobb RE, Sun N, Zhao H. Directed evolution as a powerful synthetic biology tool. *Methods*. 2012; Epub ahead of print

17. Pfleger BF, Pitera DJ, Smolke CD, Keasling JD. Combinatorial engineering of intergenic regions in operons tunes expression of multiple genes. *Nat Biotechnol*. 2006;24:1027−1032.

18. Wang HH, Isaacs FJ, Carr PA, et al. Programming cells by multiplex genome engineering and accelerated evolution. *Nature*. 2009;460:894−898.

19. Alper H, Fischer C, Nevoigt E, Stephanopoulos G. Tuning genetic control through promoter engineering. *Proc Natl Acad Sci USA*. 2005;102:12678−12683.

20. Blount BA, Weenink T, Vasylechko S, Ellis T. Rational diversification of a promoter providing fine-tuned expression and orthogonal regulation for synthetic biology. *PloS one*. 2012;7:e33279.

21. Jensen PR, Hammer K. The sequence of spacers between the consensus sequences modulates the strength of prokaryotic promoters. *Appl Environ Microbiol*. 1998;64:82−87.

22. Elowitz MB, Leibler S. A synthetic oscillatory network of transcriptional regulators. *Nature*. 2000;403:335−338.

23. Purnick PE, Weiss R. The second wave of synthetic biology: from modules to systems. *Nat Rev Mol Cell Biol*. 2009;10:410−422.

24. Lucks JB, Qi L, Mutalik VK, Wang D, Arkin AP. Versatile RNA-sensing transcriptional regulators for engineering genetic networks. *Proc Natl Acad Sci USA*. 2011;108:8617−8622.

25. Lucks JB, Qi L, Whitaker WR, Arkin AP. Toward scalable parts families for predictable design of biological circuits. *Curr Opin Microbiol*. 2008;11:567−573.

26. Zhan J, Ding B, Ma R, et al. Develop reusable and combinable designs for transcriptional logic gates. *Mol Syst Biol*. 2010;6:388.

27. Mandell JG, Barbas CF, III. Zinc finger tools: custom DNA-binding domains for transcription factors and nucleases. *Nucleic Acids Res*. 2006;34:W516−523.

28. Gonzalez B, Schwimmer LJ, Fuller RP, et al. Modular system for the construction of zinc-finger libraries and proteins. *Nat Protoc*. 2010;5:791−810.

29. Blancafort P, Magnenat L, Barbas CF, III. Scanning the human genome with combinatorial transcription factor libraries. *Nat Biotechnol*. 2003;21:269−274.

30. Boch J, Scholze H, Schornack S, et al. Breaking the code of DNA binding specificity of TAL-type III effectors. *Science*. 2009;326:1509−1512.

31. Zhang F, Cong L, Lodato S, et al. Efficient construction of sequence-specific TAL effectors for modulating mammalian transcription. *Nat Biotechnol.* 2011;29:149–153.

32. Liu CC, Qi L, Yanofsky C, Arkin AP. Regulation of transcription by unnatural amino acids. *Nat Biotechnol.* 2011;29:164–168.

33. Liu CC, Schultz PG. Adding new chemistries to the genetic code. *Annu Rev Biochem.* 2010;79:413–444.

34. Qi L, Lucks JB, Liu CC, Mutalik VK, Arkin AP. Engineering naturally occurring trans-acting non-coding RNAs to sense molecular signals. *Nucleic Acids Res.* 2012;40:5775–5786.

35. Isaacs FJ, Dwyer DJ, Collins JJ. RNA synthetic biology. *Nat Biotechnol.* 2006;24:545–554.

36. Salis HM, Mirsky EA, Voigt CA. Automated design of synthetic ribosome binding sites to control protein expression. *Nat Biotechnol.* 2009;27:946–950.

37. Isaacs FJ, Dwyer DJ, Ding C, et al. Engineered riboregulators enable post-transcriptional control of gene expression. *Nat Biotechnol.* 2004;22:841–847.

38. Callura JM, Cantor CR, Collins JJ. Genetic switchboard for synthetic biology applications. *Proc Natl Acad Sci USA.* 2012;109:5850–5855.

39. Mutalik VK, Qi L, Guimaraes JC, Lucks JB, Arkin AP. Rationally designed families of orthogonal RNA regulators of translation. *Nat Chem Biol.* 2012;8:447–454.

40. Dixon N, Duncan JN, Geerlings T, et al. Reengineering orthogonally selective riboswitches. *Proc Natl Acad Sci USA.* 2010;107:2830–2835.

41. Dixon N, Robinson CJ, Geerlings T, et al. Orthogonal riboswitches for tuneable coexpression in bacteria. *Angew Chem.* 2012;51:3620–3624.

42. Win MN, Smolke CD. Higher-order cellular information processing with synthetic RNA devices. *Science.* 2008;322:456–460.

43. Khalil AS, Collins JJ. Synthetic biology: applications come of age. *Nat Rev Genet.* 2010;11:367–379.

44. Mutalik VK, Guimaraes JC, Cambray G, et al. Composition and quality of irregular genetic elements controlling transcription and translation. *Nature Methods* 2012; in press.

45. Liu CC, Qi L, Lucks JB, et al. A converter from translational to transcriptional control. 2012; doi:10.1038/nmeth.2184.

46. Anderson JC, Voigt CA, Arkin AP. Environmental signal integration by a modular AND gate. *Mol Syst Biol.* 2007;3:133.

47. Yokobayashi Y, Weiss R, Arnold FH. Directed evolution of a genetic circuit. *Proc Natl Acad Sci USA.* 2002;99:16587–16591.

48. McAdams HH, Arkin A. Towards a circuit engineering discipline. *Curr Biol.* 2000;10:R318–320.

49. Gorgens JF, van Zyl WH, Knoetze JH, Hahn-Hagerdal B. The metabolic burden of the PGK1 and ADH2 promoter systems for heterologous xylanase production by *Saccharomyces cerevisiae* in defined medium. *Biotechnol Bioeng.* 2001;73:238–245.

50. Martin VJ, Pitera DJ, Withers ST, Newman JD, Keasling JD. Engineering a mevalonate pathway in *Escherichia coli* for production of terpenoids. *Nat Biotechnol.* 2003;21:796–802.

51. Esvelt KM, Carlson JC, Liu DR. A system for the continuous directed evolution of biomolecules. *Nature.* 2011;472:499–503.

52. Gabrielsson J, Lindberg NO, Lundstedt T. Multivariate methods in pharmaceutical applications. *J Chemometr.* 2002;16:141–160.

53. Hibbert DB. Experimental design in chromatography: a tutorial review. *J Chromatogr B Analyt Technol Biomed Life Sci.* 2012;910:2–13.

54. Bora N, Bawa Z, Bill RM, Wilks MD. The implementation of a design of experiments strategy to increase recombinant protein yields in yeast (review). *Methods Mol Biol.* 2012;866:115–127.

55. Flaherty P, Jordan MI, Arkin AP. Robust design of biological experiments. In: Weiss Y, Schoelkopf B, Platt J, eds. *Advances in Neural Information Processing Systems (NIPS)*. Cambridge, Massachusetts: MIT Press; 2006.

56. Box GEP, Wilson KB. On the experimental attainment of optimum conditions. *J R Stat Soc Ser B-Stat Methodol.* 1951;13:1–45.

57. Arkin A. Setting the standard in synthetic biology. *Nat Biotechnol.* 2008;26:771–774.

58. Canton B, Labno A, Endy D. Refinement and standardization of synthetic biological parts and devices. *Nat Biotechnol.* 2008;26:787–793.

59. Clancy K, Voigt CA. Programming cells: towards an automated 'Genetic Compiler.' *Curr Opin Biotechnol.* 2010;21:572–581.

Computational and Theoretical Tools in Synthetic Biology

CHAPTER 5

Theoretical Considerations for Reprogramming Multicellular Systems

Joseph Xu Zhou and Sui Huang
Institute for Systems Biology, Seattle, WA, USA

INTRODUCTION

The physicist Richard Feynman once said: 'What I cannot create, I do not understand.' After more than 60 years of advances in molecular biology, it has through molecular dissection provided a solid knowledge base of the central elementary molecular processes of life. Now synthetic biology launches a new era in which we manipulate systems to change their fundamental properties. This goes beyond the singular perturbations used to probe a system feature in traditional 'analytical' biology. In the most extreme scenario synthetic biology reaches into the realm of 'making' livable systems de novo. The purpose is two-fold: to better understand biological systems or, even going beyond Feynman's dictum, to harness it for a practical purpose. It is clear that our experience from engineering in man-made technology, such as designing cars or computers, offers ample methodology for synthetic biology. But in doing so we also need to be aware of fundamental differences between biological and artificial systems, notably in view of the question of how the encoded blueprints of programs (software, DNA) are translated into system behaviors. Specifically, here we confront the question of how do genes, which are arranged as a linear code in the genomic DNA and interact with each other in a gene regulatory network, give rise to complex behaviors poised at the optimum between stability and flexibility?

One essential feature of complex multicellular organisms is their development (coming into being) from a single cell (fertilized egg) into a multicellular organism comprised not only of a large number of cells, but of discretely distinct types of cells that through ordered arrangement in space form tissues and organs.[1] Development of the cell type repertoire of >1000s or so cell types found in the human body follows a hierarchical scheme of cellular differentiation characterized by a binary branching genealogy. Since cell types represent discrete, stable entities, it has long been recognized that they correspond to attractor states of molecular dynamic networks that govern the differentiation of the various cell lineages.[2-5]

Early conceptualization of developmental cell behavior characterized by the natural discontinuity of phenotype, the discrete binary decisions of less mature multipotent cells into two lineages, and the rare but observed switching between the discrete phenotypes used a landscape picture with valleys that correspond to the attractor states.[6] This 'epigenetic

Synthetic Biology. DOI: http://dx.doi.org/10.1016/B978-0-12-394430-6.00005-4

landscape' that C. Waddington sensibly proposed in the 1940s can be considered a pivotal conceptual intermediate for understanding the one-to-many mapping between the blueprint (genome) and the variety of cell phenotypes. The epigenetic landscape is akin to a high-dimensional potential-like surface with multiple potential wells that represent nonequilibrium (meta)stable steady-states and whose topography is encoded in the genome.[7]

With regard to devices or systems that exhibit multiple distinct states, in engineered systems, orderly state transitions follow a predesigned and deterministic process between explicitly designed a priori discrete states, e.g. the on−off state of a light signal. By contrast biological systems exist in a continuous space that nevertheless exhibit discretely distinct behaviors and achieve the same state through many different state transition paths. More concretely, changes of system states (= phenotypes) in living cells which appear as quasi-discontinuous switching are driven by noise-induced state transitions that can be biased if not explicitly steered by biological signals.[8,9] Hence, the various states must be attractor states for the latter to ensure discreteness and stability in a continuum of possibilities and allow for multiple entry (converging) paths to each state. Complex multicellular organisms are thought to have evolved the needed functionality by 'apposition' of new attractors to the existing system. In multicellular systems this has led to developmental paths from the egg cell attractor to the attractor states of the stem cells of the various tissues, and from these stem cells to the mature adult cell types, giving rise to a particular form of the landscape that is encoded by the genome.

Evolution thus acts on the genome to shape the particular landscape topography. The attractors are located on the landscape to allow the developing cell to access them in a particular, controlled order − when needed during development and in homeostasis. Access to an attractor is controlled locally, based on fine-tuned transition probabilities and susceptibility to regulatory signals, such as hormones. Thus, while nondeterministic processes, such as molecular noise and bifurcations,[10] are in general avoided in engineered systems, biological systems exploit noise to drive the unfolding of the genome-encoded system of constraints into a highly complex structure. They are poised between being robust to noise (staying in the same attractor despite it) and being sensitive to noise (switching to an accessible nearby attractor in response to it). In a cartoonish but useful simplification, we can view development as the noise-driven successive occupation of the attractor states, which can represent intermediate or terminal states − all encoded in the genome.[11] Thus the cellular states are predestined yet subject to randomness.

Here we are concerned with the modification of the developmental paths between attractors with the ultimate goal to control attractor state transitions − which amounts to changing cell types. This has become known as 'cell reprogramming.' While the switch from one cell type to another of a closely related lineage has long been practiced in cell biology research,[12] and can be achieved by manipulating just one 'fate-determining' gene, it is only a recent demonstration of drastic changes of cell phenotype, the reprogramming of adult cells back to an embryonic stem-cell-like state, referred to as iPS (induced pluripotent state)[13] that has convinced the reductionist defenders of orthodoxy that cell-type identities are, after all, not carved in stone but a dynamic entity. Such reprogramming can be robustly achieved by manipulating multiple transcription factors although it remains a stochastic process.[5,14]

The complete reversion of a cell's phenotype, which in a first approximation can be regarded as defined by the expression pattern of the tens of thousands of genes of the genome, warrants the shift of the notion of explorative manipulation for analysis to the realm of constructive, hence biosynthetic reengineering of a cell phenotype for practical purposes. In fact, the prospect of manipulating cell types at will, starting from easily available, proliferating cells (such as certain blood cells, liver cells, or fibroblasts) to more challenging therapeutically desired cells, has fostered the hope that such universal cell phenotype reprogramming capacity will benefit regenerative medicine, where to repair tissue deficits or correct organ dysfunction one requires specialized cells of certain types in large quantities (such as heart muscle cells, skin stem cells, or pancreas insulin-producing cells).[15]

However, the design of reprogramming experiments remains largely empirical, typically relying on brute-force trial and error and qualitative 'educated guesses' to find how to manipulate the regulatory genes in order to achieve a desired attractor state transition. Here we present a formal framework for this purpose. It is based on our increasing understanding of how gene-regulatory networks produce the stable network attractor states that determine the cell-type-specific gene expression pattern, and will permit us to quantitate the 'height of the barriers' that separate the attractors and compute the most efficient path from one attractor to another.

We organize this chapter as follows: the second section will introduce the very basic concept of dynamical systems and of the epigenetic or quasi-potential landscape in a qualitative manner for experimental biologists. Engineers, computational, and physical biologists can fast-forward to the central question of how to define a quasi-potential landscape and to calculate transition dynamics that is addressed in detail in the third section. In the fourth section a general step-by-step description of the principles for putting the theory into practice is given, considering the paucity of information on gene regulatory networks available. The fifth section offers examples for blood cell reprogramming and pancreas beta-cell reprogramming. Conclusions and outlook end this chapter in the final section.

CONCEPTUAL FRAMEWORK: GENE REGULATORY NETWORKS, NETWORK STATES, AND CELL TYPES

The Gene Regulatory Network and Gene Expression Patterns as Network State

A gene regulatory network (GRN)[16] consists of the regulatory interactions through which regulatory genes control the expression behavior of their target genes. Since regulatory genes are themselves also target genes, this leads to a network replete with feedback loops that exhibits characteristic dynamics of the global gene expression changes. The gene expression patterns and their temporal behavior are thus predestined by the network structure which has evolved a 'wiring diagram' to produce the 'meaningful,' stable gene expression states that govern cell phenotypes.

More specifically, the GRN controls cell differentiations during development. A GRN in which fate-determining transcription factors (TFs) regulate each other drives the development of tissues by orchestrating the activation or suppression of the appropriate genes across the genome to establish the stable steady-state gene expression patterns that specify a given cell type with its biological functions. These stable gene expression patterns – defined by the N gene of the genome, are the aforementioned attractor states. They guarantee the stability of the cell-type-specific expression patterns.[5,17,18] The recent integrated analysis of gene expression profiles of tens of thousands of genes across the genome based on microarrays has provided evidence that cell types indeed represent high-dimensional attractor states of the dynamics of GRNs.[5,19,20] If the cell-type-specific genomic expressions are attractors, then they are indeed 'preprogrammed' by the genome acting as a blueprint because it specifies the particular 'wiring diagram' which is defined by the invariant properties of which gene (conditionally) regulates which one. The GRN is thus directly encoded in the genome, since molecular interactions and their dependence on cooperative action (conditional binding, allostery, etc.) are defined by protein and promoter DNA sequences. Yet, while this establishes the genome as a blueprint, hence warranting the use of engineering metaphors in synthetic biology, the occupation of the attractor states is modulated by noise and environmental signal in a way that is unique to biological systems.

The Toggle-Switch as an Example for a Gene Regulatory Network

In the following we present a simplifed, qualitative introduction to network dynamics using the well-known example of $N = 2$-gene network to demonstrate key concepts of network dynamics, attractors and landscape. The mutual inhibition of two opposing transcription factors (TFs), which we call X_1 and X_2, is one of the simplest GRNs (Fig. 5.1A). We are now interested in how the network state S, which is defined by the measurable gene expression

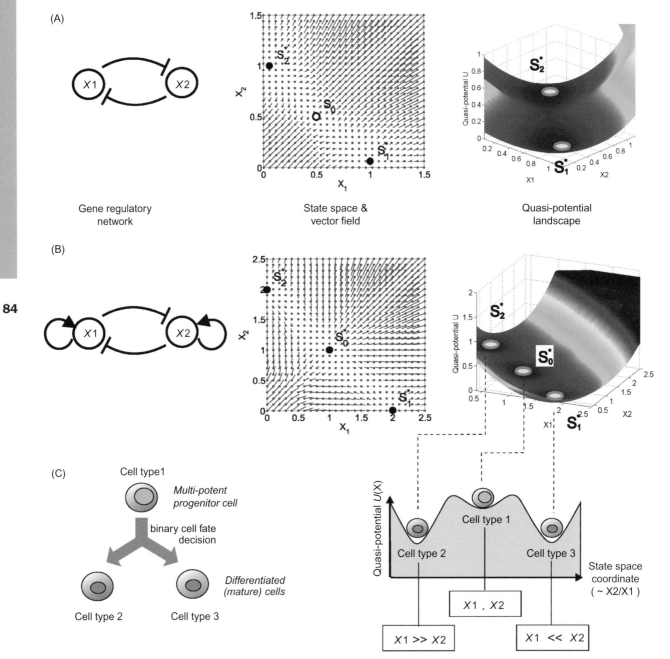

84

FIGURE 5.1

Topology, state spaces with the vector field and quasi-potential landscapes of gene-regulatory networks (circuits). For the two gene circuits (A) and (B), various representations are given from left to right: The gene circuit topology, the state space with the vector field, and the quasi-potential landscape. Circuit (A) represents the cross-inhibition and bistability, with two attractors S_1^* and S_2^* which are separated by a saddle point S_0. Circuit (B) has two positive-feedback loops which stabilize the central steady-state S_0, converting it from unstable to stable. The latter can serve to represent a metastable progenitor state which can differentiate to two differentiated cell fates, represented by attractors S_1^* ($X_1 \gg X_2$) and S_2^* ($X_2 \gg X_1$), as shown in (C).

pattern jointly established by the levels of X_1 and X_2, changes over time: $S(t) = [X_1(t), X_2(t)]$. The change of S is visualized as the movement of the state S in the two-dimensional *state space* spanned by the X_1 and the X_2 axis. The dynamics of this 2-gene network is constrained because X_1 and X_2 alter their levels in a coordinated manner — as prescribed by the network interactions (explained in Fig. 5.1A). As a consequence, most (X_1, X_2)-configurations, or states S, of the network are not stable but experience a 'force' of change until all the regulatory forces are balanced. In this case of mutual repression the dynamics is such that one typically finds two discretely distinct stable stationary states, S_1^* and S_2^* (asterisks indicate 'stability'). These stationary states are stable *attractors* because in the *state space* of all possible gene expression patterns they attract nearby states (Fig. 5.1A).[21] As argued above they represent the two alternative cell types. The term *'bistability'* is used to describe such dynamical behaviour in which the very same circuit can chose to be in either one of the two attractor states — depending on which 'basin of attraction' in the state space was its initial state, $S(t = 0) = [X_1(t_0), X_2(t_0)]$.[22]

Bistability and the associated inverse expression patterns of X_1 and X_2 in the two attractor states have been used to explain binary cell fate decisions in which an immature uncommitted multipotent cell can make a decision to commit to either one of two lineages ('fates'). For instance, in a type of blood cell progenitor that faces the decision to become a cell of the erythroid or the myeloid lineage (roughly, red versus white blood cell lineages), the erythroid lineage exhibits the $GATA1^{HIGH}/PU.1^{LOW}$ expression pattern, whereas the reciprocal myeloid lineage, has the inverse $GATA1^{LOW}/PU.1^{HIGH}$ pattern. The model can also incorporate another feature observed in many of such bistable networks: the two antagonist fate-determining factors not only repress each other (Fig. 5.1A), but often also seem to activate their own expression (Fig. 5.1B). This constitutes a different network architecture, hence a different dynamical behavior. Various mathematical models of the cross-inhibition—auto-stimulation circuit motif indicate that this departure from the classical bistable circuit structure (Fig. 5.1A) typically leads to the *stabilization* of the symmetrical bipotent state S_0, which becomes a locally stable attractor.[23] Thus, we actually have a tristable system (Fig. 5.1B). A cell differentiation corresponds to the process in which a progenitor cell at central state (X_1^{MID}/X_2^{MID}) transitions to either the left attractor with X_1^{HIGH}/X_2^{LOW} expression pattern or the right attractor with X_1^{HIGH}/X_2^{LOW} expression pattern (Fig. 5.1C).

We have here without equations given the intuition of network dynamics and the attractor states it produces. The above two-gene system, widely used in nature to control binary cell fate decisions, produces three attractor states, and thus constitutes a toy model for a multistable system. State transitions during cell fate commitment and subsequent differentiation can be viewed as transitions between attractor states — which so far are simply attracting points in a state space. But in which direction do spontaneous, noise-driven attractor transitions occur? This question of relative (meta)stability of attractors is of central biological significance, for it determines natural cell fate choice by multipotent progenitor cells and the ease of artificially induced fate switches (reprogramming). To conceptualize and quantitatively define attractor transitions and their rates, it is instrumental to introduce the notion of 'barriers' (height) between attractors and the 'relative depths' of attractors. This is the reason for the concept of potential-like landscapes which has been proposed even before the specific notion of gene networks and of attractor states in their dynamics.

THE QUASI-POTENTIAL LANDSCAPE

Biological systems of course are not just 2-gene circuits but high-dimensional dynamical systems that exhibit a large number of stable steady-states (attractors). Dealing with systems with more than two attractors, which is rather uncommon in synthetic systems, is where the landscape metaphor particularly comes in handy. This will prompt us to leave the domain of intuitive explanations as presented above and enter the realm of abstract formalism. To stay focused, however, the practical utility for cell-type modulation will guide us in the following discourse.

The stable steady-states (attractors) in the complex GRN represent so-called 'dissipative structures' that are sustained steady-states far from thermodynamic equilibrium, because their maintenance requires continuing intake of free energy and energy dissipation.[24] Transitions between attractor states correspond to cell phenotype switches and are affected by the 'relative depth' or, equivalently, 'relative barrier height' that separates these attractors. These metaphoric concepts are not arbitrary but become necessary in multiattractor systems when one has to consider attractor transitions which are frequent and integral to the behaviors of complex biological systems. Such behavior where attractor exit is the norm and not the feared extreme are rarely encountered in engineered systems where one is interested in the stability around a single attractor state. In dynamical systems theory, such 'local' stability is evaluated by linear stability analysis with multiple attractors that are 'agnostic of each other.' Thus, the landscape notion arises when one confronts the problem of nonlocal stability and wishes to relate a set of attractors to each other with respect to their 'relative stability' and the associated transition rates. If we want to manipulate biological functionality by prescribing specific transitions from one attractor to another, one needs to understand the landscape topography associated with a given GRN. This goes beyond traditional dynamical systems analysis that focuses on existence and local stability of attractors. In the following sections, we will introduce a mathematical framework to analyze 'relative stability' of multistable dynamic systems.

The Absence of Integrability and the Essence of a Potential Function

In a multistable system with M attractor states $\mathbf{x}_1^*, \mathbf{x}_2^*, \ldots \mathbf{x}_M^*$ (where $\mathbf{x}_i^*$ is the N-dimensional state vector that defines the position of attractor i), one thus wishes to obtain some sense of the 'relative depth,'[25] or more precisely, of the ordering of the attractor states with respect to their (meta)stability through some energy-like quantity, $U_1, U_2, U_3 \ldots U_M$, associated with each steady-state. This becomes relevant when we design biological processes associated with trajectories that go through a certain set of attractors. Such a potential energy-like function $U(\mathbf{x})$ for any point $\mathbf{x}$ in the state space of the system could be used to determine 'potential differences' that could inform about the probability and direction of transitions between attractor states in a noisy or perturbed system.

Let us consider a deterministic system (network) of N variables x_i (e.g. the activity of interacting genes) whose values describe the cell state $x(t) = (x_1(t), x_2(t), \ldots, x_N(t))^T$ and whose dynamics result from how each gene influences the activity of other genes (as invariantly predetermined by the gene regulatory network that is hard-coded in the genome). Such dynamics is described by the first-order ordinary differential equations (ODEs) which in general are nonlinear:

$$\frac{dx_1}{dt} = F_1(x_1, x_2, \cdots, x_N)$$

$$\frac{dx_2}{dt} = F_2(x_1, x_2, \cdots x_N) \tag{5.1}$$

$$\vdots$$

or in vector form:

$$dx/dt = \mathbf{F}(\mathbf{x})$$

$\mathbf{F}(\mathbf{x})$ represents the equivalent of 'forces' acting to change the system state $\mathbf{x}(x_1, x_2, \ldots, x_N)$ in this inertia-free system. If we have a gradient system then a potential function U can be obtained by integration, i.e. where there exists a function $U^{int}(x)$ with the following properties:

$$\frac{\partial U^{int}}{\partial x_1} = -F_1(x), \quad \cdots, \quad \frac{\partial U^{int}}{\partial x_i} = -F_i(x), \quad \cdots, \quad \frac{\partial U^{int}}{\partial x_n} = -F_n(x) \tag{5.2}$$

One can obtain a potential function $U^{int}(x)$ by successive integration,

$$dU^{int}(x) = -(F_1(x)dx_1 + \cdots + F_i(x)dx_i + \cdots + F_N(x)dx_N),$$

$$U^{int}(x) = -\int F_1(x)dx_1 + \cdots + F_i(x)dx_i + \cdots + F_N(x)dx_n \tag{5.3}$$

If U exists then the driving force is the gradient of U. However, unlike the equilibrium system, the transition rate for $x_A \rightarrow x_B$ is not path-independent here. Its spontaneity is determined by the potentials of attractor states and saddle points between them. The transition probability $P_{x_A \rightarrow x_B}$ is related to $\Delta U^{int}_{AS}(x) = U^{int}(x_S) - U^{int}(x_A)$ and the transition follows the *least action path*,[26] as discussed below. Here, x_S is the saddle point between two attractors x_A and x_B, as shown in Figure 5.2.

Biological dynamical systems with $N > 1$ are typically nonintegrable, nongradient systems, i.e. a function U cannot be obtained by integration, and a pure potential function in general does not exist. Yet, there is validity in a formal notion of an equivalent quantity. It is loosely referred to as 'quasi-potential.' Such a function would allow us to compute transition trajectories if properly defined such that it can serve as a quantity for the

87

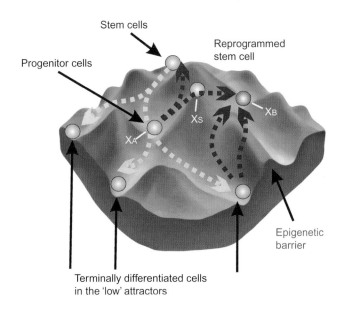

FIGURE 5.2

The quasi-potential landscape as a formal concept for cell differentiation and reprogramming. The figure shows a schematic quasi-landscape that is computed from the normal decomposition of the dynamic system, which is constructed from a gene regulatory network. Here, for representation, the high-dimensional state space ($N \gg 2$ genes) is compressed and projected into a two-dimensional plane, whereas the 'elevation' represents the quasi-potential U associated with every state. The landscape captures the global dynamics, enabling the comparison of 'relative depth' of different attractors, which is equivalent to Waddington's epigenetic landscape. X_A and X_B are two cell attractors in the landscape while the saddle point X_S is in between them. The downhill movement (in yellow) of the ball represents the differentiation process from stem cells to progenitor cells, and to terminally differentiated cells. The transition paths (in red) of the ball represent the cell reprogramming which require external stimulus and do not happen often.

ordering of (metastable) attractor states of the system in Eq. 5.1. Specifically, for U to be of meaning for this purpose we require that U:

(i) satisfies $\dfrac{dU}{dt} < 0$ for $x \neq x^*$ and $\dfrac{dU}{dt} = 0$ for $x = x^*$ (where x^* are the stable steady-states, which can be either fixed points or limit cycle) according to the stability theory of Lyapunov;[27]

(ii) is related to the Freidlin-Wentzell potential V that in turn expresses the 'the least action path' (LAP) between two states and hence, captures the barrier height U_{AS}. Thus, U permits the computation of the transition rate.[26] Note that (*i*) expresses the stability property of attractor state in a nonlocal sense; and (*ii*) refers to the relationship of any two points.

The Quasi-Potential Function for Nonintegrable Systems

Since high-dimensional, nonequilibrium systems generally are not gradient systems, i.e. Eq. 5.1 is usually not satisfied:

$$\frac{d\mathbf{x}}{dt} = \mathbf{F}(x) \neq -\nabla U \tag{5.4}$$

By contrast, one can enforce a partial notion of a quasi-potential U, if we write the driving force as a sum of two terms:

$$\frac{d\mathbf{x}}{dt} = \mathbf{F}(x) = -\nabla U + \mathbf{F}_r \tag{5.5}$$

where $\mathbf{F}_r$ is the 'remainder' beyond the component of the driving force that takes the form of a potential gradient. Thus, we decompose the nongradient vector field $\mathbf{F}(\mathbf{x})$, which needs to be finite and smooth (at least twice differentiable), into two components: one that is the gradient of some 'potential-like' function U and the second that represents the remainder of the driving forces. The question then is: what is the physical meaning of these terms? Given that there are infinite ways of decomposing a vector field into a sum of two fields, uniqueness of decomposition must come from imposing constraints through which we can incorporate the physical meaning. Our objective here is to find a decomposition such that the quasi-potential difference ΔU that manifests the 'height' of the barrier between the two attractors represents exactly the associated state transitions, whereas the 'remainder' of the driving force $\mathbf{F}_r$ will not contribute to the efforts needed for the transition.

The Normal Decomposition

One imposed constraint is that the gradient term is perpendicular to the 'remainder' force $\mathbf{F}_\perp$ (hereafter the notation U^{norm}, $\mathbf{F}_\perp$ instead of the general U, $\mathbf{F}_r$ implies that these two vector field components are perpendicular to each other):

$$\mathbf{F} = -\nabla U^{\mathrm{norm}} + \mathbf{F}_\perp \tag{5.6}$$

Then, one can show that U^{norm} is a Lyapunov function [27] and satisfies our condition (i) above. This is shown below for a two-dimensional system by taking its time derivative along any trajectory driven by the dynamic system in Eq. 5.1:

$$
\begin{aligned}
\frac{dU^{\mathrm{norm}}}{dt} &= \frac{\partial U^{\mathrm{norm}}}{\partial x}\dot{x} + \frac{\partial U^{\mathrm{norm}}}{\partial y}\dot{y} \\
&= \frac{\partial U^{\mathrm{norm}}}{\partial x}\left(-\frac{\partial U^{\mathrm{norm}}}{\partial x} + F_\perp^x\right) + \frac{\partial U^{\mathrm{norm}}}{\partial y}\left(-\frac{\partial U^{\mathrm{norm}}}{\partial y} + F_\perp^y\right) \\
&= -\left(\frac{\partial U^{\mathrm{norm}}}{\partial x}\right)^2 - \left(\frac{\partial U^{\mathrm{norm}}}{\partial y}\right)^2 + (\nabla U^{\mathrm{norm}}, \mathbf{F}_\perp)
\end{aligned}
\tag{5.7}
$$

If the gradient of U^{norm} is normal to the remainder force $F_\perp$, i.e.:

$$(\nabla U^{\text{norm}}, \mathbf{F}_\perp) = 0, \tag{5.8}$$

then the function U^{norm} will decrease monotonically with time during the process of reaching the attractors, thus satisfying Lyapunov's condition for metastability:

$$\frac{dU^{\text{norm}}}{dt} = -\left(\frac{\partial U^{\text{norm}}}{\partial x}\right)^2 - \left(\frac{\partial U^{\text{norm}}}{\partial y}\right)^2 < 0 \tag{5.9}$$

We can therefore see that the condition for the normal decomposition has an obvious physical meaning in that for $(\nabla U^{\text{norm}}, F_\perp) = 0$, U^{norm} corresponds to a Lyapunov function U^{norm} of dynamical systems and can represent the global (meta)stability as opposed to local (linear) stability.

Thus, we can decompose any sufficiently smooth vector field into a conservative potential field U^{norm} and the remaining forces $\mathbf{F}_\perp$:

$$\begin{aligned}
\mathbf{F} &= -\nabla U^{\text{norm}} + \mathbf{F}_\perp \\
(-\nabla U^{\text{norm}}, \mathbf{F}_\perp) &= 0
\end{aligned} \tag{5.10}$$

If the physical interpretation is that if $U^{\text{norm}}(\mathbf{x})$ represents a landscape over the state space region x, then for a ball in an attractor state that is perturbed to exit with least 'energy' against the system's 'driving force' $F(x)$ that keeps it in the attractor, we can decompose the field such that $\mathbf{F}_\perp$ will NOT contribute to this process, but only forces from the gradient field U^{norm} can contribute. Based on Freidlin-Wentzell's large deviation theory of a stochastic process discussed below,[26] U^{norm} can thus be used to compute the transition rate in this nonequilibrium dynamic system.

Importantly, the normal potential U^{norm} can also be directly 'read off' the system equations without time-stepping solution. We can calculate the potential field U^{norm} as follows:

$$(\nabla U^{\text{norm}}, F + \nabla U^{\text{norm}}) = 0 \tag{5.11}$$

This can be written in a component format called the Hamilton-Jacob equation:

$$\frac{\partial U^{\text{norm}}}{\partial x_1} \cdot \left(F_{x_1} + \frac{\partial U^{\text{norm}}}{\partial x_1}\right) + \cdots + \frac{\partial U^{\text{norm}}}{\partial x_i} \cdot \left(F_{x_i} + \frac{\partial U^{\text{norm}}}{\partial x_i}\right) + \cdots + \frac{\partial U^{\text{norm}}}{\partial x_n} \cdot \left(F_{x_n} + \frac{\partial U^{\text{norm}}}{\partial x_n}\right) = 0 \tag{5.12}$$

The Hamilton-Jacob equation is a nonlinear partial differential equation, which usually has no analytical solutions. However, U^{norm} can be solved numerically using the iterative Newton-Raphson method after boundary conditions are specified for a real problem.[28,29]

The Freidlin-Wentzell Theory of Large Deviation in Multistable Systems

For a dynamic system governed by deterministic forces $\mathbf{F}(x, t)$:

$$\frac{dx0}{dt} = \mathbf{F}(x0, t) \tag{5.13}$$

Let us now consider that the system is under a stochastic perturbation $\xi(t)$:

$$\frac{d\mathbf{x}}{dt} = \mathbf{F}(\mathbf{x}, t) + \varepsilon \cdot \xi(t) \tag{5.14}$$

If ε is sufficiently small, the perturbed system will converge to the original dynamical system, i.e. $\|x - x0\| \to 0$. However, if the perturbation is a random process with small average amplitude but with occasional large excursions, the perturbed dynamic system will behave differently. Freidlin and Wentzell[26] proposed a large deviation theory of stochastic process as a theoretical framework to analyze the behavior of the dynamical system with multiple attractors. Supposing that our system satisfies the Langevin dynamics, the governing equations are described by the following ODEs:

$$
\begin{cases}
\dot{x}_1 = f_1(x_1, \ldots, x_n) + \xi_1(t) \\
\quad\vdots \\
\dot{x}_i = f_i(x_1, \ldots, x_n) + \xi_i(t) \\
\quad\vdots \\
\dot{x}_n = f_{n1}(x_1, \ldots, x_n) + \xi_n(t)
\end{cases}
\tag{5.15}
$$

Suppose now that a ball is perturbed to go from attractor state $\mathbf{x}_A^* \to \mathbf{x}_B^*$, one defines an action function V_{AB} to measure the 'energy' barrier to be overcome for this transition:

$$
V_{AB} = \frac{1}{2}\min\left\{ \int_{tA}^{tB} \left[\sum_{i=1}^{n} ||\dot{x}_i - f_i(\mathbf{x})||^2 \right] dt \right\}
\tag{5.16}
$$

Here the action function V_{AB} is defined as a time integral of the square of the 'remainder' of the dynamic equations (deviation from deterministic trajectory) over the whole trajectory $X(t)$ from attractor x_A to x_B. If a ball is only driven by the 'forces' specified in the deterministic part of the ODEs (Eq. 5.14), it will correspond to 'free fall' in the 'gravity field' and the action is zero. If the ball is perturbed against the 'forces,' the remainder term $||\dot{x}_i - f_i(\bar{x})||^2$ will not be zero (in all of this metaphoric picture one is reminded that we are in an inertia-free world). We integrate the forces over the whole trajectory and obtain the total action when a ball switches from attractor x_A to x_B. Based on the variational principle, there exists a unique minimum integral, namely the action function V_{AB}, which is an objective measure of the difficulty for a state switching in a nonequilibrium dynamic system.

The Relationship Between the Freidlin-Wentzell Action Function and the Normal Decomposition

Although the Freidlin-Wentzell potential V and the normal potential U^{norm} are defined in different ways, they are mathematically related. For a dynamical system, $\mathbf{F} = -\nabla U^{norm} + \mathbf{F}_\perp$, $(-\nabla U^{norm}, \mathbf{F}_\perp) = 0$. We can rewrite the Freidlin-Wentzell potential as:

$$
\begin{aligned}
V_{AB} &= \frac{1}{2}\min\left\{ \int_{tA}^{tB} \left[||\dot{\mathbf{X}} - \mathbf{F}|| \right]^2 dt \right\} \\
&= \frac{1}{2}\min\left\{ \int_{tA}^{tB} \left[||\dot{\mathbf{X}} + \nabla U^{norm} - \mathbf{F}_\perp|| \right]^2 dt \right\} \\
&= \frac{1}{2}\min\left\{ \int_{tA}^{tB} \left[||\dot{\mathbf{X}} - \mathbf{F}_\perp - \nabla U^{norm}|| \right]^2 dt + 4\int_{tA}^{tB} \left[(\dot{\mathbf{X}} - \mathbf{F}_\perp) \cdot \nabla U^{norm} \right] dt \right\} \\
&= \frac{1}{2}\min\left\{ \int_{tA}^{tB} \left[\dot{\mathbf{X}} - \mathbf{F}_\perp - \nabla U^{norm} \right]^2 dt + 4\int_{tA}^{tB} \left[\dot{\mathbf{X}} \cdot \nabla U^{norm} \right] dt \right\} \\
&= \frac{1}{2}\min\left\{ \int_{tA}^{tB} \left[||\dot{\mathbf{X}} - \mathbf{F}_\perp - \nabla U^{norm}||^2 \right] dt + 4(U_B^{norm} - U_A^{norm}) \right\} \\
&= 2(U_B^{norm} - U_A^{norm})
\end{aligned}
\tag{5.17}
$$

During the process of exiting an attractor, the *least action path* (LAP) of a ball follows the governing equation:

$$
\mathbf{F} = \nabla U^{norm} + \mathbf{F}_\perp
\tag{5.18}
$$

As long as we are within the same attractor, the Freidlin-Wentzell potential is exactly twice as large as the potential from the normal decomposition, $V_{AB} = 2(U_B^{norm} - U_A^{norm})$.

When a 'ball' transitions from one attractor to another, the Freidlin-Wentzell potential only accounts for the uphill 'energy,' which is two times that of U^{norm}. Once it goes over the saddle point (point X_s in Fig. 5.2), the Freidlin-Wentzell potential V is zero for the remaining 'free-fall' path. The same applies for a ball that transitions through *many* attractors in between: the Freidlin-Wentzell potential is equal to the sum of all uphill potential between two points. All downhill paths contribute nothing to the Freidlin-Wentzell potential.

HOW TO OBTAIN A TRAJECTORY ON THE QUASI-POTENTIAL LANDSCAPE FOR TRANSITION BETWEEN TWO ATTRACTORS

The landscape notion offers a new vista that captures the intuition of state transitions by displaying this process as a jump between valleys. This picture has recently become fashionable to illustrate cell-type reprogramming and epigenetic regulation.[15,30] However, the link to the underlying gene regulatory network and the theoretical description of its global dynamics has not been articulated.

To demonstrate how the formal conceptualization of the landscape as a product of the GRN allows us in principles to build a landscape and determine the transition trajectories; we describe below five steps which one has to take in a typical setting in biology considering the paucity of data about the system determinants.

STEP 1: BUILD THE GENE REGULATORY NETWORK (GRN) FROM INCOMPLETE INFORMATION

To modulate the cell phenotype by triggering an attractor switch, we obviously need to know the structure of the underlying dynamical system, i.e. the gene regulatory network. Unfortunately, only for a few biological systems is there sufficient information of a GRN available that permits the straightforward translation of GRN specification into a complete dynamical system in the form of Eq. 5.1. Much effort is currently spent in such 'system identification' based on data integration, inference from gene expression profile dynamics, or direct molecular characterization of gene regulatory interactions using high-throughput experimentation such as ChipSeq.[31,32] Nevertheless, the obtained information pertains to the network architecture and remains sketchy. Not a single gene regulatory function that maps the inputs on the promoter to the output (gene expression), as needed to write the systems equations (Eq. 5.1), is even remotely known for any mammalian gene. Hence most mathematical modeling approaches that treat GRNs as dynamical systems have to use an educated guess from qualitative experiments (e.g. overexpression studies to determine whether a regulator is inhibiting or activating) based on information in the literature from isolated experiments. Given the sparseness of information, mathematical models have been limited to small subnetworks ('circuits') of a handful of genes.

STEP 2: TRANSLATE THE GRN INTO MATHEMATICAL FORMULA AND IDENTIFY FUNCTIONING ATTRACTORS IN GENETIC LANDSCAPE

Assuming a GRN has been constructed for a given biological system, we can use a mathematical formalism, such as Boolean network or ODEs, to breathe life into the static GRN structure by translating it into mathematical equations that animate the dynamics of the GRN. The central challenge is to define the 'transfer function' at each network node: to map the values of the inputs to a gene X, that is the presence and activity of its upstream regulators at a given time into its change of the expression level, $X(t)$. As mentioned above, almost no explicit molecular-level knowledge exists for the nature of the transfer functions in mammals, and thus they must be inferred from individual experiments that describe causal relationships.

Boolean networks in which gene expression X takes the value 1 or 0 (on or off) are one of the simplest mathematical formalisms to model the network dynamics, and have been historically widely used to gain insight into the most fundamental features of complex system dynamics.[22,33-35] They evade the problem of having to determine the values of parameters that are part of the description of the interactions, but face the same problem as ODEs in that the Boolean function that integrates and maps all the inputs of a gene at time t to its future expression value (output at $t + 1$) are virtually unknown. However, the qualitative nature of a Boolean network allows them to more readily capture the information obtained from qualitative overexpression/deletion experiments.[36-39] However, since in deterministic Boolean networks unstable steady-states do not exist the landscape has no saddles or hilltops, but consists of disjointed attractors.[21]

By contrast, ODE models are the most widely used formalisms and benefit from a large body of mathematical theory of dynamical systems that offer analytical tools to study the existence and stability of steady-states. Certainly, ODEs suffer from the need to define the functional form of regulatory relationships, i.e., how the expression state of upstream regulatory genes map into the expression state change of the target gene and determine the values of coefficients in the ODEs equations. One common work-around is the assumption of a universal sigmoidal transfer function (e.g. Hill function) that maps input into output.[40] This is justified on the grounds that the mechanism of transcriptional gene activation by the regulatory factors requires the assembly of multiprotein complexes on the promoter, which typically generates cooperativity. But even without the latter, the stochastic nature of gene activation and other physicochemical factors in the cell leads to sigmoidal input–output.[41]

The next challenge is how to integrate all the sigmoidal transfer functions of the individual inputs, given that for no mammalian promoter, the promoter logics, let alone of the transfer functions are known. In this respect a variety of ad hoc argued approaches have been used. Duff et al. recently compared several ODE approaches for the same network for the case for a hematopoietic system. The case of pancreas cell reprogramming below offers another example.[10,42-46]

STEP 3: DEFINE THE BIOLOGICAL TRANSFORMATION OF INTEREST AS A TRAJECTORY BETWEEN ATTRACTORS IN THE QUASI-POTENTIAL LANDSCAPE

Many useful manipulations of biological behaviors consist of imposing a change along trajectories that cause cells or tissues to exit the current (possibly maladaptive) stable state and enter into a benign attractor that may even actively contribute to a vital function of the organism. For instance, a novel therapy envisaged for diabetes patients is to convert alpha cells in the pancreatic islet into the distinct but developmentally neighboring insulin-secreting pancreas beta-cells.[47] This transition between two adjacent attractors can be externally triggered by the ectopic expression of a combination of regulatory genes or possibly by small molecules that modulate the activities of the appropriate set of genes. Such manipulations amount to the perturbation of a set of network nodes, which moves the state of a cell in state space. The ease of exit from an attractor is dictated by the gene network via the ODEs and depends on the direction of the perturbation ('push') as computed in the next step.

STEP 4: CALCULATE THE LEAST ACTION PATH (LAP) FOR THE TRANSITION BETWEEN CELL ATTRACTORS

Once the systems equations are defined and the relevant attractors (departure and destination states) are identified, we can determine the nature of the perturbation such that it follows the course of the LAP for the desired phenotype switch. This can be computed using the Freidlin-Wentzell formalism based on the same knowledge of network specification that has allowed us to compute the attractor states.

Because the Freidlin-Wentzell action function is usually too complicated to be found analytically, Eq. 5.16 can be rewritten in discrete form to find an approximate solution, as shown below:

$$V_{AB}(X(t)) = \Delta t \sum_{k=1}^{M-1} \sum_{i=1}^{n} \left[\left\| \frac{x_i^{k+1} - x_i^k}{\Delta t} - \frac{1}{2}(f_i^{k+1} - f_i^k) \right\|^2 \right] \tag{5.19}$$

Here the total time over the trajectory $X(t)$ is divided into $M-1$ equal parts Δt. The time integral of the action function is approximated with the sum of actions in the small time segments. The rate $\frac{dx}{dt}$ is approximated with the first-order difference equation $\frac{x_i^{k+1} - x_i^k}{\Delta t}$. To find the least-action trajectory in discrete form, initially two attractors are connected with a straight line and the conjugate gradient method (CG) is used to minimize the action function $V_{AB}(X(t))$.[48]

STEP 5: DESIGN THE PERTURBATIONS NEEDED TO DRIVE THE TRANSITION BASED ON THE LAP

When the LAP is calculated among the designated attractors, one will note that some genes never change their expression values while others are significantly modified as the cell moves along the LAP. Thus, the LAP serves as the first filter to select the genes whose expression needs to be modified to most efficiently drive the transition and which genes can be left unchanged. Besides identifying the list of genes that change during the state transition, the LAP also provides information on the detailed temporal profile of the expression behavior of each gene. Ideally, if we can find a 'perturbation recipe' which exactly modifies gene expression levels such that they collectively follow the LAP, this perturbation would be the most efficient way to induce that state transition. However, in reality this can currently hardly be realized because of a series of problems associated with experimentation, including intrinsic noise of the cell state (which changes the position of $x(t)$), time delays between manipulation (transfection, drug treatment) and desired gene expression change, the lack of precise control of expression levels, etc.

Thus, if the LAP computation is to benefit reprogramming it will be in cases where the cursory initial direction of the computed LAP trajectory deviates significantly from what is intuitive. Such a deviation could be implemented qualitatively to improve transition efficiency.

EXAMPLE: STATE TRANSITION IN BLOOD CELL AND PANCREAS CELL DIFFERENTIATION AND REPROGRAMMING

The ODE Model of Blood Cell Differentiation and Reprogramming

Here, we use our example of blood cells for which a qualitative description was made earlier to demonstrate how to find the Freidlin-Wentzell least action path. Based on the cross-inhibition and self-activation between GATA1 and PU.1 (Fig. 5.1B), the governing equations can be written as:

$$\begin{cases} \dfrac{dx}{dt} = F_1(x, y) = a_1 \dfrac{x^n}{\theta_{a_1}^n + x^n} + b_1 \dfrac{\theta_{b_1}^n}{\theta_{b_1}^n + y^n} - k_1 x \\[4mm] \dfrac{dy}{dt} = F_2(x, y) = a_2 \dfrac{y^n}{\theta_{a_2}^n + y^n} + b_2 \dfrac{\theta_{b_2}^n}{\theta_{b_2}^n + x^n} - k_2 y \end{cases} \tag{5.20}$$

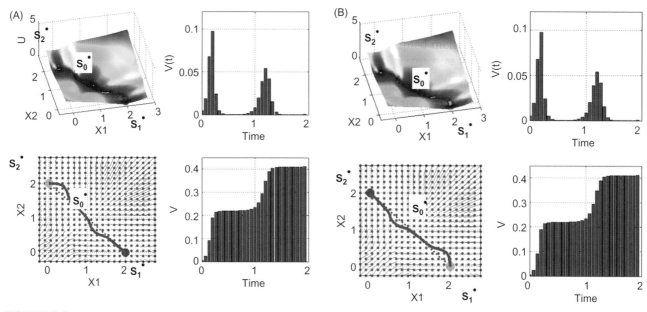

FIGURE 5.3

Least action paths on the quasi-potential landscape are computed based on the Freidlin-Wentzell theory. Least action path for attractor transition ('transdifferentiation') between the two cell attractors of the example in Figure 5.1B. (A) The least action path for transition from attractor $S_2^* \to S_1^*$. Green point is the starting point, red point is the end point. $V(t)$ is the action function at every time step. V is the accumulative action function at each time t. (B) The least action path for transition from attractor $S_1^* \to S_2^*$, which is different from that of path $S_2^* \to S_1^*$ because this is a nonlinear dynamic system. The Wentzell action function is the same as for the transition from $S_2^* \to S_1^*$ because attractors S_1^* and S_2^* have the same 'height' in the quasi-potential U^{norm}.

Since the dynamics of this network is structurally quite robust, we can use simple, symmetric values of parameters to demonstrate the network dynamics. In this example, $n = 4$, $k_1 = k_2 = 1$, $\theta_{a_1} = \theta_{a_2} = \theta_{b_1} = \theta_{b_2} = 0.5$, $a_1 = a_2 = b_1 = b_2 = 1$. The quasi-potential function U^{norm} from the normal decomposition is shown in Figure 5.3A. It has three attractors designated S_0^* (progenitor state), S_1^* (erythroid lineage exhibits the GATA1^{HIGH}/PU.1^{LOW} pattern), and S_2^* (myeloid exhibits the GATA1^{LOW}/PU.1^{HIGH} pattern). If we want to perturb the blood cell system to cause a transition from the myeloid to the erythroid lineage, the associated Wentzell action functions and LAP are calculated from the numerical minimization based on Eq. 5.19, as shown in Figure 5.3B. Note that the least action path myeloid → erythroid is different from the reverse path erythroid → myeloid. Such rather counterintuitive path irreversibility is a common characteristic of nonlinear biological systems.

The Specific Pancreas Development GRN and the ODE Model

In the second example, we return to pancreas cell reprogramming to illustrate the above principles of how to calculate a state transition given the partial knowledge of the underlying GRN. We first model the normal differentiation of the main pancreas cell lineages — as cell behavior that takes place on the quasi-potential landscape: the exocrine and the endocrine cells, including the β, δ, and α islet cells. Using ODEs to describe the mutual regulatory influence of 10 TFs involved in the differentiation of these cells during pancreas development, we present a minimal model that qualitatively captures known interactions and is able: (i) to recapitulate normal pancreas cell differentiation; (ii) to predict the temporal changes of key TFs during the development of particular cell lineages; and (iii) to predict the outcome of perturbations and hence, to help design new recipes of reprogramming experiments.[46]

There are three pairs of opposing (mutually inhibiting) TFs that control binary decisions at the three levels of pancreas differentiation. The first level of binary branching, between the exocrine and endocrine lineage, is governed by the opposing pair of the TFs, Ptf1a <-->Ngn3.[46] The second-level branching is governed in a similar manner by the Pax4 <-->Arx circuit which determines the β/δ versus the α-cell lineage, respectively.[46] The fate-determining TFs for the third branching into the β-cells versus the δ cells have not fully been characterized, but MafA has been shown to bias the decision to establishing the β-cells, whereas little information is available for the fate-determining factor for the δ cells. Thus, for modeling purposes and for maintaining symmetry, we use a placeholder for the δ-cell determining factor, called 'δ factor' and then assume a third pair of opposing TFs: MafA $<->\delta$ factor, governing the determination of β versus δ cells, respectively (Fig. 5.4A). Below are the ODEs for the network of regulatory influences described above:

$$Pdx1: \quad \dot{x}_1 = a_s \frac{x_4^n + x_7^n + x_8^n + x_{10}^n}{1 + x_4^n + x_7^n + x_8^n + x_{10}^n} - k \cdot x_1 + \xi_1(t)$$

$$Ptf1a: \quad \dot{x}_2 = a \frac{x_{10}^n}{1 + x_{10}^n + x_3^n} - k \cdot x_2 + \xi_2(t)$$

$$Ngn3: \quad \dot{x}_3 = a \frac{x_{10}^n}{1 + x_{10}^n + x_2^n} - k \cdot x_3 + \xi_3(t)$$

$$Pax6: \quad \dot{x}_4 = a \frac{\eta^n \cdot x_3^n + x_4^n}{1 + \eta^n \cdot x_3^n + x_4^n} - k \cdot x_4 + \xi_4(t)$$

$$Pax4: \quad \dot{x}_5 = a_e \frac{\eta_m^n \cdot x_1^n \cdot \eta^n \cdot x_3^n}{1 + \eta_m^n \cdot x_1^n \cdot \eta^n \cdot x_3^n + x_6^n} - k \cdot x_5 + \xi_5(t)$$

$$Arx: \quad \dot{x}_6 = a_e \frac{\eta_m^n \cdot x_1^n \cdot \eta^n \cdot x_3^n}{1 + \eta_m^n \cdot x_1^n \cdot \eta^n \cdot x_3^n + x_5^n} - k \cdot x_6 + \xi_6(t)$$

$$MafA: \quad \dot{x}_7 = a \frac{\eta_m^n \cdot x_1^n \cdot \eta^n \cdot x_3^n \cdot x_5^n + x_7^n}{1 + \eta_m^n \cdot x_1^n \cdot \eta^n \cdot x_3^n \cdot x_5^n + x_7^n + x_8^n} - k \cdot x_7 + \xi_7(t)$$

$$\delta gene: \quad \dot{x}_8 = a \frac{\eta_m^n \cdot x_1^n \cdot \eta^n \cdot x_3^n \cdot x_5^n + x_8^n}{1 + \eta_m^n \cdot x_1^n \cdot \eta^n \cdot x_3^n \cdot x_5^n + x_7^n + x_8^n} - k \cdot x_8 + \xi_8(t)$$

$$Brn4: \quad \dot{x}_9 = a \frac{\eta^n \cdot x_6^n + x_9^n}{1 + \eta^n \cdot x_6^n + x_9^n} - k \cdot x_9 + \xi_9(t)$$

$$Hnf6: \quad \dot{x}_{10} = a_s \frac{1}{1 + \eta^n (x_2 + x_7 + x_8 + x_9)^n \cdot x_{11}^n} - k \cdot x_{10} + \xi_{10}(t)$$

$$Maturity: \quad \dot{x}_{11} = m$$

(5.21)

Here, the variables x_1 to x_{10} represent the expression levels of the 10 key genes involved in pancreas cell differentiation. Equations represent the regulatory relationship among these 10 genes, as shown in Figure 5.4B. Each equation has three terms which capture the: (i) production of the gene expression in response to the upstream TFs; (ii) the linear degradation; and (iii) contribution of the stochastic fluctuations to gene expression. The production rate for the gene expression change is, as explained in Step 2 above, a sigmoidal function of its upstream regulators. Hill function-like forms with uniform exponents n ($n = 4$) were used (more details can be found in[46]).

Figure 5.4C shows the three 'branchings' of gene expression trajectories during normal pancreas cell differentiation. The expression of *Hnf6* starts at a high level and gradually decays, which activates the first switch between *Ptf1a* and *Ngn3*. The nondeterministic property of stochastic dynamics allows the state trajectory to split into two cell lineages,

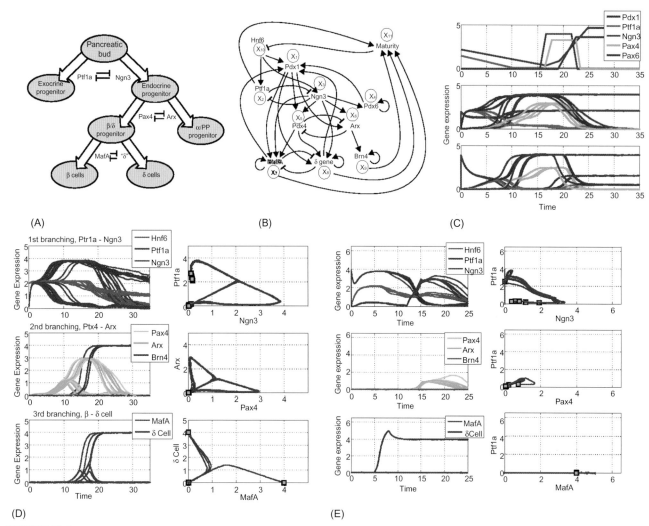

(A) (B) (C)

(D) (E)

FIGURE 5.4

The dynamic model of GRN to explain pancreatic cell differentiations and reprogramming. (A) The normal branching developmental paths for major pancreas cell types and the gene circuit modules involved in the three branching steps. (B) The underlying gene regulatory network for pancreatic cell differentiation integrating the gene circuit modules. (C) Temporal gene expression profile during pancreatic cell differentiation. The first panel is from experimental observations of both gene expression levels and timing, while the second and third panels show model simulation results from the master model and an alternative model with the inhibitory effects of Pdx1 upon *Ngn3* and *Ptf1a*. (D, E) Time course and state space trajectories for gene expression profile dynamics during normal pancreatic cell differentiation (D) and cell reprogramming with the recipe of overexpressing *Pdx1*, *Ngn3* and *MafA*(E). [46]

corresponding to the two new attractors: the exocrine and the endocrine progenitors. If cells follow the endocrine linage, they become *Ngn3* positive which dominates over *Ptf1a* ('lower branching lines' in Fig. 5.4D, right panel). High *Ngn3* triggers the second switch, embodied by the branching governed by *Pax4* and *Arx*. Later *Ngn3* decreases, reflecting its observed transient expression character, as *Hnf6* is down-regulated with maturation. Endocrine progenitors at this stage can differentiate into either α cells or β/δ cell progenitors. Cells fated to the trajectory with high *Arx* (and low *Pax4*) turn on the *Arx* target gene *Brn4*, which is a marker gene for the α phenotype and whose expression persists. By contrast, in cells fated to the trajectory with dominating *Pax4*, this TF then triggers the last switch that controls the branching between *MafA* and the δ *cell gene* which subsequently will activate the respective effector genes, producing the distinct β and δ cells.

Simulation and Prediction of Reprogramming

It was recently reported that overexpression of $Pdx1^+$, $Ngn3^+$, and $MafA^+$ could reprogram exocrine pancreas cells to the endocrine insulin-producing β cells[47] — a clinically desired transition because of its potential in cell therapy of diabetes. To model such reprogramming, we first describe the virus-mediated ectopic gene overexpression with temporal additional gain terms in the corresponding equations in our GRN model (Eq. 5.21). Figure 5.4E presents the gene expression time profiles and trajectories in the relevant phase planes during cell reprogramming using overexpression of the genes $Pdx1^+$, $Ngn3^+$, and $MafA^+$. An exocrine cell starts with high expression of Ptf1a. In the model, reprogramming is implemented by the extra production terms for Pdx1, Ngn3, and MafA during a certain time window. We see that the cell switches its expression pattern from that with a high-Ptf1a to one with a high-Ngn3 which subsequently triggers the cell to go through the Pax4-Arx branch point to finally reach the steady-state of the β cell. It should be noted that some α cells are also produced in this process because of the stochasticity, and such outliers that 'go off in the wrong direction' are often observed in reprogramming experiments.

Our model also predicts that Ngn3's role in reprogramming can be enhanced by the inhibition of Ptf1a directly. This is important because an inhibitory perturbation of a network node is technically much easier (e.g., via RNAi technology, small molecules) than an activating perturbation. Since Ptf1a and Ngn3 are cross-inhibitory, when Ptf1a expression level is suppressed, Ngn3 expression will increase. Our simulations also show that combining the perturbation of the two network nodes, Ptf1 (inhibition) and Ngn3 (activation) can synergistically enhance efficiency of β cell reprogramming.

In conclusion, this example demonstrates how a gene regulatory network model, built with qualitatively reported interaction schemes from the literature, which mostly represent incomplete 'causal networks' rather than GRNs, govern cell-type diversification and differentiation. Our results show that with a minimum of knowledge of the constraints imposed by the gene network topology, pancreas cell differentiation can be explained as the transitions among different cell attractors.

CONCLUSION AND OUTLOOK

We have demonstrated the principles of manipulating the phenotype of cells — by actuating switches between entire cell phenotypes and how such engineering can be informed by the underlying gene regulatory network that governs normal development of these cell types. The perturbations force cells to achieve a physiological, predestined phenotype, but going there through a 'road not taken' during normal development. It is in this new rationale of predictive model-based whole phenotype manipulation across uncharted terrain in gene expression state space that warrants the placement of such manipulative approaches into the domain of 'synthetic biology.'

Controlling transitions between high-dimensional attractors in rugged 'epigenetic landscapes' is an elementary capacity that will help the design and engineering of more complex biological systems.

The state transitions between attractors on a landscape is not just a helpful metaphor but the direct mathematical manifestation of network dynamics that involve deterministic constraints imposed by regulatory interactions encoded by the genome and modulated and driven by external signals and gene expression noise. At the moment little specific information about these genomic interactions is known, so that educated guessing is a substantial ingredient in the modeling. But it turns out that, as modelers of biological systems have long realized, a profound property of robust complex systems is that the qualitative constraints of interactions, independently of quantitative details, capture much of the realized effective dynamics, thus already granting practical utility to an imperfect and incomplete model.

At the academic level, much as a hypothetical generic model of a geographic landscape that we draw in the classroom teaches us about the basic principles behind the formation of mountains, valleys, and lakes, but does not help us to navigate the real world to move from one specific place to another due to the absence of information of the specific geography, so are we, with regard to gene regulatory networks and cell behaviour, only at the classroom stage of generic cartoonish models. We do not have all the exact information to compute the specific potential-landscape topography that could guide us from point (cell state) A to B. But the formal reduction of valleys and mountains in this landscape that Waddington already proposed, to the first principles of systems dynamics of a network provides us now with a solid conceptual framework for the current practice of reprogramming, much of which is still dominated by brute-force trial-and-error efforts in perturbing the dynamics of gene-regulatory networks. With sufficient information, as to be expected in the near future, it may be possible using the theories presented here, to reconstruct the real-world regulatory epigenetic landscape of model organisms. It can then serve as a specific road map to design efficient strategies for reprogramming of any desired cell type at will with high efficiency by computing the optimal starting point and paths.

References

1. Wolpert L, Tickle C. *Principles of Development*, 4th ed. Oxford, UK: Oxford University Press; 2011.
2. Delbruck M. Genetik der bakteriophagen. *Klin Wochenschr*. 1949;27:109.
3. Jacob F, Monod J. Genetic regulatory mechanisms in the synthesis of proteins. *J Mol Biol*. 1961;3:318−356.
4. Kauffman SA. Metabolic stability and epigenesis in randomly constructed genetic nets. *J Theor Biol*. 1969;22:437−467.
5. Huang S. Reprogramming cell fates: reconciling rarity with robustness. *Bioessays*. 2009;1:546−560.
6. Waddington CH. The epigenotype. *Endeavour*. 1942;1:18−20.
7. Pisco A, Fouquier d'Herouel A, Huang S. 'Epigenetics:' many meanings − one common concept. *DNA Cell Biol*. in review.
8. Eldar A, Elowitz MB. Functional roles for noise in genetic circuits. *Nature*. 2010;467:167−173.
9. Huang S. Cell lineage determination in state space: a systems view brings flexibility to dogmatic canonical rules. *PLoS Biol*. 2010;8:e1000380.
10. Huang S, Guo Y-PP, May G, Enver T. Bifurcation dynamics in lineage-commitment in bipotent progenitor cells. *Dev Biol*. 2007;305:695−713.
11. Muñoz-Descalzo S, De Navascues J, Arias AM, Munoz-Descalzo S. Wnt-Notch signalling: an integrated mechanism regulating transitions between cell states. *Bioessays*. 2012;34:110−118.
12. Graf T. Historical origins of transdifferentiation and reprogramming. *Cell Stem Cell*. 2011;9:504−516.
13. Okita K, Ichisaka T, Yamanaka S. Generation of germline-competent induced pluripotent stem cells. *Nature*. 2007;448:313−317.
14. Hanna JH, Saha K, Jaenisch R. Pluripotency and cellular reprogramming: facts, hypotheses, unresolved issues. *Cell*. 2010;143:508−525.
15. Zhou Q, Melton DA. Extreme makeover: converting one cell into another. *Cell Stem Cell*. 2008;3:382−388.
16. Smith J, Theodoris C, Davidson EH. A gene regulatory network subcircuit drives a dynamic pattern of gene expression. *Science*. 2007;318:794−797.
17. Huang S, Eichler G, Bar-Yam Y, Ingber DE, Yam YB. Cell fates as high-dimensional attractor states of a complex gene regulatory network. *Phys Rev Lett*. 2005;94:1−4.
18. Macarthur BD, Ma'ayan A, Lemischka IR, Ma A. Systems biology of stem cell fate and cellular reprogramming. *Nat Rev Mol Cell Biol*. 2009;10:672−681.
19. Chang HH, Hemberg M, Barahona M, Ingber DE, Huang S. Transcriptome-wide noise controls lineage choice in mammalian progenitor cells. *Nature*. 2008;453:544−547.
20. Enver T, Pera M, Peterson C, Andrews PW. Stem cell states, fates, and the rules of attraction. *Cell Stem Cell*. 2009;4:387−397.
21. Huang S, Kauffman S. Complex gene regulatory networks − from structure to biological observables: cell fate determination. *Encyclopedia of Complexity and Systems Science*. 2009.

22. Kaplan D, Glass L. *Understanding Nonlinear Dynamics*. New York: Springer; 1995.

23. Zhou JX, Huang S. Understanding gene circuits at cell-fate branch points for rational cell reprogramming. *Trends Genet*. 2011;27:55−62.

24. Prigogine I. *The End of Certainty*. New York NY: The Free Press; 1997.

25. Nicolis G. Dissipative systems. *Rep Prog Phys*. 1986;49:873−949.

26. Freidlin MI, Wentzell AD. *Random Perturbations of Dynamical Systems*. New York, NY: Springer-Verlag; 1984.

27. Lyapunov AM. The general problem of the stability of motion. *Int J Control*. 1992;55:531−534.

28. Polyanin AD, Zaitsev VF. Nonlinear partial differential equations. *SIAM Rev*. 1969;11:7−19.

29. Lee ES, Phillip PC. A generalized Newton-Raphson method for nonlinear partial differential equations — packed-bed reactors with axial mixing. *Chem Eng Sci*. 1966;21:143−157.

30. Yamanaka S. Elite and stochastic models for induced pluripotent stem cell generation. *Nature*. 2009;460:49−52.

31. Kerenyi MA, Orkin SH. Networking erythropoiesis. *J Exp Med*. 2010;207:2537−2541.

32. MacQuarrie KL, Fong AP, Morse RH, Tapscott SJ. Genome-wide transcription factor binding: beyond direct target regulation. *Trends Genet*. 2011;27:141−148.

33. Huang S. Gene expression profiling, genetic networks, and cellular states: an integrating concept for tumorigenesis and drug discovery. *J Mol Med*. 1999;77:469−480.

34. Shmulevich I, Kauffman SA. Activities and sensitivities in Boolean network models. *Phys Rev Lett*. 2004;93:48701.

35. Kauffman SA. *The Origins of Order*. New York: Oxford University Press; 1993.

36. Huang S, Ingber DE. Shape-dependent control of cell growth, differentiation, and apoptosis: switching between attractors in cell regulatory networks. *Exp Cell Res*. 2000;261:91−103.

37. Albert R. The topology of the regulatory interactions predicts the expression pattern of the segment polarity genes in *Drosophila melanogaster*. *J Theor Biol*. 2003;223:1−18.

38. Espinosa-Soto C, Padilla-Longoria P, Alvarez-Buylla ER. A gene regulatory network model for cell-fate determination during *Arabidopsis thaliana* flower development that is robust and recovers experimental gene expression profiles. *Plant Cell*. 2004;16:2923−2939.

39. Balleza E, et al. Critical dynamics in genetic regulatory networks: examples from four kingdoms. *PLoS One*. 2008;3:e2456.

40. Glass L, Kauffman SA. The logical analysis of continuous, non-linear biochemical control networks. *J Theor Biol*. 1973;39:103−129.

41. Andrecut M, Halley JD, Winkler DA, Huang S. A general model for binary cell fate decision gene circuits with degeneracy: indeterminacy and switch behavior in the absence of cooperativity. *PLoS One*. 2011;6:e19358.

42. Duff C, Smith-Miles K, Lopes L, Tian T. Mathematical modelling of stem cell differentiation: the PU.1-GATA-1 interaction. *J Math Biol*. 2012;64:449−468.

43. Roeder I, Glauche I. Towards an understanding of lineage specification in hematopoietic stem cells: a mathematical model for the interaction of transcription factors GATA-1 and PU.1. *J Theor Biol*. 2006;241:852−865.

44. Chickarmane V, Troein C, Nuber UA, Sauro HM, Peterson C. Transcriptional dynamics of the embryonic stem cell switch. *PLoS Comput Biol*. 2006;2:e123.

45. MacArthur BD, Please CP, Oreffo ROC. Stochasticity and the molecular mechanisms of induced pluripotency. *PloS One*. 2008;3:e3086.

46. Zhou JX, Brusch L, Huang S. Predicting pancreas cell fate decisions and reprogramming with a hierarchical multi-attractor model. *PLoS One*. 2011;6:e14752.

47. Zhou Q, Brown J, Kanarek A, Rajagopal J, Melton DA. In vivo reprogramming of adult pancreatic exocrine cells to beta-cells. *Nature*. 2008;455:627−632.

48. Hestenes MR, Stiefel E. Methods of conjugate gradients for solving linear systems 1. *J Res Natl Bur Stand*. 1952;49:409−436.

CHAPTER 6

Computational Protein Design for Synthetic Biology

Florian Richter[1] and David Baker[2]
[1]Institut für Biologie, Humboldt-Universität zu Berlin, Berlin, Germany
[2]Department of Biochemistry, University of Washington, Seattle, Washington, USA

INTRODUCTION

The objective of (computational) protein design and engineering is to create proteins with functions that are not available in natural proteomes. In many synthetic biology applications, just like in natural organisms, proteins are the workhorses that carry out the actual desired functions. But since natural proteins and the functionalities they exhibit were evolved in facilitating the survival and maintenance of cells and organisms, the synthetic biologist's toolbox is limited to functions necessary for that purpose. And while the set of naturally available proteins is already very large and can be used to design cells and organisms with novel properties, synthetic biology would benefit tremendously from transcending this barrier and being able to design proteins with functional properties that so far have not naturally evolved. For example, if one wants to engineer a strain of *Escherichia coli* that produces a certain small molecule of interest, one is dependent on the existence of an enzymatic synthesis pathway for said molecule. However, if the molecule of interest is artificial, such as a drug or biofuel candidate, it is unlikely that a natural synthesis pathway exists. In this case, a novel enzyme needs to be designed that catalyzes the desired reaction.

In a way, a synthetic biologist without the capability to create custom-tailored proteins is like an architect who is limited to using only naturally occurring materials like mud, wood, and stones to erect buildings. And while buildings with these materials are good enough for certain applications, the development of more advanced materials like brick, steel, and glass immensely increased the size and type of possible buildings. Similarly, once proteins with new functions can be reliably engineered, synthetic biology will take a huge leap forward.

Computational protein design (CPD) is by no means the only method available to engineer proteins. Other approaches such as directed evolution, which is discussed elsewhere in this book, have been employed to obtain impressive results. Computational design does, however, have some unique advantages that allow it to address problems not amenable to directed evolution, as we will demonstrate in this chapter. Conversely, if sufficient high-throughput assays for the function of interest exist, directed evolution is better suited to improve proteins starting from a threshold level of initial activity. Thus, computational design and directed evolution are perfectly complementary, and we anticipate that these two methods will often be used hand-in-hand when designing new proteins for real-life applications.

101

Synthetic Biology. DOI: http://dx.doi.org/10.1016/B978-0-12-394430-6.00006-6

In this chapter, we will first delineate where we expect CPD to have the biggest potential impact on synthetic biology. Then, we will give an overview of the general models and algorithms most often used in computational design. Next, we will give an introduction to the design of novel and specificity-changed binding proteins and enzymes, together with a brief description of the specialized computational algorithms used for these problems. Then, we will give examples of computational thermostabilization of proteins, followed by a brief overview of the design of novel protein folds. Finally, we will compare the relative strengths and weaknesses of computational design versus directed evolution, and finish with an outlook on where computational protein design could have the most imminent impact on synthetic biology.

The Potential Impact of CPD on Synthetic Biology

CPD is still a relatively young technique, and so far most synthetic applications rely on reusing and recombining existing natural proteins as building blocks.[1] However, as we will show in this chapter, the successes achieved with CPD over the last decade forecast the types of synthetic biology applications and devices that CPD will help enable. We anticipate an impact of CPD in six ways:

1. The design of novel protein–protein and protein–small molecule interactions will allow for the manipulation of signaling cascades to modify gene expression in response to designed, unnatural stimuli.
2. The design of protein–protein interactions will also allow for the creation of tailor-made proteins that bind to protein targets and can either inhibit or elicit responses not related to gene expression in a target organism.
3. The design of novel catalytic activities will allow for novel biosynthetic pathways for compounds of interest, and will also enable the creation of synthetic organisms that break down environmental pollutants or toxins.
4. The design of protein–small molecule interactions will allow for the creation of novel biosensors for compounds of interest.
5. The design of self-assembling proteins could lead to the creation of novel biomaterials, such as delivery containers for drugs or conductive fibers or sheets for bioenergy applications.
6. CPD enables the thermostabilization of proteins with relative ease, and therefore could contribute to increased robustness of synthetic biology applications.

For some of these applications, successes using computationally designed proteins have already been reported (Table 6.1). For example, regarding point 2, CPD has been used to create proteins that inhibit viral infection or the build-up of amyloid fibrils.[2] Regarding point 3, one case has been reported where a computationally designed enzyme could be used in a novel biosynthetic pathway,[3] and another novel enzyme was designed to break down a component representative of a class of pollutants.[4] For point 6, several examples have been reported where proteins have been stabilized, leading to higher expression or increased half-lives.[5] And for the types of applications where no examples have been reported yet, it is conceivable that designs can be achieved with the required functionality using computational algorithms very similar to the ones used to obtain the successful results. We are thus hopeful that CPD will have a broad impact on synthetic biology, and will put applications within reach that could not be created otherwise.

METHODS OVERVIEW

The term computational protein design (CPD) as used in this chapter describes the design of amino acid sequences based on computational structural modeling of the to-be-designed protein. While some results have been reported with design algorithms that are not based

TABLE 6.1 Computationally Designed Proteins

Synthetic Biology Application	Protein Design Task	Inputs Required	Representative Examples
Redesign of cell signaling	Protein–protein interface redesign	Crystal structure of complex	19
Interfering with target proteins	Protein–protein interface redesign, de novo protein interface design	Crystal structure of target for de novo design, structure of complex for redesign	2, 16, 17, 24
New catalytic activity/ metabolic pathways	Enzyme redesign or de novo enzyme design	Crystal structure of enzyme with substrate for redesign, theozyme for de novo design	3, 4, 37, 52
Small molecule binders and sensors	Same as for enzyme design, without catalytic machinery	Crystal structure of wild-type receptor with ligand	None yet
Material design	Protein–protein interface redesign or de novo protein–protein interface design	Crystal structure of scaffold protein	None yet
Stability increase	Monomer design	Crystal structure of protein of interest	5, 59

on an underlying structural model of the protein,[6] these approaches will not be described in this chapter.

The computational methods and algorithms typically used in CPD can broadly be divided into three categories:

1. Side-chain placement algorithms that, given a model of the protein backbone, select a set of amino-acid side-chain conformations compatible with that backbone. Since the amino acid identities of the selected side-chain set can be any of the 20 natural amino acids (or even unnatural ones), the sequence design happens at this stage.

2. Backbone conformation-generating algorithms, whose purpose is to generate models of backbone conformations according to the requirements of the specific design task. The backbone models generated by these algorithms are usually passed to side-chain placement algorithms for sequence design.

3. Rigid-body placement algorithms, which are used to place two protein models or a protein and a small molecule model in a relative spatial orientation to each other. These algorithms are often the first step when designing binding or catalytic proteins, where one has to design a functional site on one of the proteins, and thus needs to first design the spatial relation of the two interacting partners.

CPD algorithms from all three categories generally employ classical molecular mechanics[7] representations of the designed system. All atom models of the protein are used, where bond lengths and angles are usually held constant to increase computational speed. Similar to molecular dynamics force fields, CPD energy functions usually contain terms for van der Waals interactions, bond-dihedral potentials, hydrogen-bonding, simplified electrostatics, and implicit solvation.[8] A peculiarity of CPD energy functions is the additional inclusion of sequence composition terms, which are parameterized to make the distributions of amino acid identities in designed sequences similar to those in natural proteins. In this section, we will focus on describing the most often used side-chain placement algorithms (category 1), since these are broadly utilized in virtually all CPD calculations. Examples of category 2 and 3 algorithms will then be presented in subsequent sections, along with the specific design tasks these algorithms are meant to address.

The problem of computational protein design was first presented in 1983 as the so-called inverse protein-folding problem.[9] Whereas the task in protein folding and structure prediction is to derive a protein's tertiary structure given the primary amino acid sequence, the objective of the inverse folding problem is, given a certain backbone tertiary structure, to find a sequence that will fold into this template. The inverse folding problem is an extension of the threading problem in homology modeling. In both cases, a side-chain placement algorithm is tasked with finding the optimal combination of side-chain conformations on the template backbone. In threading, only one amino acid, namely that from the wild-type sequence of the threaded protein, is allowed at each residue position, whereas in design, all amino acids can be considered at each residue position. For example, for the full sequence design of a relatively small 100 residue long backbone, the side-chain placement algorithm is tasked with finding the optimal sequence out of the astronomically large number of $20^{100} \approx 10^{130}$ possible sequences.

Virtually all side-chain placement algorithms approach this problem by first discretizing side-chain conformational space into a library of so-called rotamers for each amino acid type, where each rotamer represents a frequently observed conformation for that amino acid. This approach can be justified by the observation that amino acid side-chains prefer a limited number of low-energy conformations in high-resolution crystal structures of natural proteins. The library of rotamers for each amino acid type can thus be derived from statistical analysis of protein crystal structures,[10] and as a simple rule of thumb, the rotamer library for a side-chain with n chi angles will contain 3^n rotamers (with 1 rotamer in staggered and 2 rotamers in *gauche* conformation for each chi angle). For example, in the rotamer library for valine, a residue with 1 chi angle contains 3 rotamers, whereas the rotamer library for lysine (4 chi angles) would contain 81 rotamers. The combined rotamer library for all 20 canonical amino acids contains 367 rotamers by this rule of thumb.

Using the rotamer concept, the side-chain placement algorithm's task of finding the optimal sequence for a given backbone can be formulated more specifically, namely as the task of finding the set of rotamers that give the lowest energy (as determined by the energy function used) when placed on the template backbone. This lowest energy conformation is often referred to as the GMEC (global minimum energy conformation). Thus, for the aforementioned 100 residue case, with 367 rotamers being allowed at every position, the side-chain placement algorithm needs to select a combination of rotamers out of $367^{100} \approx 10^{256}$ total possibilities. Even for a smaller problem, such as the redesign of a 20 residue binding site, there are $367^{20} \approx 10^{51}$ possible combinations. The simplest imaginable side-chain placement algorithm would be a brute-force approach that simply enumerates all possible rotamer combinations, scores each of them with the energy function, and remembers the GMEC. However, such an enumerative algorithm is evidently impractical considering the large number of possible solutions for even small design problems. Assuming that assembling and scoring one conformation takes a millisecond on modern computer hardware, an enumerative algorithm would take 10^{48} seconds to find the GMEC for the above-presented hypothetical 20 residue binding site design problem, which is roughly 31 orders of magnitude longer than the estimated age of the universe. The most important aspect of a viable side-chain placement algorithm is thus its ability to reduce the combinatorial complexity of the problem and select a low-energy rotamer combination within a short amount of time. An in-detail comparison of several algorithms developed for this purpose was done by Mayo et al.[11] Today, the most commonly employed ones are Monte Carlo algorithms,[12] and the so-called FASTER algorithm.[13]

COMPUTATIONAL DESIGN OF PROTEIN–PROTEIN INTERACTIONS

Protein–protein interactions are involved in a large number of cellular processes from signal transduction to differentiation to apoptosis and others. Being able to create new or modify existing protein–protein complexes in a rational fashion would thus endow the synthetic biologist with the capability to change cellular behavior at will. Potential synthetic biology applications include the rewiring of signal-transduction pathways to turn on a reporter gene in response to an environmental stimulus, or the design of proteins that bind to functional sites on (and thus inhibit) target proteins.

In this section we will describe methods that are commonly used for the design of protein–protein interactions, and introduce examples of several studies done in this regard so far. The goals in CPD of protein interactions can broadly be divided into two areas: redesigning existing interactions towards higher affinity or changed specificity; and the design of novel binding interactions. Depending on the specific problem, either the sequence of both binding partners in the complex may be modified by the design algorithm ('two-sided design'), or the algorithm is only allowed to design the sequence of one partner while leaving the other partner constant ('one-sided design').

Computational Redesign of Protein–Protein Interactions

There are usually two motivations to redesign an existing protein–protein interface: (1) increasing the affinity to make the interface more stable; and (2) changing the specificity of the interface, meaning to redesign the interface in such a way that the affinity of one pair of desired binding partners is retained, while the affinity towards another, competing binding partner is reduced. In both cases, a structural model of the to-be-redesigned complex, preferably a crystal structure, needs to be available, since this serves as an input for the CPD calculations. When the goal is to increase the affinity, the computational workflow used is usually some iterative combination of rigid-body docking algorithms developed for virtual protein docking[14] and general side-chain placement algorithms, while also taking into account a set of empirical rules governing affinity. When the goal is to modify specificity, these docking and side-chain placement algorithms need to be augmented by specialized algorithms that take into account and penalize the competing states.

There are several representative examples of increasing affinity by computational design. Roberts et al.[15] presented a study where a peptide inhibitor of a PDZ domain involved in cystic fibrosis was redesigned for higher affinity towards its target. Starting from an NMR structure of the natural ligand–PDZ domain complex, three mutations were introduced into the sequence to obtain a hexameric inhibitor that had 170-fold increased activity compared to the natural ligand. Lippow et al.[16] applied computational design to the problem of antibody affinity maturation, which is a field of broad therapeutic significance. In their work, the authors increased the affinity of two antibodies: one, a lysozyme-binding model antibody, by 140-fold through mutation of four residues; the other, an epidermal growth factor receptor binding therapeutic antibody, by 10-fold through mutation of three residues. Computational design was also used by Haidar et al.[17] to introduce four mutations into a solubilized T-cell receptor, increasing the affinity toward its cognate peptide MHC complex by 100-fold. The redesigned receptor is potentially better suited than the wild-type for diagnostics applications.

There has also been significant progress in the field of specificity redesign in the last several years. Perhaps the most challenging aspect of this problem from the standpoint of computational design is the need to incorporate 'negative design' into the calculation, meaning to consider the unwanted, competing stages and design a sequence that disfavors these. Another side effect of this requirement is that the designed sequences might not have the highest possible affinity against the target of interest, since the identities of the interface positions are not just determined by how well they interact with the target state, but also how well they discriminate against the unwanted states. In virtually all CPD algorithms, the designed sequence is a product of the side-chain placement algorithm, but most of these algorithms, i.e. Monte Carlo schemes and FASTER (see above) have been developed to optimize the energy function for a single state, and extending them to consider multiple states is not trivial. Therefore, algorithms specifically developed for this purpose need to be employed.

An elegant and very general approach to this problem was presented by Leaver-Fay et al.[18] In this approach, the input to the calculation is a set of backbone models, together with an algebraic rule of how to sum the scores (as determined by the energy function) of a certain sequence on each backbone model into one value. During the calculation, the sequence search is done by a genetic algorithm, and the value that is being optimized is the user-specified sum over the scores of all considered states instead of the score of a single state. When calculating the sum, absolute as well as relative scores of states can be considered in positive (for the unwanted states) or negative (for the desired states) fashion. In their work, the authors used this general framework to design sequences that in silico have favorable interaction scores for a set of desired states while having unfavorable scores for another set of competing, unwanted states. As this approach is still very recent, no experimental data on sequences designed with it exist yet, but it does represent the most general and comprehensive theoretical approach to the problem.

Several illustrative demonstrations of successful specificity redesign have been reported in recent years, some of them with direct applications in synthetic biology contexts. In a textbook example of two-sided design, Kapp et al.[19] designed an orthogonal GTPase/GEF pair, which could be used as a valuable tool to study cell signaling, as well as serve as a component in a synthetic signaling pathway. Starting from crystal structures of the GTPase Cdc42 and its activator GEF, intersectin, the authors first identified positions in Cdc42 at which mutations would interfere with intersectin binding without disrupting any of the known interfaces with other binding partners or the active site. After identifying one position (Phe-56), a cognate position on intersectin was identified where a salt-bridge could be introduced if Phe-56 was concurrently mutated to Arg. The mutated GEF stimulated nucleotide exchange in the mutated GTPase, but not in the wild-type, while the mutated GTPase could be activated by the mutated GEF but not the wild-type, demonstrating the orthogonality of the new pair. The new pair retained (albeit lower than wild-type) signaling activity in vivo. Grigoryan et al.[20] introduced a computational framework that explicitly considers competing states. In this study, the authors set out to design proteins that interact with individual members of a class of transcription factors known as bZIPs. This class comprises about 20 families, which share extensive structural and sequence similarity, making the design of inhibitors specific for only one subset very challenging. To address this problem, the authors developed an algorithm that first designs a sequence with highest possible affinity for the target, and then in a second step modifies the found sequence to increase the gap between the target state and the nearest competing state. Forty-six designs were characterized, 10 of which interacted more strongly with their intended target than with any competitor. However, to make the used algorithm computationally tractable, a scoring function specifically developed for this class of proteins had to be used, making this approach not easily extensible to other problems. In another study, Yosef et al.[21] redesigned the calcium-dependent second-messenger protein calmodulin towards preferably binding only one of its two major interaction partners. One of the designs, containing six mutations, had significantly reduced affinity towards the undesired partner, but retained affinity for the desired partner, resulting in a 900-fold specificity switch. Several more examples of successful computational redesign of protein–protein interactions have been described in a recent review.[22]

Computational Design of Novel Protein Interactions

Complementary to redesigning existing protein interfaces, CPD can be used for the de novo design of protein interactions. Generally, the objective in a de novo interface design problem is to design a protein that binds a target protein of interest, starting from a structural model (preferably a crystal structure) of the target protein alone. The computational workflow used to tackle this problem can roughly be divided into three steps:

1. Choosing a 'scaffold' protein, i.e. a protein that can serve as the backbone template that the binding sequence can be designed onto.
2. Finding a productive relative spatial orientation of the target protein and the scaffold.
3. Designing the amino acid sequence of the scaffold (and potentially the target) to stabilize this spatial orientation.

The first two steps represent computational challenges unique to this problem. In step 1, the scaffold protein can either be taken from a library of existing proteins for which the crystal structure is known, or it can be designed de novo using category 2 CPD algorithms. In step 2, after a scaffold has been decided upon, category 3 (rigid-body placement) algorithms then need to be used to place it in a spatial orientation towards the target, thus giving an initial model of the complex. Essentially, the 'global' shape of the complex and the location of the binding interfaces on the two partners are determined in step 2. This model is then passed to side-chain placement algorithms (or to the specialized algorithms developed for protein interface design as described above) to design the novel amino acid sequence.

Arguably, step 2 is the most critical stage in this workflow, because it is necessary to find a spatial orientation that is 'designable,' i.e. where a sequence can be designed such that the resulting binding interface has sufficient size, shape complementarity, and interactions for the complex to be energetically favorable compared to the unbound state and thus lead to a high-affinity interaction. Moreover, many applications require that a particular region of the target protein is part of the binding interface, meaning the rigid-body placement algorithm needs to be able to orient the scaffold to optimally interact with the desired surface patch on the target. In recent years, several impressive examples of successful de novo protein interface design have been reported, using different approaches for steps 1 and 2. These different approaches can be divided into two groups: general approaches that can in principle be used to design a binder for any arbitrary target protein; and more specialized approaches that take advantage of certain structural properties of the target.

Perhaps the most general method for scaffold selection and placement is Fleishman et al.'s so-called 'Hotspot-design' method,[23] shown in Figure 6.1, which was recently used to design proteins that bind to a conserved region of influenza hemagglutinin and inhibit this protein's ability to undergo conformational changes that underlie influenza infectiousness.[24] This method is based on the observation that in many natural protein–protein interfaces, the affinity is mostly mediated by a small subset of the residues making up the interface. When making a series of single-point mutants of a binding protein, in which residues belonging to the interface have been mutated to alanine, and measuring the resulting binding affinities, usually only a subset of variants will have significantly reduced affinities. The residues that were replaced by alanine in these variants are thus the most important interface residues, and are often referred to as 'hotspot' residues.[25] From this observation, Fleishman et al. reasoned that a viable strategy to design novel interfaces would be to first place a small number of side-chains in favorable locations on the target interface, and then use these proto-hotspot residues as anchors that the rest of the interface is designed around. In the Hotspot-design method, a small (usually on the order of 1−3) set of disembodied amino acids is placed in the vicinity of the target-protein surface patch in an energetically favorable fashion. For example, if there are exposed hydrophobics on the target surface, an aromatic amino acid might be chosen as the hotspot-residue, or in case unsatisfied hydrogen-bonding atoms are present, an amino acid with complementary hydrogen-bonding functionality could be picked. The exact location for each hotspot residue can either be set explicitly or found with the help of ligand-docking approaches.[26] Once the hotspot residues have been placed, a candidate scaffold protein is placed such that the

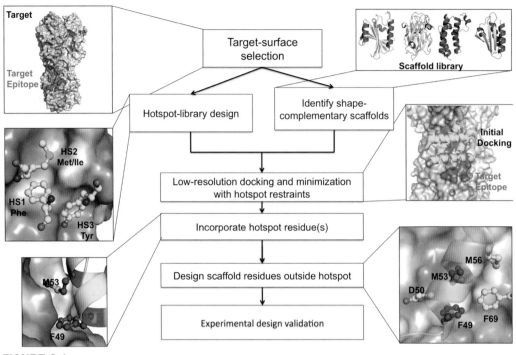

FIGURE 6.1

'Hotspot-design' method to design novel protein–protein interactions.

backbone atoms of the hotspot residues are superimposable onto the backbone atoms of any one residue of the candidate scaffold, without the scaffold backbone overlapping with the target protein. To achieve this, protein–protein docking algorithms are used, where the target protein is held constant and the candidate scaffold is docked. To guide sampling, coordinate restraints between the candidate scaffold and the hotspot residues are imposed during the docking calculation. After the scaffold has been properly placed, all residues in the vicinity of the target protein (except the hotspot residues) are redesigned with side-chain placement algorithms to obtain the final designed sequence. In practice, after the hotspot residues have been placed, the docking and design procedure is carried out millions of times, with the candidate scaffolds picked from a library of several hundred small, monomeric, globular proteins for which high-resolution crystal structures are available, the resulting designs ranked by energy, and a few dozen of the highest-ranking designs are considered for expression.

The most impressive example of a successful execution of the hotspot design strategy is represented by Fleishman et al.'s designed protein that binds to an evolutionary conserved surface region of influenza hemagglutinin, referenced above. Hemagglutinin is a protein on the flu surface that plays a vital role in the virus' capacity to infect cells, undergoing a conformational change when the virus binds to a host cell and becomes endocytosed. After inspecting a co-crystal structure of hemagglutinin with a broadly neutralizing antibody, the authors placed hotspot residues in the vicinity of the antibody binding site, and were subsequently able to design two proteins that bind hemagglutinin, one with a K_D of 200 nM and one weaker. After affinity maturation, the binding affinity was increased to a K_D of 22 nM and 38 nM for the two designs, and the designed proteins inhibited the conformational change that hemagglutinin undergoes during infection. X-ray crystallographic analysis confirmed that the designed proteins bound in the intended location and orientation. It is well conceivable that such designed proteins can play a role as therapeutics in the future.

Another general approach for steps 1 and 2 is represented by Jha et al.'s[27] so-called DDMI method ('Docking, sequence Design, and gradient-based Minimization'). In this method, first protein—protein docking methods are used to place a candidate scaffold (taken from a library) and the desired target in close proximity to each other. During the docking phase, the energy function is augmented with restraints that direct the scaffold protein towards the desired binding site on the surface of the target protein. At the end of the docking phase, a rough score filter is used to determine whether the candidate complex will be discarded or subjected to the next sequence design phase of the protocol. During this phase, the docked complex model is then subjected to iterative rounds of sequence design and gradient-based minimization, and resulting sequences are selected for expression based on a combination of energy function scores and structural parameters. The authors used their method to generate six designs against PAK1, a human kinase, the best of which bound with an affinity of 100 μM to the desired target.

Several other approaches for steps 1 and 2 that rely on the target protein to have certain structural features exist. While none of them are as general as the hotspot design or DDMI approach, they have been shown to yield high-affinity binders against targets having the required features. Two cases of exploiting solvent-exposed edges of beta-sheets to form a binding site have been reported. This strategy makes use of the fact that such edges have accessible unsatisfied hydrogen-bonding functionalities, which can be bound relatively easily by any scaffold that also has a beta-sheet with an exposed edge, thus forming an intermolecular beta-sheet upon complex formation. The residues surrounding this beta-sheet anchored interface are then redesigned to further increase the affinity and modulate specificity beyond that provided by the sheet-contacts alone. Stranges et al.[28] used this approach to turn a monomeric protein into a symmetric homodimer. Starting from a library of proteins with exposed beta-strands, the authors generated initial models consisting of two copies of the monomer interacting through the strand, followed by, like in most design protocols, cycles of sequence design and minimization. Four designs were experimentally characterized, the best of which had a K_D of 1 μM, and a crystal structure showed that the interface was virtually super-imposable onto the design model. Sievers et al. used a very similar approach to design peptides that inhibit amyloid formation.[2]

The strong interactions that some amino acids can make with metals can also be harnessed to design protein interfaces. The same way in which a metal site can significantly stabilize a protein structure can also be used to stabilize an interface. The coordination spheres of metals usually contain between four and six ligand sites. Thus, if a site containing half of a metal coordination sphere can be designed on the surface of a protein, two copies of this protein could then dimerize and form the complete metal site. Der et al.[29] demonstrated the feasibility of this approach by designing a zinc-mediated homodimer from a monomeric scaffold protein. The design had a very tight interface ($K_D < 30$ nM) in the presence of zinc, but in the absence thereof, the K_D increased ∼ 100-fold, indicating the importance of the metal. Interestingly, as demonstrated by Salgado et al.,[30] this approach can also be used as a stepping stone to design interfaces not dependent on metal. In their example, the authors first created a minimalist interface featuring a zinc-binding site, and then used CPD to create additional interactions across the interface, resulting in a design where binding was independent of metal presence.

Finally, another viable strategy to create novel protein binders against a given target is the so-called grafting of functional epitopes. This method relies on the availability of an already existing binder for the target of interest, and preferentially a crystal structure of the complex between the two. In this approach, the binding residues of the existing binder are transferred onto another scaffold, thus endowing this scaffold with binding affinity for the target of interest. This method is useful in cases where, for example, the original scaffold is difficult to express or problematic in other ways. Further, as demonstrated by Sia et al.,[31] protein grafting can lead to higher binding affinities in cases where the binding epitope is a short, relatively unstructured peptide that is grafted onto a more rigid scaffold. Grafting strategies do not necessarily require computational modeling of the system, however, as shown by Azoitei et al.,[32] computational algorithms allow for the atomically exact modeling of the connection between the grafted epitope and the new scaffold. In their study, the authors took a discontinuous epitope (i.e. an epitope consisting of two backbone segments) from the HIV protein gp120 that a broadly neutralizing antibody, b12, had been raised against, and, after scanning a large scaffold library for complementarity to both parts of the epitope, then transplanted both segments onto a different scaffold in the same relative orientation. After creating a library (also guided by computational design) around the grafted epitope, a design was obtained that bound b12 with similar affinity as the original viral protein. In theory, when challenging the immune system with these designs, they could raise antibodies with similar binding properties as b12, and thus this strategy could be used to design vaccines.

In summary, several methods have been developed recently for the de novo design of protein−protein interactions. Which method is most suited in a given situation must be decided on a case-by-case basis. If the target of interest has certain structural features that can be exploited to create a binding site, then approaches like beta-strand assembly or metal-mediated interfaces can be considered. If other binders against the target protein already exist, epitope grafting or the hotspot design method are good strategies. If neither is available, DDMI can be used, or ligand-docking algorithms can be used to generate inputs for hotspot design.

COMPUTATIONAL DESIGN OF CATALYTIC ACTIVITY

Of equal importance in biology as protein−protein interactions, the ability of enzymes to accelerate chemical transformations is involved in virtually every biological process. And just like being able to rationally engineer protein binding, being able to rationally design new or modify existing catalytic activity would tremendously expand the synthetic biologist's horizon. Potential synthetic biology applications are the redesign of metabolic pathway enzymes to yield novel biofuels, or the de novo design of enzymes with catalytic activities not available in the repertoire of natural enzymes, which would allow for these nonbiological transformations to be included in biological contexts and pathways.

In this section, we would like to highlight the advances made so far in the computational design of catalytic activity. Broadly speaking, there are three types of problems in the realm of computational enzyme design. These are, in order of increasing difficulty:

1. Redesign of an existing enzyme active site to process a different substrate (so-called 'specificity redesign').
2. Redesign of an existing enzyme active site to catalyze a different type of chemistry ('reactivity redesign').
3. De novo design of novel catalytic activity into a previously inert scaffold.

Arguably, (re-)designing an enzyme is more challenging than designing a binding protein. For the latter, it is enough to present a rigid surface that is complementary in shape to the design target, and usually only one target needs to be recognized. An enzyme, however, at a minimum needs to be able to bind the substrate(s) with sufficient affinity, then stabilize the transition state with even higher affinity, but then not have too high affinity for the product. In addition to that, often residues in the enzyme active site show large pKa shifts or form covalent intermediates with the substrate(s), and thus the enzyme's active site environment needs to modulate active-site residue reactivity. This necessity to concurrently satisfy several, sometimes conflicting criteria has led to enzymes being referred to as 'masters of compromise'. Further, while the biophysical principles governing binding are generally understood, and significant insight into the principles underlying enzymes' catalytic power has been obtained,[33] there are still open questions about the relative importance of certain phenomena for catalysis, such as the trade-off between active-site dynamics and preorganization.[34] However, developing an algorithm to approach a certain CPD problem necessitates understanding of the biophysical factors playing a role in the problem, and if this understanding is incomplete, the resulting algorithms might be lacking critical features. Thus, while impressive early results have been achieved in computational enzyme design, the procedure is still less reliable than binding or thermostability design.

Another computational challenge unique to enzyme design is that to accurately model chemical reactivity, a quantum-level treatment of the system is required, for example to assess the effect of the active site electronic environment on the energy levels of the substrate's molecular orbitals. All CPD energy functions employ a classical physical representation of the system however, and since billions of different sequences are usually considered by the side-chain placement algorithm, using a more accurate quantum-level model would be prohibitively slow. Thus, the energy function is essentially blind with respect to the catalytic competence of the designed active site. The most-often-used workaround for this problem is based on restraining the identity and allowed geometry for a handful of the active site residues.[35] In this section, we will introduce methods used and successful examples for each of the three cases.

Computational Redesign of Enzyme Specificity

The objective of specificity redesign usually is to take an existing enzyme and change it such that it transforms a different substrate. Generally, the catalytic residues of the enzyme active site (i.e. the residues that mediate the chemical steps of the reaction), and consequentially the type of chemistry and mechanism that the enzyme performs, remain unchanged. Only those active site residues that play a role in binding the substrate are modified by the design algorithm. Thus, the new substrate needs to be similar to the enzyme's natural substrate, in that it must feature the same reactive moieties that are acted upon by the enzyme. The new substrate is only allowed to differ in the nonreacting parts of the molecule.

As for most other CPD projects, a structural model, optimally a crystal structure, of the to-be-designed system is required as the input. For specificity redesign, the optimal starting point is a crystal structure of the enzyme of interest in complex with its native substrate, or with a substrate or transition state analogue. Additionally, a basic understanding of the catalytic mechanism and knowledge of the most important active site residues is required in order to prevent the design algorithm from mutating away from these critical residues. The first step is to generate a model of the new desired substrate. If this substrate features rotatable bonds, an ensemble of possible conformations also needs to be generated. Usually, this ensemble has restricted diversity in the moiety of the target substrate that resembles the wild-type substrate, but full diversity in the differing regions. The target substrate ensemble is then superimposed onto the wild-type substrate, such that the shared moiety is in the same region of the active site and making identical contacts to the catalytic residues. Next, the side-chain placement algorithm is used to design a sequence that accommodates the new substrate. In this step, besides sampling the identity and conformational diversity of the active site residues, the substrates conformational ensemble is also sampled. Usually, only the subset of residues that contact the differing moiety of the target substrate is allowed to mutate. After running the side-chain placement algorithm, the resulting structure is usually refined through gradient-based minimization, and these two steps are iterated several times. If a stochastic side-chain placement algorithm is used, the calculation is carried out several hundreds or thousands of times, leading to a large number of models. The best scoring models are selected, and depending on the throughput of the available experimental setup, anywhere between a handful to hundreds of sequences are experimentally tested for the desired activity.

Several examples of successful computational redesign of enzyme specificity have been published in recent years, with applications in different fields. Chen et al. set out to redesign the specificity of the phenylalananine adenylation domain of the gramicidin S synthetase A (GrsA-PheA).[36] This domain is part of the nonribosomal peptide synthetase (NRPS), which is a large multidomain enzyme complex that assembles peptides in an assembly-line manner. Many of the product peptides have antimicrobial properties and are thus of pharmacological interest. Being able to redesign the specificity of individual NPRS domains could yield novel, unnatural peptides with potentially improved properties. In their work, Chen et al. succeeded in redesigning GrsA-PheA to accept several different substrates instead of the native substrate phenylalanine, namely leucine, arginine, aspartate, glutamate, and lysine. Their most successful variant, a redesign for leucine, had a 2168-fold increased preference for leucine over phenylalanine compared to the wild-type enzyme, while maintaining about one-sixth of the catalytic proficiency of the native enzyme.

Another example of enzyme specificity redesign is Ashworth et al.'s redesign of the DNA cleavage site of the homing endonuclease I-MsoI.[37] Homing endonucleases are DNA-cutting enzymes that recognize target sites of ~15 nucleotides in length, as opposed to the ~6 nt cut-sites of restriction enzymes, and cut these with high specificity. Since the cleavage sites are fairly long, most homing endonucleases' cut-sites only occur once per genome, meaning that homing endonucleases are potentially valuable tools for genome engineering applications. Ashworth et al. succeeded in changing the specificity of I-MsoI for one base pair in its recognition sequence, creating a variant that cleaves the new target site 10^4-fold more efficiently than the wild-type enzyme, while having activity comparable to the wild-type enzyme with good discrimination against the original cleavage site. Ashworth et al.'s work represents an important first step towards the ultimate goal of being able to design an endonuclease for any cleavage site. Potential synthetic biology applications include the manipulation of specific genetic loci in living organisms, such as the introduction of new traits in plants,[38] or the genomic engineering of entire mosquito populations.[39]

Murphy et al. used computational design to change the specificity of a human guanine deaminase[40] towards accepting ammelide instead of guanine as a substrate. The achieved specificity switch was $2.5*10^6$-fold, albeit the designed enzyme had significantly reduced activity. Ammelide is a structural intermediate between guanine and cytosine. There are no human cytosine deaminases known. Therefore, if the specificity switch could be extended towards cytosine, the resulting enzyme could find application as a prodrug-activating enzyme with low immunogenicity. To achieve their results, Murphy et al. used a more advanced computational algorithm, where, similar to the hotspot-design method, first a disembodied side-chain was placed in ideal relation to the new substrate, and then a segment of nearby backbone was remodeled to support this desired side-chain placement.

Lippow et al. succeeded in turning a galactose 6-oxidase into a novel glucose 6-oxidase.[3] This designed enzyme could serve as the starting point in a designed efficient metabolic pathway for the synthesis of the value-added chemical D-glucaric acid in *E. coli*. Because no crystal structure of the wild-type enzyme with substrate was available, the authors first had to create a model for galactose in the wild-type active site through in-silico docking. Since a medium-throughput plate screening assay was available, the authors ran the design algorithm several thousand times to yield 2379 unique sequences, and then devised a strategy to create a library of 10^4 clones that encompassed the sequence diversity generated by the computational algorithm. The resulting library was screened and 402 hits were found (3.8% hit rate), one of which had a 400-fold increased activity for glucose and, though still having better activity with galactose, the preference for galactose over glucose was decreased 13 000-fold.

Computational Redesign of Enzyme Reactivity

The redesign of enzyme reactivity, while still considered redesign, represents a more drastic intrusion into the redesigned enzyme than mere specificity redesign. In this approach, the active site of an existing enzyme is redesigned, or 'repurposed,' to catalyze a different type of reaction. Arguably, the more similar the two reaction types are the easier the task becomes. A class of enzymes inherently suited for this approach is metalloenzymes. In many metalloenzymes, the metal carries out the essential chemical step, while the surrounding active-site amino acids serve to bind and orient the metal and the other (usually organic) substrate(s) in the proper orientation to each other, while also activating relevant functional groups of the organic substrate. Depending on what type of organic functional group is bound proximal to the metal, the same metal center can catalyze different chemistries. For example, metals such as zinc have an inherent affinity for water molecules. An H_2O molecule has reduced pKa when bound to zinc, and the resulting ZnOH species is a more potent nucleophile than unactivated H_2O. Depending on the electrophile present, the zinc bound OH^- could then either undergo a hydrolytic or a nucleophilic addition reaction.

In the inaugural example of this approach, Khare et al. redesigned an adenosine deaminase into an organophosphate hydrolase.[4] The wild-type enzyme, which was part of a set of enzymes featuring mononuclear Zn-sites with at least one of the Zn-coordination sites not occupied by a side-chain, contains a Zn-binding site that activates a water molecule for nucleophilic attack onto the amino group of adenosine. In their work, the authors first placed the design substrate, a model organophosphate featuring an activated leaving group, in the active site such that the Zn was coordinating to the phosphate's keto-oxygen, thus rendering the phosphorus atom more susceptible to nucleophilic attack by a water molecule. Next, the side-chain placement algorithm was used to redesign the surrounding active site residues to accommodate the new substrate. During this step, the Zn-coordinating residues were held constant. The resulting design featured eight mutations and hydrolyzed the model substrate with a k_{cat}/K_M of 4 $M^{-1} s^{-1}$, and after three rounds of directed evolution, a variant with a total of 13 mutations and a k_{cat}/K_M of $\sim 10^4 M^{-1} s^{-1}$ for the designed substrate was obtained. Further analysis of the importance of each of the mutations indicated that a minimal set of four mutations was absolutely required to confer hydrolytic activity for the target substrate. This suggests that obtaining a comparable result through directed evolution alone would necessitate screening of an enormous library containing all possible quadruple mutations of the protein's active site.

113

Computational De Novo Design of Enzyme Activity

The third and most difficult type of problem in computational enzyme design is the design of catalytic activity from scratch. In this case, both a catalytic mechanism as well as a protein site to carry it out need to be devised. The currently most viable approach, depicted in Figure 6.2, consists of developing a so-called theozyme for the reaction of interest, and then trying to graft this theozyme into a scaffold protein. A theozyme, or 'theoretical enzyme,' is a three-dimensional model of a minimal active site necessary to catalyze the desired reaction.[41] Usually it consists of a model of the energetically highest transition state on the reaction pathway, together with a set of disembodied amino acids placed around it that are meant to stabilize this state or perform chemical transformations on it. For example, if negative charge develops on a certain substrate atom over the course of the reaction, a strategy to stabilize this build-up and thus accelerate the reaction would be to place a positively charged side-chain, such as an arginine or lysine, next to this substrate atom. If the substrate gets deprotonated, a protic amino acid, such as a glutamate, aspartate, or histidine could be placed next to the mobile proton in the theozyme. Several strategies to develop a theozyme for a given reaction of interest have been devised. The most comprehensive and most difficult one represents an approach using quantum-mechanical

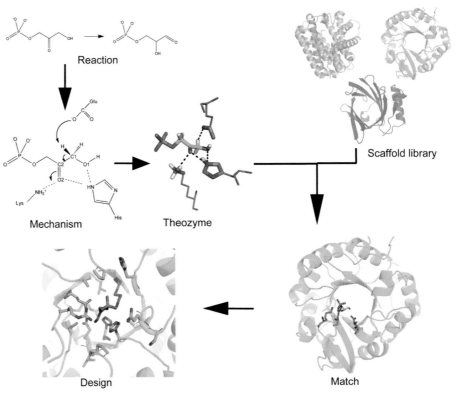

FIGURE 6.2
De novo enzyme design workflow. A theozyme for the reaction of interest is placed into a scaffold and the resulting structure designed.

modeling as described by Zhang et al.[42] Other, more ad hoc, approaches relying on chemical intuition or biological precedent will be described below.

Once a theozyme has been devised, the next step in the process is to graft it into a protein scaffold. In this step, called matching, a library of scaffold protein structures is searched for attachment sites for the theozyme. A theozyme can be attached in a scaffold if the theozyme ligand can be placed in a scaffold cavity, and if simultaneously every theozyme side-chain can be grafted onto a scaffold backbone position, such that the desired relative geometric orientation between theozyme side-chains and theozyme ligand is maintained, with no clashes between the theozyme components and the scaffold backbone. Thus, the rigid-body orientation between the substrate and the to-be-designed enzyme is determined at this stage, and hence matching algorithms fall into category 3 of computational protein design algorithms.

Several matching algorithms have been proposed, such as a simple enumerative algorithm that places each side-chain sequentially,[43] an algorithm that places the ligand for each theozyme side-chain in parallel and then employs six-dimensional geometric hashing to determine ligand positions that can make all desired theozyme contacts,[44] or an algorithm that scans through all pairwise combinations of scaffold residues to determine geometric overlap with theozyme side-chains.[45] Independent of the matching algorithm used, the result of the matching stage is a set of so-called matches, which are models featuring the desired minimal active site in a protein cavity. However, the theozyme usually only contains a handful of side-chains (2−4), while in the matches usually several dozen side-chains are within the shell contacting the placed ligand. This means that only a subset of the side-chains making up the new active site have been assigned their ideal identities at this stage. To determine sequence identities for the remaining active site residues, standard side-chain placement algorithms are run in the next stage of the designing process. Usually several iterations of sequence design and gradient-based minimization of the designed sequence are carried out. During this stage, the original theozyme residues are not allowed to change their identities anymore, and the energy function is augmented with terms that ensure that the theozyme residues stay in their desired theozyme geometry.[35] If stochastic side-chain placement algorithms are used, this stage is usually carried out several dozen times for each match. Depending on the initial number of matches, several 100−10 000 design models are generated. The models are then usually filtered and ranked according to several criteria reflecting the degree of realization of the ideal theozyme geometry, the affinity of the designed site for the new ligand, and the structural integrity of the scaffold.

The current strategy of designing a novel catalyst by placing a theozyme into a known protein structure is rooted in two observations. First, many enzymological studies have shown that the catalytic prowess of natural enzymes is usually caused by a small number of catalytic residues, while many other residues that are also found in the active site play a far less important role in catalysis.[46,47] Knocking out the catalytic residues, i.e. replacing them with chemically inert alanine, will have a huge effect on efficiency of the enzyme, while changing other binding site residues to Ala will often have no influence on the activity at all. One conclusion from this observation is that to design an active site, it is most important to properly position a handful of key catalytic residues. Second, computational protein design as a technique is still far from perfect, and, despite Kuhlman et al.'s de novo fold design work,[48] it is still unclear whether current algorithms and force fields are reliably capable of designing sequences that fold into a given protein architecture or fold. Further, as of yet there are no algorithms that would design a reasonable protein backbone conformation starting from a theozyme only. Placing a theozyme into an already known structure and then designing only the surrounding site, which usually means mutating less than 10% of the scaffold residues, circumvents both of these problems. Since the backbone structure is given by the scaffold there is no need to come up with a new one, and since the final designed sequence is highly homologous to the wild-type scaffold sequence, one can reasonably assume that the designed protein will be expressed and folded.

Several examples of successful de novo computational enzyme design have been accomplished in recent years. These include two reports[49,50] of designed catalysts for the Kemp elimination, a well-studied model reaction, one set of proteins catalyzing a retro-aldol reaction through a similar mechanism as natural aldolases,[51] and one study reporting the design of a 'Diels-alderase'[52] catalyzing the Diels-Alder reaction, which is a carbon−carbon bond forming reaction for which no natural enzymes exist. The Kemp elimination is an isoxazole ring opening reaction that can be initiated by deprotonation of the carbon adjacent to the azole nitrogen, and the substrate used in both studies featured a benzyl ring adjacent to the isoxazole moiety. In the transition state, negative charge accumulates on the isoxazole oxygen. Thus, for the Kemp eliminases, in both studies the theozyme consisted of three elements: a base (either a Glu/Asp or His) to deprotonate the relevant carbon; a hydrogen-bond donor to stabilize the isoxazole oxygen; and an aromatic side-chain (Phe/Tyr/Trp) to stack against the substrate's phenyl group and thus aid with binding the substrate. In the work by Roethlisberer et al.,[49] a scaffold library was then scanned for matches to this theozyme, and after designing the sequence and ranking the resulting matches, 59 designs were selected for expression. Eight of these designs showed detectable activity, and for each of the designs, replacing the catalytic base by a nonprotic residue resulted in a significant decrease of the catalytic activity. In the second study by Privett et al.[50] a slightly different approach was followed. Instead of searching through a number of scaffolds, the authors identified an aspartate in a hydrophobic cleft of a xylanase, and chose this residue to be the catalytic base. The substrate was placed in a position to interact with this residue, and the surrounding pocket was then designed to accommodate the substrate in this position. The resulting design had activity comparable to the designs reported by Roethlisberger et al.

Jiang et al.[51] succeeded in designing proteins catalyzing a retro-aldol reaction, which is a multistep transformation capable of breaking carbon—carbon bonds. Several natural enzymes exist that accelerate the complementary reverse reaction. The most critical active site element of these natural aldolases is a strategically placed lysine that carries out a nucleophilic attack on a ketone moiety of the substrate to form a Schiff-base covalent adduct and thus initiate the reaction. Several other protic residues act in concert with the lysine to carry out a number of proton shuffling steps necessary to complete the reaction. Consequently, Jiang et al. picked this lysine as the central element of the theozyme used for the retroaldolase designs. The authors used three variations of the theozyme: one featuring the lysine as the only amino acid but with explicit water molecules to carry out the proton transfers; the other two more similar to the natural active sites, with different combinations of protic side-chains supporting the lysine. Out of a total of 72 designs that were experimentally characterized, 32 showed activity, albeit at far lower efficiencies than natural enzymes. Somewhat surprisingly, designs based on the first, simple theozyme showed a higher rate of success than those featuring the more complicated theozyme.

Siegel et al. presented the first example of an enzyme catalyzing a nonnatural reaction, in designing the first biocatalyst for a Diels-Alder reaction,[52] which is a cycloaddition reaction that simultaneously forms two carbon—carbon bonds and four stereocenters, and is thus highly useful in organic synthesis. This study demonstrates that, in principle, the catalytic repertoire of proteins is not limited to the types of reactions observed in nature, and thus suggests new possibilities for the biotechnological application of enzymes, such as the incorporation of novel steps into biosynthesis routes. In their study, the authors employed quantum-mechanical methods to compute the relative orientation of the two reacting molecules in the transition state and to devise the placement for protein functional groups to lower the activation barrier of the cycloaddition. Out of a total of 82 designs, two showed low activity, and for one of them, the activity could be improved 100-fold through five point mutations. In addition, the resulting variant showed high selectivity for the designed stereo-configuration of the product.

In summary, a number of impressive breakthroughs have been achieved with computational enzyme design over the last few years, suggesting that this method might play an important role in future synthetic biology applications. What is important to understand however, is that the current computational methods still lack an accurate modeling of the quantum-mechanical aspects of reactivity, and thus computational design of catalytic activity requires an exquisite understanding of the reaction of interest. For a successful project, the user needs to determine which aspects of a designed active site to enforce and which parts to allow CPD to determine. A future avenue of research is the incorporation of more advanced quantum-mechanical modeling protocols into the currently used design codes. Another caveat is that for successful design of catalytic activity, i.e. discrimination between ground and transition states, placement of the active site functional groups with sub-angstrom precision is required, whereas in the design of a binding interface, somewhat more 'wiggle room' is allowed. For these two reasons, de novo enzyme design will remain a challenging endeavor in the foreseeable future.

PROTEIN THERMOSTABILIZATION BY COMPUTATIONAL DESIGN

The earliest application of CPD was the design of mutations to increase the stability of a protein of interest. As side-chain placement algorithms were developed, their ability to predict stabilizing mutations in naturally occurring proteins was used as the first experimental verification of the methodology.[53] It should be noted that, like for other CPD applications discussed so far, an experimental structure of the protein of interest needs to be available in order to stabilize it by CPD. If this is the case, CPD can be used as a valuable tool to create thermostable variants of the protein of interest for synthetic biology applications. There are three main reasons why the problem of protein stabilization is very amenable to CPD, and thus became CPD's first application:

1. Protein stability is correlated with the number of intramolecular interactions occurring in a protein, and thus can be approximated relatively well by the energy function. When comparing structures of related mesophilic and thermophilic proteins, the thermophilic variant will often have tighter packing in the hydrophobic core, as well as more hydrogen bonds and salt bridges on the surface. The simplest strategy to stabilize a protein is thus to introduce as many additional packing and hydrogen bonding interactions as possible. Fortunately, these types of interactions can be relatively well approximated by CPD energy functions through their van der Waals and hydrogen bonding terms.

2. Native proteins tend to be only marginally stable, and therefore there are usually many possible mutations that improve stability. A surprising fact of protein biochemistry is that many natural proteins are only marginally stable, with a ΔG of folding between -5 and -20 kcal/mole. By comparison, a single hydrogen bond can contribute between -1 and -4 kcal/mole to stability,[54] meaning that overall protein stability is often equivalent to only a few interactions. Explanations for this at-first paradoxical observation have been offered in detail elsewhere,[55] but this fact suggests that the stability of most proteins can be increased.

3. Thermostable variants of proteins can be obtained that have virtually unchanged backbone structure compared to their mesophilic counterpart. When comparing variants of a certain enzyme adopted for different temperatures, it is often found that the structures are virtually identical, despite often having only low sequence identity.[56] This perhaps surprising finding has favorable implications for the design of thermostable proteins: after all, if stability (and catalytic activity) at different temperatures can be realized with the same tertiary structure in many cases, then it should also be possible to create thermostable versions of other mesophilic proteins without significantly changing the backbone structure.

These three observations in combination lay out a straight-forward approach to stabilize a protein through computational design: starting from the structure of the to-be-stabilized mesophilic protein, use the side-chain placement algorithm to introduce as many favorable additional intramolecular interactions as possible. The backbone can stay fixed throughout the calculation. The presumption that the starting protein is only marginally stable implies that potentially a number of different mutations beneficial for stability can be made, and CPD energy functions can usually identify these. And indeed, there are many examples in the literature where this straight-forward approach succeeded in designing variants with improved stability, often with retained functional properties.

In an early example, Malakauskas et al.[57] were able to increase the stability of a model protein by 4.3 kcal/mol and shift the T_M by more than 20°C while maintaining (albeit reduced) affinity to a binding partner. In this example, seven mutations were introduced that mostly increased hydrophobic packing. In a comprehensive study by Dantas et al.,[58] nine globular proteins were completely redesigned. On average, only 35% of the wild-type sequence was retained in the designed variants. Six of these had a T_M above that of the wild-type protein. Computational thermostabilization of a functional protein was first described by Korkegian et al.,[5] who identified a set of three mutations that increased the T_M of yeast cytosine deaminase, an enzyme with potential applications in prodrug therapy, by 10°C while retaining the wild-type catalytic efficiency. Recent work by Borgo et al.[59] introduced an improved computational protocol that explicitly focuses on packing defects in the hydrophobic core and identifies mutations that are most likely to fix these defects. Using this protocol, the authors were able to increase stability of one of the unsuccessful cases from Dantas et al.'s test set by 2.3 kcal/mol.

COMPUTATIONAL DESIGN OF (NOVEL) PROTEIN FOLDS

One of the first and longest-standing objectives in CPD is that of full-sequence design for a given backbone conformation. Several breakthroughs in this area have been achieved. However, most of the studies done so far were only concerned with finding a sequence that folds into the target conformation, without regard to any specific function of interest. Therefore this application of CPD is of limited interest for synthetic biology, and we will only cover it briefly.

The approach of designing a novel protein structure can be subdivided into two stages: (1) generating a three-dimensional model of the desired backbone conformation and (2) designing a sequence for that backbone conformation. During stage 1, the protein is often modeled as consisting of a poly-alanine or poly-valine chain, and during stage 2, the earlier-mentioned standard side-chain placement algorithms are used to find a sequence compatible with the new backbone.

Perhaps the earliest major breakthrough in CPD was Mayo's full sequence design for a 28-residue zinc finger.[60] In this study, the authors took the backbone from a previously solved crystal structure as the template for stage 2, and designed a sequence that was 21% identical to the template's original sequence. The de novo designed sequence was shown to fold into the desired structure by NMR. While the authors in this early study did not design a novel fold and bypassed stage 1 by using a known backbone conformation, this study still represents an impressive early demonstration that side-chain placement algorithms are capable of designing a sequence for a whole protein fold. The first example of a completely de novo designed protein fold was presented by Kuhlman et al. in 2003.[48] In this study, the authors first devised a 106-residue protein fold topology not observed in nature. The novel α/β topology, which was devised as a back-of-the-envelope sketch, consisted of a five-strand antiparallel β-sheet packaged against two α-helices, with a hydrophobic core between these three secondary structure elements. Algorithms developed for protein folding[61] were then used to create three-dimensional models of the desired topology (with poly-Val as sequence). Next, an iterative protocol of side-chain placement algorithms and gradient-based minimization was used to design a sequence compatible with the backbone model. The expressed sequence was well-folded and exceptionally stable, and a crystal structure showed close agreement with the design model.

The aforementioned approach developed by Kuhlman et al. represents a general algorithm to design a sequence that folds into a desired fold (assuming the fold is 'designable'), starting from nothing more than a very rough, back-of-the-envelope draft of the fold. It should be noted that other rational/computational approaches to design novel protein structures have been developed, but most of these are only suitable to design a certain class of protein fold and are not as general as Kuhlman et al.'s approach. The most prominent example is probably the work by Harbury, DeGrado and others,[62] who succeeded in designing alpha-helix bundle assemblies not seen in nature. Their approach relies on exploiting a unique and well-understood property of a protein-alpha helix: amphiphatic alpha helices can be designed by assigning alternating hydrophobic and hydrophilic amino acids to the seven positions in the heptad-repeat unit that helices are made up of. These amphiphatic helices can then be designed to assemble into helix bundles by choosing amino acids complementary to each other at the positions of the heptad that form the hydrophobic interface (knobs-into-holes approach). Most recently this technique was used to design an alpha helix bundle consisting of six helices (bundles containing this number of helices are not known in nature) that featured a water-filled pore on the inside.[63]

A common property of de novo designed folds so far is that most of them have fairly compact, convex shapes, and therefore these folds do not have any cavities or pockets that could be used as binding or active sites to mediate a certain function. Designing folds with concave features and cavities that could harbor functional sites is arguably harder, as the fold needs to be stable enough to not collapse around and bury the cavity. Computational algorithms to generate backbone templates that feature cavities and are at the same time designable have yet to be developed.

COMPLEMENTARITY WITH DIRECTED EVOLUTION

Before the onset of the computational methods described in this chapter, the design and engineering of protein function was mainly carried out by directed evolution (DE). Many powerful high-throughput screening techniques (HTS) have been developed to address the various engineering challenges, such as a number of display strategies for evolving protein–protein interactions (e.g. phage-, yeast-, *E. coli*-, and ribosome-, or mRNA-display) and growth selections, microfluidic/in vitro compartmentalization systems, or plate screens for the evolution of new catalysts. The details and relative advantages of these different approaches have been comprehensively reviewed elsewhere.[64] Computational approaches and DE are by no means mutually exclusive. Since each of these two methods offers its own unique advantages but also has shortcomings, they are perfectly complementary, and we anticipate that these methods will usually be applied hand-in-hand in future design efforts. This prognosis is supported by the fact that many of the designed proteins described in this chapter have been optimized by DE.

Perhaps currently the biggest shortcoming of CPD methods is the inaccuracy of the energy function, both in terms of estimating the precise energetics of a candidate-designed sequence and in terms of translating this estimated value into a precise estimate for the functional parameter of interest, i.e. a reaction rate or a dissociation constant. However, energy functions are certainly accurate enough to identify the usually extremely small subset of sequence space that is considered compatible with a function of interest, and, as described earlier in this chapter, current algorithms are fast enough to search through sequence spaces of size 10^{130} and above.

The sequence throughput of DE methods can approach 10^{12} for some of the in vitro display techniques, $\sim 10^8$ for in vivo display approaches and growth selections, and 10^4 for plate screening methods, and is thus far below that of CPD. Considering that most active or binding sites are comprised of one or two dozen residues (i.e. a sequence space of $\sim 10^{26}$ for a 20-residue site), no currently available HTS method can cover more than a tiny subset of possible sequences in a functional design problem. In addition, experimental work is usually much more laborious and resource-intensive than computation. However, DE allows for the selection of the best variant from the sequence pool based on the actual activity of interest instead of based on an (inaccurate) energy function value.

Thus, an ideal way to combine CPD and DE (assuming that an HTS assay is available) is to use computation to come up with initial designs, and then use those initial designs that show measurable activity as a foothold in sequence space and starting point for successive rounds of DE. This strategy was used for several of the cases described here, and the reported successes could not have been achieved with either CPD or DE alone. In Fleishman et al.'s[24] influenza binder, the affinity of the initial computational designs was increased 10-fold. In Khare et al.'s[4] designed organophosphate hydrolase, the best variant after DE was four orders of magnitude more active than the original design, but to obtain activity by DE alone, a very large library containing all possible quadruple mutations of the scaffold's active site would have to be screened. In Azoitei et al.'s two-loop graft,[32] the authors directly compared a computation-informed library to a naïve library, and were able to isolate much tighter binders from the former. Further, in the future, with both advanced DNA synthesis and deep sequencing technologies becoming more readily available, we expect that approaches like the one presented by Lippow et al.,[3] where computation is used to design a library instead of a single sequence, will become more widely used.

CONCLUSION AND OUTLOOK

The recent advances in computational protein design highlighted in this chapter suggest the likely areas in which synthetic biology could be impacted by CPD most immediately. Generally, CPD should have an advantage over directed evolution in cases where the sequence space related to the design problem is too large for library construction and/or there is no high-throughput assay. Of the different available techniques discussed here, the redesign of protein–protein interactions is currently the most robust, and thus most readily applicable. This puts within reach the redesign of cell-signaling pathways (as presented by Kapp et al.[19]) or improved protein therapeutics that bind their targets with picomolar affinity (as presented by Lippow et al.[16]).

De novo design of binders against a given epitope on a target of interest, while shown to work in several examples, is probably still a few years away from being routinely used for practical applications and arbitrary targets. One problem in this area that is not yet solved is, in cases where no existing binder is known, how to identify 'bindable' epitopes on the target of interest and corresponding productive 'anchoring' points on the new binder. For example, if the hotspot-design strategy is used, it is not yet clear how best to identify potential hotspot residues if only a structure of the target in its apo form is known. However, in case a crystal structure of the target in complex with an already existing binder is known, but that binder happens to have undesirable biochemical properties (i.e. size, immunogenicity), the existing de novo design algorithms can be used out of the box to try to design a novel binder starting from a more favorable scaffold. Specificity redesign of enzymes, i.e. for bioremediation of pollutants or modification of existing metabolic pathways, should also be achievable with the currently available methods. A further complication for pathway engineering lies in the fact that, in the pathway of interest, possibly all enzymes downstream of the first redesigned enzyme would also have to be redesigned to transform the novel metabolite(s). De novo design of enzymes, being perhaps the hardest problem in CPD, will likely require several more years of research effort before it can reliably be applied for practical applications, and should be deployed in combination with directed evolution to increase the (probably low) activity of the initial designs.

References

1. Purnick PEM, Weiss R. The second wave of synthetic biology: from modules to systems. *Nat Rev Mol Cell Biol.* 2009;10:410−422.

2. Sievers SA, Karanicolas J, Chang HW, et al. Structure-based design of non-natural amino-acid inhibitors of amyloid fibril formation. *Nature.* 2011;475:96−100.

3. Lippow SM, Moon TS, Basu S, et al. Engineering enzyme specificity using computational design of a defined-sequence library. *Chem Biol.* 2010;17:1306−1315.

4. Khare SD, Kipnis Y, Greisen PJ, et al. Computational redesign of a mononuclear zinc metalloenzyme for organophosphate hydrolysis. *Nat Chem Biol.* 2012;8:294−300.

5. Korkegian A, Black ME, Baker D, Stoddard BL. Computational thermostabilization of an enzyme. *Science.* 2005;308:857−860.

6. Russ WP, Lowery DM, Mishra P, Yaffe MB, Ranganathan R. Natural-like function in artificial WW domains. *Nature.* 2005;437:579−583.

7. Leach AR. *Molecular Modelling: Principles and Applications.* Englewood Cliffs, NJ: Prentice-Hall; 2001.

8. Boas FE, Harbury PB. Potential energy functions for protein design. *Curr Opin Struct Biol.* 2007;17:199−204.

9. Pabo C. Molecular technology. Designing proteins and peptides. *Nature.* 1983;301:200.

10. Dunbrack RL. Rotamer libraries in the 21st century. *Curr Opin Struct Biol.* 2002;12:431−440.

11. Voigt CA, Gordon DB, Mayo SL. Trading accuracy for speed: a quantitative comparison of search algorithms in protein sequence design. *J Mol Biol.* 2000;299:789−803.

12. Holm L, Sander C. Fast and simple Monte Carlo algorithm for side chain optimization in proteins: application to model building by homology. *Proteins: Struct Funct Genet*. 1992;14:213−223.

13. Desmet J, Spriet J, Lasters I. Fast and accurate side-chain topology and energy refinement (FASTER) as a new method for protein structure optimization. *Proteins: Struct Funct Bioinf*. 2002;48:31−43.

14. Janin J. Protein−protein docking tested in blind predictions: the CAPRI experiment. *Mol Biosyst*. 2010;6:2351−2362.

15. Roberts KE, Cushing PR, Boisguerin P, Madden DR, Donald BR. Computational design of a PDZ domain peptide inhibitor that rescues CFTR activity. *PLoSComput Biol*. 2012;8(4):e1002477.

16. Lippow SM, Wittrup KD, Tidor B. Computational design of antibody-affinity improvement beyond in vivo maturation. *Nat Biotechnol*. 2007;25:1171−1176.

17. Haidar JN, Pierce B, Yu Y, Tong W, Li M, Weng Z. Structure-based design of a T-cell receptor leads to nearly 100-fold improvement in binding affinity for pepMHC. *Proteins*. 2009;74:948−960.

18. Leaver-Fay A, Jacak R, Stranges PB, Kuhlman B. A generic program for multistate protein design. *PLoS ONE*. 2011;6(7):e20937.

19. Kapp GT, Liu S, Stein A, et al. Control of protein signaling using a computationally designed GTPase/GEF orthogonal pair. *Proc Natl Acad Sci USA*. 2012;109:5277−5282.

20. Grigoryan G, Reinke AW, Keating AE. Design of protein-interaction specificity gives selective bZIP-binding peptides. *Nature*. 2009;458:859−864.

21. Yosef E, Politi R, Choi MH, Shifman JM. Computational design of calmodulin mutants with up to 900-fold increase in binding specificity. *J Mol Biol*. 2009;385:1470−1480.

22. Karanicolas J, Kuhlman B. Computational design of affinity and specificity at protein−protein interfaces. *Curr Opin Struct Biol*. 2009;19:458−463.

23. Fleishman SJ, Corn JE, Strauch E-M, Whitehead TA, Karanicolas J, Baker D. Hotspot-centric de novo design of protein binders. *J Mol Biol*. 2011;413:1047−1062.

24. Fleishman SJ, Whitehead TA, Ekiert DC, et al. Computational design of proteins targeting the conserved stem region of influenza hemagglutinin. *Science*. 2011;332:816−821.

25. Bogan AA, Thorn KS. Anatomy of hot spots in protein interfaces. *J Mol Biol*. 1998;280:1−9.

26. Davis IW, Baker D. RosettaLigand docking with full ligand and receptor flexibility. *J Mol Biol*. 2009;385:381−392.

27. Jha RK, Leaver-Fay A, Yin S, et al. Computational design of a PAK1 binding protein. *J Mol Biol*. 2010;400:257−270.

28. Stranges PB, Machius M, Miley MJ, et al. Computational design of a symmetric homodimer using β-strand assembly. *Proc Natl Acad Sci USA*. 2011;108:20562−20567.

29. Der BS, Machius M, Miley MJ, Mills JL, Szyperski T, Kuhlman B. Metal-mediated affinity and orientation specificity in a computationally designed protein homodimer. *J Am Chem Soc*. 2012;134:375−385.

30. Salgado EN, Ambroggio XI, Brodin JD, Lewis RA, Kuhlman B, Tezcan FA. Metal templated design of protein interfaces. *Proc Natl Acad Sci USA*. 2010;107:1827−1832.

31. Sia SK, Kim PS. Protein grafting of an HIV-1-inhibiting epitope. *Proc Natl Acad Sci USA*. 2003;100:9756−9761.

32. Azoitei ML, Correia BE, Ban Y-EA, et al. Computation-guided backbone grafting of a discontinuous motif onto a protein scaffold. *Science*. 2011;334:373−376.

33. Garcia-Viloca M, Gao J, Karplus M, Truhlar DG. How enzymes work: analysis by modern rate theory and computer simulations. *Science*. 2004;303:186−195.

34. Bhabha G, Lee J, Ekiert DC, et al. A dynamic knockout reveals that conformational fluctuations influence the chemical step of enzyme catalysis. *Science*. 2011;332:234−238.

35. Richter F, Leaver-Fay A, Khare SD, Bjelic S, Baker D. De novo enzyme design using rosetta3. *PLoS ONE*. 2011;6(5):e19230.

36. Chen C-Y, Georgiev I, Anderson AC, Donald BR. Computational structure-based redesign of enzyme activity. *Proc Natl Acad Sci USA*. 2009;106:3764−3769.

37. Ashworth J, Havranek JJ, Duarte CM, et al. Computational redesign of endonuclease DNA binding and cleavage specificity. *Nature*. 2006;441:656−659.

38. Gao H, Smith J, Yang M, et al. Heritable targeted mutagenesis in maize using a designed endonuclease. *Plant J*. 2010;61:176−187.

39. Windbichler N, Menichelli M, Papathanos PA, et al. A synthetic homing endonuclease-based gene drive system in the human malaria mosquito. *Nature*. 2011;473:212−215.

40. Murphy PM, Bolduc JM, Gallaher JL, Stoddard BL, Baker D. Alteration of enzyme specificity by computational loop remodeling and design. *Proc Natl Acad Sci USA*. 2009;106:9215−9220.

41. Tantillo DJ, Chen J, Houk KN. Theozymes and compuzymes: theoretical models for biological catalysis. *Curr Opin Chem Biol.* 1998;2:743–750.

42. Zhang X, DeChancie J, Gunaydin H, et al. Quantum mechanical design of enzyme active sites. *J Org Chem.* 2008;73:889–899.

43. Hellinga HW, Richards FM. Construction of new ligand binding sites in proteins of known structure. I. Computer-aided modeling of sites with pre-defined geometry. *J Mol Biol.* 1991;222:763–785.

44. Zanghellini A, Jiang L, Wollacott AM, et al. New algorithms and an in silico benchmark for computational enzyme design. *Protein Sci.* 2006;15:2785–2794.

45. Malisi C, Kohlbacher O, Höcker B. Automated scaffold selection for enzyme design. *Proteins.* 2009;77:74–83.

46. Miller BG. The mutability of enzyme active-site shape determinants. *Protein Sci.* 2007;16(9):1965–1968.

47. Bartlett GJ, Porter CT, Borkakoti N, Thornton JM. Analysis of catalytic residues in enzyme active sites. *J Mol Biol.* 2002;324:105–121.

48. Kuhlman B, Dantas G, Ireton GC, Varani G, Stoddard BL, Baker D. Design of a novel globular protein fold with atomic-level accuracy. *Science.* 2003;302:1364–1368.

49. Röthlisberger D, Khersonsky O, Wollacott AM, et al. Kemp elimination catalysts by computational enzyme design. *Nature.* 2008;453:190–195.

50. Privett HK, Kiss G, Lee TM, et al. Iterative approach to computational enzyme design. *Proc Natl Acad Sci USA.* 2012;109:3790–3795.

51. Jiang L, Althoff EA, Clemente FR, et al. De novo computational design of retro-aldol enzymes. *Science.* 2008;319:1387–1391.

52. Siegel JB, Zanghellini A, Lovick HM, et al. Computational design of an enzyme catalyst for a stereoselective bimolecular Diels-Alder reaction. *Science.* 2010;329:309–313.

53. Dahiyat BI. In silico design for protein stabilization. *Curr Opin Biotechnol.* 1999;10:387–390.

54. Gao J, Bosco DA, Powers ET, Kelly JW. Localized thermodynamic coupling between hydrogen bonding and microenvironment polarity substantially stabilizes proteins. *Nat Struct Mol Biol.* 2009;16:684–690.

55. Wintrode PL, Arnold FH. *Temperature adaptation of enzymes: lessons from laboratory evolution. Evolutionary Protein Design.* San Diego, CA: Academic Press; 2001 [cited 2012 Jan 30]. pp. 161–225.

56. Russell RJ, Gerike U, Danson MJ, Hough DW, Taylor GL. Structural adaptations of the cold-active citrate synthase from an Antarctic bacterium. *Structure.* 1998;6:351–361.

57. Malakauskas SM, Mayo SL. Design, structure and stability of a hyperthermophilic protein variant. *Nat Struct Mol Biol.* 1998;5:470–475.

58. Dantas G, Kuhlman B, Callender D, Wong M, Baker D. A large scale test of computational protein design: folding and stability of nine completely redesigned globular proteins. *J Mol Biol.* 2003;332:449–460.

59. Borgo B, Havranek JJ. Automated selection of stabilizing mutations in designed and natural proteins. *Proc Natl Acad Sci USA.* 2012;109:1494–1499.

60. Dahiyat BI, Mayo SL. De novo protein design: fully automated sequence selection. *Science.* 1997;278:82–87.

61. Simons KT, Kooperberg C, Huang E, Baker D. Assembly of protein tertiary structures from fragments with similar local sequences using simulated annealing and Bayesian scoring functions. *J Mol Biol.* 1997;268:209–225.

62. Hill RB, Raleigh DP, Lombardi A, DeGrado WF. De novo design of helical bundles as models for understanding protein folding and function. *Acc Chem Res.* 2000;33:745–754.

63. Zaccai NR, Chi B, Thomson AR, et al. A de novo peptide hexamer with a mutable channel. *Nat Chem Biol.* 2011;7:935–941.

64. Jäckel C, Kast P, Hilvert D. Protein design by directed evolution. *Annu Rev Biophys.* 2008;37:153–173.

Computer-Aided Design of Synthetic Biological Constructs with the Synthetic Biology Software Suite

Katherine Volzing, Konstantinos Biliouris, Patrick Smadbeck and Yiannis Kaznessis
University of Minnesota, Minneapolis, MN, USA

123

INTRODUCTION

Robustly controlled biomolecular interactions are the key to maintaining optimal cellular function. DNA−DNA, DNA−protein, and protein−protein interactions are instrumental in governing the underlying intracellular behaviors that define living organisms.[1,2] Over the last 60 years, molecular resolution studies have revealed the inner workings of biomolecular interactions. Specific protein domains and amino acid sequences that drive protein−DNA binding and protein multimerization are now well known for many molecules. In addition, the particular DNA base pairs that are required for transcription factor binding, transcription, and translation initiation and termination have also been identified for numerous operons.[3−8]

This increasingly comprehensive understanding of natural cell function at the molecular level is propelling engineering efforts to utilize the control mechanisms and their downstream targets as molecular biology tools. To date numerous nonnatural, or synthetic, systems have been designed and built experimentally, employing this knowledge of molecular interaction mechanisms. Many of these systems couple an extracellular signal (or input) to a unique intracellular target (or output), as is the case of synthetic receptor kinases.[9−14] Other significant efforts have focused on building systems to study or address previously unmet needs in gene expression regulation. Gene expression can be controlled at the level of transcription,[15,16] mRNA stability,[17] translation,[18] or protein stability.[19,20] While all four of these approaches are very powerful, the following discussion will focus on the development of synthetic systems that regulate gene transcription.

Two pioneering systems constructed to achieve novel gene expression control are the bistable switch and the repressillator.[21,22] Both systems are based on synthetic promoter

Synthetic Biology. DOI: http://dx.doi.org/10.1016/B978-0-12-394430-6.00007-8

sequences constructed from operator sites, or parts, of well-characterized naturally occurring *E. coli* promoters and their complementary regulatory proteins. Over the last decade many additional systems have been designed and built based on a similar methodology.[16,23–35] For example, one family of systems, the AND gates, is a set of synthetic promoters composed of multiple operator sites. The sites are organized in such a way that the resulting promoters follow AND logic in response to inducer molecules.[16]

Although our understanding of synthetic biological systems is growing, how to optimally construct and tune them still remains challenging. A major challenge that researchers face throughout the development of new synthetic systems is finding which parts are most suitable for constructing a particular system with a specific, targeted phenotypic behavior.[36] The complexity of synthetic constructs may quickly overwhelm the intuition of even the most seasoned of practitioners.

Mathematical modeling has been widely used in explaining the behavior of various synthetic systems, and can contribute significantly to their full characterization. As the interactions and dynamics governing the biological systems are significantly nonlinear, and therefore arduous to be predicted experimentally, building such systems benefits from modeling. Mathematical modeling of synthetic biological systems has therefore become a necessary ally of experimental work.[37–41] Remarkable examples of synthetic systems that have been computationally described include protein production in an oscillatory or bistable manner,[21,22,42,43] tight regulation of gene expression,[15,44] and logical gate-like behavior.[16,45] Importantly, we are now transitioning from relatively simple biological constructs to complex synthetic biological systems which integrate several simple systems and are capable of robustly performing multiple user-defined functions.[46] These functions have also been captured using mathematical modeling and include, but are not limited to, cells that efficiently synchronize with one another[47] and cells that kill or rescue each other.[48–50] Synthetic biology technologies resulting from these systems have culminated in important applications in producing biofuels with high efficiency, in targeting and destroying cancer cells, in preventing and treating infections, and in developing inexpensive vaccines.[51–53]

To date, a gamut of software products have been developed that allow synthetic biologists to model the behavior of biological systems of interest. These packages include, but are not limited to, SynBioSS,[54,55] CellDesigner,[56] GenNetDes,[57] COPASI,[58] TinkerCell,[59] and others.[60,61] Each of the aforementioned products has its own limitations and advantages. In what follows, we discuss the Synthetic Biology Software Suite (SynBioSS) and its various components. We first revisit two sets of experimentally constructed synthetic biological switches to illustrate the usefulness of SynBioSS and its significance in designing and characterizing synthetic biological systems. The first is a set of AND gates and the second is a set of inducible activators. Combined, these systems offer a wide spectrum of regulating devices: a repressor; a derepressor; an activator; and a deactivator. All these result in varying, well-characterized protein expression amounts.

SYNTHETIC LOGICAL-AND GATES AND PROTEIN DEVICES
AND Gates

The AND gate systems were built using components of the lactose and tetracycline operons, and then tested both experimentally and computationally.[16] Inducible synthetic promoters were constructed by placing the operator sites *lacO* and *tetO* downstream, between and upstream of the *E. coli* −35 and −10 consensus sequences. Three operator sites were included for each promoter, one *lacO* and two *tetO*, or two *lacO* and one *tetO*. A strong ribosomal binding site (RBS) and a transcriptional start site were included to complete the functional promoter. This construction, as illustrated in Figure 7.1, provides the AND gate

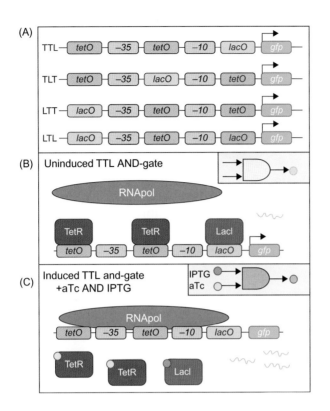

FIGURE 7.1

AND gates system design. (A) Illustration of the TTL, TLT, LTT, and LTL AND gates controlling *gfp* expression. The *tetO* and *lacO* operator sites are located upstream of the −35, between the −35 and −10, and downstream of the −10 consensus sequences. (B) Uninduced TTL AND gate. In the absence of inducers (IPTG and aTc), both TetR and LacI bind their operators (*tetO* and *lacO*) and prevent *gfp* transcription by RNApol. (C) Induced TTL AND gate. In the presence of IPTG and aTc, inducer-bound TetR and LacI proteins dissociate from their operators. RNApol can access the −35 and −10 binding sites of the free promoter and transcribe downstream *gfp*.

character of inducible gene expression. Endogenous expression of the LacI and TetR repressor proteins was exploited for transcription repression (via binding their complementary operator sequences *lacO* and *tetO*, respectively) in the absence of inducer molecules. In the presence of the appropriate inducers, LacI and TetR undergo a conformational change and dissociate from their operator sequences. While expression from the lactose operon is induced by Isopropyl β-D-1-thiogalactopyranoside (IPTG), and expression from the tetracycline operon is induced by anhydrotetracycline (aTc), the character of these synthetic promoters allows transcription to occur only in the presence of both inducers, IPTG and aTc. In the presence of only one inducer, only one type of repressor protein dissociates from the promoters and, therefore, gene expression is still repressed by the second repressor type. However, in the presence of both inducers, IPTG binds LacI while aTc binds TetR. Both interactions result in a conformational change driving the repressors to dissociate from their complementary operators. Upon the release of LacI and aTc, gene transcription can occur.

Using these promoters, AND-logic gene expression control is achieved. While none of the AND gates has a perfectly digital behavior, the promoters containing two *tetO* operator sites (and one *lacO*) do produce an output signal, GFP expression, that increases in strength with increasing aTc and IPTG concentrations. The robustness of the AND gate systems is largely dependent upon both the placement of the operator sites relative to the −35 and −10 sequences, as well as the ratio of *tetO* to *lacO* sites. The results shown in Figure 7.2 illustrate these dependencies.

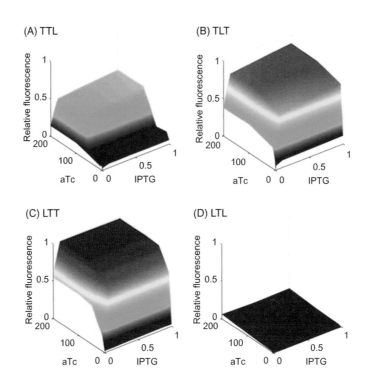

FIGURE 7.2

AND gate behavior. (A) TTL AND gate. AND-logic behavior is observed for the TTL AND gate as expression only occurs with aTc and IPTG. (B, C) TLT and LTT AND gates. AND-logic is less apparent as the *lacO* operator site is moved farther upstream in the synthetic promoter. Background expression is high even in the absence of IPTG. (D) LTL AND gate. Very low induction is observed from promoters with two *lacO* operator sites even with high aTc and IPTG concentrations.

It has been previously demonstrated that transcription regulation by TetR:*tetO* is tight, while regulation by LacI:*lacO* is significantly leakier in the absence of inducer molecules. Additionally, regulation by LacI:*lacO* has been shown to be largely affected by its location within the promoter.[35] Unfortunately, the AND gates with two *lacO* operator sites were minimally inducible even at high IPTG concentrations. This was an artifact of very high intracellular LacI concentrations in the experimental systems (*E. coli*). However, the promoters with two *tetO* operator sites showed informative GFP expression dynamics. These promoters allowed very little background GFP expression in the absence of aTc, while they did exhibit a range of background GFP in the absence of IPTG. These behaviors capture this leakiness and position dependence of LacI:*lacO*. Taken together, the behaviors of the AND gates further support previous observations which proposed that the location of *lacO* significantly influences the promoters' leakiness, while *tetO* robustly regulates expression from multiple locations.[35] These characteristics highlight the modularity limitations of biomolecules and parts of biomolecules. However, despite their limitations, the AND gates are applicable new molecular biology tools with which one can regulate target gene expression given two input signals. They have also provided the field with new insights regarding the modularity of biomolecules and part of biomolecules. As mentioned above, the dynamics of the AND gates were also characterized using computer simulations. These simulations are briefly discussed in the next section.

proTeOn and proTeOff

The proTeOn and proTeOff systems were also built to control target gene expression in *E. coli*. These systems were built and characterized both experimentally and computationally.[15] In contrast to the AND gates, both the promoter sequence and the protein molecules of proTeOn and proTeOff were engineered. Additionally, for both systems, gene expression is up-regulated by the synthetic PROTEON and PROTEOFF transactivators, rather than controlled by derepression upon induction. Also, unlike the TETON protein of the original TetOn system, multiple domains of the PROTEON and PROTEOFF proteins must bind to their complementary operator sequences of the synthetic promoter prior to recruiting RNA polymerase (RNApol) and activating transcription. With these added degrees of complexity, molecular dynamic (MD) simulations were employed to guide the experimental construction of both systems. First, the structures of the synthetic promoter and the synthetic proteins were optimized and their geometrical constraints were satisfied in the simulations. Next, the two systems were built experimentally and tested. Operator sites from the tetracycline operon and the lux operon were integrated along with −35 and −10 RNApol binding sites, a transcriptional start site, an mRNA enhancer sequence, and a strong RBS to construct the synthetic promoter as shown in Figure 7.3. In parallel, the PROTEON and PROTEOFF proteins were built by linking the inducible DNA-binding protein rTetR or TetR to the transactivator domain of LuxR with a flexible peptide linker. The association between the synthetic promoter and PROTEON is also illustrated in Figure 7.3.

aTc-dependent gene expression control is achieved with both the proTeOn and proTeOff systems as depicted in Figure 7.4. For proTeOn, in the absence of aTc, the inducible DNA binding domain (rTetR) does not bind its operator (*tetO*)and the PROTEON protein does not up-regulate the target gene. Upon activation with aTc, rTetR binds the inducer, undergoes a conformational change, and binds *tetO*. This binding brings the LuxR transactivator domain near its operator site allowing it to bind the *luxbox* and up-regulate gene transcription through RNApol recruitment to the promoter.

In contrast, for proTeOff, in the absence of aTc, its inducible DNA binding domain (TetR) binds *tetO*. This binding brings its LuxR transactivator domain close to its operator site, permitting it to bind *luxbox* and up-regulate transcription through RNApol recruitment. After

127

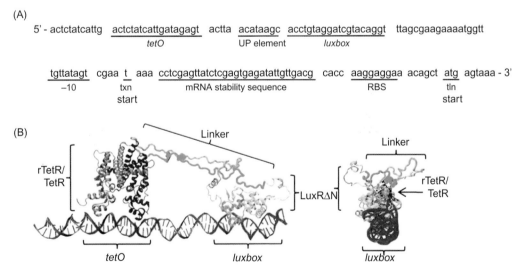

FIGURE 7.3

proTeOn and proTeOff systems' design. (A) Synthetic promoter sequence. Both PROTEON and PROTEOFF bind and recruit RNApol to this synthetic promoter sequence. Moving from 5′ to 3′: the rTetR/TetR protein domain binds *tetO*, RNApol binds the UP element and the −10 region, LuxR binds the *luxbox*, the mRNA stability sequence stabilizes the mRNA transcript, and ribosomes bind the RBS of this resulting mRNA message. (B) proTeOn and proTeOff molecular model. Both proTeOn and proTeOff are designed to assemble as shown. The inducible DNA binding domain (rTetR or TetR in blue) binds the *tetO* operator (purple), and the transcription activator domain (LuxR in orange) binds the *luxbox* (red). The two domains bind their operators along the same face of the DNA double helix and are connected (TetR/rTetR's C-terminus to LuxR's N-terminus) by a linker peptide (green).

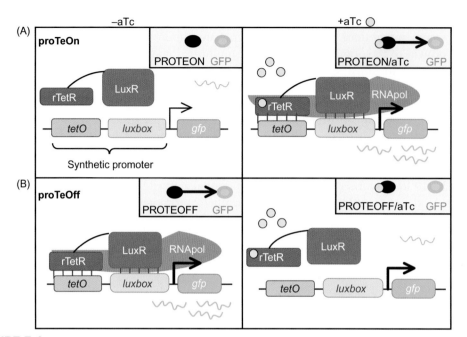

FIGURE 7.4

proTeOn and proTeOff behavior. (A) proTeOn behavior and logic. In the absence of aTc, rTetR does not bind *tetO* and PROTEON does not upregulate *gfp*. Upon activation with aTc, rTetR binds the inducer, undergoes a conformational change and binds *tetO*, bringing LuxR near its operator site to bind *luxbox* and up-regulate *gfp* transcription through RNApol recruitment to the promoter. (B) proTeOff behavior and logic. In the absence of aTc, TetR binds to *tetO*, brings LuxR near its operator site to bind *luxbox* and up-regulate *gfp* transcription through RNApol recruitment. After the addition of aTc, TetR binds to the small molecule, undergoes a conformational change, and releases *tetO*. Upon dissociation of TetR:*tetO*, the LuxR:*luxbox* interaction is destabilized and upregulation by RNApol recruitment terminated.

the addition of aTc, TetR binds the small molecule, undergoes a conformational change and releases *tetO*. Upon dissociation of TetR:*tetO*, the LuxR:*luxbox* interaction is destabilized and up-regulation by RNApol recruitment is terminated.

proTeOn and proTeOff both up-regulate gene expression in the presence and absence of aTc, respectively. Flow cytometry results illustrate the phenotypes of both systems in Figure 7.5. proTeOn achieves steady-state levels of gene overexpression, 20-fold above uninduced samples, approximately 5 hours post-induction with aTc, and maintains them through long periods. Also, the PROTEON protein is rapidly transcribed, translated, folded, and becomes functional when under the control of a free promoter.[15] Given these characteristics, applications of proTeOn would be appropriate in systems when quick bursts or long-term gene up-regulation is desired. On the other hand, proTeOff readily activates target gene expression in the absence of aTc, whereas upon administration of the latter, up-regulation is significantly reduced. This decrease is achieved just 1 hour post-treatment with aTc, and expression is maintained at a low level, 10-fold below untreated samples, through long periods. The PROTEOFF protein is also rapidly transcribed, translated, folded, and functional when under the control of a free promoter.[15] Thus, the proTeOff system can be applied to up-regulate gene expression for both short and long periods. Continuous long-term up-regulation can be achieved, as well as up-regulation with intermittent periods of low expression, using a range of aTc concentrations. Taken together, the proTeOn and proTeOff systems are new tools that can be used to robustly control gene expression in prokaryotic systems. As mentioned above, stochastic simulations were performed to capture the behavior of both systems and further characterize the governing interactions. These computational analyses are detailed in the next section.

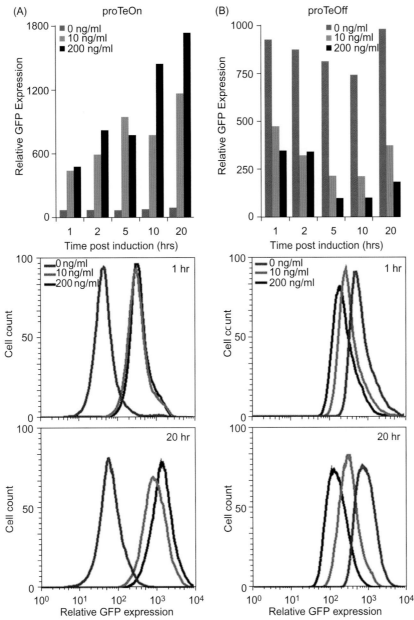

FIGURE 7.5

proTeOn and proTeOff phenotype. Mean GFP expression and expression distribution as analyzed by flow cytometry. (A) Cells expressing PROTEON were induced with 0, 10, and 200 ng/ml aTc. With 10 and 200 ng/ml aTc, proTeOn up-regulates GFP six-fold 1 hour post induction. Steady-state expression is achieved by 5 hours and 10 hours with 10 and 200 ng/ml respectively, and maintained through 20 hours. Maximum overexpression is 10- and 20-fold above uninduced controls with 10 and 200 ng/ml respectively. (B) In the absence of aTc, cells expressing PROTEOFF up-regulate GFP expression. With 10 and 200 ng/ml aTc, expression is reduced to half that of the untreated samples by 1 hour. Steady-state expression is reached by 5 hours with both 10 and 200 ng/ml and maintained through 20 hours post-treatment. Maximum expression with 0 ng/ml aTc is up to 10-fold above the aTc-treated samples.

THE SYNTHETIC BIOLOGY SOFTWARE SUITE

SynBioSS is a software package that endows researchers with the ability to easily and quickly model and simulate the behavior of both naturally occurring and synthetic biological systems, circumventing the need for previous coding experience. It is publicly available at http://synbioss.sourceforge.net/ and consists of three main parts: (1) the SynBioSS Wiki;

129

(2) the SynBioSS Designer; and (3) the SynBioSS Desktop Simulator. Even though the three SynBioSS parts function independently of one another, information with regards to kinetics and biomolecular interactions underlying the biological systems of interest may flow bidirectionally between any two components of SynBioSS. Figure 7.6 illustrates a snapshot of the main window of the three SynBioSS components.

The first element of SynBioSS, SynBioSS Wiki, is a database used for storing information related to biological system components, their interplay, and the corresponding biological information. SynBioSS Wiki is openly available at https://www.synbioss.org/wiki/, and its development is still in progress. Specifically, the user can use SynBioSS Wiki to store information associated with the function of a DNA sequence, the behavior of a protein, and the kinetic parameters and reactants/products of a biochemical reaction. It enables users to not only add new information regarding their synthetic (or natural) systems, but also to

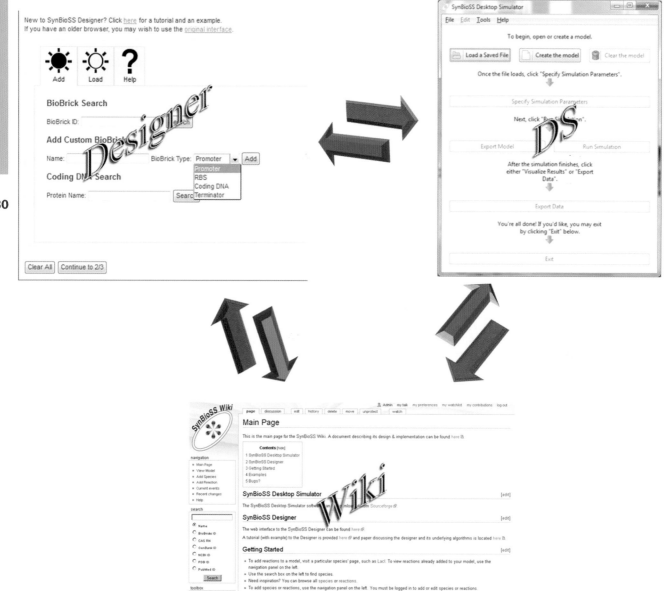

FIGURE 7.6

Screenshots of the three SynBioSS components: SynBioSS Designer; SynBioSS Wiki; and SynBioSS DS. Information regarding the molecular parts and their interactions may flow bidirectionally between any two SynBioSS components.

either edit or use information that has already been added for biological constructs. This information can subsequently be used by the SynBioSS Desktop Simulator (SynBioSS DS) to simulate the behavior of synthetic systems of interest. A more detailed description of SynBioSS Wiki can be found at http://synbioss.sourceforge.net/wp-content/uploads/supplement.pdf.

SynBioSS Designer constitutes the second component of SynBioSS, and is the means for automatically producing a reaction network that captures the interplay between different molecular components of a biosystem, or among different biosystems.[54,55] These molecular components include promoters, operators, repressors, and effector molecules. SynBioSS Designer carries information about these molecular components and is linked to the database of BioBrick Standard Biological Parts. In short, 'BioBricks' are open source molecular biology elements (promoter components, genes, terminators, proteins, etc.) that have been functionally well-characterized and can be used to design and build synthetic biological constructs.[62,63] Once the user has defined the molecular elements from the BioBricks database to use in building their system of interest, SynBioSS Designer implements the available information to produce a reaction network describing the molecular interactions associated with these elements.

With SynBioSS Wiki and Designer, synthetic biologists can now efficiently develop a model describing a synthetic construct by just inputting to SynBioSS Designer the desired molecular parts participating in this construct's function. SynBioSS Designer's output is then a reaction cascade describing the behavior of the synthetic system. The output can be saved in a Systems Biology Markup Language (SBML) format, making the simulation of the network accessible using a software package. Users can then simply load the model into the SynBioSS DS to calculate the evolution of the system's behavior as outlined below. Thus, this software suite greatly simplifies modeling synthetic biological systems.

Even though the SynBioSS Designer package has rendered the development of synthetic biology models quite trivial, it currently still has the following limitations:[55]

1. Numerous molecular parts are not well-defined, and therefore there are no kinetic parameters associated with them. In this case, SynBioSS Designer uses default values increasing the assumptions underlying the model which, in turn, decreases its accuracy.
2. The SynBioSS Designer modeling approach is focused on biology's central dogma neglecting other mechanisms, such as antisense regulation, transcriptional inhibition, and RNA silencing, that may be important for the simulated phenotype. The results could therefore be inaccurate in cases where these mechanisms significantly influence the behavior of the system.
3. The desired molecular parts required for constructing the synthetic system may not be available in the BioBricks database. In this case, SynBioSS Designer cannot produce a reaction network, as it currently only retrieves information from this database.

(More details about the SynBioSS Designer can be found in reference.[55])

As mentioned previously, the third component of SynBioSS, SynBioSS DS, simulates the behavior of the biological systems of interest. Unlike other software packages, SynBioSS DS accounts for the stochasticity which is omnipresent in gene expression.[64–66] It possesses a windows graphical user interface (GUI) where the user can provide all of the biochemical reactions and the conditions describing the synthetic system of interest. The conditions of the system include the cell volume, time interval between the saved points, the mean cell division time, and other fundamental information associated with the model. The algorithm behind the GUI then implements a fusion of a stochastic-discrete and stochastic-continuous approach[67] to calculate the evolution of each species participating in the synthetic system. Subsequently, the results are stored in an. nc or .xls format so that the user can process them.

AND Gate Models

SynBioSS has been shown to play an instrumental role in designing various synthetic biological AND gates.[16] As described above, these AND gates consist of molecular parts from the tetracycline and lactose operons, and switch on gene expression if and only if two inducer molecules, namely aTc and IPTG, are present. See above for details regarding these experimental systems (and reference[16]). As the molecular components of these synthetic systems are well-characterized and found in the BioBricks database, SynBioSS can be used to develop a complete model of the systems (see references [16,55,68]) and can then be used to accurately predict their behavior under a wide range of conditions. Therefore, the simulations can inform the experimental design of different and more complex AND gates and systems containing AND gates. Let us now discuss how SynBioSS can aid in the design and characterization of two synthetic biological switches that have been recently developed.

proTeOn and proTeOff Models

Using the SynBioSS DS, two new biological switches were recently designed, built, and simulated. As discussed in the previous section, these switches, proTeOn and proTeOff, robustly switch the expression of a target gene on and off in prokaryotic cells in an aTc-dependent fashion. Their experimental design and characterization are described in the previous section, and can be found in reference[15]. Briefly, proTeOn up-regulates gene expression from basal to high levels in response to aTc. On the other hand, proTeOff up-regulates gene expression only in the absence of aTc. Basal expression is achieved in the presence of aTc with this system. Even though similar gene expression regulation devices have been designed previously for eukaryotic systems,[28–31] to the author's knowledge this is the first time that two similar prokaryotic switches have been developed. Furthermore, these switches exhibit the exact opposite functionality from one another relative to the presence and absence of aTc. This was accomplished for proTeOn and proTeOff with only a single amino acid difference between the regulatory proteins PROTEON and PROTEOFF.

Pairing the experimental efforts with computation, proTeOn and proTeOff were initially built and tested in vivo, and then the strength of their governing biomolecular interactions were quantified in silico. To do this, a detailed reaction network was developed, capturing the biomolecular interplay between the components comprising proTeOn and proTeOff. The model was simulated stochastically using the SynBioSS DS and validated against the experimental data. The important feature of this model is its high level of detail as it captures most of the biological processes (molecule transport, transcription activation and repression, gene transcription and mRNA translation, and protein folding and degradation).

The model describing the behavior of proTeOn and proTeOff consists of 25 species and 32 reactions. The value of only five key kinetic parameters differentiates the proTeOn from the proTeOff model. In other words, the same model can accurately capture the behavior of both genetic switches by modifying the five unique kinetic parameters (for details see [15]).

The majority of the kinetic parameters used in the model were adopted from the literature. However, as a number of the synthetic molecular parts used were novel, some parameters did not exist previously. These parameters were first estimated as values close to similar systems' parameters and then adjusted such that the computational results matched with the experimental phenotype.

The procedure to simulate the behavior of these two synthetic biological switches is as follows:

1. Open the main SynBioSS DS window.
2. Click on 'Create the Model' to open a new window (in case you have previously developed and saved the model, you can load it by clicking on 'Load a Saved File').

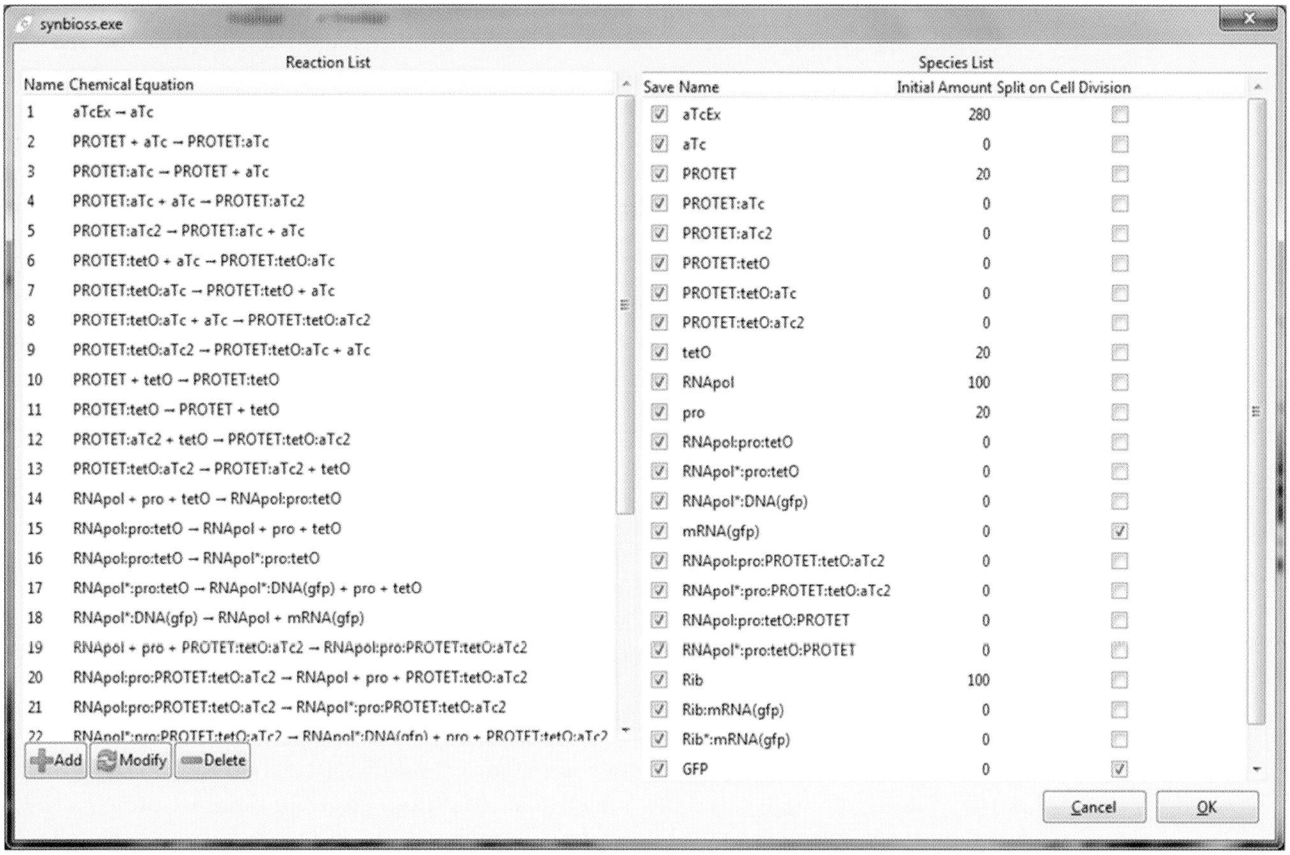

FIGURE 7.7

Screenshot of the SynBioSS DS including the reaction network and corresponding species underlying the proTeOn and proTeOff models.

133

3. In the new window, click on 'Add' to add a reaction, 'Modify' to modify an existing reaction, and 'Delete' to delete an existing reaction. For the purposes of this study, the 32 biochemical reactions that describe the behavior of proTeOn and proTeOff were added to build the representative reaction network. These reactions are shown in Figure 7.7 in the 'Reaction List' column, and account for the process of transcription, translation, protein folding, degradation, and transport. To provide the reaction type (e.g. first order, second order, Michaelis-Menten, etc.) for each reaction, click on 'Rate Law.' In this window, you are also asked to define the reaction parameter(s) and reactant(s)/product(s), along with their corresponding stoichiometric coefficients.

4. Click on the boxes to the left of the species in the 'Save' column (Fig. 7.7) to choose the species you would like to save. Next, click on the boxes to the right of the species in the 'Split on Cell Division' column to choose the species whose amount is halved during cell division. Finally, provide the initial number of molecules for each species in the 'Initial Amount' column. After providing all the conditions of the system, click on 'OK.' In the proTeOn and proTeOff model all the above conditions were set in compliance with the experimental work.

5. The next step is to provide the simulation parameters. Click on 'Specify Simulation Parameters' to open a new window. Here, click on 'Number of Simulation Trajectories to Generate' to provide the number of trials to consider. One hundred thousand trials have typically been used such that the simulation results can be compared with the experimental phenotype of 100 000 cells. Next click on 'End Time (seconds)' to define the simulation time. Seventy-two thousand has been a typical simulation time used to compare

the computational results with the experimental data collected 6 hours post-induction. Finally, by clicking on 'Show Advanced Options' other parameters such as the cell volume, integration step, time between saved points, and division time (if any) can be defined. In many cases, the cell volume has been set equal to 10^{-15} L, as the simulations refer to *E. coli* cells, and the cell division time and standard deviation were set to 30 ± 5 min. More advanced simulation parameters available by clicking on 'Show More Advanced Options' include the type of algorithm used for the simulation, the tolerance, and the population threshold under which continuous approaches are replaced by discrete simulation approaches. For the proTeOn and proTeOff simulation the default values of these parameters were used.

6. Next, click on 'Export Model' to save your model for future implementation. You are provided with the option to save it in .xml or. nc format.

7. Finally, click on 'Run Simulation' to run the simulation.

8. Once your simulation is finished, click on the highlighted 'Export Data' to save the simulation results for post-processing. You are given the option to save the data either as an xls. or .nc file.

Simulating synthetic biological systems stochastically using the SynBioSS DS provides the user with the opportunity to compare computational with experimental results at both the population level and the single-cell level. Thus, the simulations capture the average behavior of the system, as well as the distribution of the protein(s) of interest over the cell population. While deterministic simulations aim to capture mean population behavior as well, they cannot provide information at single-cell resolution.

Figure 7.8 illustrates experimental and computational results of 100 000 cells carrying the proTeOn device. In both cases, the intracellular amount of green fluorescence protein (GFP) was monitored. The upper panel of Figure 7.8 compares the average behavior of the cells, as provided by flow cytometry measurements and SynBioSS DS, at 1, 5, and 10 hours post aTc induction. The lower panel of Figure 7.8 shows computational and experimental results at the single-cell level, the distribution of GFP over 100 000 cells. Overall, both experiments and simulations report that cells carrying proTeOn exhibit tight aTc-dependent gene expression regulation. The higher the inducer concentration is, the greater the increase in gene expression. Interestingly, this protein up-regulation is maintained in the cells for a long period post-induction. As evident, the SynBioSS DS simulation results agree well with the experimental phenotype over different inducer concentrations, and for different post-induction times.

Similarly to proTeOn, the upper panel of Figure 7.9 depicts the simulation and experimental data of the mean GFP value of 100 000 cells possessing the proTeOff system, and the lower panel of Figure 7.9 depicts the distribution of GFP across 100 000 cells. As these four plots show, proTeOff efficiently maintains basal gene expression levels even with low inducer concentrations. However, it robustly up-regulates expression in the absence of inducer. Intriguingly, gene expression drops to low levels within a very short time of aTc administration, and it remains at basal levels for a long period. It should be stressed that the results provided by SynBioSS DS are consistent with the experimental measurements. As discussed previously, the model describing proTeOff differs from the proTeOn model by only the five kinetic parameters that differ between the two experimental systems.

SynBioSS DS is a powerful tool that provides simulation results that accurately describe the experimental phenotype of synthetic biological systems. Demonstrating that the proTeOn and proTeOff models match the in vivo system's behavior well allows for further quantification of the parameters related to the biological constructs, and provides a guide for building more complex synthetic systems from these components. Specifically, the binding strength between the PROTEON and PROTEOFF protein and *tetO* operator can be quantified both when the former is free and when it is bound to aTc. In addition, the

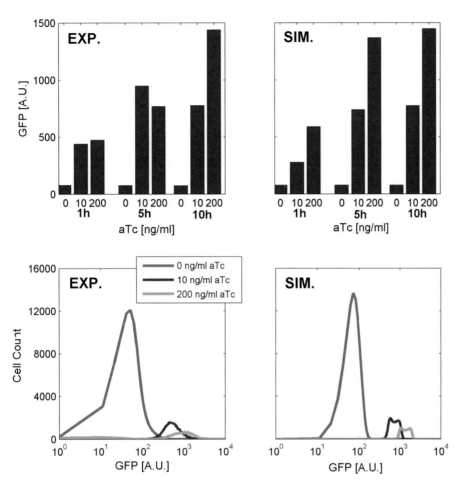

FIGURE 7.8

Behavior of the proTeOn system. Upper panel: Experimental (left) and simulation (right) results of the average cell behavior 1, 5, and 10 hours post-induction with 0, 10, and 200 ng/ml aTc. Lower panel: Experimental (left) and simulation (right) results of the average cell behavior 10 hour post-induction with 0, 10, and 200 ng/ml aTc.

affinity of RNApol for the synthetic promoter can be determined both when the latter is free and when it is bound to PROTEON or PROTEOFF. The corresponding dissociation constants are detailed in Table 7.1.

The simulations demonstrate that the binding strength of RNApol for the synthetic promoter increases 22 times upon binding of PROTEON to *tetO*. In addition, they show that binding of PROTEOFF to *tetO* increases the binding strength of RNApol to DNA by 14 times. On the other hand, upon aTc treatment, the binding strength of PROTEOFF protein to the synthetic promoter is reduced by 14 times. The kinetic parameters applied to obtain these results can now also be added to the database of kinetic information characterizing important biomolecular interactions.

CONCLUSION

We have shown how SynBioSS nicely complements experimental work to accurately describe the behavior of synthetic biological systems and quantify the underlying interactions. Therefore, it can be used to further predict, guide, and test the behaviors of such systems quickly and economically. Integrating this kind of computational tool with

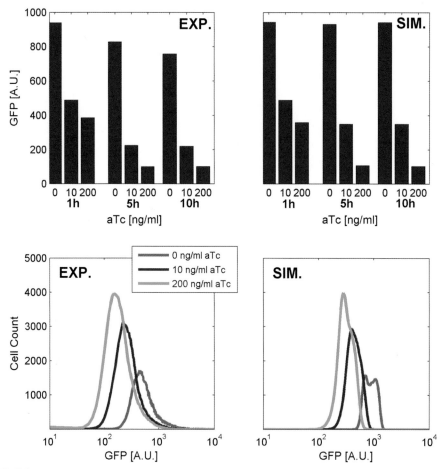

FIGURE 7.9

Behavior of the proTeOff system. Upper panel: Experimental (left) and simulation (right) results of the average cell behavior 1, 5, and 10 hours post-induction with 0, 10, and 200 ng/ml aTc. Lower panel: Experimental (left) and simulation (right) results of the average cell behavior 1 hour post-induction with 0, 10, and 200 ng/ml aTc.

TABLE 7.1 Dissociation Constants of Biomolecular Interactions Underlying proTeOn and proTeOff (Units: Second Order Reactions: M, Third Order Reactions: M^2)[15]

Biomolecular Interactions	Binding Affinity	
	proTeOn	**proTetOff**
PROTET + tetO ↔ PROTET:tetO	$2.5 \cdot 10^{-5}$	$2.5 \cdot 10^{-10}$
PROTET:aTc2 + tetO ↔ PROTET:tetO:aTc2	$2.5 \cdot 10^{-10}$	$2.5 \cdot 10^{-2}$
RNApol + pro + tetO ↔ RNApol:pro:tetO	$3.7 \cdot 10^{-9}$	$3.7 \cdot 10^{-9}$
RNApol + pro + PROTET:tetO ↔ RNApol:pro:tetO:PROTET	$1.67 \cdot 10^{-10}$	$2.56 \cdot 10^{-10}$

experimental work will be increasingly important as synthetic systems become larger and increasingly nonlinear. However, accurate implementation of mathematical models will depend upon the availability of quality kinetic data for the species or parts used. High-quality, cross-disciplinary work between computer science, mathematics, and the biological sciences is now needed to drive continuous progress in this emerging field of synthetic biology.

Acknowledgments

This work was supported by a grant from the National Institutes of Health (American Recovery and Reinvestment Act grant GM086865) and a grant from the National Science Foundation (CBET-0644792). Partial funding for this work was also provided through a grant from the Synthetic Ecology Program of the Biotechnology Institute and from the Presidential Initiative on Biocatalysis at the University of Minnesota.

References

1. Alon U. *An Introduction to Systems Biology: Design Principles of Biological Circuits.* Chapman and Hall London, UK: CRC Press; 2007.

2. Stelling J, Sauer U, Szallasi Z, Doyle III FJ, Doyle J. Robustness of cellular functions. *Cell.* 2004;118 (6):675–685.

3. Hillen W, Berens C. Mechanisms underlying expression of TN10 encoded tetracycline resistance. *Annu Rev Microbiol.* 1994;48(1):345–369.

4. Ramos JL, Martínez-Bueno M, Molina-Henares AJ, et al. The TetR family of transcriptional repressors. *Microbiol Mol Biol Rev.* 2005;69(2):326–356.

5. González JE, Keshavan ND. Messing with bacterial quorum sensing. *Microbiol Mol Biol Rev.* 2006;70 (4):859–875.

6. Qin N, Callahan SM, Dunlap PV, Stevens AM. Analysis of LuxR regulon gene expression during quorum sensing in *vibrio fischeri. J Bacteriol.* 2007;189(11):4127–4134.

7. Wilson C, Zhan II, Swint-Kruse L, Matthews K. The lactose repressor system: paradigms for regulation, allosteric behavior and protein folding. *Cell Mol Life Sci.* 2007;64(1):3–16.

8. Lewis M. The lac repressor. *C R Biol.* 2005;328(6):521–548.

9. Utsumi R, Brissette RE, Rampersaud A, Forst SA, Oosawa K, Inouye M. Activation of bacterial porin gene expression by a chimeric signal transducer in response to aspartate. *Science.* 1989;245 (4923):1246–1249.

10. Feng X, Baumgartner JW, Hazelbauer GL. High- and low-abundance chemoreceptors in escherichia coli: differential activities associated with closely related cytoplasmic domains. *J Bacteriol.* 1997;179 (21):6714–6720.

11. Weerasuriya S, Schneider BM, Manson MD. Chimeric chemoreceptors in *Escherichia coli*: signaling properties of tar-tap and tap-tar hybrids. *J Bacteriol.* 1998;180(4):914–920.

12. Repik A, Rebbapragada A, Johnson MS, Haznedar JÖ, Zhulin IB, Taylor BL. PAS domain residues involved in signal transduction by the Aer redox sensor of *Escherichia coli. Mol Microbiol.* 2000;36(4):806–816.

13. Appleman JA, Chen LL, Stewart V. Probing conservation of HAMP linker structure and signal transduction mechanism through analysis of hybrid sensor kinases. *J Bacteriol.* 2003;185(16):4872–4882.

14. Salis H, Tamsir A, Voigt C. Engineering bacterial signals and sensors. *Contrib Microbiol.* 2009;16:194–225.

15. Volzing K, Biliouris K, Kaznessis YN. proTeOn and proTeOff, new protein devices that inducibly activate bacterial gene expression. *ACS Chem Biol.* 2011;6(10):1107–1116.

16. Ramalingam KI, Tomshine JR, Maynard JA, Kaznessis YN. Forward engineering of synthetic biological AND gates. *Biochem Eng J.* 2009;47(1–3):38–47.

17. Win MN, Smolke CD. A modular and extensible RNA-based gene-regulatory platform for engineering cellular function. *PNAS.* 2007;104(36):14283–14288.

18. Goldfless SJ, Belmont BJ, Paz DMA, Liu JF, Niles JC. Direct and specific chemical control of eukaryotic translation with a synthetic RNA–protein interaction. *Nucleic Acids Res.* 2012;40(9):1–12.

19. Culyba EK, Price JL, Hanson SR, et al. Protein native-state stabilization by placing aromatic side chains in N-glycosylated reverse turns. *Science.* 2011;331(6017):571–575.

20. Popp MW, Dougan SK, Chuang T-Y, Spooner E, Ploegh HL. Sortase-catalyzed transformations that improve the properties of cytokines. *PNAS.* 2011;108(8):3169–3174.

21. Elowitz MB, Leibler S. A synthetic oscillatory network of transcriptional regulators. *Nature.* 2000;403 (6767):335–338.

22. Gardner TS, Cantor CR, Collins JJ. Construction of a genetic toggle switch in *Escherichia coli. Nature.* 2000;403:339–342.

23. Becskei A, Seraphin B, Serrano L. Positive feedback in eukaryotic gene networks: cell differentiation by graded to binary response conversion. *EMBO J.* 2001;20(10):2528–2535.

24. Becskei A, Serrano L. Engineering stability in gene networks by autoregulation. *Nature.* 2000;405 (6786):590–593.

25. Guet CC, Elowitz MB, Hsing W, Leibler S. Combinatorial synthesis of genetic networks. *Science.* 2002;296 (5572):1466−1470.

26. Kobayashi H, Kærn M, Araki M, et al. Programmable cells: interfacing natural and engineered gene networks. *PNAS.* 2004;101(22):8414−8419.

27. Hasty J, McMillen D, Collins JJ. Engineered gene circuits. *Nature.* 2002;420(6912):224−230.

28. Gossen M, Bujard H. Tight control of gene expression in mammalian cells by tetracycline-responsive promoters. *PNAS.* 1992;89(12):5547−5551.

29. Gossen M, Freundlieb S, Bender G, Muller G, Hillen W, Bujard H. Transcriptional activation by tetracyclines in mammalian cells. *Science.* 1995;268(5218):1766−1769.

30. Deuschle U, Meyer WK, Thiesen HJ. Tetracycline-reversible silencing of eukaryotic promoters. *Mol Cell Biol.* 1995;15(4):1907−1914.

31. Baron U, Gossen M, Bujard H. Tetracycline-controlled transcription in eukaryotes: novel transactivators with graded transactivation potential. *Nucleic Acids Res.* 1997;25(14):2723−2729.

32. Urlinger S, Helbl V, Guthmann J, Pook E, Grimm S, Hillen W. The p65 domain from NF-κB is an efficient human activator in the tetracycline-regulatable gene expression system. *Gene.* 2000;247 (1−2):103−110.

33. Akagi K, Kanai M, Saya H, Kozu T, Berns AA. Novel tetracycline-dependent transactivator with E2F4 transcriptional activation domain. *Nucleic Acids Res.* 2001;29(4):e23.

34. Ryu J-R, Olson LK, Arnosti DN. Cell-type specificity of short-range transcriptional repressors. *PNAS.* 2001;98 (23):12960−12965.

35. Lanzer M, Bujard H. Promoters largely determine the efficiency of repressor action. *PNAS.* 1988;85 (23):8973−8977.

36. Ellis T, Wang X, Collins JJ. Diversity-based, model-guided construction of synthetic gene networks with predicted functions. *Nat Biotechnol.* 2009;27(5):465−471.

37. De Jong H. Modeling and simulation of genetic regulatory systems: a literature review. *J Comput Biol.* 2002;9(1):67−103.

38. Hasty J, McMillen D, Isaacs F, Collins JJ. Computational studies of gene regulatory networks: in numero molecular biology. *Nat Rev Genet.* 2001;2(4):268−279.

39. Kaznessis YN. Models for synthetic biology. *BMC Syst Biol.* 2007;1(1):47.

40. Kaznessis Y. Multi-scale models for gene network engineering. *Chem Eng Sci.* 2006;61(3):940−953.

41. Kaznessis Y. Mathematical models in biology: from molecules to life. *Wiley Interdiscip Rev Syst Biol Med.* 2011;3 (3):314−322.

42. Tuttle L, Salis H, Tomshine J, Kaznessis Y. Model-driven designs of an oscillating gene network. *Biophys J.* 2005;89(6):3873−3883.

43. Stricker J, Cookson S, Bennett M, Mather W, Tsimring L, Hasty J. A fast, robust and tunable synthetic gene oscillator. *Nature.* 2008;456(7221):516−519.

44. Beisel C, Bayer T, Ho K, Smolke C. Model-guided design of ligand-regulated RNAi for programmable control of gene expression. *Mol Syst Biol.* 2008;4:224.

45. Anderson J, Voigt C, Arkin A. Environmental signal integration by a modular AND gate. *Mol Syst Biol.* 2007;3:133.

46. Brenner K, You L, Arnold F. Engineering microbial consortia: a new frontier in synthetic biology. *Trends Biotechnol.* 2008;26(9):483−489.

47. Prindle A, Samayoa P, Razinkov I, Danino T, Tsimring L, Hasty J. A sensing array of radically coupled genetic biopixels. *Nature.* 2011;481:39−44.

48. Biliouris K, Babson D, Schmidt-Dannert C, Kaznessis YN. Stochastic simulations of a synthetic bacteria−yeast ecosystem. *BMC Sys Biol.* 2012;6:58.

49. You L, Cox R, Weiss R, Arnold F. Programmed population control by cell−cell communication and regulated killing. *Nature.* 2004;428(6985):868−871.

50. Balagadde F, Song H, Ozaki J, et al. A synthetic *Escherichia coli* predator-prey ecosystem. *Mol Syst Biol.* 2008;4:187.

51. Ruder W, Lu T, Collins J. Synthetic biology moving into the clinic. *Science.* 2011;333(6047):1248−1252.

52. Anderson J, Clarke E, Arkin A, Voigt C. Environmentally controlled invasion of cancer cells by engineered bacteria. *J Mol Biol.* 2006;355(4):619−627.

53. Lee S, Chou H, Ham T, Lee T, Keasling J. Metabolic engineering of microorganisms for biofuels production: from bugs to synthetic biology to fuels. *Curr Opin Biotechnol.* 2008;19(6):556−563.

54. Hill A, Tomshine J, Weeding E, Sotiropoulos V, Kaznessis Y. SynBioSS. The synthetic biology modeling suite. *Bioinformatics.* 2008;24(21):2551−2553.

55. Weeding E, Houle J, Kaznessis Y. SynBioSS designer: a web-based tool for the automated generation of kinetic models for synthetic biological constructs. *Brief Bioinform*. 2010;11(4):394−402.

56. Funahashi A, Morohashi M, Kitano H. CellDesigner: a process diagram editor for gene-regulatory and biochemical networks. *BIOSILICO*. 2003;1:159−162.

57. Rodrigo G, Carrera J, Jaramillo A. Genetdes: automatic design of transcriptional networks. *Bioinformatics*. 2007;23(14):1857−1858.

58. Hoops S, Sahle S, Gauges R, et al. COPASI — a complex pathway simulator. *Bioinformatics*. 2006;22(24):3067−3074.

59. Chandran D, Bergmann FT, Sauro HM. Tinkercell: modular CAD tool for synthetic biology. *J Biol Eng*. 2009;3:19.

60. Marchisio M, Stelling J. Computational design of synthetic gene circuits with composable parts. *Bioinformatics*. 2008;24(17):1903−1910.

61. Kurata H, Masaki K, Sumida Y, Iwasaki R. CADLIVE dynamic simulator: direct link of biochemical networks to dynamic models. *Genome Res*. 2005;15:590−600.

62. Canton B, Labno A, Endy D. Refinement and standardization of synthetic biological parts and devices. *Nat Biotechnol*. 2008;26(7):787−793.

63. Shetty RP, Endy D, Knight Jr TF. Engineering biobrick vectors from biobrick parts. *J Biol Eng*. 2008;2:5.

64. Raser JM, O'Shea EK. Noise in gene expression: origins, consequences, and control. *Science*. 2005;309(5743):2010−2013.

65. Elowitz M, Levine A, Siggia E, Swain P. Stochastic gene expression in a single cell. *Science*. 2002;297(5584):1183.

66. Kaern M, Elston T, Blake W, Collins J. Stochasticity in gene expression. *Nat Rev Genet*. 2005;6:451−464.

67. Salis H, Kaznessis Y. Accurate hybrid stochastic simulation of a system of coupled chemical or biochemical reactions. *J Chem Phys*. 2005;122:054103.

68. Kaznessis Y. SynBioSS-aided design of synthetic biological constructs. *Methods Enzymol*. 2011;498:137.

Computational Methods for Strain Design

**Sang Yup Lee[1,2,3], Seung Bum Sohn[1,2], Yu Bin Kim[1,2], Jae Ho Shin[1], Jin Eyun Kim[1]
and Tae Yong Kim[2]**
[1]Institute for the BioCentury, KAIST, Daejeon, Republic of Korea
[2]Bioinformatics Research Center, KAIST, Daejeon, Republic of Korea
[3]Department of Bio and Brain Engineering and Bioinformatics Research Center, KAIST,
Daejeon, Republic of Korea

INTRODUCTION

To fully harness the synthetic capacity of biological systems, it is necessary to go beyond
natural genetic circuits and pathways for the design of engineered cells.[1-5] Early methods
were designed to assemble simple genetic components into the biological network such that
the cells would perform a distinct function atypical from the wild-type phenotype.[3-5] With
the development towards the construction of large-scale systems with complex functions,
integration of complex synthetic biology components into the cell is critical. However,
despite the increase in knowledge of the functions of the cellular components within these
systems, there is still much that remains to be illuminated, because of the complex nature
of cells, resulting in uncertainties and partial description of the biological network that the
synthetic version was designed to represent. To achieve such a comprehensive integration
of synthetic biology components in a large-scale system, computational methods would
be essential to represent the dynamics that the various components have with each other.[6,7]

Computational approaches have contributed greatly in consolidating and enhancing the
predictive capabilities of synthetic biology for characterizing biological systems and their
components. The use of computational methods allows for the computational modeling
and simulation of biological systems for better understanding of the complexities within
these systems. The results of computational prediction of cellular behaviors are then
validated against actual observations of the system and discrepancies between the model
and the observations result in the refining of the model to account for the differences.
This process of simulation and validation is repeated iteratively until the model is able to
accurately represent the system of interest. With the availability of computational methods,
the analysis of data from observations allows for a more rapid and acceptable approach in
designing synthetic biological networks for different environmental and/or genetic
perturbations.[2,7,8]

An important application of synthetic biology is the development of strategies to
engineering microorganisms for the purpose of producing high-valued compounds.
Metabolic engineering has been the main tool for altering the metabolic network to direct

141

Synthetic Biology. DOI: http://dx.doi.org/10.1016/B978-0-12-394430-6.00008-X

the metabolic flux towards producing the target compound. With synthetic biology tools, in conjunction with systems metabolic engineering strategies, the engineering of microorganisms can achieve a high level of efficiency and control to optimize the production of the target compound.[2,9–12] In this chapter, we shall explore the fundamentals of computational synthetic biology, the tools and methods in predicting and developing synthetic networks at the genetic and pathway levels, and finally discuss the necessity for a systems-level analysis of the synthetic network to gauge the effect it has on the overall phenotype of the host cell.

FUNDAMENTAL COMPONENTS OF SYNTHETIC BIOLOGY

Synthetic biology designs synthetic biological networks for the purpose of introducing specified functions and new capabilities, such as genetic counters used to count the number of arabinose pulses, and sensors for detecting boundaries between light and dark.[13–17] Towards this goal, regulation of genes and their physiological functions have been identified (e.g. riboswitch, promoter regulations, quorum sensing) and modified for various applications.[8,17–19] One such application is the programming of living cells with genetic parts for strain design. To conceptualize the complex regulation of cellular functions, the analogy to electronic circuits and computer programing has been widely used.[13–17] This comparison is fitting when observing the logic and decision nodes of gene expression and control in response to various stimuli and perturbations. Genetic programming is engineering of the genetic code to implement and configure de novo regulation of cellular functions.

Boolean logic operators (e.g. NOT, AND, NAND, OR, NOR) are commonly employed to simulate decision nodes of various biological functions, such as promoter regulation, riboswitch, and quorum sensing (Fig. 8.1).[7,13–15,17,20] Promoter regulation is the most commonly engineered biological function in synthetic biology.[17,19–22] Riboswitches are another function and usually consist of two functional groups: an aptamer and an expression platform.[17] In riboswitches, the aptamer selectively binds to the target ligand and the expression platform modulates the gene expression. A tandem riboswitch utilizes two riboswitches to create NOR, AND, and NAND operators. The riboswitches are influenced independently, and do not interact with each other despite the close proximity of the active sites. Biological sensors capable of detecting variations in chemical or photon concentrations and quorum sensing are also employed to send signals to the cell and influence cellular functions and responses to stimuli.[17] In this section, design of synthetic biological tools to function as counters, signaling systems to perform specific functions, and oscillators are discussed.

Genetic Counters

Genetic counters retain a memory of events or objects, each as a distinct state, and allow for the accounting of tightly controlled states for the maintenance of metabolism and cellular growth.[15] Two examples of a genetic counter developed for synthetic biology are the riboregulated transcriptional cascade (RTC) counter and the DNA invertase cascade (DIC).[15] The RTC counter utilizes a transcriptional cascade and has been employed in translational regulation. Two types of the RTC genetic counter have been developed: the two-counter, a counter that can count up to two; and the three-counter, a counter that can count up to three. These RTC counters were designed to count the number of arabinose pulses (two or three for the respective counters) experienced by the cell. The two-counter is driven by the constitutive promoter $P_{Ltet0-1}$ leading to the expression of the T7 RNA polymerase (RNAP). This RNAP then binds to the T7 promoter and transcribes the green fluorescent protein (GFP) gene.[15] The expression for the T7 RNAP and the GFP genes are further regulated by riboregulators (ribreg), which silence and activate the post-transcriptional expression of the

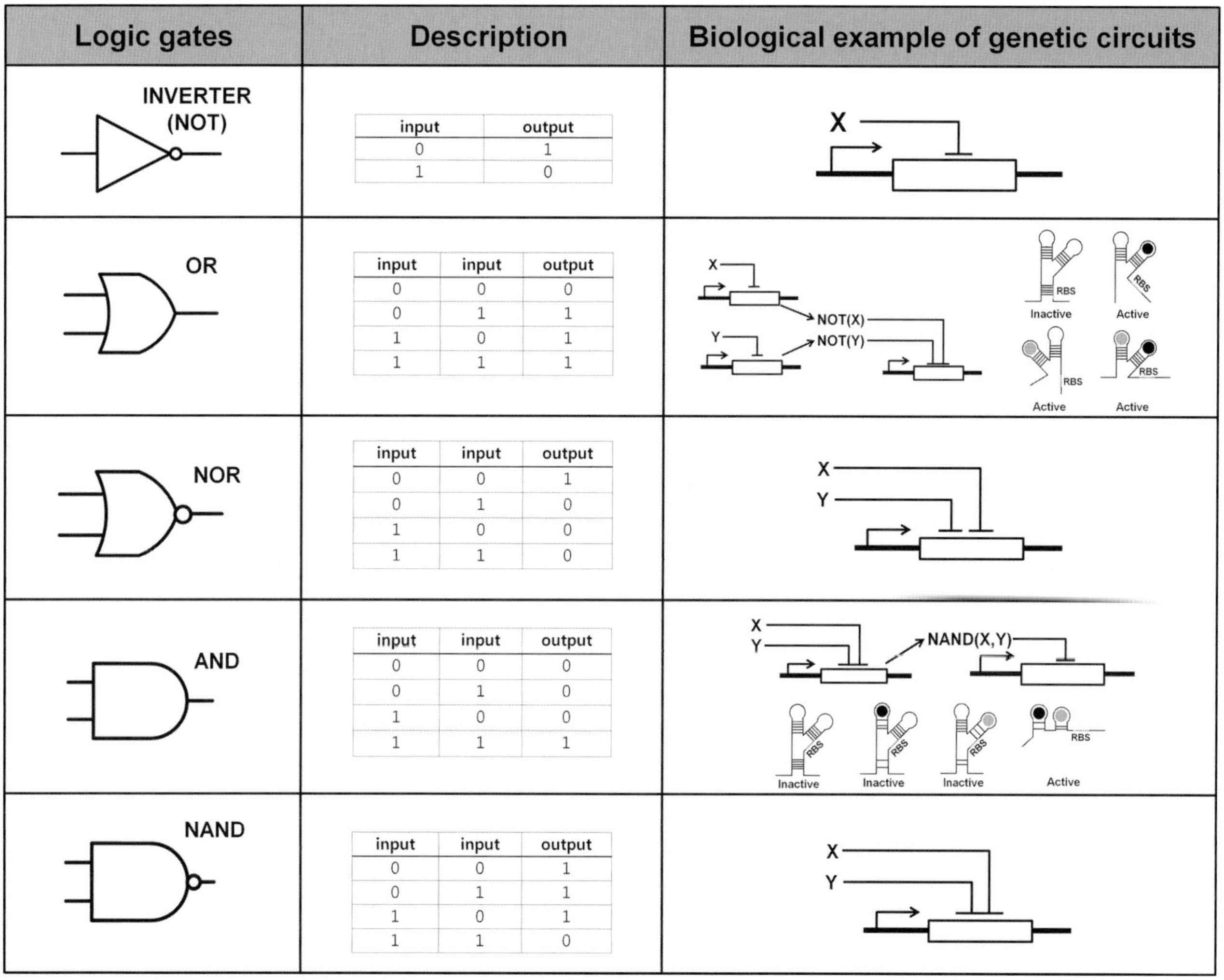

Logic gates	Description	Biological example of genetic circuits

INVERTER (NOT)

input	output
0	1
1	0

OR

input	input	output
0	0	0
0	1	1
1	0	1
1	1	1

NOT(X)
NOT(Y)
RBS — Inactive / Active

NOR

input	input	output
0	0	1
0	1	0
1	0	0
1	1	0

AND

input	input	output
0	0	0
0	1	0
1	0	0
1	1	1

NAND(X,Y)
RBS — Inactive / Inactive / Inactive / Active

NAND

input	input	output
0	0	1
0	1	1
1	0	1
1	1	0

FIGURE 8.1

Summary of Boolean logic gates (operators) used in synthetic biology to design genetic circuits. On the left, the symbols of the logic gates are listed and the corresponding Boolean description is presented next to each symbol. The biological examples of each logic gate are presented, where X and Y are the inputs and the output is the expression of the end gene. An example of NOT gates manifested in biological systems is the simple inhibition of a gene expression by a single inhibitor. NOR and NAND gates utilize two inhibitors where one is sufficient to inhibit gene expression (NOR), or both are required to be present for gene expression inhibition (NAND). OR gates are represented by a nested operator, NAND(NOT(X),NOT(Y)), where the inhibitors X and Y inhibit the expression of two other inhibitors that are needed simultaneously to block the expression of the output. AND gates also use a nested operation of NOT(NAND(X,Y)), where X and Y are both needed to inhibit the inhibitor to the target output gene, thereby allowing it to be expressed.

genes (RNAP and GFP). Thus, with the presence of the riboregulators, the transcription of the genes is silenced, even for RNAP with the constitutive $P_{Lteto-1}$ promoter: [$P_{Lteto-1}$(ON)-ribreg(ON)-T7RNAP(OFF)]; [P_{T7}(OFF)-ribreg(ON)-GFP(OFF)]. To relieve the repression, a short transactivating, noncoding RNA (taRNA), driven by the arabinose P_{BAD} promoter, is expressed and binds to the riboregulator and activates gene expression. When the cell encounters the first pulse of arabinose, the transcription of RNAP is activated and is accumulated within the cell: [$P_{Lteto-1}$(ON)-ribreg(OFF)-T7RNAP(ON)]. Slight increase in the expression of the GFP is encountered as well, but not enough for a significant fluorescent signal due to insufficient levels of RNAP. When the cell encounters a second pulse of arabinose, the RNAP accumulated from the initial arabinose pulse leads to the expression of GFP and a significant fluorescent signal is generated: [P_{T7}(ON)-ribreg(OFF)-GFP(ON)]. The three-counter was constructed by adding a T3 RNAP and a T3 promoter, where the T3 RNAP is linked to the T7 promoter and the GFP is linked to the T3 promoter: [$P_{Ltet0-1}$-ribreg-T7RNAP]; [P_{T7}-ribreg-T3RNAP]; [P_{T3}-ribreg-GFP].[15]

The DIC counter uses recombinase that can invert DNA between the respective and oppositely oriented recombination target (FRT) sites, for example the recombinases Cre and Flp_e with the FRT$loxP$ and flp_e respectively, to construct a counter module.[15] This counter constructs a single invertase memory module (SIMM) that is placed in sequence with each other, each with different FRT sites, and the promoter for each consecutive module is oriented in the opposite direction within the previous module. The first module, whose promoter is oriented in the forward direction, becomes activated by a pulse, resulting in the expression of the recombinase and the inversion of the module. When the module becomes inverted, the inverted promoter is reoriented to the forward direction, and is able to activate the expression of the next module in the sequence when a second pulse is experienced. The components of the SIMM consists of an inverted promoter (P_{inv}), a recombinase gene (*rec*), an *ssrA* tag (this causes rapid protein degradation to prepare the system for the next pulse), a terminator (Term), and two recombination target sites for *rec* placed at both ends of the module. The sequence of these components between the recombination target sites is: P_{inv}-*rec*-ssrA-Term. The number of SIMM used determines the number of pulses the system is able to count. For instance, two SIMMs, one with the Cre recombinase and the other with the Flp_e recombinase, with the overall sequence of P_{BAD}-SIMM (*flp_e*)-SIMM (*cre*)-gfp, and the inverted promoter in each SIMM is P_{BAD}, the system is able to count three pulses of arabinose before a green fluorescent signal is detected. In this example, the first pulse of arabinose flips the first SIMM (*flp_e*), so that the inverted promoter (P_{BAD}) is able to activate the next SIMM and so on in a cascade, resulting in the production of GFP.[15]

Cellular Signaling System

The edge detection program employs a signal processing system to synthetically engineer the cell to detect the edges or changes in the environment.[13] This edge detection program was demonstrated using a light image sensing program to produce a signal and produce pigment in response to light by *Escherichia coli*. The Boolean logic program for detecting the edge between light and dark is: [IF NOT light → produce signal] and[IF signal AND NOT (NOT light) → produce pigment].[13] Thus, for cells at the boundary between light and dark, pigment is produced. This program was implemented as a genetic circuit by employing a chimeric light-sensitive protein, Cph8, a cell−cell communication signal 3-oxohexanoyl-homoserine lactone (AHL), and β-galactosidase. The first component is the light detection module which consists of a NOT gate that inhibits the expression of the signal protein in the presence of light. In the presence of light, the gene *luxI*, which produces AHL (IF signal), and the gene *cI*, which encodes for a transcriptional repressor from phage λ, are repressed. AHL complexed with the constitutively expressed transcriptional factor luxR in the absence of *cI*(NOT(NOT light)) allows for the expression from the $P_{lux-\lambda}$ promoter, which in turn, leads to the expression of the β-galactosidase gene *lacZ* in cells located at the boundary of light and dark, resulting in the cleavage of a substrate in the media to produce the black pigment. This system was also simulated computationally by using a sigmoidal function (f_{light}) for the [NOT light] gate and a Shea-Ackers formalism (f_{logic}) for the [IF signal AND NOT(NOT light)] gate.

A more complex signaling system is the predator−prey ecosystem that employs two different strains to control the populations of the two organisms.[20] To demonstrate this system, two different *E. coli* strains were engineered to communicate with each other through quorum sensing to control gene expression and survival of the two strains via engineered bidirectional gene circuits. Two quorum sensing molecules were employed: N-3-oxododecanoyl homoserine lactone, 3OC12HSL; and 3-oxohexanoyl-homoserine lactone, 3OC6HSL.[20] In the predator *E. coli* strain, the suicide gene, *ccdB*, is constitutively expressed and the cell dies off. However, the prey *E. coli* strain synthesizes and releases the quorum sensing molecule, 3OC6HSL, which activates the expression of the antidote gene *ccdA*, which blocks the activity of *ccdB*, thereby saving the predator *E. coli*. As the predator *E. coli* population increases, the concentration of another quorum sensing molecule, 3OC12HSL, increases. This molecule induces the expression of *ccdB* in the prey *E. coli*, resulting in the death of prey cells. Consequently, the predator *E. coli* cannot survive without its 'prey,' which is then killed off by the predator when the predator population achieves a certain cell density. Eventually steady-states in the populations of both strains are achieved with predator *E. coli* killing off the prey, and the prey *E. coli* saving the predator.[20]

Genetic Oscillators

Oscillators are important characteristics of biological systems as many cellular functions display oscillating behaviors, such as cell cycle regulation and circadian rhythms. These oscillators can be constructed using simple components already present in the cell, with additional features to confirm and evaluate the synthetic oscillatory behavior. One such oscillating network was developed in *E. coli* using a three transcriptional repressor system (repressilator) that is not found in any natural biological clocks.[23] This oscillating network was used to monitor the state of the individual *E. coli* cells through the periodic synthesis of GFP.[23]

Three different types of transcriptional oscillator have also been constructed and evaluated in vitro: (1) a negative feedback oscillator regulated by excitatory and inhibitory RNA signals; (2) a positive-feedback loop to extend and modulate the oscillatory regime of the negative feedback oscillator; and (3) a three-switch ring oscillator that is analogous to the repressilator.[8] Mathematical modeling was employed using kinetic models in the design and analysis of the three oscillators. Using the mathematical model, the robustness of the regulatory oscillator to different conditions was analyzed, which allowed better understanding of oscillator's principles of operation within the kinetic models.

Implementing Synthetic Biology Components Through Engineering of Cellular Components

The implementation of these genetic circuits is achieved through the engineering of different native components to display the numerous decision nodes needed in constructing the genetic circuits. One such tool that allows for the implementation of synthetic biology genetic circuits is RNA programming.[18] RNA has become an important tool in synthetic biology with increased knowledge of their functional roles in cellular physiology, beyond being a bridge between DNA and proteins. Small RNA molecules have been found to perform regulatory functions through attenuating genetic expression. Through the design of functional synthetic RNA molecules that exhibit complex functions, cellular circuits can be engineered to perform high-level biological functions.[18]

Small hairpin RNAs (shRNA) are small molecules of RNA with tight hairpins that have been used to silence gene expression through ligand control of RNA interferences (RNAi).[18] One of the limitations of employing shRNA as a regulatory control element is the lack of predictive tools in optimizing the design of shRNA sequences in order to not only alter and tune the system's response to perturbations, but also to broaden the implementation of shRNA to a wide range of systems. The shRNA switch platform was developed to provide ligand control of RNAi in mammalian cell lines, which provides a fine-tuning, multi-input control, and model-guided design of regulatory systems.[18] The platform utilizes a strand displacement strategy where ligand binding, RNAi activation, and translation of the binding interaction are isolated into individual domains within the platform. Standard RNA folding algorithms were used to establish quantitative sequence-to-function relationships. By demonstrating combinatorial tuning strategies and multi-input control, the switch optimized the dynamic range within a specified context, allowing extension of utility of RNAi as regulatory tools and a valuable component for building complex biological systems.[18]

Orthogonal ribosome and mRNA pairs are duplicated molecules in an organism that are capable of processing information in parallel with their progenitors without any crosstalk, which have been developed to create a regulatory network controlled by Boolean logic operators.[19] The development of orthogonal pairs of ribosomes and mRNA requires an extensive and exhaustive screening of ribosomes that are capable of specifically translating the orthogonal mRNA with a high efficiency. The ribosomes and mRNA are then evolved through a duplication strategy developed to evolve highly specific and active orthogonal pairs using positive and negative selection. With orthogonal ribosomal and mRNA pairs, Boolean AND and OR gates were constructed.[19] In addition to engineering orthogonal ribosomes, the design of synthetic ribosome binding sites (RBSs) has been investigated.[19] The major goal of designing synthetic RBSs is to alter and optimize protein expression levels within the cell where orthogonal ribosomes are not used to synthesize proteins. The design of RBSs was performed using a predictive method that interconverts between the DNA sequence of the RBSs and their function inside a genetic system, allowing for the optimization of the system at a genetic level. Combination of a biophysical model of translation initiation with an optimization algorithm allows predictions of the sequence of a synthetic RBS sequence that provides target translation initiation rate on a proportional scale. This design method provides prediction of reusing identical RBS sequences in different genetic contexts that can result in different protein expression levels.

COMPUTATIONAL PREDICTION TOOLS FOR SYNTHETIC BIOLOGY COMPONENTS

Biomolecular Design

Biomolecular design is a significant aspect of synthetic biology, as demonstrated by the highlights and development of new tools in biomolecular design applied towards protein engineering to create proteins with modified or novel functions, engineering of controllable ribosome binding sites with a specific translation initiation rate, and the design of synthetic promoter sequences with specific transcription rates.[19,21,24,25] Rosetta is one molecular modeling tool widely used in protein engineering.[25] Some features of Rosetta include structure prediction of proteins through ab initio calculation (RosettaAbInitio), protein–protein (RosettaDock) and protein–small molecule (RosettaLigand) docking simulations, and DNA–protein interaction evaluation (RosettaDNA).[25]

RBS Designer and RBS Calculator are two tools used to design synthetic RBSs to optimize the translation of natural or synthetic gene constructs. These tools incorporate thermodynamic characteristics of the RBS binding to specific coding sequences (CDS) and the molecular interaction between the ribosomes and mRNA (Table 8.1).[19,21]

DNA Assembly

In a large genetic combinatorial space, construction of a defined set of DNA sequences is a powerful method for synthetic biology. However, the process of assembling more than a few small DNA sequences can be time-consuming, costly, and error-prone. To reduce the time and cost, as well as improve the efficiency of DNA assembly, a number of standardized and automatable DNA assembly methods or algorithms have been developed for synthetic biology.[26–30] Gene Designer is an easy and fast tool for building artificial DNA segments by a graphic user interface (GUI) that provides facile manipulation of genetic elements. Using this tool, synthetic new genes can be visually assembled by combining DNA traits.

TABLE 8.1 Computational Tools in Designing Synthetic Biological Components

Synthetic system	Website	Refs
DNA assembly		
GeneDesigner	https://www.dna20.com/genedesigner2	29
GeneDesign	http://www.genedesign.org	28
j5	http://j5.jbei.org	26
Asmparts	http://soft.synth-bio.org/asmparts.html	30
Gene Composer	http://www.genecomposer.net	27
Biomolecular design		
Rosetta	http://www.rosettacommons.org	25
RBSDesigner	http://rbs.kaist.ac.kr	21
RBS Calculator	https://salis.psu.edu/software	19
Optimizer	http://genomes.urv.cat/OPTIMIZER	24
Genetic circuit design		
Genetdes	http://soft.synth-bio.org/genetdes.html	31
RoVerGeNe	http://iasi.bu.edu/~batt/rovergene/rovergene.htm	7
OptCircuit	http://maranas.che.psu.edu/research_circuits.htm	32
GEC	http://lepton.research.microsoft.com/webgec	16
Biojade	http://web.mit.edu/jagoler/www/biojade	36
GenoCAD	http://www.genocad.org	33
SynBioSS	http://synbioss.sourceforge.net	34
TinkerCell	http://www.tinkercell.com	35
Pathway prediction		
EnzMatcher	stand-alone software	1
KEGG PathPred	http://www.genome.jp/tools/pathpred	39
BNICE	http://systemsbiology.northwestern.edu/bnice	2
UM-BBD Pathway prediction	http://umbbd.msi.umn.edu/predict	38
OptStrain	framework (No website)	42
DESHARKY	http://soft.synth-bio.org/desharky.html	43
RetroPath	http://www.issb.genopole.fr/~faulon/retropath.php	12
FMM	http://fmm.mbc.nctu.edu.tw/index.php	40
CarbonSearch	http://www.kavrakilab.org/atommetanet	41

147

Furthermore, it is possible to optimize the codons of open reading frames for protein expression for a specific host organism.[29]

GeneDesign is a web-based tool that provides modules, algorithms, restriction enzyme library, and batch processing capabilities. Each module can be in charge of different tasks, such as reverse translation, codon juggling, and building block design.[28] For example, after a target protein has been characterized and its presence inside a novel gene circuit has been confirmed, the protein can be reverse translated back to DNA, whose sequence is then modified and engineered to optimally function in the host organism which the gene circuit is to be incorporated into.

Another tool that can automate the design of DNA assembly protocol is j5, a web-based software tool.[26] The design process of j5 includes the optimization of costs, design specification rule enforcement, strategies in hierarchical assembly to minimize assembly errors, and the manual or automated construction of scarless combinatorial DNA libraries. The design algorithms are generally compatible with other DNA construction methodologies, and can be used to complement other DNA assembly tools (Table 8.1).

Genetic Circuit Design

A number of computational methods have already been developed to facilitate system design. To date these methods have concentrated mainly on the design of small transcriptional circuits. Some have been packaged into downloadable software tools. Common to most of the methods developed so far is the use of ordinary differential equations (ODEs) to model the dynamics of the system (though some can also handle stochastic dynamics). These computational methods differ in how the networks are parameterized, how the dynamics are approximated, and how the optimization is formulated.

The Genetdes method attempts to design optimal transcriptional networks and kinetic parameters with targeted behaviors.[31] The optimization is performed using simulated annealing with moves in the model space including synthetic genes and regulatory interactions, and the change of kinetic parameters. The Robust Verification of Gene Networks (RoVerGeNe) method inputs an existing network and generates a set of constraints to express the desired behavior of the network.[7] Approximating the regulation terms in the dynamical equations using piecewise linear functions, the analysis of the network becomes more efficient compared to a full nonlinear ODE model. The Linear Temporal Logic (LTL) is used to express the desired network behavior, and RoVerGeNe uses abstraction and model checking for a given set of parameters to determine whether the parameters are able to satisfy the constraints of the desired behavior. If the parameters are valid for the given constraints, then a specific set of parameters are selected to represent the specified behavior. The MATLAB code for RoVerGeNe is available online (Table 8.1).[7]

The OptCircuit method uses multiple sources of literature on promoters, protein molecules, and inducers.[32] By doing so, systems are constructed to maximize an objective function derived from the desired dynamics. Here the full dynamics is approximated under the assumption that fast reactions (rate constants on the order of seconds) are in pseudoequilibrium. The optimization is formulated as a mixed integer dynamic optimization problem and can be applied to both system topology and kinetic parameters. This method has been used to design a toggle switch, a genetic decoder, and a concentration band detector.[32]

A different approach is to develop languages that can be compiled into sequences of standard biological parts. The Genetic Engineering of living Cells (GEC) method is an elaborate approach that allows the expression of logical interactions between biological parts using a programming language.[16] The language can also serve as an explicit proposal and to guide the emerging standard of biological parts which so far has been related to biological, rather than logical, properties of parts.[16] GenoCAD is an example of a context-free grammar that enforces a set of production rules that ensure that the user will produce a biologically valid construct.[33] GenoCAD is a web-based tool used to design artificial genes, protein expression vectors, and genetic networks composed of multiple functional genetic parts.[33] By using simple GUI, complex constructs composed of dozens of functional blocks can be designed within minutes. It also provides the genetic sequence of the construct designed from its comprehensive libraries of genetic parts. SynBioSS Designer is another web-based tool available for the design of DNA constructs.[34] With the input of the molecular parts involved in gene expression and regulation, SynBioSS Designer automatically generates a complete network of biomolecular reactions as output.[34]

GUI enables convenient construction of a genetic circuit by dragging and dropping components on a canvas.[33–37] BioJADE was the first genetic circuit tool using GUI, and interactively utilizes the BioBrick collection.[36] Using the GUI, BioJADE allows designers to specify the circuit, tune it, and perform simulations of the circuit behaviors. It also provides connections between databases and simulations of the designed circuits through the Distributed Flexible and User eXtensible (D-FLUX) protocol. One of the standard simulators in D-FLUX is TABASCO, which performs genome-scale simulation of transcription and translation in the cell at the molecular and single base pair level.[37]

Although GUI is convenient and makes it easy to design genetic circuits, there are some limitations in that the user is limited to only what is available in the software. The TinkerCell tool allows users to create and analyze synthetic networks using third-party C and Python programs with an extensive C and Python application programming interface (API). As an open-source tool, it allows for the development and implementation of new tools that can be used in TinkerCell for those who have a background in computer programing (Table 8.1).[35]

COMPUTATIONAL TOOLS FOR PATHWAY PREDICTION

Advancing beyond the level of local circuit design in synthetic biology, the engineering of biological systems at the enzyme or pathway level have been investigated for metabolic engineering.[1,2,12,38—42] Pathway engineering can occur in all forms, from mixing and matching pathway components into an ideal host to the prediction and design of novel or de novo pathways that have not been identified in biological systems. While engineering existing pathways and combining them to create new ones result in the efficient production of desired molecules, this has limited our options to enzymatic reactions that are currently known. This has led to the development of tools focused on the discovery of de novo pathways, partially or as a whole, consisting of nonnatural enzymes that can expand the known database of metabolism. With an ever-growing database and knowledge of bioinformatics and metabolic pathways, it is often easy to think that the easiest way to engineer genes for strain design is to use the enzymes with known functions and well-characterized properties. The approach of identifying de novo pathways of nonnatural enzymes, defined as enzymes that have not yet been characterized and are not commonly found in known organisms, employs a method that scans the realm of all known chemical reactions rather than known enzymatic reactions. This allows for the discovery of chemical reactions that can potentially occur in a cell but are overlooked because the enzyme that would catalyze such a reaction has not been characterized.

Enzymes have evolved through millions of years to carry out specific metabolic reactions. However, many of these enzymes are promiscuous and have broad substrate specificities. In vitro characterization of an enzyme may not fully explain the enzyme's behavior or function in vivo, thus logical steps can be taken to hypothesize that an enzyme can possibly perform the same function with multiple similar substrates. Therefore, heterologous expression of enzymes that have been modified to efficiently utilize alternate substrates can possibly lead to the production of nonnatural metabolites and the discovery of a new metabolic pathway.

To facilitate the discovery of de novo pathways and their enzymes, a retrosynthetic analysis is employed. Furthermore, algorithms have been developed to automate and efficiently perform the analysis and present potential biochemical reactions to explore. Although de novo pathway construction for strain design is limited by the biology of enzymes, it also opens new opportunities for finding nonnatural metabolite-forming pathways that can be ultimately useful for metabolic engineering.

Algorithms and computational tools that have been developed to address possible biochemical pathways explore a wide range of chemical properties, including but not limited to the thermodynamic feasibility of metabolic reactions under physiological conditions. Computational tools such as the University of Minnesota Pathway Prediction System (UM-PPS), DESHARKKY, the web-based PathPred from KEGG, Biochemical Network Integrated Computational Explorer (BNICE), and EnzMatcher are examples of the tools available for developing strategies in designing novel pathways.[1,2,38,39,43]

The combinatorial explosion resulting in the addition of rule sets is also addressed in the publically available UM-PPS, in which certain restrictions such as aerobic likelihood, relative reasoning, super rules and increasing rule stringency are introduced to meet the need of users that are not looking for too many possible answers.[38] The UM-PPS is part of the University of Minnesota Biocatalysis/Biodegradation Database (UM-BBD), a database dedicated to information regarding microbial degradation of xenobiotic chemical compounds. Chemical compounds are described in SMILES format and based on a set of rules that can be found on the UM-PPS webpage, potential pathways for the degradation of the compound may be predicted. A limitation of this method is the dependence on biotransformation rules based on known reactions that can be found in the database or supported by literature.[38]

DESHARKY is another pathway prediction tool that generates a candidate pathway and presents it in the context of the host organism where these novel pathways are to be implemented.[43] Additionally, amino acid sequences of the enzymes from the closest organisms phylogenetically related to the host organism are presented. To achieve this, DESHARKY uses the Monte Carlo algorithm to predict a potential pathway route connecting two compounds. This tool, however, reconstructs potential pathways from databases of known enzymatic reactions, and thus is limited in its capability in the bioconversion of chemical compounds.[43]

The KEGG database also has a pathway prediction tool available on its website, PathPred.[39] PathPred utilizes only the metabolic reactions that can be found in the KEGG database to predict metabolic pathways between two biochemical compounds. Also, the minimum number of metabolic steps required for the biotransformation can be specified. Again, as with the UM-PPS tool and DESHARKY, the KEGG pathway prediction tool is limited to known metabolic reactions and chemical compounds found in the database.[39]

One pathway prediction tool that is not limited to known enzymatic/metabolic reactions to generate pathways between two biochemical compounds is BNICE, in which the chemical bonds in the reactants that undergo changes during a chemical reaction are represented as a bond−electron matrix (BEM).[2] The BEM is then added with another matrix that represents a 'reaction operator.' The reaction matrix is a representative of the 'generalized enzyme reactions' described by the third-level classification of the Enzyme Commission (EC) numbers. The method screens out unlikely candidates and predicts the most favorable pathway by calculating the thermodynamic energy changes along the pathway. Although BNICE identifies all possible pathways in a given target molecule from a starting molecule, the novel pathways found using BNICE are devoid of 'combinatorial explosion,' because it weeds out thermodynamically infeasible reactions.[2]

Another pathway prediction tool that is not limited to known enzymatic/metabolic reactions to generate pathways between two biochemical compounds is the standalone program Enzyme Matcher (EnzMatcher), a retrosynthesis framework with a prioritization scoring algorithm developed at the Korea Advanced Institute of Science and Technology (KAIST).[1] This tool employs a set of rules that are not dependent on known metabolic reactions. Instead, the tool employs a set of rules to determine all possible biochemical reactions that may occur for each functional group found in the chemical compound. By using a rule set based on all possible biochemical reactions rather than known metabolic reactions, EnzMatcher is capable of predicting not only known metabolic reactions that are found in databases, but also reactions that are biochemically feasible but not characterized or identified in biological systems. The results from EnzMatcher analysis are then prioritized based on binding site covalence, chemical similarity of the compounds, thermodynamic favorability and pathway distance (Table 8.1).[1] Although none of the tools aforementioned have been complemented by experimental validation for their capability of predicting feasible pathways, the effort to solve such problems is complemented by the development of tools for 'custom' designing enzymes to catalyze chemical reactions that have not yet been observed in nature.

With the identification of metabolic reactions that have not been previously characterized by enzymes in known organisms, enzymes that are capable of similar chemical bioconversions need to be engineered to perform the desired bioconversion of a substrate to the desired product. This is to fill in the biochemical 'gaps' found in the pathway prediction programs, since the programs do not pinpoint which specific enzyme is needed in the novel pathway. One way to fill in such gaps is to engineer an enzyme for carrying out a particular nonnatural reaction of interest. A successful case of such engineering enzymes by directed evolution is the engineering of an (R)-selective transaminase for sitagliptin synthesis.[44] By using computational tools for active site analysis, the substrate binding site was evaluated. The enzyme was then subjected to multiple rounds of mutations to expand the binding site in order to accommodate a truncated homologue of sitagliptin. Further engineering of the enzyme ultimately led to an efficient biocatalyst for sitagliptin synthesis. Similarly, Baker and colleagues have computationally designed enzymes and experimentally demonstrated de novo reactions. Nonnatural enzymatic retro-aldol,[45] Kemp-elimination,[46] and Diels-Alder reactions[47] are successful stories of designing biocatalysts using the Rosetta methodologies assisted with RosettaMatch and RosettaDesign to in silico design a novel enzyme, to select for the ideal protein backbones and optimize the matches. These examples deserve particular attention, since the enzymes developed for each study carry out chemical reactions that have never been observed in nature. Although the aforementioned biocatalysts need to be in specific in vitro reaction conditions, the computational methods for designing enzymes can be of great assistance to synthetic biology for stain design, because without exploring the enzyme evolutions, pathway prediction algorithms alone may not be sufficient for actually producing recombinant strains for high titer production of chemicals from renewable feedstock.

COMPUTATIONAL TOOLS FOR STRAIN OPTIMIZATION

In addition to these computational tools for synthetic biology components, the in silico genome-scale metabolic model is another aspect in strain design. Employing constraints-based flux balance analysis (FBA), an insight into cellular metabolism under various genetic and/or environmental perturbed conditions can be obtained and strategies into modifying the metabolic network through genetic manipulations can be developed.[48] To this end, various algorithms have been developed to dissect and understand genotype—phenotype relationships and to represent strategies for improving production of the desired bioproducts.[49]

In order to evaluate the physiological features of strains under gene knockout conditions, minimization of metabolic adjustment (MOMA) and OptKnock algorithms have been developed.[50,51] In MOMA, the metabolic fluxes under knockout mutant strain are assumed to lead the minimal flux redistribution with respect to those of the wild-type. Therefore, MOMA results represent a unique flux distribution which has very similar fluxes to the wild-type strain.[50] For example, to further strain optimization, Park et al. used MOMA to improve the L-valine production in E. coli.[11] The triple knockout genes (aceF, mdh, and pfkA) were identified by using MOMA simulation, which allowed a 45.5% increase in L-valine production, and also as high as 0.378 g of L-valine per gram of glucose.

OptKnock was also developed to identify knockout gene targets.[51] OptKnock included two competing objective functions such as cellular growth and biochemical production. This bilevel optimization algorithm leads to the overproduction of desired biochemicals by adjusting the metabolic fluxes under gene knockout conditions. The redistributed fluxes were then expected to improve production of the target metabolites which are essential components for cellular growth. Recently, OptKnock was successfully demonstrated for the production of 1,4-butanediol(BDO), a nonnatural chemical, in E. coli.[52] OptKnock results represented a promising strategy, removing four genes (adhE, pfl, ldh, and mdh). This strategy contributed to designing a strain in which overproduction of BDO was the method to balance redox and enable anaerobic growth.

An effort to identify gene amplification targets was also considered a meaningful method for strain optimization, as well as gene deletion targets. Flux scanning based on enforced objective flux (FSEOF) was able to identify gene amplification targets by scanning all the metabolic fluxes in the metabolic model and selecting fluxes which increase when flux toward desired product formation was enforced as an additional constraint.[9] In other words, gene amplification targets represent the fluxes increasing gradually from their initial value of the wild-type strain to a value close to the maximum theoretical yield. This algorithm was applied to overproduction of lycopene in *E. coli*.[9] Although amplification gene targets through FSEOF algorithm contributed to increase lycopene production, final strain optimization was further reflected by the FSEOF and MOMA simulation results. Flux variability scanning based on enforced objective flux (FVSEOF) with grouping reaction (GR) constraints was also used to identify gene amplification targets.[53] FVSEOF scan changes in the variabilities of metabolic fluxes in response to an artificially enforced objective flux of product formation so that putrescine production was experimentally increased by overexpression of each identified amplification gene.

SYNTHETIC BIOLOGY FOR SYSTEMS-LEVEL METABOLIC ENGINEERING

Development of synthetic biology tools is driven by the applications in which these tools are employed, ranging from the characterization of genetic circuits to the metabolic engineering of microorganisms for production of high-value industrial compounds. However, while synthetic biology tools are powerful in their ability to control and design cellular functions at the local level, they are limited in determining how the changes will affect the overall cellular phenotype. Therefore, these tools are combined with other established tools that would expand their focus from the genetic level to the systems level (Fig. 8.2). One such tool is the genome-scale metabolic models that are widely available for various host organisms, including widely studied and employed host systems in biology and biotechnology, such as *E. coli*, *Saccharomyces cerevisiae*, and even *Homo sapiens*.[54−56] By employing the genome-scale metabolic models, researchers are able to simulate the effect synthetic biology tools have at the systems level.

For instance, systems metabolic engineering of microorganisms for the production of high-value compounds would seek to achieve high productivity and yield of the target product while maintaining a sufficiently high growth rate as well as other objectives, including but not limited to minimal byproduct formation. Strategies employing many of the synthetic biology tools have been investigated towards this goal. One example is the identification of novel pathways for the production of 3-hydroxypropionate (3HP) by employing BNICE.[10] Here, all candidate pathways that were thermodynamically feasible were identified and discussed. However, although the metabolic reactions identified may be thermodynamically feasible, they would be irrelevant if the pathways failed to achieve high production levels of 3HP for large-scale production. To investigate this, the pathways were incorporated into a genome-scale metabolic model of *E. coli* and the maximum theoretical yield of 3HP was examined. Through the use of the genome-scale metabolic model, it was found that the cofactors of the novel pathways can influence the maximum yield of 3HP (e.g. ATP and NADH). Furthermore, other tools and methods that have been established in the systems metabolic engineering of host organisms for the production of a target compound can be employed to complement the introduction of the novel pathways.[10]

Another example of employing systems-level strategies in conjunction with a synthetic pathway prediction tool is the engineering of *E. coli* to produce 1,4-butanediol (14BDO).[52]

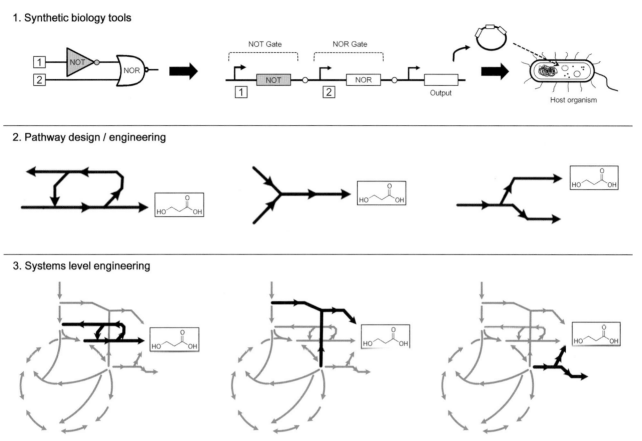

FIGURE 8.2

Overall scheme of systems metabolic engineering from (i) the genetic level to (ii) the pathway level and then to (iii) the level of the metabolic network. At the genetic level, synthetic biology tools are employed to create modifications in the cell, in this case a regulatory NOT and NOR system for the expression of a gene. This modification is geared towards the design and engineering of metabolic pathways such that the metabolic flux funnels towards the production of a target metabolite. From the list of possible pathways that can be used to produce the target compound, a systems level analysis is performed to determine the optimal route that would not be detrimental to the overall system (e.g. retarded cellular growth or generation of toxic byproducts).

153

This compound is not synthesized by any known organism and thus, a complete biosynthetic pathway has not been characterized. The in-house pathway prediction tool Sim Pheny Biopathway Predictor was used to identify all potential pathways for 14BDO synthesis based on known chemistry and not enzymatic reactions. More than 10 000 potential pathways were identified and then ranked based on thermodynamics, theoretical yields, number of steps that are characterized, number of uncharacterized steps, and total number of steps required for 14BDO synthesis. Ultimately, pathways starting from tricarboxylic acid cycle intermediates and passing through the intermediate 4-hydroxybutyrate were selected for producing 14BDO.[52]

CONCLUDING REMARKS

Synthetic biology tools have come a long way since their inception, and consequently computational methods and tools have been developed to facilitate the design of synthetic biology components and the development of new synthetic biology tools. However, the complexity of biological systems and the large gaps in our knowledge of these systems show that there is still much more we need to do before achieving a true synthetic cell. An understanding of the mechanisms of the various complex cellular components and how they are all interconnected to manifest into a fully dynamic and robust living organism is essential to synthesize such a cell. The advancement of high-throughput technologies has certainly aided in the understanding of these mechanisms. However, the lack of efficient methods to extract new information from the overwhelming volume of new data has bottlenecked the advancement. With the development of new synthetic biology components and tools targeting specific cellular components/functions, the high-throughput data can be translated into new knowledge and information. In addition to modifying genetic components of the cell, a systems-level analysis of what these modifications will do to the physiology of the cell is necessary to allow the cell to optimally perform the task for which is has been engineered.

Acknowledgments

This work was supported by the Technology Development Program to Solve Climate Changes on Systems Metabolic Engineering for Biorefineries (NRF-2012-C1AAA001-2012M1A2A2026556) of the Ministry of Education, Science and Technology (MEST) through the National Research Foundation (NRF). Further support by the Intelligent Synthetic Biology Center (2011-0031963) of the Global Frontier Project and the World Class University Program (R32-2009-000-10142-0) of MEST through the NRF is appreciated.

References

1. Cho A, Yun H, Park JH, Lee SY, Park S. Prediction of novel synthetic pathways for the production of desired chemicals. *BMC Syst Biol.* 2010;4:35.

2. Hatzimanikatis V, Li C, Ionita JA, Henry CS, Jankowski MD, Broadbelt LJ. Exploring the diversity of complex metabolic networks. *Bioinformatics.* 2005;21(8):1603–1609.

3. Clancy K, Voigt CA. Programming cells: towards an automated 'Genetic Compiler.' *Curr Opin Biotechnol.* 2010;21(4):572–581.

4. Khalil AS, Collins JJ. Synthetic biology: applications come of age. *Nat Rev Genet.* 2010;11(5):367–379.

5. Medema MH, van Raaphorst R, Takano E, Breitling R. Computational tools for the synthetic design of biochemical pathways. *Nat Rev Microbiol.* 2012;10(3):191–202.

6. Lee JW, Kim TY, Jang YS, Choi S, Lee SY. Systems metabolic engineering for chemicals and materials. *Trends Biotechnol.* 2011;29(8):370–378.

7. Batt G, Yordanov B, Weiss R, Belta C. Robustness analysis and tuning of synthetic gene networks. *Bioinformatics.* 2007;23(18):2415–2422.

8. Kim J, Winfree E. Synthetic *in vitro* transcriptional oscillators. *Mol Syst Biol.* 2011;7:465.

9. Choi HS, Lee SY, Kim TY, Woo HM. *In silico* identification of gene amplification targets for improvement of lycopene production. *Appl Environ Microbiol.* 2010;76(10):3097–3105.

10. Henry CS, Broadbelt LJ, Hatzimanikatis V. Discovery and analysis of novel metabolic pathways for the biosynthesis of industrial chemicals: 3-hydroxypropanoate. *Biotechnol Bioeng.* 2010;106(3):462–473.

11. Park JH, Lee KH, Kim TY, Lee SY. Metabolic engineering of *Escherichia coli* for the production of L-valine based on transcriptome analysis and *in silico* gene knockout simulation. *Proc Natl Acad Sci USA.* 2007;104 (19):7797–7802.

12. Carbonell P, Planson AG, Fichera D, Faulon JL. A retrosynthetic biology approach to metabolic pathway design for therapeutic production. *BMC Syst Biol.* 2011;5:122.

13. Tabor JJ, Salis HM, Simpson ZB, et al. A synthetic genetic edge detection program. *Cell.* 2009;137 (7):1272–1281.

14. Basu S, Gerchman Y, Collins CH, Arnold FH, Weiss R. A synthetic multicellular system for programmed pattern formation. *Nature.* 2005;434(7037):1130–1134.

15. Friedland AE, Lu TK, Wang X, Shi D, Church G, Collins JJ. Synthetic gene networks that count. *Science.* 2009;324(5931):1199–1202.

16. Pedersen M, Phillips A. Towards programming languages for genetic engineering of living cells. *J R Soc Interface.* 2009;6(Suppl 4):S437–S450.

17. Sharma V, Nomura Y, Yokobayashi Y. Engineering complex riboswitch regulation by dual genetic selection. *J Am Chem Soc.* 2008;130(48):16310–16315.

18. Beisel CL, Bayer TS, Hoff KG, Smolke CD. Model-guided design of ligand-regulated RNAi for programmable control of gene expression. *Mol Syst Biol.* 2008;4:224.

19. Salis HM, Mirsky EA, Voigt CA. Automated design of synthetic ribosome binding sites to control protein expression. *Nat Biotechnol.* 2009;27(10):946–950.

20. Balagadde FK, Song H, Ozaki J, et al. A synthetic *Escherichia coli* predator–prey ecosystem. *Mol Syst Biol.* 2008;4:187.

21. Na D, Lee D. RBSDesigner: software for designing synthetic ribosome binding sites that yields a desired level of protein expression. *Bioinformatics.* 2010;26(20):2633–2634.

22. Kim D, Rossi J. RNAi mechanisms and applications. *Biotechniques.* 2008;44(5):613–616.

23. Elowitz MB, Leibler S. A synthetic oscillatory network of transcriptional regulators. *Nature.* 2000;403 (6767):335–338.

24. Puigbo P, Guzman E, Romeu A, Garcia-Vallve S. OPTIMIZER: a web server for optimizing the codon usage of DNA sequences. *Nucleic Acids Res.* 2007;35(Web Server issue):W126–W131.

25. Richter F, Leaver-Fay A, Khare SD, Bjelic S, Baker D. *De novo* enzyme design using Rosetta3. *PLoS One.* 2011;6(5):e19230.

26. Hillson NJ, Rosengarten RD, Keasling JD. j5 DNA assembly design automation software. *ACS Synth Biol.* 2012;1(1):14–21.

27. Lorimer D, Raymond A, Walchli J, et al. Gene composer: database software for protein construct design, codon engineering, and gene synthesis. *BMC Biotechnol.* 2009;9(1):36.

28. Richardson SM, Nunley PW, Yarrington RM, Boeke JD, Bader JS. GeneDesign 3.0 is an updated synthetic biology toolkit. *Nucleic Acids Res.* 2010;38(8):2603–2606.

29. Villalobos A, Ness JE, Gustafsson C, Minshull J, Govindarajan S. Gene designer: a synthetic biology tool for constructing artificial DNA segments. *BMC Bioinformatics.* 2006;7:285.

30. Rodrigo G, Carrera J, Jaramillo A. Asmparts: assembly of biological model parts. *Syst Synth Biol.* 2007;1(4):167–170.

31. Rodrigo G, Carrera J, Jaramillo A. Genetdes: automatic design of transcriptional networks. *Bioinformatics.* 2007;23(14):1857–1858.

32. Dasika MS, Maranas CD. OptCircuit: an optimization based method for computational design of genetic circuits. *BMC Syst Biol.* 2008;2:24.

33. Czar MJ, Cai Y, Peccoud J. Writing DNA with GenoCAD. *Nucleic Acids Res.* 2009;37(Web Server issue): W40–W47.

34. Weeding E, Houle J, Kaznessis YN. SynBioSS designer: a web-based tool for the automated generation of kinetic models for synthetic biological constructs. *Brief Bioinform.* 2010;11(4):394–402.

35. Chandran D, Bergmann FT, Sauro HM. TinkerCell: modular CAD tool for synthetic biology. *J Biol Eng.* 2009;3:19.

36. Goler JA. BioJADE: a design and simulation tool for synthetic biological systems. 2004.

37. Kosuri S, Kelly JR, Endy D. TABASCO: A single molecule, base-pair resolved gene expression simulator. *BMC Bioinformatics.* 2007;8:480.

38. Ellis LB, Gao J, Fenner K, Wackett LP. The University of Minnesota pathway prediction system: predicting metabolic logic. *Nucleic Acids Res.* 2008;36(Web Server issue):W427–W432.

39. Moriya Y, Shigemizu D, Hattori M, et al. PathPred: an enzyme-catalyzed metabolic pathway prediction server. *Nucleic Acids Res.* 2010;38(Web Server issue):W138–W143.

40. Chou CH, Chang WC, Chiu CM, Huang CC, Huang HD. FMM: a web server for metabolic pathway reconstruction and comparative analysis. *Nucleic Acids Res.* 2009;37(Web Server issue): W129–W134.

41. Heath AP, Bennett GN, Kavraki LE. Finding metabolic pathways using atom tracking. *Bioinformatics.* 2010;26 (12):1548–1555.

42. Pharkya P, Burgard AP, Maranas CD. OptStrain: a computational framework for redesign of microbial production systems. *Genome Res.* 2004;14(11):2367–2376.

43. Rodrigo G, Carrera J, Prather KJ, Jaramillo A. DESHARKY: automatic design of metabolic pathways for optimal cell growth. *Bioinformatics.* 2008;24(21):2554–2556.

44. Savile CK, Janey JM, Mundorff EC, et al. Biocatalytic asymmetric synthesis of chiral amines from ketones applied to sitagliptin manufacture. *Science.* 2010;329(5989):305.

45. Jiang L, Althoff EA, Clemente FR, et al. *De novo* computational design of retro-aldol enzymes. *Science.* 2008;319(5868):1387–1391.

46. Rothlisberger D, Khersonsky O, Wollacott AM, et al. Kemp elimination catalysts by computational enzyme design. *Nature.* 2008;453(7192):190–195.

47. Siegel JB, Zanghellini A, Lovick HM, et al. Computational design of an enzyme catalyst for a stereoselective bimolecular Diels-Alder reaction. *Science.* 2010;329(5989):309–313.

48. Lewis NE, Nagarajan H, Palsson BO. Constraining the metabolic genotype–phenotype relationship using a phylogeny of *in silico* methods. *Nat Rev Microbiol.* 2012;10(4):291–305.

49. Kim TY, Sohn SB, Kim YB, Kim WJ, Lee SY. Recent advances in reconstruction and applications of genome-scale metabolic models. *Curr Opin Biotechnol.* 2012;23(4):617–623.

50. Segre D, Vitkup D, Church GM. Analysis of optimality in natural and perturbed metabolic networks. *Proc Natl Acad Sci USA.* 2002;99(23):15112–15117.

51. Burgard AP, Pharkya P, Maranas CD. Optknock: a bilevel programming framework for identifying gene knockout strategies for microbial strain optimization. *Biotechnol Bioeng.* 2003;84(6):647–657.

52. Yim H, Haselbeck R, Niu W, et al. Metabolic engineering of *Escherichia coli* for direct production of 1,4-butanediol. *Nat Chem Biol.* 2011;7(7):445–452.

53. Park JM, Park HM, Kim WJ, Kim HU, Kim TY, Lee SY. Flux variability scanning based on enforced objective flux for identifying gene amplification targets. *BMC Syst Biol.* 2012;6:106.

54. Ma H, Sorokin A, Mazein A, et al. The Edinburgh human metabolic network reconstruction and its functional analysis. *Mol Syst Biol.* 2007;3:135.

55. Feist AM, Henry CS, Reed JL, et al. A genome-scale metabolic reconstruction for *Escherichia coli* K-12 MG1655 that accounts for 1260 ORFs and thermodynamic information. *Mol Syst Biol.* 2007;3:121.

56. Mo ML, Palsson BO, Herrgard MJ. Connecting extracellular metabolomic measurements to intracellular flux states in yeast. *BMC Syst Biol.* 2009;3:37.

SECTION III

Applications in Synthetic Biology

Design and Application of Synthetic Biology Devices for Therapy

Boon Chin Heng and Martin Fussenegger
ETH-Zurich, Basel, Switzerland

INTRODUCTION

Synthetic biology is the rational and systematic design/construction of biological systems with desired functionality.[1−4] This is achieved through applying engineering and computational principles within the field of molecular biology.[1−4] As such, synthetic biology represents an eclectic fusion of multiple disciplines. The debut of synthetic biology took place in the year 2000, when the first synthetic gene circuits, a toggle switch[5] and an oscillator,[6] were constructed in bacterial systems. Since then, the field has progressed extremely rapidly, with increasingly complex synthetic gene regulatory circuits being placed into mammalian cells,[7,8] fungi,[9,10] and viruses,[11,12] in addition to prokaryotic bacterial systems.[5,6,13] This, in turn, has spawned a diverse array of potential therapeutic applications, ranging from drug screening and discovery to cancer treatment and even fabrication of novel biomaterials.[14,15]

In recent years, the pace of progress in synthetic biology has been further accelerated by the gradual adoption of a bioinformatics and computer-aided design approach to the construction of synthetic gene circuits.[16−20] A number of software tools have been specifically developed for synthetic biology applications. These include SynBioSS (Synthetic Biology Software Suite),[21] TinkerCell,[22] and BioNetCAD.[23] Such programs provide databases of the different potential building blocks of synthetic biological systems, and enable modeling and simulation of hypothetical gene circuits in silico before conducting actual experiments in vitro and in vivo.

This chapter will start by providing an overview of the target organisms and the molecular toolkit available for synthetic biology applications. This will be followed by an extensive review of the potential therapeutic applications of synthetic biology in various aspects of human health and disease. Finally, we will critically examine future challenges and safety issues associated with the application of synthetic biology to clinical practice.

TARGET ORGANISMS AND CELL TYPES FOR THERAPEUTIC APPLICATIONS OF SYNTHETIC BIOLOGY

Human cells and tissues appear to be the most direct targets for the engineering of therapeutic gene circuits,[14,24,25] particularly for cancer therapy, to correct aberrant metabolic conditions,

Synthetic Biology. DOI: http://dx.doi.org/10.1016/B978-0-12-394430-6.00009-1

and for application in regenerative medicine. However, tinkering with the biological functions of bacteria and viruses can also have useful clinical applications. The diverse population of commensal and symbiotic microorganisms associated with the human body, known as the human microbiome, forms a complex ecosystem that plays an important role in regulating human physiology, health, and disease states.[26,27] The fact that the human body naturally tolerates the various species of microorganisms associated with it makes these microorganisms ideal targets for synthetic biology applications. Indeed, synthetic gene circuits have been engineered in human microbiome-associated bacterial species to confer host resistance to infectious diseases such as cholera,[28] as well as to kill cancerous cells.[29] Besides bacterial cells, synthetic gene circuits can also be deployed in viruses for therapeutic applications. Of particular interest are the bacteriophages, which are viruses that specifically infect bacterial cells. The T7 phage has been engineered to express the bacterial biofilm-degrading enzyme dispersin B,[30] while a synthetic gene circuit that interferes with the SOS response network in bacterial cells has been engineered in the M13 phage.[31]

MOLECULAR TOOLKIT FOR SYNTHETIC BIOLOGY
Synthetic Gene Circuits Encoded by Recombinant DNA
GENE TARGETING AND GENOME EDITING TECHNOLOGIES

Transfection of recombinant DNA into mammalian or bacterial cells is an essential prerequisite for the deployment of synthetic gene circuits. The major challenge is to achieve efficient and site-specific integration of the transfected recombinant DNA into the host genome. Transient transfection with nonintegrating plasmid DNA imposes severe limitations on long-term clinical applications. Moreover, there is a low probability of random and nonsite-specific integration of the plasmid DNA into the host genome, which could cause insertional mutagenesis of host genes. This could potentially lead to cancer, and hence invoke serious safety concerns.[32] Although the use of viral vectors can efficiently integrate recombinant DNA into the host genome, again, the problem is the random and nonsite-specific integration of recombinant DNA into the cellular host genome.[33,34]

The development of site-specific recombinase (SSR) technology was the first attempt at site-specific manipulation of genomic DNA. SSR systems such as Cre-LoxP,[35] Flp-FRT,[36] Dre,[37] and PhiC31[38] can be used to delete, insert, or invert a segment of DNA flanked by specific recombination sites within genomic DNA, and have now been incorporated into viral vectors.[39,40] More recently, there have been advances in the rational design of zinc finger (ZF) nucleases for site-specific insertion of recombinant DNA.[41−44] The nonspecific Fok1 domain of ZF nucleases can be coupled to transcription activator-like effectors (TALEs) to form novel genome-editing tools that enable precise integration of recombinant DNA into target chromosomal locations.[45]

It would also be of clinical interest to be able to transiently insert and remove transgenic elements in a precise and site-specific manner, without leaving any permanent modification to genomic DNA. For example, some synthetic biology applications may require temporary expression of synthetic gene circuits within cells for only a limited period of time. The PiggyBac transposon system can be particularly useful for this purpose.[46,47] Wilson et al.[48] demonstrated that PiggyBac integration and excision within human genomic DNA is very precise, without leaving any 'footprint' mutations at the site of transposon excision. The same study[48] also mapped a total of 575 PiggyBac integration sites within human genomic DNA to demonstrate the nonrandom site-selectivity of PiggyBac transposon integration. Such useful properties of the PiggyBac transposon system have been utilized for reprogramming mammalian somatic cells to pluripotent stem cells through the transient expression of four transgenes (c-Myc, Klf4, Oct4, and Sox2), without leaving any permanent genetic alteration to the reprogrammed cells.[49,50]

TABLE 9.1 Different Types of Synthetic Gene Circuits

Type of Synthetic Gene Circuit	References
Switches	Gardner et al.[5]
	Atkinson et al.[54]
	Stupak and Stupa[55]
	Boczko et al.[56]
	Greber et al.[57]
	Ghim and Almas[58]
	Kramer and Fussenegger[59]
	Isaacs et al.[60]
	Lou et al.[61]
	Kramer et al.[62]
	Xiong and Ferrell[63]
	Ptashne[64]
	Zordan et al.[65]
	Huang et al.[66]
	Ajo-Franklin et al.[67]
Oscillators	Elowitz and Leibler[6]
	Atkinson et al.[54]
	Goodwin[68]
	Stricker et al.[69]
	Fung et al.[70]
	Tigges et al.[71]
	Tigges et al.[72]
Filters	Austin et al.[73]
	Hooshangi et al.[74]
	Basu et al.[75]
	Sohka et al.[76]
	Muranaka and Yokobayashi[77]
	Weber et al.[78]
	Greber and Fussenegger[79]
Communication modules	You et al.[80]
	Kobayashi et al.[81]
	Balagadde et al.[82]
	Danino et al.[83]
Logic gates	Anderson et al.[84]
	Guet et al.[85]
	Rackham et al.[86]
	Rinaudo et al.[87]
	Stojanovic et al.[88]
	Win and Smolke[89]

To date, most synthetic gene networks have been constructed using the classical restriction−digestion-based molecular cloning method that has several inherent limitations. Multiple steps are often required, which makes this classical cloning technique both labor- and time-intensive. Moreover, the use of restriction enzymes would leave a restriction site-scar between annealed DNA fragments. Molecular cloning with specific restriction enzymes can also be potentially hindered by the presence of multiple restriction sites within either the target gene sequence or the destination vector backbone. These limitations may be overcome through newly developed restriction-enzyme free molecular cloning techniques that enable scarless, sequence-independent multipart DNA assembly, such as SLIC (sequence and ligase independent cloning),[51] Gibson DNA assembly,[52] and CPEC (circular polymerase extension cloning).[53] All that is required is for the target gene sequence to be PCR amplified with oligonucleotide primers that have 5′ termini sequence homology (at least 25 bp) to the corresponding ends of the destination vector. In the case of SLIC and Gibson DNA assembly, exonuclease activity is used to generate complementary overhangs on the target DNA fragment and the linearized destination vector, which are then annealed in the absence (SLIC) or presence (Gibson DNA assembly) of ligase. Another major difference between the two techniques is that T5 exonuclease is utilized in the case of Gibson DNA assembly, whereas SLIC utilizes T4 DNA polymerase which displays 3′ exonuclease activity in the absence of dNTPs.[51,52] By contrast, in the case of CPEC, no exonuclease activity is utilized to generate overhangs, and neither are oligonucleotide primers utilized.[53] Instead, both the target gene sequence and the linearized destination vector are melted into single strands and subsequently annealed to each other, thus allowing both DNA fragments to prime each other in the presence of Phusion polymerase.[53]

TYPES OF SYNTHETIC GENE CIRCUITS

Overview

The application of engineering and computing principles in synthetic biology has helped develop a diverse variety of synthetic gene circuits with a panoply of different functions. It is convenient to compare the functioning of synthetic gene circuits with analogous electronic devices. These are summarized in Table 9.1, and can be broadly classified into genetic switches,[5,54–67] oscillators,[6,54,68–72] filters,[73–79] communication modules,[80–83] and other miscellaneous synthetic gene circuits such as various digital logic gates.[84–89] Each of these is discussed here in turn.

Genetic Switches

In electronics, a switch is a device that allows conditional transition from one state to another, in response to an input signal. Similarly, synthetic genetic switches are artificial regulatory networks that allow cells to undergo conditional transition between gene expression states. Although it may appear simple and straightforward to genetically engineer cells to switch on/off gene expression in response to metabolic, physical, or cytokine-induced stimuli, the challenge is to achieve a robust bistable transition from one state to another without the tendency to flip randomly between states as a result of fluctuations inherent in gene expression. This challenge may be overcome through the incorporation of positive and negative feedback loops.[5,54–57] Construction of bistable genetic toggle switches has been reported in bacteria,[5,54,55] as well as in yeast[56] and mammalian cells.[57] Subsequently, more complex genetic switches with tunable,[58] hysteresis,[59,60] and gating[61] functions have also been created. Besides control of transcription and translation, genetic switches for epigenetic regulation have also been built.[62]

Perhaps one of the most important applications of genetic switches in synthetic biology is to function as cellular memory elements. Cellular memory can be defined as a protracted response to previous exposure to a transient stimulus. The best example of cellular memory in nature is cell fate decision-making during the process of differentiation, whereby a stem/progenitor cell makes a permanent and irreversible decision to commit to a particular somatic lineage. Memory elements of gene regulatory networks are characterized by two major feedback motifs — mutual inhibition and autoregulatory positive feedback.[63–66] Examples of synthetic memory elements displaying mutual inhibition and autoregulatory positive feedback include genetic toggle switches[5,55–57] and hysteresis networks,[59,60] respectively. Ajo-Franklin et al.[67] constructed a high-fidelity memory device in yeast based on transcriptionally controlled autoregulatory positive feedback. This synthetic gene circuit allowed cells to heritably retain an induced state after responding to a transient stimulus. This was achieved by expression of an 'auto-feedback' transcription factor that binds to its own promoter upon exposure to a transient stimulus. The result is that the 'auto-feedback'

162

transcription factor continues to activate its own expression even when the stimulus is later absent, resulting in cellular memory.

Oscillators

In nature, cells exhibit a variety of oscillatory functions such as the cell cycle and circadian clock. Genetic oscillators are of particular interest to synthetic biology because these can be employed to precisely synchronize and time the expression of proteins and RNA for both therapeutic and nontherapeutic applications. The first synthetic genetic oscillators were reported in prokaryotic systems. The simplest example, reported by Goodwin[68] in 1963, just comprised a single gene (LacI) that represses itself. This was followed by the construction of more complex repressilators, which are regulatory networks of more than one gene, with each gene repressing its successor in the cycle.[6] The first synthetic repressilator was developed by Elowitz and Leibler.[6] This involved a closed loop of three genes consecutively repressing each other, with LacI repressing tetR, which represses λcI, which in turn represses LacI gene, thus completing the loop.

Both the Goodwin oscillator and repressilators function through repression of gene function. The next step in oscillator design would be to incorporate both activation and repression of gene expression. This was realized by the construction of an amplified negative-feedback oscillator by Atkinson et al.[54] This consisted of one gene (NRI) promoting its own transcription via a positive self-feedback loop, which also promoted the transcription of a second gene (LacI), which in turn repressed expression of the first gene (NRI), forming a negative feedback loop. Subsequently, more complex synthetic gene oscillators with tunable function,[69] and which utilizes metabolic flux as a control factor in system-wide oscillation,[70] were constructed in *E. coli*.

The first synthetic genetic oscillator to be constructed in mammalian systems was reported by Tigges et al.[71] This is, in fact, an amplified negative-feedback oscillator that contains a delay in the negative-feedback loop. The resultant effect is that the oscillations in gene expression are undamped. This system is composed of two genes, with both sense and antisense transcription occurring from one of the genes. The sense transcript is translated into a protein (tTA) that promotes its own transcription and that of a second gene (PIT), which in turns activates antisense transcription from the first gene (tTA). This antisense transcript will then repress translation of the sense transcript through hybridization, constituting a negative feedback loop. Following up on this study, Tigges and colleagues constructed a low-frequency variant of this oscillator.[72] To date, these are the only synthetic genetic oscillators to be reported in mammalian systems.

Filters

In synthetic biology, filters and bandpasses are required for noise control, as well as for controlling the output of specific signal modes. The study of Austin et al.[73] with *E. coli* cultures demonstrated that negative feedback is able to shape noise spectrum. In the study of Hooshangi et al.,[74] low-pass filtering of input signals was achieved in a synthetic transcriptional cascade within bacterial cells. This conferred some robustness to fluctuations in input signal conditions, that is, noise. The first synthetic bandpass filter in bacterial systems was originally reported by Basu et al.,[75] in which one cell population (receiver) was programmed to respond to a defined concentration range of acyl-homoserine lactone (AHL) produced by another cell population (sender). Subsequently, Sohka et al.[76] developed a bacterial bandpass filter with tunable function. Tunability of the bandpass filter was achieved by utilizing three different β-lactamase genes that conferred different levels of ampicillin resistance.[76] A more complex bacterial bandpass filter incorporating metabolite-responsive riboswitches was reported by Muranaka and Yokobayashi.[77] This comprised of

163

tandem riboswitches that could function as the Boolean logic gates AND and NAND in response to two chemical inputs – theophylline and thiamine pyrophosphate (TPP).

The first synthetic bandpass filter in mammalian cells was reported by Weber et al.,[78] which displayed specific modulation of target gene expression within a defined range of biotin concentrations (input signal). This system involved BirA-mediated ligation of biotin to a biotinylation signal-containing VP16 transactivation domain, which triggered heterodimerization of chimeric VP16 to a streptavidin-linked tetracycline repressor (TetR). With increasing biotin concentration up to 20 nM, there is gradual activation of tetR-specific promoters until maximal induction is achieved at a biotin concentration range of 20 nM to 10 μM. Above a biotin concentration of 10 μM, expression of the target gene is shut off. Yet another mammalian bandpass filter responsive to a defined range of tetracycline concentrations was reported by Greber and Fussenegger.[79] This consisted of a high-detect and low-detect switch that repressed gene expression at high and low tetracycline concentrations, respectively.

Communication Modules

Synthetic gene circuits that enable communication between cells have also been constructed, and are often referred to as communication modules. By utilizing elements of a bacterial quorum-sensing system from the marine bacterium *Vibrio fischeri*, You and colleagues[80] were able to program an *E. coli* population to maintain a cell density lower than that imposed by environmental constraints; that is, a limited supply of oxygen and nutrients as well as the accumulation of toxic metabolites. This was achieved by engineering the expression of the LuxI/LuxR quorum-sensing system from *V. fischeri* into *E. coli*. The LuxI protein synthesizes acyl-homoserine lactone (AHL), which diffuses into the surrounding milieu and neighboring cells. At above a certain threshold concentration, AHL binds and activates the LuxR repressor protein, which in turn induces expression of a 'suicide' gene.

Utilizing the same LuxI/LuxR quorum-sensing system from *V. fischeri*, Kobayashi et al.[81] were able to program *E. coli* populations to activate synthesis of specific proteins when the cell population exceeded a certain critical threshold density. Additionally, the same study also engineered *E. coli* populations to form a biofilm in response to DNA-damaging chemicals by utilizing elements of the bacterial SOS signaling response to DNA damage.[81] In yet another study, Basu et al.[75] programmed bacterial cells to generate intricate two-dimensional patterns through intercellular communication by splitting the LuxI/LuxR quorum-sensing system into two distinct 'sender' and 'receiver' subpopulations of bacterial cells. By using quorum sensing to regulate expression of 'killer' and 'antidote' proteins in two distinct *E. coli* populations, Balgadde et al.[82] were able to construct a synthetic ecosystem resembling a canonical predator–prey relationship. More recently, Danino et al.[83] used elements of the quorum-sensing machineries of both *V. fischeri* and *Bacillus thuringiensis* to generate synchronized oscillations of gene expression within bacterial populations.

Miscellaneous Synthetic Gene Circuits

Other miscellaneous synthetic gene circuits analogous to complex electronic devices have also been constructed. This includes a diverse array of synthetic gene circuits that correspond to various digital logic gates.[84–89]

FUNCTIONAL PROTEINS AND RNA COMPONENTS OF SYNTHETIC GENE CIRCUITS

For synthetic gene circuits to be effective, they must encode for a variety of functional proteins and RNA that exert control over diverse biological processes within cells. Proteins encoded by synthetic gene circuits can be broadly classified into membrane components and intracellular components.

Membrane components generally serve as functional receptors for extracellular molecular or physical signals such as metabolites, cytokines, light, and electrical pulses. This will be discussed in greater detail in the next section. Commonly utilized membrane protein receptors in synthetic biology belong to two major families — the G-protein coupled receptors (GPCR)[90] and the tyrosine kinase family.[91]

Intracellular components can function to relay signals from activated membrane receptors and influence gene expression at the transcriptional or translational level. Very often, intracellular components are already present endogenously within the cells, and need not be encoded by the synthetic gene circuit; such as the G-proteins associated with the GPCR. Nevertheless, the transcriptional control element of the synthetic gene circuit must be responsive to secondary messengers (Ca^{2+} influx, cAMP) activated by the transgenically expressed membrane receptors. Among the more common transcriptional control elements utilized in synthetic biology are NFAT,[92] which responds to Ca^{2+} influx, and CRE,[93] which responds to elevated cAMP levels. Alternatively, the intracellular components encoded by the synthetic gene circuits may respond directly to small molecules that diffuse into the cells and activate or repress gene expression. Perhaps the best-known example is the tetracycline tetO-TetR system developed for mammalian cells.[94]

In recent years, there has also been increased interest in synthetic gene circuits that code for functional RNA molecules.[95] Besides being short-lived transmitters of information from DNA to proteins, RNA molecules can potentially serve as functional molecules in their own right, in a manner similar to proteins. Examples of the functionality of RNA molecules are their roles as short interfering RNA (siRNA) and microRNA in RNA interference,[96] and in directly binding specific ligands to influence gene expression; that is, RNA aptamer domains of riboswitches.[97,98] Although either microRNA or siRNA mediates RNA interference in nature, the expression of short hairpin RNA (shRNA) is preferable for synthetic biology applications. The reason is that shRNA offers silencing longevity, better delivery options, and lower costs compared with siRNA or microRNA.[99] The cellular machinery cleaves the shRNA hairpin structure into siRNA, which then binds to the RNA-induced silencing complex (RISC) to effect gene silencing.[99]

RNA structural motifs that can bind to specific ligands (aptamers) have also aroused much interest in the field of synthetic biology.[95] Of particular interest are riboswitches,[98] which are cis-acting structural motifs within the untranslated portion of an mRNA molecule, and consist of an aptamer domain linked to an expression platform. Upon binding to its specific ligand, the aptamer domain effects a structural change in the expression platform, which in turn affects mRNA transcription, translation, and splicing.[98] A ribozyme is an RNA molecule with a well-defined tertiary structure that enables it to act like a protein enzyme in catalyzing biochemical and metabolic reactions within a cell. Because some ribozymes, such as the hammerhead ribozyme and the hairpin ribozyme, cleave RNA, there has been increasing interest in utilizing them for antiretroviral therapy.[100,101] An intriguing possibility is to conjugate ligand-specific aptamer domains to ribozymes to effect their catalytic function. Indeed, this has already been achieved with hammerhead ribozymes.[102,103]

INDUCTION OF SYNTHETIC GENE CIRCUITS

Synthetic gene circuits can be engineered to be responsive to a diverse array of inductive stimuli, such as antibiotics,[104–106] orally ingestible food supplements such as vitamins, amino acids and flavorings,[107–109] specific drugs and medications,[110,111] skin-lotion-based chemicals that can penetrate the skin,[112,113] endogenous metabolites, hormones and redox states,[114–116] gaseous chemicals (for example, involving smell),[117,118] and physical stimuli such as light and electrical pulses.[119,120] These are summarized in Table 9.2.

TABLE 9.2 Different Inductive Stimuli for Synthetic Gene Circuits

Inductive Stimuli for Synthetic Gene Circuit	References
Antibiotics	Reeves et al.[104] Jiang et al.[105] Tonack et al.[106]
Food supplements	Hartenbach et al.[107] Weber et al.[108]
Drugs and medications	Weber et al.[109] Chong et al.[110] Tascou et al.[111]
Skin-penetrating chemicals	Valenta et al.[112] Gitzinger et al.[113]
Endogenous metabolite, hormone, or redox state	Kemmer et al.[114] Kemmer et al.[115]
Gaseous chemicals, i.e. smell	Weber et al.[116] Weber et al.[118] Weber et al.[119]
Physical stimuli, i.e. light, electrical pulse	Weber et al.[119] Ye et al.[120]

One of the first inductive stimuli which synthetic gene circuits were engineered to respond to was the presence/absence of antibiotics, best exemplified by the TET_{ON} and the TET_{OFF} system developed for mammalian cells.[104–106] Despite this achievement, the use of antibiotics for the induction of synthetic gene circuits has limited therapeutic application. For example, the systemic administration of antibiotics may kill off symbiotic bacterial populations within the patient (for example, in the gut), as well as encourage the development of antibiotic resistance.[121]

To allay safety concerns for therapeutic applications, it may be advantageous to design synthetic gene circuits that are responsive to nontoxic and orally ingestible food supplements such as vitamins, amino acids, and flavorings. Indeed, synthetic gene circuits responsive to the amino acid arginine,[107] vitamin H (Biotin),[108] and strawberry flavor 2-phenyl ethyl butyrate[109] have been reported. Nevertheless, it must be noted that because vitamins and amino acids are found naturally within the human body and are implicated in key metabolic processes, ingestion of high doses may interfere with the patient's own metabolism.

For certain therapeutic applications, it may be advantageous to engineer synthetic gene circuits to be responsive to specific drugs and medications administered to the patient. Chong et al.[110] developed a synthetic gene network to control adenoviral vector replication based on the response to the immunosuppressive drug rapamycin. In another study by Tascou et al.,[111] a synthetic gene circuit responsive to the antidiabetic drug rosiglitazone was constructed.

Besides oral and intravenously administered substances, synthetic gene circuits can also be engineered to be responsive to skin-lotion-based chemicals that can penetrate the skin. This is best exemplified by phloretin, an anti-bacterial plant defense metabolite found naturally within apples, and commonly utilized as a penetration enhancer for skin-based drug delivery.[112] Gitzinger et al.[113] constructed a synthetic gene circuit responsive to phloretin by using a flavonoid-specific biosensor (TtgR) that interacts with phloretin and binds to a specific operator sequence (O_{TtgR}).

In addition to exogenous stimuli, it may be necessary in some therapeutic applications to engineer synthetic gene circuits to respond to endogenous stimuli in situ within the human body itself. For example, a high level of uric acid in the blood circulation is associated with tumor lysis syndrome and gout. Kemmer et al.[114] constructed a synthetic gene circuit that senses endogenous uric acid levels through a *Deinococcus radiodurans*-derived protein, and which subsequently triggers elimination of excessive uric acid through a secretion-engineered *Aspergillus flavus* urate oxidase. Another example of a synthetic gene circuit responding to endogenous signals in situ is a designer gene network for coordinating bovine artificial insemination by the ovulation-triggered release of implanted sperm.[115] This involves transgenic expression of cellulase in response to a lutenizing hormone surge associated with ovulation. The cellulase enzyme then breaks down the cellulose microcapsules containing the spermatozoa, thus effecting their release for fertilization at the time of ovulation. Because certain diseases are characterized by changes in redox state, it may also be therapeutically relevant to engineer synthetic gene circuits to be responsive to the redox state within cells. Weber et al.[116] constructed a genetic redox sensor for mammalian cells based on the *Streptomyces coelicolor* redox control system.

In addition to their response to soluble molecules, synthetic gene circuits can be engineered to be responsive to molecular signals in the gas phase. The pioneering study of Weber et al.[117] achieved gas-inducible transgene expression in mammalian cells by reengineering the acetaldehyde-inducible $AlcR-P_{alcA}$ transactivator–promoter interaction found within *Aspergillus nidulans*. By utilizing this synthetic gene circuit, together with the conversion of ethanol to acetyldehyde, a further study[118] established a synthetic airborne cell communication network.

Besides response to molecular signals, it may be useful in some therapeutic applications for synthetic gene circuits to be able to respond to physical stimuli. For example, it may be necessary to engineer synthetic gene circuits to respond to electrical signals in order to interface between implanted electronic/electrical devices and biological systems. Weber et al.[119] created an electro-genetic transcription unit by using electrical power to convert ethanol into acertaldehyde, which in turn interacted with the acetyldehyde-inducible gene circuit discussed earlier.[117] Gene transcription can be adjusted according to the intensity of an applied direct current, as well as to the amplitude or frequency of an applied alternating current. Utilizing such an electro-genetic transcription unit, modulation of the beating frequency of primary heart cells was demonstrated.[119] A more recent study by the Fussenegger group[120] used a light-inducible synthetic gene circuit to control the transgenic expression of GLP-1 in a type II diabetic mouse model. Using this system, the attenuation of glycemic excursions and the enhancement of blood-glucose homeostasis were achieved in diabetic mice.

Alternatives to Recombinant DNA: Synthetic/Modified RNA, Proteins, and Other Analogue Molecules

The use of recombinant DNA in synthetic biology applications carries a risk of permanent genetic alteration to cells, which in turn raises serious safety concerns. To address these safety concerns, an alternative may be to utilize synthetic/modified RNA, proteins and other analogue molecules (for example, peptide nucleic acid, locked nucleic acid) in place of recombinant DNA for some, but not all, synthetic biology applications. Nevertheless, it must be noted that because these molecules are not integrated into the cellular genome, and are eventually degraded, their effect is only transient. Hence, these molecules are only suited for synthetic biology applications that require just a transient change in gene/protein expression.

For transient gene silencing through RNA interference, it may be advantageous to utilize nucleic acid analogues, such as peptide nucleic acid (PNA)[122] and locked nucleic acid (LNA),[123] rather than siRNA or microRNA. This is due to the greater stability and superior hybridization capacity of PNA and LNA compared with RNA, which in turn translates into a more potent and longer-lasting gene silencing effect.[122,123] Another example of a synthetic biology application requiring transient changes in gene/protein expression is in cellular reprogramming and the determination of cell lineage fate. Reprogramming of somatic cells into an embryonic stem-cell-like state (induced pluripotent stem cells) through transfection of cell-permeable proteins[124] and mRNA[125] without the use of recombinant DNA has already been achieved. This will be discussed in greater detail below in the section 'Stem cells and regenerative medicine.'

THERAPEUTIC APPLICATIONS OF SYNTHETIC BIOLOGY
Cancer Therapy

Cancer therapy remains a persistent and formidable clinical challenge because common treatment modalities such as surgical excision, radiotherapy, and chemotherapy inevitably lead to some destruction of healthy tissues within the patient. Synthetic biology can potentially allow more selective targeting and destruction of cancer cells while avoiding the destruction of healthy tissues. Indeed, a number of novel synthetic gene circuits have been engineered in mammalian cells,[126] viruses,[127] and bacteria[29,128] to achieve this purpose.

Chen and colleagues[126] were able to engineer a ribozyme-based synthetic gene circuit into tumor-targeting T-cells to enable rapid proliferation in response to small-molecule drugs. This was achieved by coupling ligand-responsive ribozyme switches to relevant T-cell growth-promoting cytokine genes such as interleukin-2 and interleukin-15.[126] At the same time, the synthetic gene circuit incorporated a 'safety-switch' in the form of the suicide gene thymidine kinase, which provides a means of ablating T-cell proliferation once it is no longer needed.[126]

Ramachandra et al.[127] utilized a genetically engineered adenovirus for cancer therapy. These investigators constructed a synthetic gene circuit that coupled adenoviral replication to the p53 pathway in human cells. In this circuit, adenoviral replication was inhibited in the presence of a normal p53 pathway, while an aberrant p53 pathway characteristic of malignant cells triggered adenoviral replication, leading to cell death.

Besides mammalian cells and viruses, bacteria can also be utilized for cancer therapy. In the study of Anderson et al.,[29] *E. coli* was conferred with the ability to invade and destroy cancer cells through expression of the invasin protein from *Yersinia pseudotuberculosis*, which was activated in the presence of a hypoxic environment characteristic of cancerous tissues. This was achieved by placing expression of the invasin protein under the control of the formate dehydrogenase promoter that is induced under anaerobic conditions. In another study by Xiang et al.,[128] invasin-expressing *E. coli* was utilized to target the colon-cancer-causing ß-1 catenin gene (CTNNB1) through RNA interference. Immune-deficient mice subcutaneously xenografted with colon cancer cells exhibited significant regression of the tumorous xenograft upon intravenous administration of genetically engineered *E. coli*. This demonstrated that localized administration was not necessary, and that cancer-invading bacteria can be utilized for selective targeting of tumorous tissues at distant sites.

Treatment and Prevention of Infectious Diseases

Human populations worldwide are being threatened by newly emerging infectious diseases,[129,130] as well as from rapidly evolving new strains of old pathogens that exhibit increasing resistance to antibiotics and other therapeutic drugs. Hence, a major focus in the field of synthetic biology is the treatment and prevention of infectious diseases.

A persistent challenge in the treatment of bacterial infections is the development of antibiotic resistance.[121] This may possibly be overcome through the disruption of molecular networks implicated in antibiotic defense mechanisms. Lu et al.[31] genetically engineered the nonlytic M13 bacteriophage to express lex A3, a repressor of the SOS response network that protects bacteria against DNA damage induced by antibiotics. Subsequent exposure of various resistant bacterial strains to this genetically engineered bacteriophage in the presence of three major classes of antibiotics (quinolones, ß-lactams, and aminoglycosides) resulted in up to a 500-fold enhancement in bacterial cell death, compared with treatment by antibiotics alone.

Another major challenge faced in combating bacterial infections is the formation of protective biofilms by pathogenic bacteria,[131–133] which shield them from antibiotics and the host immune system. In another study by Lu et al.,[30] the lytic T7 phage was genetically engineered to express the enzyme dispersin B, which readily degrades the biofilm matrix. Degradation of the biofilm matrix by dispersin B allowed exposure of unprotected bacterial cells to infection and subsequent lysis by the genetically engineered T7 phage.

Naturally occurring commensal and symbiotic bacterial species within the human body can be genetically engineered to prevent and fight infectious pathogens. Of particular interest is *E. coli*, which is probably the most abundant microbe found within the human body. Duan and March[28] genetically engineered *E. coli* to express both cholera autoinducer-1 (CAI-1) and autoinducer-2 (CAI-2), both of which are naturally utilized by *Vibrio cholerae* for quorum sensing. When the population density of *Vibrio cholerae* is high, elevated levels of CAI-1 and CAI-2 secreted by the bacterial cells inhibit the secretion of the cholera toxin. Hence, in this manner, the genetically engineered *E. coli* that produce CAI-1 and CAI-2 can mitigate the virulence of *Vibrio cholerae* infections through the inhibition of cholera toxin production. Rao et al.[134] genetically engineered a highly colonizing probiotic strain of *E. coli* (Nissle 1917) to secrete HIV-gp41-hemolysin A hybrid peptides that block HIV fusion and subsequent entry into target cells. Subsequently, it was demonstrated that the genetically engineered *E. coli* was capable of colonizing various tissues such as the rectum, vagina, and small intestine of mice for prolonged durations up to several months, while actively secreting the HIV fusion inhibitor peptide.[134]

Besides targeting infectious pathogens per se, it may also be possible to take a synthetic biology approach to target their insect vectors. Windbichler et al.[135] attempted to reduce the ability of mosquitoes to transmit malaria by disrupting the genes that encode malaria vector capability within the mosquito genome. This was achieved with a synthetic gene drive comprised of the homing endonuclease gene I-SceI together with aberrant mosquito regulatory genes that reduced malaria vector capability. The homing endonuclease induced double-stranded DNA breaks, which in turn activated the endogenous DNA repair system within the mosquito cells. The homologous chromosome containing the synthetic gene drive was utilized as a template for repair, which in turn caused both of the chromosomes to carry the synthetic gene drive. In this manner, the synthetic gene drive that reduced malaria vector capability could be transmitted rapidly within mosquito populations.

Vaccine development could also benefit from a synthetic biology approach. Amidi et al.[136] were able to encapsulate a reconstituted bacterial transcription and translation network, together with DNA encoding a model antigen within synthetic liposome vesicles, and utilize these to provoke a humoral immune response in mice. This can provide a safer alternative to attenuated live antigens without the potential of becoming virulent.

In another study by Coleman et al.,[137] an attenuated poliovirus vaccine was created by exploiting species-specific codon bias. This is based on the principle that while several different codon sequences can code for the same amino acid, each individual species exhibits bias for the adjacent codons that it can translate efficiently into protein. Hence, by switching synonymous codons that encoded the poliovirus capsid protein, the translation efficiency was greatly reduced, resulting in an attenuated poliovirus with reduced virulence that could be utilized as a live vaccine.

Treatment of Metabolic Disorders

The use of synthetic gene circuits to correct aberrant metabolic conditions is another major area of interest in synthetic biology. Obesity has recently become a major health concern in developed countries, due primarily to lifestyle and dietary factors. Dean et al.[138] incorporated a synthetic gene circuit encoding the glyco-oxylate shunt pathway into mice liver cells, resulting in increased fatty acid oxidation. Upon being placed on a high-fat diet, the mice expressing the glyco-oxylate shunt pathway in their liver cells showed increased resistance to diet-induced obesity, as well as reduced plasma triglyceride and cholesterol levels.

A metabolic disorder closely associated with obesity is type II diabetes, the incidence of which is increasing rapidly in developed countries.[139,140] In a groundbreaking study by the Fussenegger group,[120] a light-responsive synthetic gene regulatory circuit expressing glucagon-like peptide I (GLP-I) was shown to be capable of attenuating glycemic excursions in a diabetic mouse model. Exogenous light stimulation was delivered through implanted fiber optic cables. In an earlier study by the Fussenegger group, aberrant uric acid metabolism associated with gout and tumor lysis syndrome was corrected by a synthetic gene circuit expressing uricase mUox, which converted the toxic urate metabolite to allantoin for easy excretion.[114]

Responsive Biomaterials and Controlled Release of Biopharmaceuticals

A newly emerging application of synthetic biology is in the fabrication of 'smart' responsive biomaterials that may be capable of controlled release of biopharmaceuticals. This is best exemplified by hydrogels composed of different high-molecular-weight polymers cross-linked by either DNA or proteins.[15]

DNA-based responsive hydrogels exploit the ability of a defined sequence of single-stranded DNA (ssDNA) to bind specifically to either its complementary ssDNA strand or to various small molecules and proteins. Tierney & Stokke[141] achieved crosslinking within a hydrogel with complementary ssDNA oligonucelotide sequences functionalized to polyacrylamide chains. Upon exposure to free ssDNA oligonucleotides with complementary sequences to the crosslinking DNA, competitive DNA hybridization took place, resulting in disruption of the crosslink and subsequent dissolution of the hydrogel.

In an alternative hydrogel configuration reported by Lin et al.,[142] polyacrylamide chains were functionalized to two different noncomplementary ssDNA oligonucleotide sequences. Crosslinking was then achieved with a third ssDNA oligonucleotide that had terminal complementary sequences to the two ssDNA oligonucleotides functionalized on the polyacrylamide chains. It was then demonstrated that the stiffness of the hydrogel could be modulated with a fourth ssDNA oligonucleotide that had a complementary sequence to the nonhybridized mid-portion of the third oligonucleotide.

The existence of well-defined DNA sequences that bind specifically to various small molecules (aptamers) and protein transcription factors (operator/promoter sequences) may be exploited to further expand the functionality of DNA-based hydrogels to enable the controlled release of biopharmaceuticals. In the study of He et al.,[143] aptamer sequences that bind specifically to small molecules, such as adenosine-5′-triphosphate and human α-thrombin, were localized within the mid-section of ssDNA that crosslinks polyacrylamide chains functionalized with complementary DNA oligonucleotide sequences. Crosslinking was achieved through hybridization at the terminal ends of the ssDNA, while the mid-section aptamer domain remained free to bind to small molecules. Subsequent exposure to free ssDNA of complementary sequences resulted in competitive DNA hybridization and displacement of the small molecule, as well as disruption of crosslinking and dissolution of the hydrogel. This effected controlled release of the small molecule. In an alternative configuration, binding of small molecules to the aptamer domain of crosslinking ssDNA could be used to effect dissolution of the hydrogel. Yang et al.[144] reported on DNA-based hydrogels that underwent dissolution in the presence of adenosine or thrombin.

The interaction of specific protein transcription factors with their cognate promoter/operator sequences has been well-studied,[145] and can be potentially exploited to further expand the functionality of DNA-based hydrogels. Of particular interest is the fact that many of these protein transcription factors also bind to various small-molecule inducer/repressors that in turn modulate their binding affinity to their cognate operator/promoter sequences. Christen et al.[146] fabricated a tetracycline-responsive DNA-based hydrogel based on the TetR-tetO interaction.[94] Polyacrylamide chains within the hydrogel were functionalized with either the TetR protein or DNA oligonucleotide sequences corresponding to tetO. Hence, exposure to tetracycline led to disruption of the TetR-tetO crosslink, resulting in dissolution of the hydrogel.

Besides DNA-based hydrogels, the use of proteins to crosslink high-molecular-weight polymers in hydrogels has also been reported. Kampf et al.[147] created a hydrogel by crosslinking polyacrylamide chains with Fm dimer proteins. In the presence of the immunosuppressive drug FK506 (tacrolimus), the Fm dimers dissociated, disrupting the crosslink and causing dissolution of the hydrogel. Subsequently, Kampf et al.[147] demonstrated that this FK506-responsive hydrogel could be utilized for the controlled release of vascular endothelial growth factor (VEGF) in a mouse model. The ability of some proteins to undergo drastic conformational change in the presence/absence of their cognate ligands can also be exploited in the fabrication of responsive hydrogels. In the study of King et al.,[148] the protein calmodulin was utilized as a crosslinker for a polyethyleneglycol (PEG)-based hydrogel. The addition or withdrawal of its cognate ligand trifluoperazine (TFP) induced drastic conformational changes, which in turn were translated into swelling or shrinking of the hydrogel.

Drug Screening and Discovery

Synthetic biology has also found increasing applications in the field of drug screening and discovery. Synthetic gene circuits engineered into mammalian cells, and incorporating suitable reporter genes, enable an easily measurable and quantifiable readout upon pharmaceutical induction of the target cellular signaling pathway. Moreover, the use of mammalian cells as screening tools also allows simultaneous evaluation of cytotoxicity and cell penetration ability, in addition to assessment of its pharmacological effect on the target signaling pathway. Mammalian cell-based screening technology can be combined with a high-throughput analyzer for rapid screening and identification of pharmacologically active compounds from numerous potential candidates extracted from natural products or chemically synthesized in the laboratory.

Ligands to the G-protein-coupled receptors (GPCR) form an important class of potential drug candidates.[149] Currently, there is a commercially available mammalian cell-based assay that specifically identifies GPCR-binding compounds (marketed by Life Technologies Inc., Gaithersburg, MD, USA, under the trademark name of Tango® assay). A synthetic gene circuit incorporating the β-lactam reporter gene is used to provide the readout for receptor binding and activation.[150]

Another important category of potential drug candidates is anticancer drugs, which must have the ability to kill malignant neoplastic cells while at the same time have minimal cytotoxic effect on normal cells. An initial first screen of potential anticancer drug candidates would be to identify compounds that selectively kill rapidly dividing cells, while having no effect on mitotically inactive cells. Gonzalez-Niccolini et al.[151] engineered a synthetic gene network into Chinese hamster ovary (CHO) cells that allowed mitosis to take place only in

171

the presence of tetracycline. G1-phase-specific growth arrest occurs when tetracycline is withdrawn due to activation of the cycline-dependent kinase inhibitor p27^{kip1} within the synthetic gene network. Nevertheless, some cells that can escape p27^{kip1}-mediated G1-phase growth arrest spontaneously arise within the cell population at a precise frequency. This somewhat recapitulates the natural development of cancer, whereby mitotically active cells spontaneously appear within a growth-arrested population. Utilizing such a cellular model, Gonzalez-Niccolini et al.[151] demonstrated that commonly used cancer drugs, such as 5-fluorouracil, doxorubicin, and etoposide, can selectively kill the spontaneously occurring mitotic cells, while having no effect on the growth-arrested cells. Such a cellular model can potentially be used as a first-line screening assay for potential anticancer drug candidates.

In recent years, newly emerging infectious bacterial diseases[129] and widespread antibiotic resistance[121] have made it imperative to screen and identify new antibiotic compounds. Aubel et al.[152] engineered a synthetic gene circuit incorporating the *Streptomyces pristinaespiralis*-derived streptogramin-responsive promoter PIP into CHO cells. In response to the streptogramin antibiotic pristinamycin I, the PIP repressor dissociates from its cognate promoter, leading to expression of the reporter gene (secreted alkaline phosphatase). Subsequently, the genetically engineered CHO cells were utilized for the screening and identification of potential antibiotic candidates against clinical pathogens, which were able to penetrate mammalian cells while exhibiting minimal cytotoxic effects.

Weber et al.[153] adopted a similar approach and developed a screening tool to identify potential antituberculosis drug candidates that are able to penetrate mammalian cells while having negligible cytotoxic effects. This is critical in view of the fact that *Mycobacterium tuberculosis* is an intracellular pathogen. Currently, the most effective therapeutic drug against *Mycobacterium tuberculosis* is ethionamide.[154] However, for ethionamide to exert its toxic effects on *Mycobacterium tuberculosis*, it must be activated by Baeyer-Villiger monooxygenase EthA within the bacterium itself.[154] Nevertheless, EthA expression is naturally repressed by another protein, EthR. Hence, inhibitors of EthR can potentiate the therapeutic efficacy of ethionamide on *Mycobacterium tuberculosis*. Therefore, the synthetic gene circuit developed by Weber et al.[153] linked the inhibition of EthR to the expression of a reporter gene (secreted alkaline phosphatase). Utilizing this setup, Weber et al.[153] were able to screen and identify several cell-penetrating compounds that can inhibit EthR, while exhibiting minimal cytotoxicity to mammalian cells.

Loose et al.[155] used linguistically based algorithms developed for syntax and grammar analysis to examine the sequences of naturally occurring antimicrobial peptides. Certain characteristic patterns were identified in the sequences of antimicrobial peptides, and this information was used to design new artificial peptides that exhibit antimicrobial activity. In this manner, the rational design of antimicrobial peptides can be achieved.

Stem Cells and Regenerative Medicine

Synthetic biology has also found increasingly numerous applications in the field of stem cells and regenerative medicine. Of particular interest is cellular reprogramming for the generation of induced pluripotent stem cells (iPSC).[156] Yamanaka and colleagues first achieved this feat in 2007 by demonstrating that differentiated somatic cells could be reprogrammed to a pluripotent embryonic stem-cell-like state through recombinant expression of four genes — KLF4, OCT4, c-MYC, and SOX2.[157] This groundbreaking study has made it possible to derive immunocompatible cells of any lineage for transplantation/transfusion therapy.

Nevertheless, the major challenge faced in cellular reprogramming for the generation of iPSC is the use of recombinant DNA, with its attendant risk of permanent genetic modification to the cellular genome. One strategy to overcome this challenge would be to transiently insert and remove transgenic elements in a precise and site-specific manner without leaving any permanent genetic modification to cellular genomic DNA. Indeed, this can be achieved with the PiggyBac transposon system.[49,50] Utilizing PiggyBac transposons with four different transgene inserts (KLF4, OCT4, c-MYC, and SOX2), Woltjen et al.[49,50] managed to successfully reprogram human fibroblasts into iPSC and excise these transgenic elements from the cellular genome, without leaving any permanent trace of genetic alteration.[49,50]

Another strategy to avoid the use of recombinant DNA would be to utilize cell-permeable recombinant proteins for cellular reprogramming.[124] The existence of specific peptide sequences that confer cell-penetrating ability is well known, and are commonly referred to as protein transduction domains.[158−160] Zhou et al.[124] fused a polyarginine (11R) protein transduction domain to four recombinant transcription factors (KLF4, OCT4, c-MYC, and SOX2), and these were utilized to successfully reprogram murine embryonic fibroblasts to iPSC. Using a similar approach, Kim et al.[161] were able to derive iPSC from human fibroblasts with recombinant transcription factors fused to polyarginine protein transduction domains.

Success in cellular reprogramming with cell-permeable proteins was quickly followed by the first successful derivation of iPSC through transfection with chemically modified mRNA.[125] There are two major challenges in utilizing RNA transfection in cellular reprogramming. The first challenge is the relative instability and short half-life of RNA within the cell. The second challenge is that there exists an innate antiretroviral response within mammalian cells against foreign RNA that triggers cellular apoptosis. Warren et al.[125] extended the stability and half-life of synthetic mRNA with a 5'-guanine cap,[162,163] while the innate antiretroviral response within mammalian cells against foreign RNA was overcome through chemical modifications to the synthetic mRNA. This included substitution of uridine with pseudouridine,[164] and substitution of cytidine with 5-methylcytidine.[165] Additionally, the B18R protein was supplemented into the culture media during RNA transfection to suppress the interferon-1 pathway that leads to cellular apoptosis upon introduction of foreign RNA into the cell.[166] This is achieved by the B18R protein acting as decoy receptor for type I interferon,[166] and is absolutely crucial for maintaining cell viability during cellular reprogramming with RNA transfection. By utilizing chemically modified mRNA corresponding to the four Yamanaka transcription factors (KLF4, OCT4, c-MYC, and SOX2), together with the supplementation of B18R within the culture medium, Warren et al.[125] were able to successfully reprogram human fibroblasts to iPSC at higher efficiencies compared with previous techniques utilizing recombinant DNA.

Although the derivation of iPSC from differentiated somatic cells is an exciting development in the field of stem cells and regenerative medicine, a pertinent question is whether it is absolutely necessary to set the developmental clock back to the embryonic state for therapeutic applications in regenerative medicine. It is possible that the developmental clock may instead be reset halfway to a less immature developmental stage that is more directly applicable to therapeutic applications, that is, the transit amplifying progenitor stage.[167,168] Some recent studies have even demonstrated direct lineage conversion from one differentiated phenotype to another through the recombinant expression of transcription factors. Most notably, murine fibroblasts have been converted directly into cardiomyocytes,[169] neurons,[170,171] and hepatocytes[172,173]; while human fibroblasts have been converted directly into neurons[174−176] and hematopoietic progenitors.[177]

Nevertheless, it is unlikely that future regenerative medicine applications would bypass the stem cell or progenitor cell stages completely. This is because terminally differentiated somatic phenotypes have limited proliferative capacity, and transplantation/transfusion therapy would certainly require large quantities of cells for individual patients. Extensive proliferation in situ or ex vivo is only possible at either the stem cell or progenitor cell stages.

CONCLUSION: CHALLENGES AND SAFETY ISSUES

As synthetic biology makes major inroads into diverse therapeutic applications, an obvious key requirement is system robustness; that is, long-term stability, tight control, and consistent performance of the engineered synthetic gene networks. A major obstacle to achieving this requirement is that there is a limit to the size of genetic programs that can be transferred into human cells with clinically approved viral and plasmid vectors, which would in turn necessitate the use of multiple vectors for piecemeal transfer of different components of a synthetic gene network into human cells. This would obviously compromise system stability, and make it difficult to achieve tight control of synthetic gene circuits.

Translation of synthetic biology into clinical applications demands stringent evaluation of toxicity and other safety issues associated with the introduction of engineered synthetic gene networks into the human body. Of particular concern is the possibility of genetically modified cells and organisms causing unanticipated adverse effects on patients, such as cancer. There is also a risk of inadvertent transmission and spread of genetically modified organisms (for example, commensal bacteria) to human populations and the environment. Hence, a top priority for synthetic biology is 'biosafety engineering;' that is, the development of safety switches or brakes within synthetic gene networks that can restrain growth and proliferation of genetically modified cells and organisms.

Currently, the synthetic biology field does not have too many studies that specifically address safety issues pertaining to the construction of synthetic gene circuits and their interaction with the human body and environment. Nevertheless, what is certainly needed is stringent regulatory oversight exercised by international agencies and local government bodies over research institutions and commercial entities involved in the creation of genetically modified cells and organisms. Perhaps an international registration system with standardized nomenclature should be set up to ensure traceability of genetically modified cells and organisms. This may be facilitated by making it mandatory for research institutions and commercial entities to uniquely tag the DNA sequence of all genetically modified cells and organisms that they develop. Compulsory registration and tagging will undoubtedly motivate research institutions and commercial entities to operate with greater accountability and transparency. Additionally, new legislation should also make provisions for compulsory biosafety engineering of all genetically modified cells and organisms, to inhibit their growth and survival outside a contained environment, once the technology for doing so is matured and widely available.

To date, a substantial portion of the synthetic biology toolkit has been derived from bacterial prokaryotic systems. Hence, another major safety concern is that heterologous proteins (particularly from prokaryotic systems) expressed by synthetic gene networks might trigger an immune response within the patient. Accordingly, there is a critical need to find new building blocks of synthetic gene networks from human systems. An expanded synthetic biology toolkit based on human genes, proteins, and RNA is unquestionably crucial for translation into clinical practice.

References

1. Nandagopal N, Elowitz MB. Synthetic biology: integrated gene circuits. *Science*. 2011;333(6047):1244–1248.
2. Agapakis CM, Silver PA. Synthetic biology: exploring and exploiting genetic modularity through the design of novel biological networks. *Mol Biosyst*. 2009;5(7):704–713.
3. Purnick PE, Weiss R. The second wave of synthetic biology: from modules to systems. *Nat Rev Mol Cell Biol*. 2009;10(6):410–422.
4. Tigges M, Fussenegger M. Recent advances in mammalian synthetic biology-design of synthetic transgene control networks. *Curr Opin Biotechnol*. 2009;20(4):449–460.

5. Gardner TS, Cantor CR, Collins JJ. Construction of a genetic toggle switch in *Escherichia coli*. *Nature*. 2000;403 (6767):339−342.

6. Elowitz MB, Leibler S. A synthetic oscillatory network of transcriptional regulators. *Nature*. 2000;403 (6767):335−338.

7. Weber W, Fussenegger M. Synthetic gene networks in mammalian cells. *Curr Opin Biotechnol*. 2010;21 (5):690−696.

8. Weber W, Fussenegger M. Engineering of synthetic mammalian gene networks. *Chem Biol*. 2009;16 (3):287−297.

9. Krivoruchko A, Siewers V, Nielsen J. Opportunities for yeast metabolic engineering: lessons from synthetic biology. *Biotechnol J*. 2011;6(3):262−276.

10. Kozma-Bognar L, Hajdu A, Nagy F. Light-regulated gene expression in yeast. *Methods Mol Biol*. 2012;813:187−193.

11. Atsumi S, Little JW. Regulatory circuit design and evolution using phage lambda. *Genes Dev*. 2004;18 (17):2086−2094.

12. Atsumi S, Little JW. A synthetic phage lambda regulatory circuit. *Proc Natl Acad Sci USA*. 2006;103 (50):19045−19050.

13. Michalodimitrakis K, Isalan M. Engineering prokaryotic gene circuits. *FEMS Microbiol Rev*. 2009;33(1):27−37.

14. Ruder WC, Lu T, Collins JJ. Synthetic biology moving into the clinic. *Science*. 2011;333(6047):1248−1252.

15. Jakobus K, Wend S, Weber W. Synthetic mammalian gene networks as a blueprint for the design of interactive biohybrid materials. *Chem Soc Rev*. 2012;41(3):1000−1018.

16. MacDonald JT, Barnes C, Kitney RI, Freemont PS, Stan GB. Computational design approaches and tools for synthetic biology. *Integr Biol (Camb)*. 2011;3(2):97−108.

17. Marchisio MA. In silico implementation of synthetic gene networks. *Methods Mol Biol*. 2012;813:3−21.

18. Blakes J, Twycross J, Romero-Campero FJ, Krasnogor N. The infobiotics workbench: an integrated in silico modelling platform for systems and synthetic biology. *Bioinformatics*. 2011;27(23):3323−3324.

19. Umesh P, Naveen F, Rao CU, Nair AS. Programming languages for synthetic biology. *Syst Synth Biol*. 2010;4 (4):265−269.

20. Alterovitz G, Muso T, Ramoni MF. The challenges of informatics in synthetic biology: from biomolecular networks to artificial organisms. *Brief Bioinform*. 2010;11(1):80−95.

21. Weeding E, Houle J, Kaznessis YN. SynBioSS Designer: a web-based tool for the automated generation of kinetic models for synthetic biological constructs. *Brief Bioinform*. 2010;11(4):394−402.

22. Chandran D, Bergmann FT, Sauro HM. Computer-aided design of biological circuits using TinkerCell. *Bioeng Bugs*. 2010;1(4):274−281.

23. Rialle S, Felicori L, Dias-Lopes C, et al. BioNetCAD: design, simulation and experimental validation of synthetic biochemical networks. *Bioinformatics*. 2010;26(18):2298−2304.

24. Chen YY, Smolke CD. From DNA to targeted therapeutics: bringing synthetic biology to the clinic. *Sci Transl Med*. 2011;3(106):106ps42.

25. Aubel D, Fussenegger M. Mammalian synthetic biology − from tools to therapies. *Bioessays*. 2010;32 (4):332−345.

26. Kau AL, Ahern PP, Griffin NW, Goodman AL, Gordon JI. Human nutrition, the gut microbiome and the immune system. *Nature*. 2011;474(7351):327−336.

27. Grice EA, Segre JA. The skin microbiome. *Nat Rev Microbiol*. 2011;9(4):244−253.

28. Duan F, March JC. Engineered bacterial communication prevents *Vibrio cholerae* virulence in an infant mouse model. *Proc Natl Acad Sci USA*. 2010;107(25):11260−11264.

29. Anderson JC, Clarke EJ, Arkin AP, Voigt CA. Environmentally controlled invasion of cancer cells by engineered bacteria. *J Mol Biol*. 2006;355(4):619−627.

30. Lu TK, Collins JJ. Dispersing biofilms with engineered enzymatic bacteriophage. *Proc Natl Acad Sci USA*. 2007;104(27):11197−11202.

31. Lu TK, Collins JJ. Engineered bacteriophage targeting gene networks as adjuvants for antibiotic therapy. *Proc Natl Acad Sci USA*. 2009;106(12):4629−4634.

32. Thomas CE, Ehrhardt A, Kay MA. Progress and problems with the use of viral vectors for gene therapy. *Nat Rev Genet*. 2003;4(5):346−358.

33. Modlich U, Baum C. Preventing and exploiting the oncogenic potential of integrating gene vectors. *J Clin Invest*. 2009;119(4):755−758.

34. Bohne J, Cathomen T. Genotoxicity in gene therapy: an account of vector integration and designer nucleases. *Curr Opin Mol Ther*. 2008;10(3):214−223.

35. Hoess RH, Abremski K. Mechanism of strand cleavage and exchange in the Cre-lox site-specific recombination system. *J Mol Biol.* 1985;181(3):351–362.

36. Schlake T, Bode J. Use of mutated FLP recognition target (FRT) sites for the exchange of expression cassettes at defined chromosomal loci. *Biochemistry.* 1994;33(43):12746–12751.

37. Anastassiadis K, Fu J, Patsch C, et al. Dre recombinase, like Cre, is a highly efficient site-specific recombinase in *E. coli*, mammalian cells and mice. *Dis Model Mech.* 2009;2(9–10):508–515.

38. Thyagarajan B, Olivares EC, Hollis RP, Ginsburg DS, Calos MP. Site-specific genomic integration in mammalian cells mediated by phage phiC31 integrase. *Mol Cell Biol.* 2001;21(12):3926–3934.

39. Berghella L, De Angelis L, Coletta M, et al. Reversible immortalization of human myogenic cells by site-specific excision of a retrovirally transferred oncogene. *Hum Gene Ther.* 1999;10(10):1607–1617.

40. Badorf M, Edenhofer F, Dries V, Kochanek S, Schiedner G. Efficient in vitro and in vivo excision of floxed sequences with a high-capacity adenoviral vector expressing Cre recombinase. *Genesis.* 2002;33(3):119–124.

41. Wright DA, Thibodeau-Beganny S, Sander JD, et al. Standardized reagents and protocols for engineering zinc finger nucleases by modular assembly. *Nat Protoc.* 2006;1(3):1637–1652.

42. Maeder ML, Thibodeau-Beganny S, Sander JD, Voytas DF, Joung JK. Oligomerized pool engineering (OPEN): an 'open-source' protocol for making customized zinc-finger arrays. *Nat Protoc.* 2009;4(10):1471–1501.

43. Sander JD, Dahlborg EJ, Goodwin MJ, et al. Selection-free zinc-finger-nuclease engineering by context-dependent assembly (CoDA). *Nat Methods.* 2011;8(1):67–69.

44. Sander JD, Maeder ML, Reyon D, Voytas DF, Joung JK, Dobbs D. ZiFiT (Zinc Finger Targeter): an updated zinc finger engineering tool. *Nucleic Acids Res.* 2010;38(Web Server issue):W462–W468.

45. Li T, Huang S, Jiang WZ, et al. TAL nucleases (TALNs): hybrid proteins composed of TAL effectors and FokI DNA-cleavage domain. *Nucleic Acids Res.* 2011;39(1):359–372.

46. Fraser MJ, Ciszczon T, Elick T, Bauser C. Precise excision of TTAA-specific lepidopteran transposons piggyBac (IFP2) and tagalong (TFP3) from the baculovirus genome in cell lines from two species of Lepidoptera. *Insect Mol Biol.* 1996;5(2):141–151.

47. Elick TA, Bauser CA, Fraser MJ. Excision of the piggyBac transposable element in vitro is a precise event that is enhanced by the expression of its encoded transposase. *Genetica.* 1996;98(1):33–41.

48. Wilson MH, Coates CJ, George Jr AL. PiggyBac transposon-mediated gene transfer in human cells. *Mol Ther.* 2007;15(1):139–145.

49. Woltjen K, Michael IP, Mohseni P, et al. piggyBac transposition reprograms fibroblasts to induced pluripotent stem cells. *Nature.* 2009;458(7239):766–770.

50. Woltjen K, Hämäläinen R, Kibschull M, Mileikovsky M, Nagy A. Transgene-free production of pluripotent stem cells using piggyBac transposons. *Methods Mol Biol.* 2011;767:87–103.

51. Li MZ, Elledge SJ. Harnessing homologous recombination in vitro to generate recombinant DNA via SLIC. *Nat Methods.* 2007;4(3):251–256.

52. Gibson DG, Young L, Chuang RY, Venter JC, Hutchison III CA, Smith HO. Enzymatic assembly of DNA molecules up to several hundred kilobases. *Nat Methods.* 2009;6(5):343–345.

53. Quan J, Tian J. Circular polymerase extension cloning of complex gene libraries and pathways. *PLoS One.* 2009;4(7):e6441.

54. Atkinson MR, Savageau MA, Myers JT, Ninfa AJ. Development of genetic circuitry exhibiting toggle switch or oscillatory behavior in *Escherichia coli*. *Cell.* 2003;113(5):597–607.

55. Stupak EE, Stupak IV. Inheritance and state switching of genetic toggle switch in different culture growth phases. *FEMS Microbiol Lett.* 2006;258(1):37–42.

56. Boczko E, Gedeon T, Mischaikow K. Dynamics of a simple regulatory switch. *J Math Biol.* 2007;55(5–6):679–719.

57. Greber D, El-Baba MD, Fussenegger M. Intronically encoded siRNAs improve dynamic range of mammalian gene regulation systems and toggle switch. *Nucleic Acids Res.* 2008;36(16):e101.

58. Ghim CM, Almaas E. Two-component genetic switch as a synthetic module with tunable stability. *Phys Rev Lett.* 2009;103(2):028101.

59. Kramer BP, Fussenegger M. Hysteresis in a synthetic mammalian gene network. *Proc Natl Acad Sci USA.* 2005;102(27):9517–9522.

60. Isaacs FJ, Hasty J, Cantor CR, Collins JJ. Prediction and measurement of an autoregulatory genetic module. *Proc Natl Acad Sci USA.* 2003;100(13):7714–7719.

61. Lou C, Liu X, Ni M, et al. Synthesizing a novel genetic sequential logic circuit: a push-on push-off switch. *Mol Syst Biol.* 2010;6:350.

62. Kramer BP, Viretta AU, Daoud-El-Baba M, Aubel D, Weber W, Fussenegger M. An engineered epigenetic transgene switch in mammalian cells. *Nat Biotechnol.* 2004;22(7):867–870.

63. Xiong W, Ferrell Jr JE. A positive-feedback-based bistable 'memory module' that governs a cell fate decision. *Nature*. 2003;426(6965):460−465.

64. Ptashne M. *A Genetic Switch: Phage Lambda Revisited*. Cold Spring Harbor, NY: Cold Spring Harbor Laboratory Press; 2004.

65. Zordan RE, Galgoczy DJ, Johnson AD. Epigenetic properties of white-opaque switching in *Candida albicans* are based on a self-sustaining transcriptional feedback loop. *Proc Natl Acad Sci USA*. 2006;103(34):12807−12812.

66. Huang G, Wang H, Chou S, Nie X, Chen J, Liu H. Bistable expression of WOR1, a master regulator of white-opaque switching in *Candida albicans*. *Proc Natl Acad Sci USA*. 2006;103(34):12813−12818.

67. Ajo-Franklin CM, Drubin DA, Eskin JA, et al. Rational design of memory in eukaryotic cells. *Genes Dev*. 2007;21(18):2271−2276.

68. Goodwin B. *Temporal Organization in Cells*. London: Academic Press; 1963.

69. Stricker J, Cookson S, Bennett MR, Mather WH, Tsimring LS, Hasty J. A fast, robust and tunable synthetic gene oscillator. *Nature*. 2008;456(7221):516−519.

70. Fung E, Wong WW, Suen JK, Bulter T, Lee SG, Liao JC. A synthetic gene-metabolic oscillator. *Nature*. 2005;435 (7038):118−122.

71. Tigges M, Marquez-Lago TT, Stelling J, Fussenegger M. A tunable synthetic mammalian oscillator. *Nature*. 2009;457(7227):309−312.

72. Tigges M, Dénervaud N, Greber D, Stelling J, Fussenegger M. A synthetic low-frequency mammalian oscillator. *Nucleic Acids Res*. 2010;38(8):2702−2711.

73. Austin DW, Allen MS, McCollum JM, et al. Gene network shaping of inherent noise spectra. *Nature*. 2006;439 (7076):608−611.

74. Hooshangi S, Thiberge S, Weiss R. Ultrasensitivity and noise propagation in a synthetic transcriptional cascade. *Proc Natl Acad Sci USA*. 2005;102(10):3581−3586.

75. Basu S, Gerchman Y, Collins CH, Arnold FH, Weiss R. A synthetic multicellular system for programmed pattern formation. *Nature*. 2005;434(7037):1130−1134.

76. Sohka T, Heins RA, Phelan RM, Greisler JM, Townsend CA, Ostermeier M. An externally tunable bacterial band-pass filter. *Proc Natl Acad Sci USA*. 2009;106(25):10135−10140.

77. Muranaka N, Yokobayashi Y. A synthetic riboswitch with chemical band-pass response. *Chem Commun (Camb)*. 2010;46(36):6825−6827.

78. Weber W, Stelling J, Rimann M, et al. A synthetic time-delay circuit in mammalian cells and mice. *Proc Natl Acad Sci USA*. 2007;104(8):2643−2648.

79. Greber D, Fussenegger M. An engineered mammalian band-pass network. *Nucleic Acids Res*. 2010;38(18):e174.

80. You L, Cox III RS, Weiss R, Arnold FH. Programmed population control by cell−cell communication and regulated killing. *Nature*. 2004;428(6985):868−871.

81. Kobayashi H, Kaern M, Araki M, et al. Programmable cells: interfacing natural and engineered gene networks. *Proc Natl Acad Sci USA*. 2004;101(22):8414−8419.

82. Balagaddé FK, Song H, Ozaki J, et al. A synthetic *Escherichia coli* predator−prey ecosystem. *Mol Syst Biol*. 2008;4:187.

83. Danino T, Mondragón-Palomino O, Tsimring L, Hasty J. A synchronized quorum of genetic clocks. *Nature*. 2010;463(7279):326−330.

84. Anderson JC, Voigt CA, Arkin AP. Environmental signal integration by a modular AND gate. *Mol Syst Biol*. 2007;3:133.

85. Guet CC, Elowitz MB, Hsing W, Leibler S. Combinatorial synthesis of genetic networks. *Science*. 2002;296:1466−1470.

86. Rackham O, Chin JW. Cellular logic with orthogonal ribosomes. *J Am Chem Soc*. 2005;127:17584−17585.

87. Rinaudo K, et al. A universal RNAi-based logic evaluator that operates in mammalian cells. *Nat Biotechnol*. 2007;25:795−801.

88. Stojanovic MN, Stefanovic D. A deoxyribozyme-based molecular automaton. *Nat Biotechnol*. 2003;21:1069−1074.

89. Win MN, Smolke CD. Higher-order cellular information processing with synthetic RNA devices. *Science*. 2008;322:456−460.

90. Milligan G. New aspects of G-protein-coupled receptor signalling and regulation. *Trends Endocrinol Metab*. 1998;9(1):13−19.

91. Biarc J, Chalkley RJ, Burlingame AL, Bradshaw RA. Receptor tyrosine kinase signaling − a proteomic perspective. *Adv Enzyme Regul*. 2011;51(1):293−305.

92. Crabtree GR, Schreiber SL. SnapShot: Ca^{2+}-calcineurin-NFAT signaling. *Cell*. 2009;138(1):210, 210.e1.

93. Hoeffler JP, Meyer TE, Yun Y, Jameson JL, Habener JF. Cyclic AMP-responsive DNA-binding protein: structure based on a cloned placental cDNA. *Science.* 1988;242(4884):1430–1433.

94. Deuschle U, Meyer WK, Thiesen HJ. Tetracycline-reversible silencing of eukaryotic promoters. *Mol Cell Biol.* 1995;15(4):1907–1914.

95. Wieland M, Fussenegger M. Ligand-dependent regulatory RNA parts for synthetic biology in eukaryotes. *Curr Opin Biotechnol.* 2010;21(6):760–765.

96. Laitala-Leinonen T. Update on the development of microRNA and siRNA molecules as regulators of cell physiology. *Recent Pat DNA Gene Seq.* 2010;4(2):113–121.

97. Isaacs FJ, Dwyer DJ, Ding C, Pervouchine DD, Cantor CR, Collins JJ. Engineered riboregulators enable post-transcriptional control of gene expression. *Nat Biotechnol.* 2004;22(7):841–847:Epub 2004 Jun 20

98. Garst AD, Edwards AL, Batey RT. Riboswitches: structures and mechanisms. *Cold Spring Harb Perspect Biol.* 2011;3(6):pii: a003533.

99. McIntyre GJ, Fanning GC. Design and cloning strategies for constructing shRNA expression vectors. *BMC Biotechnol.* 2006;6:1.

100. Mitsuyasu RT, Merigan TC, Carr A, et al. Phase 2 gene therapy trial of an anti-HIV ribozyme in autologous CD34+ cells. *Nat Med.* 2009;15(3):285–292.

101. Hotchkiss G, Maijgren-Steffensson C, Ahrlund-Richter L. Efficacy and mode of action of hammerhead and hairpin ribozymes against various HIV-1 target sites. *Mol Ther.* 2004;10(1):172–180.

102. Soukup GA, Breaker RR. Engineering precision RNA molecular switches. *Proc Natl Acad Sci USA.* 1999;96 (7):3584–3589.

103. Link KH, Guo L, Ames TD, Yen L, Mulligan RC, Breaker RR. Engineering high-speed allosteric hammerhead ribozymes. *Biol Chem.* 2007;388(8):779–786.

104. Reeves PJ, Kim JM, Khorana HG. Structure and function in rhodopsin: a tetracycline-inducible system in stable mammalian cell lines for high-level expression of opsin mutants. *Proc Natl Acad Sci USA.* 2002;99 (21):13413–13418.

105. Jiang W, Zhou L, Breyer B, et al. Tetracycline-regulated gene expression mediated by a novel chimeric repressor that recruits histone deacetylases in mammalian cells. *J Biol Chem.* 2001;276(48):45168–45174.

106. Tonack S, Patel S, Jalali M, et al. Tetracycline-inducible protein expression in pancreatic cancer cells: effects of CapG overexpression. *World J Gastroenterol.* 2011;17(15):1947–1960.

107. Hartenbach S, Daoud-El Baba M, Weber W, Fussenegger M. An engineered L-arginine sensor of *Chlamydia pneumoniae* enables arginine-adjustable transcription control in mammalian cells and mice. *Nucleic Acids Res.* 2007;35(20):e136.

108. Weber W, Bacchus W, Daoud-El Baba M, Fussenegger M. Vitamin H-regulated transgene expression in mammalian cells. *Nucleic Acids Res.* 2007;35(17):e116.

109. Weber W, Schoenmakers R, Keller B, et al. A synthetic mammalian gene circuit reveals antituberculosis compounds. *Proc Natl Acad Sci USA.* 2008;105(29):9994–9998.

110. Chong H, Ruchatz A, Clackson T, Rivera VM, Vile RG. A system for small-molecule control of conditionally replication-competent adenoviral vectors. *Mol Ther.* 2002;5(2):195–203.

111. Tascou S, Sorensen TK, Glénat V, et al. Stringent rosiglitazone-dependent gene switch in muscle cells without effect on myogenic differentiation. *Mol Ther.* 2004;9(5):637–649.

112. Valenta C, Cladera J, O'Shea P, Hadgraft J. Effect of phloretin on the percutaneous absorption of lignocaine across human skin. *J Pharm Sci.* 2001;90(4):485–492.

113. Gitzinger M, Kemmer C, El-Baba MD, Weber W, Fussenegger M. Controlling transgene expression in subcutaneous implants using a skin lotion containing the apple metabolite phloretin. *Proc Natl Acad Sci USA.* 2009;106(26):10638–10643.

114. Kemmer C, Gitzinger M, Daoud-El Baba M, Djonov V, Stelling J, Fussenegger M. Self-sufficient control of urate homeostasis in mice by a synthetic circuit. *Nat Biotechnol.* 2010;28(4):355–360.

115. Kemmer C, Fluri DA, Witschi U, Passeraub A, Gutzwiller A, Fussenegger M. A designer network coordinating bovine artificial insemination by ovulation-triggered release of implanted sperms. *J Control Release.* 2011;150 (1):23–29.

116. Weber W, Link N, Fussenegger M. A genetic redox sensor for mammalian cells. *Metab Eng.* 2006;8 (3):273–280.

117. Weber W, Rimann M, Spielmann M, et al. Gas-inducible transgene expression in mammalian cells and mice. *Nat Biotechnol.* 2004;22(11):1440–1444.

118. Weber W, Daoud-El Baba M, Fussenegger M. Synthetic ecosystems based on airborne inter- and intra-kingdom communication. *Proc Natl Acad Sci USA.* 2007;104(25):10435–10440.

119. Weber W, Luzi S, Karlsson M, et al. A synthetic mammalian electro-genetic transcription circuit. *Nucleic Acids Res.* 2009;37(4):e33.

120. Ye H, Daoud-El Baba M, Peng RW, Fussenegger M. A synthetic optogenetic transcription device enhances blood-glucose homeostasis in mice. *Science*. 2011;332(6037):1565−1568.

121. Alanis AJ. Resistance to antibiotics: are we in the post-antibiotic era? *Arch Med Res*. 2005;36(6):697−705.

122. Nielsen PE. Gene targeting and expression modulation by peptide nucleic acids (PNA). *Curr Pharm Des*. 2010;16(28):3118−3123.

123. Grünweller A, Hartmann RK. Locked nucleic acid oligonucleotides: the next generation of antisense agents? *BioDrugs*. 2007;21(4):235−243.

124. Zhou H, Wu S, Joo JY, et al. Generation of induced pluripotent stem cells using recombinant proteins. *Cell Stem Cell*. 2009;4(5):381−384.

125. Warren L, Manos PD, Ahfeldt T, et al. Highly efficient reprogramming to pluripotency and directed differentiation of human cells with synthetic modified mRNA. *Cell Stem Cell*. 2010;7(5):618−630.

126. Chen YY, Jensen MC, Smolke CD. Genetic control of mammalian T-cell proliferation with synthetic RNA regulatory systems. *Proc Natl Acad Sci USA*. 2010;107(19):8531−8536.

127. Ramachandra M, Rahman A, Zou A, et al. Re-engineering adenovirus regulatory pathways to enhance oncolytic specificity and efficacy. *Nat Biotechnol*. 2001;19(11):1035−1041.

128. Xiang S, Fruehauf J, Li CJ. Short hairpin RNA-expressing bacteria elicit RNA interference in mammals. *Nat Biotechnol*. 2006;24(6):697−702.

129. Maudlin I, Eisler MC, Welburn SC. Neglected and endemic zoonoses. *Philos Trans R Soc Lond B Biol Sci*. 2009;364(1530):2777−2787.

130. Childs JE, Gordon ER. Surveillance and control of zoonotic agents prior to disease detection in humans. *Mt Sinai J Med*. 2009;76(5):421−428.

131. Vlassova N, Han A, Zenilman JM, James G, Lazarus GS. New horizons for cutaneous microbiology: the role of biofilms in dermatological disease. *Br J Dermatol*. 2011;165(4):751−759.

132. Cogan NG, Gunn JS, Wozniak DJ. Biofilms and infectious diseases: biology to mathematics and back again. *FEMS Microbiol Lett*. 2011;322(1):1−7.

133. Chen L, Wen YM. The role of bacterial biofilm in persistent infections and control strategies. *Int J Oral Sci*. 2011;3(2):66−73.

134. Rao S, Hu S, McHugh L, et al. Toward a live microbial microbicide for HIV: commensal bacteria secreting an HIV fusion inhibitor peptide. *Proc Natl Acad Sci USA*. 2005;102(34):11993−11998.

135. Windbichler N, Menichelli M, Papathanos PA, et al. A synthetic homing endonuclease-based gene drive system in the human malaria mosquito. *Nature*. 2011;473(7346):212−215.

136. Amidi M, de Raad M, Crommelin DJ, Hennink WE, Mastrobattista E. Antigen-expressing immunostimulatory liposomes as a genetically programmable synthetic vaccine. *Syst Synth Biol*. 2011;5 (1−2):21−31.

137. Coleman JR, Papamichail D, Skiena S, Futcher B, Wimmer E, Mueller S. Virus attenuation by genome-scale changes in codon pair bias. *Science*. 2008;320(5884):1784−1787.

138. Dean JT, Tran L, Beaven S, et al. Resistance to diet-induced obesity in mice with synthetic glyoxylate shunt. *Cell Metab*. 2009;9(6):525−536.

139. Shamseddeen H, Getty JZ, Hamdallah IN, Ali MR. Epidemiology and economic impact of obesity and type 2 diabetes. *Surg Clin North Am*. 2011;91(6):1163−1172.

140. Ryan JG. Cost and policy implications from the increasing prevalence of obesity and diabetes mellitus. *Gend Med*. 2009;6(suppl 1):86−108.

141. Tierney S, Stokke BT. Development of an oligonucleotide functionalized hydrogel integrated on a high resolution interferometric readout platform as a label-free macromolecule sensing device. *Biomacromolecules*. 2009;10(6):1619−1626.

142. Lin DC, Yurke B, Langrana NA. Inducing reversible stiffness changes in DNA-crosslinked gels. *J Mater Res*. 2005;20:1456−1464.

143. He X, Wei B, Mi Y. Aptamer based reversible DNA induced hydrogel system for molecular recognition and separation. *Chem Commun (Camb)*. 2010;46(34):6308−6310.

144. Yang H, Liu H, Kang H, Tan W. Engineering target-responsive hydrogels based on aptamer-target interactions. *J Am Chem Soc*. 2008;130(20):6320−6321.

145. Wang J, Lu J, Gu G, Liu Y. In vitro DNA-binding profile of transcription factors: methods and new insights. *J Endocrinol*. 2011;210(1):15−27.

146. Christen EH, Karlsson M, Kampf MM, et al. Conditional DNA−protein interactions confer stimulus-sensing properties to biohybrid materials. *Adv Func Mat*. 2011;21:2861−2867.

147. Kampf MM, Christen EH, Ehrbar M, et al. Gene therapy technology-based biomaterial for the trigger-inducible release of biopharmaceuticals in mice. *Adv Func Mat*. 2010;20(15):2534−2538.

148. King WJ, Toepke MW, Murphy WL. Facile formation of dynamic hydrogel microspheres for triggered growth factor delivery. *Acta Biomater*. 2011;7(3):975−985.

149. Congreve M, Langmead C, Marshall FH. The use of GPCR structures in drug design. *Adv Pharmacol*. 2011;62:1−36.

150. Life Technologies Inc. Tango™ GPCR Assay System. Accessible at the following website: <http://www.invitrogen.com/site/us/en/home/Products-and-Services/Applications/Drug-Discovery/Target-and-Lead-Identification-and-Validation/g-protein_coupled_html/GPCR-Cell-Based-Assays/Tango.html> Date Accessed 30.01.12.

151. Gonzalez-Nicolini V, Fux C, Fussenegger M. A novel mammalian cell-based approach for the discovery of anticancer drugs with reduced cytotoxicity on nondividing cells. *Invest New Drugs*. 2004;22(3):253−262.

152. Aubel D, Morris R, Lennon B, et al. Design of a novel mammalian screening system for the detection of bioavailable, noncytotoxic streptogramin antibiotics. *J Antibiot (Tokyo)*. 2001;54(1):44−55.

153. Weber W, Schoenmakers R, Keller B, et al. A synthetic mammalian gene circuit reveals antituberculosis compounds. *Proc Natl Acad Sci USA*. 2008;105(29):9994−9998.

154. Fraaije MW, Kamerbeek NM, Heidekamp AJ, Fortin R, Janssen DB. The prodrug activator EtaA from *Mycobacterium tuberculosis* is a Baeyer-Villiger monooxygenase. *J Biol Chem*. 2004;279(5):3354−3360.

155. Loose C, Jensen K, Rigoutsos I, Stephanopoulos G. A linguistic model for the rational design of antimicrobial peptides. *Nature*. 2006;443(7113):867−869.

156. Ebben JD, Zorniak M, Clark PA, Kuo JS. Introduction to induced pluripotent stem cells: advancing the potential for personalized medicine. *World Neurosurg*. 2011;76(3−4):270−275.

157. Takahashi K, Tanabe K, Ohnuki M, et al. Induction of pluripotent stem cells from adult human fibroblasts by defined factors. *Cell*. 2007;131(5):861−872.

158. Lindgren M, Hällbrink M, Prochiantz A, Langel U. Cell-penetrating peptides. *Trends Pharmacol Sci*. 2000;21(3):99−103.

159. Joliot A. Transduction peptides within naturally occurring proteins. *Sci STKE*. 2005;2005(313):pe54.

160. van den Berg A, Dowdy SF. Protein transduction domain delivery of therapeutic macromolecules. *Curr Opin Biotechnol*. 2011;22(6):888−893.

161. Kim D, Kim CH, Moon JI, et al. Generation of human induced pluripotent stem cells by direct delivery of reprogramming proteins. *Cell Stem Cell*. 2009;4(6):472−476.

162. Jiao X, Xiang S, Oh C, Martin CE, Tong L, Kiledjian M. Identification of a quality-control mechanism for mRNA 5'-end capping. *Nature*. 2010;467(7315):608−611.

163. Yisraeli JK, Melton DA. Synthesis of long, capped transcripts in vitro by SP6 and T7 RNA polymerases. *Methods Enzymol*. 1989;180:42−50.

164. Kariko K, Muramatsu H, Welsh FA, et al. Incorporation of pseudouridine into mRNA yields superior nonimmunogenic vector with increased translational capacity and biological stability. *Mol Ther*. 2008;16:1833−1840.

165. Kormann MS, Hasenpusch G, Aneja MK, et al. Expression of therapeutic proteins after delivery of chemically modified mRNA in mice. *Nat Biotechnol*. 2011;29(2):154−157.

166. Symons JA, Alcamí A, Smith GL. Vaccinia virus encodes a soluble type I interferon receptor of novel structure and broad species specificity. *Cell*. 1995;81(4):551−560.

167. Heng BC, Cao T. Could the transit-amplifying stage of stem cell differentiation be the most suited for transplantation purposes? *Med Hypotheses*. 2005;65(2):412−413.

168. Heng BC, Richards M, Ge Z, Shu Y. Induced adult stem (iAS) cells and induced transit amplifying progenitor (iTAP) cells − a possible alternative to induced pluripotent stem (iPS) cells? *J Tissue Eng Regen Med* 2010;4(2):159−162.

169. Ieda M, Fu JD, Delgado-Olguin P, et al. Direct reprogramming of fibroblasts into functional cardiomyocytes by defined factors. *Cell*. 2010;142(3):375−386.

170. Vierbuchen T, Ostermeier A, Pang ZP, Kokubu Y, Südhof TC, Wernig M. Direct conversion of fibroblasts to functional neurons by defined factors. *Nature*. 2010;463:1035−1041.

171. Caiazzo M, Dell'Anno MT, Dvoretskova E, et al. Direct generation of functional dopaminergic neurons from mouse and human fibroblasts. *Nature*. 2011;476(7359):224−227.

172. Huang P, He Z, Ji S, et al. Induction of functional hepatocyte-like cells from mouse fibroblasts by defined factors. *Nature*. 2011;475(7356):386−389.

173. Sekiya S, Suzuki A. Direct conversion of mouse fibroblasts to hepatocyte-like cells by defined factors. *Nature*. 2011;475:390−393.

174. Pang ZP, Yang N, Vierbuchen T, et al. Induction of human neuronal cells by defined transcription factors. *Nature*. 2011;476(7359):220−223.

175. Pfisterer U, Kirkeby A, Torper O, et al. Direct conversion of human fibroblasts to dopaminergic neurons. *Proc Natl Acad Sci USA*. 2011;108(25):10343−10348.

176. Yoo AS, Sun AX, Li L, et al. MicroRNA-mediated conversion of human fibroblasts to neurons. *Nature*. 2011;476(7359):228−231.

177. Szabo E, Rampalli S, Risueño RM, et al. Direct conversion of human fibroblasts to multilineage blood progenitors. *Nature*. 2010;468(7323):521−526.

Drug Discovery and Development via Synthetic Biology

Ryan E. Cobb, Yunzi Luo, Todd Freestone and Huimin Zhao
University of Illinois at Urbana-Champaign, Urbana, IL, USA

INTRODUCTION

Since the expression of the first synthetic gene in 1977, the pharmaceutical sector has been revolutionized with the manufacturing of protein therapeutics such as hormones, vaccines, antibodies, blood factors, and therapeutic enzymes. The industry is poised for another revolution as advancements in synthetic biology allow the production of bioactive compounds in nonnative hosts. Molecules from natural sources have proved to be an important source of new pharmaceuticals, and newly discovered natural products have been shown to be promising in the treatment of cardiovascular disease, cancer, and infectious diseases.[1] Sources for such compounds are equally diverse, including microbes, plants, marine sources, and even insects. Because natural products of interest often have complex structures, their chemical synthesis can be impractical. Additionally, they are formed in only low concentrations in their natural producer, preventing sustainable and economical extraction. For these reasons, optimizing natural product pathways in genetically tractable and cultivatable hosts have been an attractive alternative.

However, heterologous production of small molecules has often been avoided for several reasons. Often the complete metabolic pathways are not known. Even if the necessary genes are known, protein expression and enzyme function may be severely diminished in the nonnative environment. Another problem is the numerous permutations that must be tested when modulating protein expression throughout a pathway. Synthetic biology offers both computational and experimental tools that help to overcome these problems. In addition, the discovery of new natural products is possible with synthetic biology, increasing the influence the field will have on future pharmaceutical development.

TOOLS FOR PATHWAY DISCOVERY AND ENGINEERING

Synthetic biology is built upon the concept of biological systems as machines that can be designed and engineered to perform a desired objective, as in the synthesis of a drug or drug candidate molecule. The success of such an approach is predicated on two key concepts: first, an extensive set of diverse parts from which to construct such machines; and second, sophisticated techniques by which to assemble and engineer these parts into working pathways. The former has been significantly aided by the rapid development of

183

Synthetic Biology. DOI: http://dx.doi.org/10.1016/B978-0-12-394430-6.00010-8

bioinformatics tools, while the latter has been enabled by a number of sophisticated and versatile experimental tools.

Computational Tools

GENE IDENTIFICATION

In nature, there exists an incredibly diverse array of species adapted to survival in all kinds of environments. In the genomes of these organisms are the templates for countless proteins which catalyze a myriad of chemical transformations, many of which could be useful for synthetic biology applications. As a result, computational tools are essential to rapidly and accurately identify the true protein-coding sequences from DNA sequence data. The earliest gene identification algorithms were developed mainly for the analysis of shorter DNA sequences in which the exact coding sequence of a protein was ambiguous. These methods were reasonably simplistic, but provided fairly accurate predictions of 'coding' versus 'noncoding' sequences. For example, the TESTCODE algorithm devised in 1982 misclassified only 5% of test sequences, but drew no conclusion for 20% of test sequences.[2]

Subsequent prediction algorithms employed more sophisticated approaches to achieve better results. For example, the GeneMark program of Borodovsky and McIninch (initially referred to as GENMARK) combined nonhomogeneous Markov chain models with Bayesian decision-making for coding sequence prediction.[3] This program also introduced simultaneous analysis of both DNA strands as a method of improving accuracy. As the sequencing of entire genomes became realized, the need for reliable gene prediction was underscored. To improve the GeneMark program for entire bacterial genomes, a hidden Markov model framework was implemented, as well as recognition of ribosome binding site sequences.[4] Further improvements came with the application of self training for new prokaryotic genome sequences,[5] and expansion to eukaryotic and viral systems.[6,7] GLIMMER represents a complementary tool for gene identification that was built on interpolated Markov models.[8,9] This tool has similarly been adapted to eukaryotic DNA,[10,11] as well as endosymbiont and metagenome DNA.[12,13] These tools and others continue to be indispensable in the identification of new and potentially interesting protein-coding sequences from the ever-expanding volume of DNA sequence information available.

PREDICTION OF GENE FUNCTION

Synthetic biologists are typically interested in proteins for the transformations that they catalyze, but sequence information alone is not enough to describe a protein's utility. Bench-top experiments both in vitro and in vivo are, of course, the best way to determine protein function. However, the vast success in DNA sequencing and coding sequence identification has provided such a wealth of putative protein targets that laboratory characterization of them all is simply not feasible. Fortunately, if two proteins have similar primary sequences, it is quite likely that they will also share similar functions. As a result, sequence alignment and homology analysis based on proteins of known function have proved vital to the accurate prediction of protein function from sequence data alone.

The earliest exercises in protein homology comparisons were carried out to evaluate evolutionary relationships rather than to predict function.[14-16] Nevertheless, these algorithms provided the groundwork upon which subsequent protein alignment tools would be built. In 1985, Lipman and Pearson noted the increasing number of protein sequences made available at the time, and that functions could be inferred by comparison to other characterized proteins. As a result, they developed the FASTP algorithm for rapid in silico comparison of a query sequence to a protein sequence database.[17] This was followed by FASTA, which featured improved sensitivity, and LFASTA, which allowed for analyses of local similarity.[18,19] Other tools followed, such as MSA and CLUSTAL W, for the high-sensitivity alignment of smaller sets of proteins.[20,21]

Currently, state-of-the-art sequence database search tools employ more sophisticated algorithms to reduce computation time and increase sensitivity to weak similarities. For example, tools such as SAM[22] and HMMER[23] employ hidden Markov models to identify sequence homology. Perhaps the most widely used search tool is the Basic Local Alignment Search Tool, or BLAST, first presented by Altschul and coworkers in 1990.[24] At its inception, BLAST offered high-sensitivity database searching at speeds much faster than any previous algorithm, and proved amenable to mathematical and statistical analysis. Subsequent versions, such as gapped BLAST and position-specific iterative (PSI) BLAST, have further improved computation time and sensitivity to weak, but still biologically relevant, similarity.[25] Functional prediction via BLAST is further enhanced through coupling with the Conserved Domain Database (CDD),[26] which integrates data from sources such as Pfam[27] and SMART[28] to identify regions of the query sequence with evolutionarily conserved functions, such as binding a metal ion or cofactor. BLAST results can also be coupled to tools such as GCView,[29] which enable analysis of the genomic context of search results to facilitate more accurate functional prediction.

In recent years, predictive algorithms have expanded beyond individual coding sequence queries to a variety of other targets. For example, IsoRankN enables the alignment of entire protein–protein interaction networks for the prediction of functional orthologues across species.[30] Tools such as PromPredict,[31] ConTra,[32] and RSAT,[33] among others, focus not on protein sequences, but on the sequences of regulatory regions such as promoters and transcription factor binding sites. Such tools have clear applications in synthetic biology, not only for the design of biosynthetic pathways, but also for synthetic gene circuits and signal transduction systems.

PATHWAY DISCOVERY, PREDICTION, AND ANALYSIS

The computational tools described above are generally applicable to any DNA or protein sequence, and as a result have proven very useful for a wide range of applications. For the synthesis of drugs and drug candidates, however, special attention is paid to those proteins that are involved in secondary metabolism. This is due to the observation that secondary metabolite natural products and their derivatives and analogues represent a substantial fraction of the drugs available today. For example, in 2007 it was reported that 72.9% of anticancer drugs and 68.9% of small molecule antiinfectives are natural products or derived therefrom.[1] As a result, a number of tools have been developed for the discovery, prediction, and analysis of secondary metabolite gene clusters (Table 10.1).

Two classes of natural products that have garnered significant research interest are polyketides and nonribosomal peptides, complex compounds that are synthesized by multimodular, assembly line megasynthases known as polyketide synthases (PKSs) and nonribosomal peptide synthetases (NRPSs), respectively. In the past 15 years, a number of computational tools have been developed not only for the identification of PKS and NRPS gene clusters from DNA sequence data, but also for the prediction of their corresponding products. Some of the earliest efforts toward in silico prediction of NRPS products focused on the specificity of adenylation domains. In 1997, de Crécy-Lagard and coworkers examined 55 adenylation domain sequences to devise rules for specificity prediction, but found that they could only come up with good predictions in 43% of cases.[34] Two years later, however, analysis of the crystal structure of the adenylation domain PheA involved in gramicidin S biosynthesis enabled two groups to provide much more accurate specificity predictions. Stachelhaus and coworkers identified 10 specificity-conferring residues, allowing 86% accuracy in specificity prediction.[35] Challis and coworkers took a very similar approach, identifying an eight-residue signature sequence.[36] More recently, a sophisticated prediction algorithm based on transductive support vector machines has been devised that also incorporates the physicochemical properties of the

TABLE 10.1 Tools for Natural Product Pathway Discovery, Prediction, and Analysis

Tool	Full Name	Description	Web Address
antiSMASH	Antibiotics & Secondary Metabolite Analysis Shell	Annotation of biosynthetic gene clusters for many secondary metabolites from genomic DNA, including polyketides, nonribosomal peptides, terpenes, aminoglycosides, aminocoumarins, indolocarbazoles, lantibiotics, bacteriocins, nucleosides, beta-lactams, butyrolactones, siderophores, melanins, and others. Includes comparative gene cluster analysis	http://antismash.secondarymetabolites.org/
ASMPKS	Analysis System for Modular Polyketide Synthesis	Identification of PKS genes from genomic DNA, prediction of domain architecture, and visualization of predicted polyketide products. Also includes a database of known polyketides	http://gate.smallsoft.co.kr:8008/pks/
Bagel2	Bacteriocin Genome Mining Tool 2	Annotation of putative bacteriocins and lantibiotics from genomic DNA. Includes a database of validated bacteriocins	http://bagel2.molgenrug.nl/
CLUSEAN	Cluster Sequence Analyzer	Identification of domains and prediction of specificities for PKS and NRPS genes	http://redmine.secondarymetabolites.org/projects/clusean
ClustScan	Cluster Scanner	Annotation and product structure prediction for PKS, NRPS, and hybrid NRPS/PKS genes from genomic DNA. Includes stereochemistry prediction	http://bioserv.pbf.hr/cms/index.php?page = clustscan/
NORINE	Nonribosomal peptides (with 'ine' as a typical ending of peptide names)	Database of more than 1100 peptides containing structural information as well as biological activity, producing organisms, and literature references	http://bioinfo.lifl.fr/norine/
NP.Searcher	Natural Product Searcher	Automated annotation and product structure prediction for PKS, NRPS, and hybrid NRPS/PKS genes from genomic DNA	http://dna.sherman.lsi.umich.edu/
NRPS-PKS	—	PKS, NRPS, and hybrid NRPS/PKS domain identification and specificity prediction, as well as a database of characterized gene clusters	http://www.nii.res.in/nrps-pks.html
NRPSpredictor2	—	Prediction of adenylation domain specificity based on 10-amino-acid Stachelhaus code and 8-angstrom signature. Includes bacterial and fungal prediction	http://nrps.informatik.uni-tuebingen.de/
PKS-NRPS Analysis Website	—	Identification of PKS and NRPS domains, as well as adenylation domain specificity prediction based on the 8-amino-acid code	http://nrps.igs.umaryland.edu/nrps/
SBSPKS	Structure Based Sequence Analysis of Polyketide Synthases	Structural modeling of PKS modules and identification of key residues in the interfaces between modular PKS subunits. Also includes all functionalities and database support of the NRPS-PKS tool	http://www.nii.ac.in/sbspks.html
SMURF	Secondary Metabolite Unique Regions Finder	Annotation of polyketide, non-ribosomal peptide, NRPS-PKS hybrid, indole alkoloid, and terpene biosynthetic gene clusters from fungal genomic DNA, including prediction of borders	http://jcvi.org/smurf/index.php

residues lining the active site to narrow down its predictions.[37] This tool is available online via NRPSpredictor2.[38] For the analysis and comparison of the nonribosomal peptide products themselves, the database NORINE was established, which currently contains over 1100 compounds.[39]

For the analysis of PKSs, the earliest computational tools focused on the identification of their constituent domains[40,41] and of the linker regions between them.[42] The purpose here was not only to aid product prediction, but also to facilitate the reconstitution of individual domains and the combinatorial biosynthesis of 'unnatural' natural products via domain swapping. Identification of the specificity-determining residues of acyltransferase domains was again enabled by crystallographic data, allowing prediction of malonyl-CoA or methylmalonyl-CoA specificity.[40,41] More recent tools have afforded more accurate predictions and increased functionality. For example, ASMPKS allows the input of entire genome sequences for analysis.[43] As with NRPSs, support vector machines have been applied to type III PKSs to afford improved prediction accuracy.[44] Finally, SBSPKS allows PKS domains and modules to be modeled and docked to better predict and engineer intersubunit contacts, as well as to predict the order of substrate channeling in gene clusters with multiple PKS open reading frames.[45]

Currently, the bulk of the state-of-the-art computational tools for analysis of secondary metabolites focus on concomitant analysis of both PKS and NRPS genes. These tools, such as NRPS-PKS,[46] ClustScan,[47] CLUSEAN,[48] NP.Searcher,[49] and the PKS-NRPS Analysis Web-site[50] continue to improve upon the accuracy and functionality of their predecessors while providing user-friendly interfaces and increased computational power. Nevertheless, other classes of secondary metabolites have seen increased attention in recent years as well. For example, the genome mining tool BAGEL2 focuses exclusively on bacteriocins, which are ribosomally synthesized antimicrobial peptides from bacteria.[51] BAGEL2 considers conserved domains, physical properties, and genomic context to identify putative bacteriocins, which can easily be missed by other genome annotation tools due to their short size (<100 amino acids). Another example is the SMURF program, which can be used to scan fungal genome sequences not only for gene clusters producing polyketides and nonribosomal peptides, but also for indole alkaloids and terpenes by identifying prenyltransferases and terpene cyclases, respectively.[52] Finally, a new program called antiSMASH promises the greatest versatility of any program to date, expanding beyond just polyketides and nonribosomal peptides to include such classes of compounds as β-lactams, lantibiotics, and siderophores, among others.[53] The identification and analysis of such diverse secondary metabolism genes will surely be of great benefit to the synthetic biologist for the design and engineering of new pathways. Of course, it must be stated that despite their increasing utility, in silico tools will never completely replace laboratory experiments for the analysis and understanding of secondary metabolite genes. Nevertheless, such tools are clearly a very valuable component of the synthetic biology toolkit.

DE NOVO PATHWAY CONSTRUCTION AND OPTIMIZATION

While the identification and analysis of genes, proteins, and pathways found in nature provide an excellent starting point, it is the goal of the synthetic biologist to go above and beyond natural biological systems to obtain improved, or even completely novel, functionalities. The former of these goals has been addressed by a number of in silico experiments and tools designed to predict genetic modifications that will improve productivity of a particular value-added compound. An early example of this was provided by Lee and coworkers in 2002, who modeled the metabolism of *E. coli* to study succinic acid synthesis and predict a more efficient pathway.[54] One of the landmark computational tools for strain optimization is OptKnock, originally described by the Maranas group in 2003.[55] OptKnock utilizes a bilevel optimization framework to reshape the metabolic network of *E. coli* such that the desired product is a necessary byproduct of growth.

The result is a set of candidate reactions for knockout. OptKnock was subsequently expanded through OptStrain, which considered not only knockouts but also addition of nonnative pathways to *E. coli*;[56] OptReg, which further allowed for up- and down-regulation of reactions;[57] and OptForce, which incorporates flux data to identify all of the necessary reaction modifications that force the system towards the target production level.[58] Other groups have since devised their own complementary optimization tools, such as CASOP, which evaluates every reaction in the host organism via weighted elementary modes,[59] and OptORF, which explicitly focuses on gene deletion and overexpression targets rather than on reactions.[60] Finally, the OptFlux framework was recently developed to make such tools easily usable by researchers without significant expertise in bioinformatic analysis.[61] While such tools have predominantly been applied to the synthesis of platform and specialty chemicals, the underlying principles are readily applicable to drugs and drug candidates as well.

In addition to the systems-level approaches described above, smaller-scale computational tools have also been devised for pathway optimization. For example, Salis and coworkers have developed a method for the in silico design of ribosome binding sites to control protein expression levels.[62] To demonstrate the utility of their approach, they rationally tuned an AND gate genetic circuit to express green fluorescent protein only in the presence of arabinose and salicylate. Subsequently, they have expanded their toolkit to include the design of diverse RBS libraries, bacterial operons, and small RNAs for control of translation initiation. Another example of a computational tool that operates at the protein level is Gene Designer, which can be used to optimize protein expression via codon optimization, as well as optimization of promoter strength and mRNA stability.[63]

One step beyond pathway optimization is the goal of complete de novo pathway design. Several tools, including BNICE, ReBiT, and UM-BBD, have been developed that propose biosynthetic connections between two given compounds. For a detailed description of these and other tools refer to Chapter 3. At present, the reported applications of de novo design software to complex natural product pathways are limited. A notable example, however, was presented by Gonzalez-Lergier and coworkers, who employed the BNICE computational framework to explore routes to novel polyketides.[64] Otherwise, general application of these tools to natural product biosynthesis has yet to be realized, in part due to the difficulty in evaluating the feasibility of a proposed pathway with a given set of enzymes. One computational tool that begins to address this problem is DESHARKY, which not only can predict routes from a host's metabolism to a desired product, but also evaluates the impact of the proposed routes via mathematical modeling.[65] As such tools continue to develop, they will no doubt prove invaluable to the engineering of designer pathways and organisms for drug biosynthesis.

Experimental Tools

ASSEMBLY OF LARGE DNA CONSTRUCTS

The biosynthesis of drugs and drug candidate molecules typically requires the coordinated actions of many gene products. As a result, the DNA constructs of interest to the synthetic biologist encompass many genes and regulatory elements, and can be quite large, ranging from tens of kilobases to hundreds of kilobases or more. Perhaps the most straightforward route to such constructs is through direct chemical synthesis. However, chemical synthesis of large constructs can introduce sequence errors and can also be cost-prohibitive, although significant strides have been made in this direction.[66] Traditionally, designer DNA constructs have been prepared through sequential digestion and ligation-based cloning techniques, but this can be a laborious, time-consuming, and inefficient process. As a result, some powerful new techniques have emerged in recent years for the assembly of large DNA constructs, up to and including entire genomes.

Early examples of assembly technology were carried out in vitro, such as the PCR-based method of Stemmer and coworkers.[67] Using a mixture of overlapping 40 nucleotide oligomers, they were able to generate a 1.1 kb β-lactamase gene via several cycles of PCR. Following standard PCR amplification of the gene, restriction digestion, and ligation into a plasmid, the assembled gene was successfully cloned and expressed in *E. coli*. They analogously demonstrated assembly of a 2.7 kb plasmid, which was maintained in *E. coli* following circularization and transformation. In 2003, a similar approach was taken to assemble the complete 5386 bp genome of φX174.[68] While the synthetic DNA was able to produce infectious virions following *E. coli* transformation, reduced infectivity was observed, indicating a rate of approximately one lethal error per 500 bp. More recently, the Golden Gate assembly method and the Gibson assembly method have been developed for high-efficiency in vitro assembly of large constructs, making possible the accurate assembly of large natural product gene clusters.[69,70] The details of these methods can be found in Chapter 3.

As an alternative to total in vitro assembly, techniques that make partial or complete use of in vivo recombination have been devised. Two examples are Sequence and Ligation-Independent Cloning (SLIC)[71] and Circular Polymerase Extension Cloning (CPEC),[72] which were covered in Chapter 3. Of particular note in natural product biosynthesis is the method of Reisinger and coworkers, who assembled the complete 31.6 kb 6-deoxyerythronolide B gene cluster.[73] They utilized a PCR-based assembly of 40 bp oligomers to generate 500 bp fragments in vitro, followed by in vivo ligation-independent cloning to generate 3−6 kb intermediate plasmids. Finally, digestion with type II restriction enzymes and ligation-based selection yielded the fully assembled gene cluster. An example of a completely in vivo assembly technique is the DNA assembler method of Shao and coworkers.[74] This method, which relies on the high homologous recombination proficiency of *Saccharomyces cerevisiae*, was demonstrated to construct an 11 kb five-gene biosynthetic pathway to the carotenoid natural product zeaxanthin. At the same time, Gibson and coworkers demonstrated in vivo assembly of the entire *Mycoplasma genitalium* genome in *S. cerevisiae* via simultaneous transformation of 25 DNA fragments of ~24 kb.[75] This impressive feat clearly demonstrates the utility of in vivo assembly for the preparation of large DNA constructs, underscoring the applicability of such techniques in the field of natural product biosynthesis.

OPTIMIZATION OF ASSEMBLED PATHWAYS

In traditional engineering disciplines, the standardization of parts enables the rapid assembly of novel devices and the reliable prediction of their functionality a priori. As the field of synthetic biology matures, efforts are also being made to standardize biological parts in the form of a standard parts registry of so-called BioBrick components.[76,77] Nevertheless, the ideal configuration of components to maximize flux through a biosynthetic pathway is often difficult or impossible to predict in the context of a complex biological system. As a result, a number of experimental tools have been developed to facilitate the optimization of an assembled pathway.

One method by which to optimize a pathway is through combinatorial assembly of isofunctional variants of each gene. These variants could either be derived from different species or could be engineered versions of a parental construct. Many of the assembly tools described in the previous section, such as Golden Gate, DNA assembler, and Gibson assembly, are largely sequence-independent, and thus can readily be applied to the combinatorial assembly of pathway libraries from diverse fragments with compatible ends. Such a library generation approach was recently demonstrated by Wingler and Cornish.[78] By combining homologous recombination with type II restriction digestion in vivo, they demonstrated the construction of a library of $>10^4$ biosynthetic pathways.

Particularly in the field of natural product biosynthesis, combinatorial assembly strategies can be useful not only for the optimization of a given pathway, but also for the diversification of a given scaffold for the exploration of 'nonnatural' derivatives. For example, Shao and coworkers employed the DNA assembler method to rapidly modify a PKS gene involved in the biosynthesis of the polyketide natural product aureothin, generating a new derivative.[79] Expanding this concept further, one can also envision combinatorial assembly as a tool to diversify the tailoring enzymes frequently found in natural product pathways, such as oxidoreductases, methyltransferases, and glycosyltransferases. By exploiting the promiscuity of such enzymes, libraries of compounds can conceivably be generated that feature diverse modifications of a conserved core structure.

As an alternative to engineering via combinatorial assembly, many methods have been devised to optimize a set of biosynthetic pathways through tuning at the transcriptional, translational, and post-translational levels. Some examples of these techniques include the engineering of promoters, ribosome binding sites, and intergenic regions, as well as the scaffolding of biosynthetic proteins to direct metabolic flux. While many of these strategies have been applied primarily to short pathways in proof-of-concept experiments, their potential for significant contribution to the optimization of complex secondary metabolite pathways is evident. For more details on these and other pathway optimization approaches, refer to Chapter 3.

APPLICATIONS

As mentioned above, synthetic biology offers exciting techniques and fascinating strategies for drug discovery and development. Advances in computational tools enable rapid identification, characterization, and modification of novel genes and pathways, while powerful experimental tools accelerate the assembly and optimization of large DNA constructs. With the help of these tools, applications of synthetic biology have largely focused on the perfection of the investigative nature of biology with the constructive nature of engineering. In other words cells are becoming recast as true programmable and customizable entities.[80] Here we will discuss the applications of synthetic biology in both discovery and production of drugs.

Discovery of Novel Compounds

METAGENOMICS APPROACH TOWARDS THE DISCOVERY OF NATURAL PRODUCTS FROM SOIL

Soil microorganisms are a rich source of natural products. However, culture-independent analyses of environmental samples indicate that traditional laboratory cultivation approaches have most likely missed the majority of bacterial natural products that exist in the environment, as only a tiny fraction of soil microbes are cultivable in the laboratory.[81,82] To address this limitation, the concept of metagenomics was proposed in 1998 to analyze genes and pathways in samples obtained directly from the environment.[83] Such a strategy explores the secondary metabolites produced by the large collections of bacteria that are known to be present in the environment, but remain recalcitrant to culturing.[84]

The isolation and subsequent examination of DNA extracted directly from naturally occurring microbial populations (environmental DNA, eDNA) is the foundation of metagenomics.[84] The basic strategy is to isolate metagenomic DNA directly from soil, clone the large pieces of DNA into a readily cultivable organism such as *E. coli*, and screen the clones for biological activity (Fig. 10.1).[83] Metagenomics is particularly appealing for novel drug discovery from soil bacteria, as the secondary metabolite biosynthetic pathways

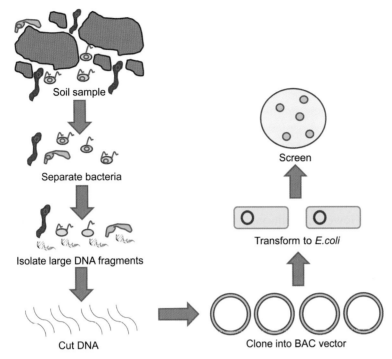

FIGURE 10.1
Process flow of the metagenomic method.

in bacteria are typically clustered together on the genome. Therefore, this method is widely used to identify new biologically active compounds.

There are two major metagenomic screening strategies: expression-dependent (function-based) and expression-independent (sequence-based). Both have been used to identify eDNA clones that produce bioactive compounds. In expression-dependent studies, the production of bioactive compounds is usually linked to the change of phenotypes generated from the eDNA libraries through a simple high-throughput assay. On the other hand, in expression-independent studies, the eDNA libraries are examined by screening methods that use probes containing conserved sequences traditionally associated with secondary metabolite biosynthesis to identify target clones. Valuable information collected from these initial high-throughput assays is subsequently tested in model cultivable heterologous hosts to confer the production of desirable compounds.[84]

There are several examples of using the function-based metagenomics library screening method. One new isocyanide-containing eDNA-derived antibiotic was isolated and characterized by Brady and coworkers.[85] In this work, eDNA clones in *E. coli* libraries were screened using top agar overlay assays to identify potential bioactive compounds by looking for clones generating inhibition zones against test microbes. In addition, a variety of long-chain *N*-acylated amino acid antibiotics were characterized by the same method.[86] Antibacterially active pigments violacein, indigo,[81,87] and turbomycins[88,89] were also found by examining pigmented eDNA clones or through randomly selected culture broth extracts.[90−92]

Compared to the function-based metagenomics library screening method, the sequence-based metagenomics library screening method offers a better way to clone natural product gene clusters from uncultivable symbionts or gene cluster families from the environment. The anticancer drug pederin was originally isolated from the beetle *Paederus fuscipes*,[93] while the biosynthetic gene cluster for pederin was identified to originate from an uncultured symbiotic

Pseudomonad by using a cosmid library constructed from metagenomic DNA.[94] In addition, Banik and coworkers[95] identified two new biosynthetic gene clusters associated with the biosynthesis of teicoplanin and vancomycin-like glycopeptide antibiotics after they screened a soil DNA cosmid library using the sequence-based method. One of these glycopeptide pathways contains three sulfotransferases, which is a class of tailoring enzymes that has rarely been found within glycopeptide biosynthesis clusters. By using the eDNA-derived sulfotransferases, seven new anionic glycopeptide congeners[87] were generated in vitro.

METAGENOMICS APPROACH TOWARDS THE DISCOVERY OF NATURAL PRODUCTS FROM THE SEA

Whereas the bacterial world represents a rich source of bioactive natural products in the soil, marine organisms provide an untapped reservoir for undiscovered, bioactive natural products as well. However, traditional chemical approaches to discover new valuable bioactive compounds are not effective because the vast majority of marine natural products have extreme chemical variability.[84] In contrast, metagenomic methods enabled the discovery of new natural products. For example, the sequence-based metagenomic library screening method was used to successfully clone a gene cluster encoding the biosynthesis of pederin-related onnamide from symbionts associated with field collected *Demospongiae* sponges.[96] Similarly, the gene cluster encoding the biosynthesis of the antitumor agent psymberin was identified from a sponge—bacteria association.[97] Later an enhanced antitumor compound derived from pederin was identified by using a recombinant *O*-methyltransferase, PedO, from the pederin biosynthesis gene cluster to methylate mycalamide A.[98]

Cyanobactins are an important example of novel bioactive compounds discovered from the sea using metagenomics. One example is from cyanobacterial symbionts associated with marine *Demospongiae* sponges. Schmidt and coworkers[99] identified several patellamide biosynthetic genes and a related cluster in an important cyanobacterium, *Trichodesmium erythraeum* IMS101, which was confirmed by heterologous expression of the whole pathway in *E. coli*. A large family of cytotoxic cyclic peptides exemplified by the patellamides was isolated from ascidians harboring the obligate cyanobacterial symbionts *Prochloron* spp. Later, the Schmidt group reported the PCR amplification of 30 genes encoding novel patellamide-like precursor peptides from uncultured *Prochloron* spp. symbionts living in consortia with marine sponges.[100,101] The discovery of new patellamide-like precursor peptides demonstrates the power of comparative analysis of closely related symbiotic pathways to direct the biological synthesis of new molecules and aids in generating additional members of the important cyclic peptide families. In addition, an assembly line responsible for the biosynthesis of diverse cyanobactins was reported.[102] By comparing five new cyanobactin biosynthetic clusters, the prenylated antitumor preclinical candidate trunkamide was produced in *E. coli* using genetic engineering.

COMBINATORIAL APPROACHES FOR NOVEL DRUG GENERATION

In the past few decades, more than 20 000 chemically diverse and biologically active compounds have been discovered from microbial sources by traditional screening efforts.[103] The successful cloning of the first antibiotic biosynthetic gene cluster[104] prompted researchers to investigate whether recombinant bacteria containing one or more genes from different organisms that make biosynthetically related metabolites could produce novel ones. The resulting combinatorial biosynthesis method, which was developed independently by Houghten and Geysen in the 1980s, allows interchanging of secondary metabolism genes between antibiotic-producing microorganisms to generate a large library of pathways for high-throughput screening.

Combinatorial biosynthesis has been particularly successful with PKS genes. Novel combinations of type I and type II PKS genes produced numerous derivatives of medically important macrolide antibiotics and unusual polycyclic aromatic compounds.[103,105–108] The initial demonstration was comprised of novel polycyclic aromatic metabolites produced by hybrid forms of the actinorhodin producing genes and tetracenomycin type II PKS genes,[109–113] resulting in rational design of new analogues.[114] A large number of novel compounds were produced through such a strategy,[115] including tetracenomycin M resulting from the combination of mithramycin and tetracenomycin genes,[116] and a novel 18-carbon polyketide made by hybrid forms of tetracenomycin and griseusin cyclases.[117] The modular PKS fosters the real excitement of using combinatorial biosynthesis for drug discovery and development. One comprehensively studied example is the 6-deoxyerythronolide B synthase (DEBS), which will be discussed in detail in the following section. Deletion, inactivation, or shuffling of domains or modules within or outside the system generated many novel compounds.[118,119] Another strategy to generate novel compounds is to change the starter or extension units.[120–133] This approach has been extended to other drugs.[134–136]

The potential of combinatorial biosynthesis was further expanded by the addition of the deoxysugar (DOS) biosynthesis genes. Usually the formation of DOSs as glycosides, which are made by the glucose-1-phosphate key metabolic intermediate, thymidine diphospho 4-keto-6-deoxyglucose or one of its derivatives,[137] generates biological activities. Since the genes involved in these DOS-producing pathways have been identified, attempts have been made to use combinatorial biosynthesis to make analogues of known antibiotic glycosides or novel metabolites.[138–141]

ACTIVATION OF SILENT GENE CLUSTERS

The increasing availability of whole-genome sequences has revised our view of the metabolic capabilities of microorganisms. Analyses of more than 500 microbial genome sequences currently in the publicly accessible databases have revealed numerous examples of gene clusters encoding enzymes similar to those known to be involved in the biosynthesis of many important natural products.[142] Examples of such enzymes include NRPSs, PKSs, and terpene synthases, as well as enzymes belonging to less thoroughly investigated families. Many of these gene clusters are hypothesized to produce novel natural products. However, most of them are cryptic biosynthetic pathways, for which the encoded natural product is unknown, or the pathway is silent and the product cannot be detected under general growth conditions. Since genomic sequencing projects are mostly focused on bacteria and fungi, cryptic pathways related to valuable secondary metabolites are often revealed from these genome databases. For example, while *Streptomyces coelicolor* was known to produce only four secondary metabolites,[143] genome analysis revealed 18 additional cryptic biosynthetic pathways. It is intriguing to note that this is not a special case, because analyses of other microbial genomes originating from myxobacteria, cyanobacteria, and filamentous fungi shows the presence of comparable or even larger numbers of cryptic pathways.[144,145] These pathways represent an untapped treasure trove, which is likely to grow exponentially in the future, uncovering many novel and possibly bioactive compounds.[146]

Various strategies have been developed to discover the products of cryptic biosynthetic gene clusters (Fig. 10.2). Bioinformatics tools are usually used to predict the putative function of a target gene cluster from sequenced genomes and metagenomes before any experiments. Several techniques have been used to activate the cryptic pathways. One is to identify the product through prediction of physicochemical properties.[147] By using this approach, 17 novel biosynthetic loci were identified from the genus *Salinispora*, and bioinformatic analysis elucidated the structure of the polyene macrolactam salinilactam A. This provides a powerful bridge between genomic analysis and traditional natural product isolation

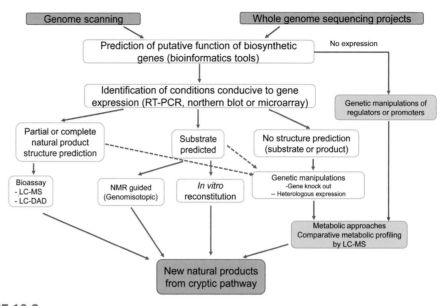

FIGURE 10.2

Techniques used to activate cryptic pathways.

studies. Another method is the genomisotopic approach.[148] This method employs a combination of genomic sequence analysis and isotope-guided fractionation to identify unknown compounds synthesized from orphan gene clusters. Identification of an orphan gene cluster which led to the discovery of orfamide A, the model for a group of bioactive cyclic lipopeptides, was done through analysis of the *Pseudomonas fluorescens* Pf-5 genome. A third approach is the in vitro reconstitution approach.[149] This strategy was applied to a terpene synthase encoded by the SCO5222 (SC7E4.19) gene of *S. coelicolor*. Incubation of the recombinant protein, SCO5222p, with farnesyl diphosphate in the presence of Mg(II) gave a new sesquiterpene, (+)-epi-isozizaene. The fourth approach is the biosynthetic gene inactivation/comparative metabolic profiling approach.[150] Gcs, a type III PKS encoded by the SCO7221 ORF of the bacterium *S. coelicolor*, required for biosynthesis of the germicidin family of 3,6-dialkyl-4-hydroxypyran-2-one natural products, was discovered through this method. The fifth approach is the genome mining approach.[151] In this approach, genes that encode tailoring enzymes from natural product biosynthesis pathways serve as indicator genes for the identification of strains that have the genetic potential to produce natural products of interest. One example is from halogenases, which are known to be involved in the synthesis of halometabolites. From PCR screening of 550 randomly selected *Actinomycetes* strains, 103 novel putative halogenase genes were identified. The sixth approach is to activate silent clusters by manipulation of regulatory genes. Homologous overexpression of a putative activator gene *apdR* led to activation of a silent PKS-NRPS gene cluster from *Aspergillus nidulans*.[145] Another recent example to activate silent fungal secondary metabolite gene clusters was done by deletion of *hdaA*, the gene encoding an *Aspergillus nidulans* histone deacetylase (HDAC).[152] Using this approach, two telomere-proximal gene clusters were transcriptionally activated, leading to the production of multiple secondary metabolites not previously known to be produced by *A. nidulans*.[153]

Production of Valuable Compounds

ERYTHROMYCIN A AND ITS DERIVATIVES

Polyketides are a diverse group of natural products with important applications in the pharmaceutical industry. They can be found in a variety of organisms including plants, fungi, and bacteria, in which they are produced by PKSs which polymerize acyl-Coenzyme As (CoAs) in a fashion similar to fatty acid synthesis.

Erythromycin A is a clinically important polyketide antibiotic produced by the Gram-positive bacterium *Saccharopolyspora erythraea*.[154] It is a macrolide antibiotic with an antimicrobial spectrum similar to that of penicillin, and is often used for people who are allergic to penicillin to treat respiratory infections, whooping cough, syphilis, Legionnaires' disease, gastrointestinal infections, and acne. After the discovery and characterization of the DEBS PKSs,[118,155−157] which synthesize the macrocyclic aglycone of the antibiotic erythromycin, the biochemical basis for these highly controlled megasynthases was gradually revealed. Meanwhile, the functional expression of the DEBS proteins and their mutants in heterologous hosts such as *S. coelicolor*[158] and *E. coli*[159] allows the rational design of biosynthetic products, and leads to the generation of diverse polyketide libraries by means of combinatorial cloning of naturally occurring and mutant PKS modules.

One notable example of rational design is the incorporation of unnatural starter units to change the chain initiation process. Chain initiation on many modular polyketide synthases is mediated by acyl transfer from the CoA thioester of a dicarboxylic acid, followed by decarboxylation in situ by KSQ, a ketosynthase-like decarboxylase domain.[160] Therefore, one approach is to replace the loading domain AT with other AT domains,[127,160,161] or incorporate synthetic oligoketide precursors into the downstream DEBS pathway.[131] Kinoshita and coworkers[162] fed several diketide and triketide substrates to a recombinant *E. coli* strain containing a variant form of the DEBS cluster from which the first elongation module was deleted. They successfully obtained macrolactone formation from diketide substrates, but not the triketide substrates. To investigate the structural basis for selectivity among triketide analogues, Cane and coworkers[163] incubated a series of *N*-acetyl cysteamine (-SNAC) esters of unsaturated triketides with DEBS module 2 + TE and identified some that were good to excellent substrates. Based on this study, Regentin and coworkers[164] used precursor-directed biosynthesis to produce different triketide lactones (R-TKLs) in a fermentation process, and found that the R group on the precursor significantly affected titer (propyl >> chloromethyl > vinyl). An overproducing strain of *S. coelicolor* expressing the DEBS module 2 + TE protein was found to be best for production of R-TKLs, and a maximum titer of 500 mg/L 5-chloromethyl-TKL was obtained using 3.5 g/L precursor. Later, rather than using the KS1⁰ system (the DEBS1 protein with an inactivated ketosynthase domain), Ward and coworkers[165] showed that removing the DEBS loading domain and first module can lead to a two-fold increase in the utilization of diketide precursors, and production of 6-dEB analogues (R-6dEB) in *S. coelicolor* and *S. erythraea* exhibited a 10-fold increase.

Besides the precursors, the extender unit can also be replaced in the DEBS system. The DEBS AT domains of each module are the primary gatekeepers for the incorporation of methylmalonyl-CoA into the polyketide chain. With replacement by heterologous AT domains that accept malonyl-, ethylmalonyl-, or methoxylmalonyl units, corresponding building blocks have been verified to be incorporated into the final polyketide products.[119,123−125,166−168]

In addition to swapping domains, 17 new heterologous genes were expressed in a 6-deoxyerythronolide B (6dEB) producer *E. coli* strain, and the resulting strain was capable of producing the potent antibiotic erythromycin C.[169] Combinatorial polyketide biosynthesis by de novo design and rearrangement of modular DEBS genes with other PKS genes allow production of novel derivatives as well.[170] Fourteen modules from eight PKS clusters were synthesized, and 154 bimodular combinations were created. Nearly half of the combinations successfully mediated the biosynthesis of a polyketide in *E. coli*, and all individual modules participated in productive bimodular combinations.

Besides production of novel derivatives, synthetic biology also offers platforms for productivity enhancement. Random mutagenesis was applied to a *S. coelicolor* strain expressing the heterologous DEBS for the production of erythromycin aglycones.[171] Several strains exhibited two-fold higher productivities and reached >3 g/L of total macrolide aglycones. By combining classical mutagenesis, recombinant DNA techniques, and process development, the fermentation achieved 1.3 g/L of 15-methyl-6-dEB, and the productivity was increased by over 100% in a scalable fermentation process. Furthermore, the introduction of an engineered mutase-epimerase pathway in *E. coli*, which provides the extender unit methylmalonyl-CoA to DEBS, enabled the five-fold higher heterologous production of 6-dEB.[172,173] Also, a deletion strain to increase biosynthetic pathway carbon flow and an engineered stain which contains an extra copy of a key deoxysugar glycosyltransferase gene both showed significant improvement in erythronolide B (EB) and 3-mycarosylerythronolide B (MEB) production, and a seven-fold increase in erythromycin A titer.[174]

PRECURSORS FOR ARTEMISININ

The biosynthetic production of artemisinin precursors is the most promising example of how synthetic biology is revolutionizing natural-product-based drug manufacturing. Artemisinin is an antimalarial agent naturally produced by sweet wormwood (*Artemisia annua*). As malaria-causing parasites are becoming increasingly resistant to commonly used therapies, more focus has been placed on artemisinin-based treatments. An estimated 225 million people are infected with malaria, and over 750 000 people die from it each year.[175] Current artemisinin extraction methods are neither economical nor sustainable enough to produce the compound for wide use. However, the progress of artemisinin biosynthesis in *E. coli* and *S. cerevisiae* over the past decade has led to titers that make fermentative production a practical and viable process.

Artemisinin is a sesquiterpene, a C15 molecule consisting of three isoprene building blocks. Two molecules of isopentenyl pyrophosphate (IPP) and one dimethylallyl pyrophosphate (DMAPP) react to form farnesyl pyrophosphate (FPP), which can be cyclized to form a number of sesquiterpenes. Amorpha-4,11-diene synthase (ADS) cyclizes FPP into amorphadiene, a bicyclic precursor of artemisinin. The gene for ADS was first discovered by probing an *A. annua* cDNA library with gene fragments derived from other sesquiterpene synthases.[176] Initial expression of ADS in *E. coli* led to a production of 0.086 mg caryophyllene equivalent/L/OD_{600} (caryophyllene is a sesquiterpene that was used for standardized measurements when amorphadiene was not commercially available).[177]

In *E. coli*, IPP and DMAPP are products of the 1-deoxy-D-xylulose 5-phosphate (DXP) pathway, while in many eukaryotes they are made via the mevalonate pathway. To decouple amorphadiene production from the native DXP pathway, Martin and coworkers introduced the *S. cerevisiae* mevalonate pathway into *E. coli*, leading to amorphadiene titers of 6.5 mg caryophyllene equivalent/L/OD_{600}, a 75-fold improvement.[177] The mevalonate pathway was split into two stages, with the first three enzymes contained on plasmid pMevT and the latter four along with FPP synthase on pMBIS (Fig. 10.3). By optimizing the codons of the pMevT genes and switching both promoters from lac to the stronger and less regulation-sensitive lacUV5, amorphadiene production rose to 221 mg/L/OD_{600}.[178] Because a considerable increase was seen when pMBIS genes were placed under lacUV5, it was determined that one of the latter-stage mevalonate enzymes was rate-limiting. Each of the pMBIS genes was individually placed on a higher copy number plasmid and coexpressed with pMBIS while cultures were supplemented with mevalonate, the substrate for the MBIS pathway stage. The strain overexpressing mevalonate kinase, encoded by *ERG12*, showed considerable production of amorphadiene, and a codon optimized *ERG12* was placed on a high copy number plasmid with ADS, leading to amorphadiene titers of 293 mg/L/OD_{600}.[178]

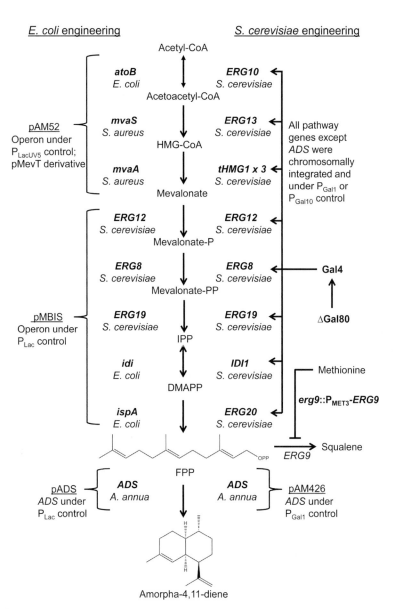

E. coli engineering   *S. cerevisiae* engineering

FIGURE 10.3

Engineered pathways in highest amorphadiene producing *E. coli* (left) and *S. cerevisiae* (right) strains. Sources of genes are in italics under gene names. Enzymes encoded by genes: *atoB* and *ERG10*, acetoacetyl-CoA thiolase; *mvaS* and *ERG13*, HMG-CoA synthase; *mvaA*, HMG-CoA reductase; *tHMG1*, truncated HMG-CoA reductase; *ERG12*, mevalonate kinase; *ERG8*, phosphomevalonate kinase; *ERG19*, mevalonate pyrophosphate decarboxylase; *idi* and *IDI1*, IPP isomerase; *ispA* and *ERG20*, farnesyl pyrophosphate synthase; *ADS*, amorphadiene synthase.

Engineering on the MevT genes has also led to considerable increase in concentrations. Studies showed that hydroxymethylglutaryl-CoA (HMG-CoA), an intermediate in the upper stage of the mevalonate pathway, inhibits cell growth.[179] Tsuruta and coworkers swapped the HMG-CoA synthase and reductase genes originally from yeast with those from *Staphylococcus aureus*, producing around 275 mg/L amorphadiene compared to 223 mg/L without the swapped genes.[180] It should be noted that titers listed are from shake flask cultures, and that improvements in fermentation process and compound separation led to amorphadiene production of over 25 g/L in *E. coli*.[180]

Considerable engineering has been done on yeast's native mevalonate pathway to give amorphadiene titers comparable with those in *E. coli*. Along with the overexpression of tHMGR and FPP synthase, a mutant allele of UPC2, a global transcription factor that up-regulates several mevalonate pathway genes, was overexpressed. A methionine repressible promoter was placed in front of squalene synthase, reducing FPP flux through the alternate pathway. These changes in yeast increased amorphadiene production from 4.4 mg/L to 153 mg/L.[181] Further engineering included expressing each mevalonate pathway gene under galactose inducible promoters, giving titers near 1.2 g/L amorphadiene in 72 hours. To allow activation of the mevalonate pathway without galactose induction, the gene for the Gal80 protein was deleted. In the presence of galactose, Gal80 is released from Gal4, allowing the latter to activate transcription of the target genes. With the deletion of *gal80*, Gal4 readily binds to activation sequences without the addition of an inducer. Even without the addition of galactose, the *ΔGAL80* strain produced amorphadiene in levels comparable to those of the galactose inducible strain.[182] As with *E. coli*, experimentation with a fed-batch process led to a considerable increase in amorphadiene production over shake-flask cultures, resulting in concentrations near 40 g/L.[182]

With the addition of the cytochrome P450 amorphadiene oxidase (AMO) to engineered strains in *E. coli* and yeast, artemisinic acid, a precursor closer to artemisinin, has been produced.[181,183] Although amorphadiene turnover was not seen upon initial expression of AMO in *E. coli*, after codon optimization, *N*-terminal engineering, and placement of AMO on a plasmid designed for P450 expression, artemisinic acid titers of over 100 mg/L were reached.[183] In yeast, similar titers were also achieved.[102] However, since amorphadiene titers were nearly 10-fold higher and the yield for semisynthesis of artemisinic acid from amorphadiene nears 50%, processes optimized for amorphadiene production were pursued.[182]

PRECURSORS FOR TAXOL

Taxol (paclitaxel) is a potent antitumor compound used to treat a variety of cancers. It is naturally produced by the Pacific yew tree (*Taxus brevifolia*). Like artemisinin, Taxol is a terpenoid that is currently derived from plant sources, which prevents wider availability of the drug and its derivatives. Synthetic pathways have been introduced into *E. coli* and yeast for the production of taxadiene, a committed intermediate of Taxol. Taxadiene forms upon the cyclization of geranylgeranyl pyrophosphate (GGPP) by taxadiene synthase (TS). GGPP is a C20 terpenoid consisting of three molecules of IPP and one of DMAPP. An additional 18 steps are required for completion of Taxol, including eight oxygenations by cytochrome P450 enzymes.[184]

DeJong and coworkers expressed in yeast the *Taxus canadensis* GGPP synthase (GGPPS) and TS along with two cytochrome P450 enzymes and an acyltransferase that convert taxadiene into taxadien-5α-acetoxy-10β-ol. Both the GGPPS and TS were truncated to remove peptides for plastid localization, and all genes except for a hydroxylase were placed under galactose inducible promoters. Taxadiene titers reached 1 mg/L and taxadien-5α-ol, the subsequent intermediate after taxadiene, was measured to be around 25 µg/L. Although immunoblots showed that the enzymes that formed the next two metabolites expressed well, the compounds could not be detected.[185] Further modifications made to the mevalonate pathway were similar to changes made for artemisinin pathway engineering. A truncated HMGR (tHMGR) was introduced that prevented negative-feedback regulation, and the mutated *UPC2* gene led to overexpression of mevalonate pathway genes. One bottleneck in the pathway was the relatively low availability of GGPP. Since the *T. canadensis* GGPPS competed with sesquiterpene synthases for FPP, a *Sulfolobus acidocaldarius* GGPPS that used only DMAPP building blocks was expressed. In addition, since *T. candaensis* taxadiene synthase showed poor expression in yeast, it was codon optimized. This final strain exhibited taxadiene production of 8.7 mg/L. GGPP amounts were also considerable at 33.1 mg/L, allowing more production of taxadiene.[186]

The first engineering for taxadiene production in *E. coli* employed overexpression of IPP isomerase, GGPPS, TS, and DXP synthase, the first enzyme of the DXP pathway. An initial titer of 1.3 mg/L taxadiene was formed.[187] Extensive pathway engineering was carried out by Ajikumar and coworkers in which the taxadiene pathway was separated into an upstream module, consisting of four of the eight DXP pathway enzymes that were reported to be rate-limiting, and a downstream module, consisting of GGPPS and TS. The expression of each module was varied by different promoters and plasmid copy numbers. Thirty-two total combinations of the modulated upstream and downstream modules were tested for taxadiene production, the maximum producing nearly 300 mg/L by shake-flask and over 1 g/L by fed-batch reactor.[188] Taxadiene-5α-ol hydroxylase (T5αOH), the P450 that oxidizes taxadiene to taxadiene-5α-ol, was engineered with its partner reductase (TCPR). Optimal taxadiene-5α-ol production (24 mg/L in shake-flask, 58 mg/L in bioreactor) was found with an *N*-terminal engineered T5αOH linked to a 74-amino-acid truncated TCPR.

CONCLUSIONS AND FUTURE PERSPECTIVES

The potential applications of synthetic biology to the pharmaceutical industry are far-reaching, from the earliest stages of compound discovery through lead optimization and industrial-scale production. As we have demonstrated in this chapter, a vast array of computational tools is now available for tasks ranging from gene and pathway discovery all the way to production host optimization. Additionally, sophisticated and facile experimental tools have been developed that make the manipulation of biosynthetic pathways and even entire genomes routine. Application of these tools and synthetic biology concepts has already led to the discovery of new and interesting natural products from diverse environments. Further, engineering and combinatorial approaches enable the biosynthesis of 'unnatural' natural products not found in nature. We have already seen the application of synthetic biology techniques make significant contributions to the production of valuable drugs such as erythromycin A, artemisinin, and taxol; and countless similar examples are likely to appear in the coming years. As a result, synthetic biology is poised to be a major driving force in the pharmaceutical industry, unlocking as-yet undiscovered drug molecules and transforming the perception of the cell from an ill-defined black box to a programmable, tunable machine.

Acknowledgments

We thank the National Institutes of Health (GM077596), the National Academies Keck Futures Initiative on Synthetic Biology, the Energy Biosciences Institute, and the National Science Foundation as part of the Center for Enabling New Technologies through Catalysis (CENTC), CHE-0650456 for financial support in our synthetic biology projects. Y. Luo also acknowledges fellowship support from the National Research Foundation of Korea (NRFK) (220-2009-1-D00033).

References

1. Newman DJ, Cragg GM. Natural products as sources of new drugs over the last 25 years. *J Nat Prod.* 2007;70:461–477.

2. Fickett JW. Recognition of protein coding regions in DNA sequences. *Nucleic Acids Res.* 1982;10:5303–5318.

3. Borodovsky M, McIninch J. GENMARK: parallel gene recognition for both DNA strands. *Compu Chem.* 1993;17:123–133.

4. Lukashin AV, Borodovsky M. GeneMark.hmm: new solutions for gene finding. *Nucleic Acids Res.* 1998;26:1107–1115.

5. Besemer J, Lomsadze A, Borodovsky M. GeneMarkS: a self-training method for prediction of gene starts in microbial genomes. Implications for finding sequence motifs in regulatory regions. *Nucleic Acids Res.* 2001;29:2607–2618.

6. Besemer J, Borodovsky M. GeneMark: web software for gene finding in prokaryotes, eukaryotes and viruses. *Nucleic Acids Res.* 2005;33:W451–454.

7. Ter-Hovhannisyan V, Lomsadze A, Chernoff YO, Borodovsky M. Gene prediction in novel fungal genomes using an ab initio algorithm with unsupervised training. *Genome Res.* 2008;18:1979−1990.

8. Delcher AL, Harmon D, Kasif S, White O, Salzberg SL. Improved microbial gene identification with GLIMMER. *Nucleic Acids Res.* 1999;27:4636−4641.

9. Salzberg SL, Delcher AL, Kasif S, White O. Microbial gene identification using interpolated Markov models. *Nucleic Acids Res.* 1998;26:544−548.

10. Majoros WH, Pertea M, Salzberg SL. TigrScan and GlimmerHMM: two open source ab initio eukaryotic gene-finders. *Bioinformatics.* 2004;20:2878−2879.

11. Salzberg SL, Pertea M, Delcher AL, Gardner MJ, Tettelin H. Interpolated Markov models for eukaryotic gene finding. *Genomics.* 1999;59:24−31.

12. Delcher AL, Bratke KA, Powers EC, Salzberg SL. Identifying bacterial genes and endosymbiont DNA with Glimmer. *Bioinformatics.* 2007;23:673−679.

13. Kelley DR, Liu B, Delcher AL, Pop M, Salzberg SL. Gene prediction with Glimmer for metagenomic sequences augmented by classification and clustering. *Nucleic Acids Res.* 2012;40:e9.

14. Fitch WM. An improved method of testing for evolutionary homology. *J Mol Biol.* 1966;16:9−16.

15. Sellers PH. On the theory and computation of evolutionary distances. *SIAM J Appl Math.* 1974;26:787−793.

16. Gotoh O. An improved algorithm for matching biological sequences. *J Mol Biol.* 1982;162:705−708.

17. Lipman DJ, Pearson WR. Rapid and sensitive protein similarity searches. *Science.* 1985;227:1435−1441.

18. Pearson WR. Using the FASTA program to search protein and DNA sequence databases. *Methods Mol Biol.* 1994;24:307−331.

19. Pearson WR, Lipman DJ. Improved tools for biological sequence comparison. *Proc Natl Acad Sci USA.* 1988;85:2444−2448.

20. Lipman DJ, Altschul SF, Kececioglu JD. A tool for multiple sequence alignment. *Proc Natl Acad Sci USA.* 1989;86:4412−4415.

21. Thompson JD, Higgins DG, Gibson TJ. CLUSTAL W: improving the sensitivity of progressive multiple sequence alignment through sequence weighting, position-specific gap penalties and weight matrix choice. *Nucleic Acids Res.* 1994;22:4673−4680.

22. Karplus K, Barrett C, Hughey R. Hidden Markov models for detecting remote protein homologies. *Bioinformatics.* 1998;14:846−856.

23. Eddy SR. Profile hidden Markov models. *Bioinformatics.* 1998;14:755−763.

24. Altschul SF, Gish W, Miller W, Myers EW, Lipman DJ. Basic local alignment search tool. *J Mol Biol.* 1990;215:403−410.

25. Altschul SF, Madden TL, Schaffer AA, et al. Gapped BLAST and PSI-BLAST: a new generation of protein database search programs. *Nucleic Acids Res.* 1997;25:3389−3402.

26. Marchler-Bauer A, Anderson JB, Chitsaz F, et al. CDD: specific functional annotation with the Conserved Domain Database. *Nucleic Acids Res.* 2009;37:D205−210.

27. Finn RD, Mistry J, Tate J, et al. The Pfam protein families database. *Nucleic Acids Res.* 2010;38:D211−222.

28. Letunic I, Doerks T, Bork P. SMART 7: recent updates to the protein domain annotation resource. *Nucleic Acids Res.* 2012;40:D302−D305.

29. Grin I, Linke D. GCView: the genomic context viewer for protein homology searches. *Nucleic Acids Res.* 2011;39:W353−356.

30. Liao CS, Lu K, Baym M, Singh R, Berger B. IsoRankN: spectral methods for global alignment of multiple protein networks. *Bioinformatics.* 2009;25:i253−258.

31. Rangannan V, Bansal M. High-quality annotation of promoter regions for 913 bacterial genomes. *Bioinformatics.* 2010;26:3043−3050.

32. Broos S, Hulpiau P, Galle J, Hooghe B, Van Roy F, De Bleser P. ConTra v2: a tool to identify transcription factor binding sites across species, update 2011. *Nucleic Acids Res.* 2011;39:W74−78.

33. Thomas-Chollier M, Defrance M, Medina-Rivera A, et al. RSAT 2011: regulatory sequence analysis tools. *Nucleic Acids Res.* 2011;39:W86−91.

34. de Crécy-Lagard V, Saurin W, Thibaut D, et al. Streptogramin B biosynthesis in *Streptomyces pristinaespiralis* and *Streptomyces virginiae*: molecular characterization of the last structural peptide synthetase gene. *Antimicrob Agents Chemother.* 1997;41:1904−1909.

35. Stachelhaus T, Mootz HD, Marahiel MA. The specificity-conferring code of adenylation domains in nonribosomal peptide synthetases. *Chem Biol.* 1999;6:493−505.

36. Challis GL, Ravel J, Townsend CA. Predictive, structure-based model of amino acid recognition by nonribosomal peptide synthetase adenylation domains. *Chem Biol.* 2000;7:211−224.

37. Rausch C, Weber T, Kohlbacher O, Wohlleben W, Huson DH. Specificity prediction of adenylation domains in nonribosomal peptide synthetases (NRPS) using transductive support vector machines (TSVMs). *Nucleic Acids Res.* 2005;33:5799–5808.

38. Rottig M, Medema MH, Blin K, Weber T, Rausch C, Kohlbacher O. NRPSpredictor2 – a web server for predicting NRPS adenylation domain specificity. *Nucleic Acids Res.* 2011;39:W362–367.

39. Caboche S, Pupin M, Leclere V, Fontaine A, Jacques P, Kucherov G. NORINE: a database of nonribosomal peptides. *Nucleic Acids Res.* 2008;36:D326–331.

40. Yadav G, Gokhale RS, Mohanty D. Computational approach for prediction of domain organization and substrate specificity of modular polyketide synthases. *J Mol Biol.* 2003;328:335–363.

41. Yadav G, Gokhale RS, Mohanty D. SEARCHPKS: a program for detection and analysis of polyketide synthase domains. *Nucleic Acids Res.* 2003;31:3654–3658.

42. Udwary DW, Merski M, Townsend CA. A method for prediction of the locations of linker regions within large multifunctional proteins, and application to a type I polyketide synthase. *J Mol Biol.* 2002;323:585–598.

43. Tae H, Kong EB, Park K. ASMPKS: an analysis system for modular polyketide synthases. *BMC Bioinformatics.* 2007;8:327.

44. Mallika V, Sivakumar KC, Jaichand S, Soniya EV. Kernel based machine learning algorithm for the efficient prediction of type III polyketide synthase family of proteins. *J Integr Bioinform.* 2010;7:143.

45. Anand S, Prasad MV, Yadav G, et al. SBSPKS: structure based sequence analysis of polyketide synthases. *Nucleic Acids Res.* 2010;38:W487–496.

46. Ansari MZ, Yadav G, Gokhale RS, Mohanty D. NRPS-PKS: a knowledge-based resource for analysis of NRPS/PKS megasynthases. *Nucleic Acids Res.* 2004;32:W405–413.

47. Starcevic A, Zucko J, Simunkovic J, Long PF, Cullum J, Hranueli D. ClustScan: an integrated program package for the semi-automatic annotation of modular biosynthetic gene clusters and *in silico* prediction of novel chemical structures. *Nucleic Acids Res.* 2008;36:6882–6892.

48. Weber T, Rausch C, Lopez P, et al. CLUSEAN: a computer-based framework for the automated analysis of bacterial secondary metabolite biosynthetic gene clusters. *J Biotechnol.* 2009;140:13–17.

49. Li MH, Ung PM, Zajkowski J, Garneau-Tsodikova S, Sherman DH. Automated genome mining for natural products. *BMC Bioinformatics.* 2009;10:185.

50. Bachmann BO, Ravel J. Methods for *in silico* prediction of microbial polyketide and nonribosomal peptide biosynthetic pathways from DNA sequence data. *Methods Enzymol.* 2009;458:181–217.

51. de Jong A, van Heel AJ, Kok J, Kuipers OP. BAGEL2: mining for bacteriocins in genomic data. *Nucleic Acids Res.* 2011;38:W647–651.

52. Khaldi N, Seifuddin FT, Turner G, et al. SMURF: genomic mapping of fungal secondary metabolite clusters. *Fungal Genet Biol.* 2010;47:736–741.

53. Medema MH, Blin K, Cimermancic P, et al. antiSMASH: rapid identification, annotation and analysis of secondary metabolite biosynthesis gene clusters in bacterial and fungal genome sequences. *Nucleic Acids Res.* 2011;39:W339–346.

54. Lee SY, Hong SH, Moon SY. *In silico* metabolic pathway analysis and design: succinic acid production by metabolically engineered *Escherichia coli* as an example. *Genome Inform.* 2002;13:214–223.

55. Burgard AP, Pharkya P, Maranas CD. Optknock: a bilevel programming framework for identifying gene knockout strategies for microbial strain optimization. *Biotechnol Bioeng.* 2003;84:647–657.

56. Pharkya P, Burgard AP, Maranas CD. OptStrain: a computational framework for redesign of microbial production systems. *Genome Res.* 2004;14:2367–2376.

57. Pharkya P, Maranas CD. An optimization framework for identifying reaction activation/inhibition or elimination candidates for overproduction in microbial systems. *Metab Eng.* 2006;8:1–13.

58. Ranganathan S, Suthers PF, Maranas CD. OptForce: an optimization procedure for identifying all genetic manipulations leading to targeted overproductions. *PLoS Comput Biol.* 2010;6:e1000744.

59. Hadicke O, Klamt S. CASOP: a computational approach for strain optimization aiming at high productivity. *J Biotechnol.* 2010;147:88–101.

60. Kim J, Reed JL. OptORF: optimal metabolic and regulatory perturbations for metabolic engineering of microbial strains. *BMC Syst Biol.* 2010;4:53.

61. Rocha I, Maia P, Evangelista P, et al. OptFlux: an open-source software platform for in silico metabolic engineering. *BMC Syst Biol.* 2010;4:45.

62. Salis HM, Mirsky EA, Voigt CA. Automated design of synthetic ribosome binding sites to control protein expression. *Nat Biotechnol.* 2009;27:946–950.

63. Welch M, Villalobos A, Gustafsson C, Minshull J. Designing genes for successful protein expression. *Methods Enzymol.* 2011;498:43–66.

64. Gonzalez-Lergier J, Broadbelt LJ, Hatzimanikatis V. Theoretical considerations and computational analysis of the complexity in polyketide synthesis pathways. *J Am Chem Soc.* 2005;127:9930–9938.

65. Rodrigo G, Carrera J, Prather KJ, Jaramillo A. DESHARKY: automatic design of metabolic pathways for optimal cell growth. *Bioinformatics.* 2008;24:2554–2556.

66. May M. Engineering a new business. *Nat Biotechnol.* 2009;27:1112–1120.

67. Stemmer WP, Crameri A, Ha KD, Brennan TM, Heyneker HL. Single-step assembly of a gene and entire plasmid from large numbers of oligodeoxyribonucleotides. *Gene.* 1995;164:49–53.

68. Smith HO, Hutchison 3rd CA, Pfannkoch C, Venter JC. Generating a synthetic genome by whole genome assembly: phiX174 bacteriophage from synthetic oligonucleotides. *Proc Natl Acad Sci USA.* 2003;100:15440–15445.

69. Engler C, Kandzia R, Marillonnet S. A one pot, one step, precision cloning method with high throughput capability. *PLoS One.* 2008;3:e3647.

70. Gibson DG, Young L, Chuang RY, Venter JC, Hutchison 3rd CA, Smith HO. Enzymatic assembly of DNA molecules up to several hundred kilobases. *Nat Methods.* 2009;6:343–345.

71. Li MZ, Elledge SJ. Harnessing homologous recombination *in vitro* to generate recombinant DNA via SLIC. *Nat Methods.* 2007;4:251–256.

72. Quan J, Tian J. Circular polymerase extension cloning of complex gene libraries and pathways. *PLoS One.* 2009;4:e6441.

73. Reisinger SJ, Patel KG, Santi DV. Total synthesis of multi-kilobase DNA sequences from oligonucleotides. *Nat Protoc.* 2006;1:2596–2603.

74. Shao Z, Zhao H. DNA assembler, an *in vivo* genetic method for rapid construction of biochemical pathways. *Nucleic Acids Res.* 2009;37:e16.

75. Gibson DG, Benders GA, Andrews-Pfannkoch C, et al. Complete chemical synthesis, assembly, and cloning of a *Mycoplasma genitalium* genome. *Science.* 2008;319:1215–1220.

76. Endy D. Foundations for engineering biology. *Nature.* 2005;438:449–453.

77. Knight T. Idempotent vector design for standard assembly of biobricks. *MIT Synthetic Biology Working Group Technical Reports*; 2003.

78. Wingler LM, Cornish VW. Reiterative recombination for the *in vivo* assembly of libraries of multigene pathways. *Proc Natl Acad Sci USA.* 2011;108:15135–15140.

79. Shao Z, Luo Y, Zhao H. Rapid characterization and engineering of natural product biosynthetic pathways via DNA assembler. *Mol Biosyst.* 2011;7:1056–1059.

80. Purnick PE, Weiss R. The second wave of synthetic biology: from modules to systems. *Nat Rev Mol Cell Biol.* 2009;10:410–422.

81. Rappe MS, Giovannoni SJ. The uncultured microbial majority. *Annu Rev Microbiol.* 2003;57:369–394.

82. Hugenholtz P, Goebel BM, Pace NR. Impact of culture-independent studies on the emerging phylogenetic view of bacterial diversity. *J Bacteriol.* 1998;180:4765–4774.

83. Handelsman J, Rondon MR, Brady SF, Clardy J, Goodman RM. Molecular biological access to the chemistry of unknown soil microbes: a new frontier for natural products. *Chem Biol.* 1998;5:R245–249.

84. Banik JJ, Brady SF. Recent application of metagenomic approaches toward the discovery of antimicrobials and other bioactive small molecules. *Curr Opin Microbiol.* 2010;13:603–609.

85. Brady SF, Clardy J. Cloning and heterologous expression of isocyanide biosynthetic genes from environmental DNA. *Angew Chem Int Ed Engl.* 2005;44:7063–7065.

86. Brady SF, Clardy J. Long-chain N-acyl amino acid antibiotics isolated from heterologously expressed environmental DNA. *J Am Chem Soc.* 2000;122:12903–12904.

87. Lim HK, Chung EJ, Kim JC, et al. Characterization of a forest soil metagenome clone that confers indirubin and indigo production on *Escherichia coli*. *Appl Environ Microbiol.* 2005;71:7768–7777.

88. Torsvik V, Goksoyr J, Daae FL. High diversity in DNA of soil bacteria. *Appl Environ Microbiol.* 1990;56:782–787.

89. Gillespie DE, Brady SF, Bettermann AD, et al. Isolation of antibiotics turbomycin A and B from a metagenomic library of soil microbial DNA. *Appl Environ Microb.* 2002;68:4301–4306.

90. Brady SF, Chao CJ, Handelsman J, Clardy J. Cloning and heterologous expression of a natural product biosynthetic gene cluster from eDNA. *Org Lett.* 2001;3:1981–1984.

91. MacNeil IA, Tiong CL, Minor C, et al. Expression and isolation of antimicrobial small molecules from soil DNA libraries. *J Mol Microb Biotech.* 2001;3:301–308.

92. Wang GYS, Graziani E, Waters B, et al. Novel natural products from soil DNA libraries in a streptomycete host. *Org Lett.* 2000;2:2401–2404.

93. Soldati M, Fioretti A, Ghione M. Cytotoxicity of pederin and some of its derivatives on cultured mammalian cells. *Experientia.* 1966;22:176–178.

94. Piel J. A polyketide synthase-peptide synthetase gene cluster from an uncultured bacterial symbiont of Paederus beetles. *Proc Natl Acad Sci USA*. 2002;99:14002−14007.

95. Banik JJ, Brady SF. Cloning and characterization of new glycopeptide gene clusters found in an environmental DNA megalibrary. *Proc Natl Acad Sci USA*. 2008;105:17273−17277.

96. Piel J, Hui D, Wen G, et al. Antitumor polyketide biosynthesis by an uncultivated bacterial symbiont of the marine sponge *Theonella swinhoei. Proc Natl Acad Sci USA*. 2004;101:16222−16227.

97. Fisch KM, Gurgui C, Heycke N, et al. Polyketide assembly lines of uncultivated sponge symbionts from structure-based gene targeting. *Nat Chem Biol*. 2009;5:494−501.

98. Zimmermann K, Engeser M, Blunt JW, Munro MH, Piel J. Pederin-type pathways of uncultivated bacterial symbionts: analysis of O-methyltransferases and generation of a biosynthetic hybrid. *J Am Chem Soc*. 2009;131:2780−2781.

99. Schmidt EW, Nelson JT, Rasko DA, et al. Patellamide A and C biosynthesis by a microcin-like pathway in *Prochloron didemni*, the cyanobacterial symbiont of *Lissoclinum patella. Proc Natl Acad Sci USA*. 2005;102:7315−7320.

100. Donia MS, Hathaway BJ, Sudek S, et al. Natural combinatorial peptide libraries in cyanobacterial symbionts of marine ascidians. *Nat Chem Biol*. 2006;2:729−735.

101. Schmidt EW, Donia MS. Cyanobactin ribosomally synthesized peptides − a case of deep metagenome mining. *Methods Enzymol*. 2009;458:575−596.

102. Donia MS, Ravel J, Schmidt EW. A global assembly line for cyanobactins. *Nat Chem Biol*. 2008;4:341−343.

103. Donadio S, Sosio M. Strategies for combinatorial biosynthesis with modular polyketide synthases. *Comb Chem High Throughput Screen*. 2003;6:489−500.

104. Malpartida F, Hopwood DA. Molecular cloning of the whole biosynthetic pathway of a Streptomyces antibiotic and its expression in a heterologous host. *Nature*. 1984;309:462−464.

105. Olano C, Mendez C, Salas JA. Post-PKS tailoring steps in natural product-producing actinomycetes from the perspective of combinatorial biosynthesis. *Nat Prod Rep*. 2010;27:571−616.

106. Rix U, Fischer C, Remsing LL, Rohr J. Modification of post-PKS tailoring steps through combinatorial biosynthesis. *Nat Prod Rep*. 2002;19:542−580.

107. Shen B, Chen M, Cheng Y, et al. Prerequisites for combinatorial biosynthesis: evolution of hybrid NRPS/PKS gene clusters. *Ernst Schering Res Found Workshop*. 2005:107−126.

108. Weissman KJ, Leadlay PF. Combinatorial biosynthesis of reduced polyketides. *Nat Rev Microbiol*. 2005;3:925−936.

109. Fu H, Alvarez MA, Khosla C, Bailey JE. Engineered biosynthesis of novel polyketides: regiospecific methylation of an unnatural substrate by the tcmO O-methyltransferase. *Biochemistry*. 1996;35:6527−6532.

110. Fu H, Hopwood DA, Khosla C. Engineered biosynthesis of novel polyketides: evidence for temporal, but not regiospecific, control of cyclization of an aromatic polyketide precursor. *Chem Biol*. 1994;1:205−210.

111. McDaniel R, Ebert-Khosla S, Fu H, Hopwood DA, Khosla C. Engineered biosynthesis of novel polyketides: influence of a downstream enzyme on the catalytic specificity of a minimal aromatic polyketide synthase. *Proc Natl Acad Sci USA*. 1994;91:11542−11546.

112. Fu H, McDaniel R, Hopwood DA, Khosla C. Engineered biosynthesis of novel polyketides: stereochemical course of two reactions catalyzed by a polyketide synthase. *Biochemistry*. 1994;33:9321−9326.

113. McDaniel R, Ebert-Khosla S, Hopwood DA, Khosla C. Engineered biosynthesis of novel polyketides. *Science*. 1993;262:1546−1550.

114. McDaniel R, Ebert-Khosla S, Hopwood DA, Khosla C. Rational design of aromatic polyketide natural products by recombinant assembly of enzymatic subunits. *Nature*. 1995;375:549−554.

115. Khosla C, Zawada RJ. Generation of polyketide libraries via combinatorial biosynthesis. *Trends Biotechnol*. 1996;14:335−341.

116. Kunzel E, Wohlert SE, Beninga C, et al. Tetracenomycin M, a novel genetically engineered tetracenomycin resulting from a combination of mithramycin and tetracenomycin biosynthetic genes. *Chem-Eur J*. 1997;3:1675−1678.

117. Zawada RJX, Khosla C. Domain analysis of the molecular recognition features of aromatic polyketide synthase subunits. *J Biol Chem*. 1997;272:16184−16188.

118. Donadio S, McAlpine JB, Sheldon PJ, Jackson M, Katz L. An erythromycin analog produced by reprogramming of polyketide synthesis. *Proc Natl Acad Sci USA*. 1993;90:7119−7123.

119. McDaniel R, Thamchaipenet A, Gustafsson C, et al. Multiple genetic modifications of the erythromycin polyketide synthase to produce a library of novel 'unnatural' natural products. *Proc Natl Acad Sci USA*. 1999;96:1846−1851.

120. Carreras CW, Pieper R, Khosla C. The chemistry and biology of fatty acid, polyketide, and nonribosomal peptide biosynthesis. *Top Curr Chem*. 1997;188:85−126.

121. McDaniel R, Kao CM, Fu H, et al. Gain-of-function mutagenesis of a modular polyketide synthase. *J Am Chem Soc*. 1997;119:4309–4310.

122. Kao CM, McPherson M, McDaniel RN, Fu H, Cane DE, Khosla C. Gain of function mutagenesis of the erythromycin polyketide synthase. 2. Engineered biosynthesis of eight-membered ring tetraketide lactone. *J Am Chem Soc*. 1997;119:11339–11340.

123. Oliynyk M, Brown MJB, Cortes J, Staunton J, Leadlay PF. A hybrid modular polyketide synthase obtained by domain swapping. *Chem Biol*. 1996;3:833–839.

124. Liu L, Thamchaipenet A, Fu H, Betlach M, Ashley G. Biosynthesis of 2-nor-6-deoxyerythronolide B by rationally designed domain substitution. *J Am Chem Soc*. 1997;119:10553–10554.

125. Ruan X, Pereda A, Stassi DL, et al. Acyltransferase domain substitutions in erythromycin polyketide synthase yield novel erythromycin derivatives. *J Bacteriol*. 1997;179:6416–6425.

126. Kuhstoss S, Huber M, Turner JR, Paschal JW, Rao RN. Production of a novel polyketide through the construction of a hybrid polyketide synthase. *Gene*. 1996;183:231–236.

127. Marsden AFA, Wilkinson B, Cortes J, Dunster NJ, Staunton J, Leadlay PF. Engineering broader specificity into an antibiotic-producing polyketide synthase. *Science*. 1998;279:199–202.

128. Dutton CJ, Gibson SP, Goudie AC, et al. Novel avermectins produced by mutational biosynthesis. *J Antibiot*. 1991;44:357–365.

129. Cane DE, Yang CC. Macrolide biosynthesis. 4. Intact incorporation of a chain-elongation intermediate into erythromycin. *J Am Chem Soc*. 1987;109:1255–1257.

130. Yue S, Duncan JS, Yamamoto Y, Hutchinson CR. Macrolide biosynthesis – tylactone formation involves the processive addition of 3 carbon units. *J Am Chem Soc*. 1987;109:1253–1255.

131. Jacobsen JR, Hutchinson CR, Cane DE, Khosla C. Precursor-directed biosynthesis of erythromycin analogs by an engineered polyketide synthase. *Science*. 1997;277:367–369.

132. Chuck JA, McPherson M, Huang H, Jacobsen JR, Khosla C, Cane DE. Molecular recognition of diketide substrates by a beta-ketoacyl-acyl carrier protein synthase domain within a bimodular polyketide synthase. *Chem Biol*. 1997;4:757–766.

133. McDaniel R, Kao CM, Hwang SJ, Khosla C. Engineered intermodular and intramodular polyketide synthase fusions. *Chem Biol*. 1997;4:667–674.

134. Behrens OK, Corse J. Biosynthesis of penicillins; new crystalline biosynthetic penicillins. *J Biol Chem*. 1948;175:793–809.

135. Daum SJ, Lemke JR. Mutational biosynthesis of new antibiotics. *Annu Rev Microbiol*. 1979;33:241–265.

136. Khaw LE, Bohm GA, Metcalfe S, Staunton J, Leadlay PF. Mutational biosynthesis of novel rapamycins by a strain of *Streptomyces hygroscopicus* NRRL 5491 disrupted in rapL, encoding a putative lysine cyclodeaminase. *J Bacteriol*. 1998;180:809–814.

137. Liu HW, Thorson JS. Pathways and mechanisms in the biogenesis of novel deoxysugars by bacteria. *Annu Rev Microbiol*. 1994;48:223–256.

138. Decker H, Haag S, Udvarnoki G, Rohr J. Novel genetically-engineered tetracenomycins. *Angew Chem Int Edit Engl*. 1995;34:1107–1110.

139. Solenberg PJ, Matsushima P, Stack DR, Wilkie SC, Thompson RC, Baltz RH. Production of hybrid glycopeptide antibiotics *in vitro* and in *Streptomyces toyocaensis*. *Chem Biol*. 1997;4:195–202.

140. Madduri K, Kennedy J, Rivola G, et al. Production of the antitumor drug epirubicin (4′-epidoxorubicin) and its precursor by a genetically engineered strain of *Streptomyces peucetius*. *Nat Biotechnol*. 1998;16:69–74.

141. Stachelhaus T, Schneider A, Marahiel MA. Engineered biosynthesis of peptide antibiotics. *Biochem Pharmacol*. 1996;52:177–186.

142. Donadio S, Monciardini P, Sosio M. Polyketide synthases and nonribosomal peptide synthetases: the emerging view from bacterial genomics. *Nat Prod Rep*. 2007;24:1073–1109.

143. Bentley SD, Chater KF, Cerdeno-Tarraga AM, et al. Complete genome sequence of the model actinomycete *Streptomyces coelicolor* A3(2). *Nature*. 2002;417:141–147.

144. Challis GL. Genome mining for novel natural product discovery. *J Med Chem*. 2008;51:2618–2628.

145. Bergmann S, Schumann J, Scherlach K, Lange C, Brakhage AA, Hertweck C. Genomics-driven discovery of PKS-NRPS hybrid metabolites from *Aspergillus nidulans*. *Nat Chem Biol*. 2007;3:213–217.

146. Gross H. Strategies to unravel the function of orphan biosynthesis pathways: recent examples and future prospects. *Appl Microbiol Biotechnol*. 2007;75:267–277.

147. Udwary DW, Zeigler L, Asolkar RN, et al. Genome sequencing reveals complex secondary metabolome in the marine actinomycete *Salinispora tropica*. *Proc Natl Acad Sci USA*. 2007;104:10376–10381.

148. Gross H, Stockwell VO, Henkels MD, Nowak-Thompson B, Loper JE, Gerwick WH. The genomisotopic approach: a systematic method to isolate products of orphan biosynthetic gene clusters. *Chem Biol*. 2007;14:53–63.

149. Lin X, Hopson R, Cane DE. Genome mining in *Streptomyces coelicolor*: molecular cloning and characterization of a new sesquiterpene synthase. *J Am Chem Soc.* 2006;128:6022−6023.

150. Song L, Barona-Gomez F, Corre C, et al. Type III polyketide synthase beta-ketoacyl-ACP starter unit and ethylmalonyl-CoA extender unit selectivity discovered by *Streptomyces coelicolor* genome mining. *J Am Chem Soc.* 2006;128:14754−14755.

151. Hornung A, Bertazzo M, Dziarnowski A, et al. A genomic screening approach to the structure-guided identification of drug candidates from natural sources. *Chembiochem.* 2007;8:757−766.

152. Shwab EK, Bok JW, Tribus M, Galehr J, Graessle S, Keller NP. Histone deacetylase activity regulates chemical diversity in Aspergillus. *Eukaryot Cell.* 2007;6:1656−1664.

153. Bok JW, Chiang YM, Szewczyk E, et al. Chromatin-level regulation of biosynthetic gene clusters. *Nat Chem Biol.* 2009;5:462−464.

154. Vannucchi V. Clinical study of a new antibiotic; erythromycin. *Riv Crit Clin Med.* 1952;52:128−136.

155. Cortes J, Haydock SF, Roberts GA, Bevitt DJ, Leadlay PF. An unusually large multifunctional polypeptide in the erythromycin-producing polyketide synthase of *Saccharopolyspora erythraea*. *Nature.* 1990;348:176−178.

156. Donadio S, Staver MJ, Mcalpine JB, Swanson SJ, Katz L. Modular organization of genes required for complex polyketide biosynthesis. *Science.* 1991;252:675−679.

157. Caffrey P, Bevitt DJ, Staunton J, Leadlay PF. Identification of DEBS 1, DEBS 2 and DEBS 3, the multienzyme polypeptides of the erythromycin-producing polyketide synthase from *Saccharopolyspora erythraea*. *Febs Lett.* 1992;304:225−228.

158. Kao CM, Katz L, Khosla C. Engineered biosynthesis of a complete macrolactone in a heterologous host. *Science.* 1994;265:509−512.

159. Pfeifer BA, Admiraal SJ, Gramajo H, Cane DE, Khosla C. Biosynthesis of complex polyketides in a metabolically engineered strain of *E. coli. Science.* 2001;291:1790−1792.

160. Long PF, Wilkinson CJ, Bisang CP, et al. Engineering specificity of starter unit selection by the erythromycin-producing polyketide synthase. *Mol Microbiol.* 2002;43:1215−1225.

161. Liou GF, Lau J, Cane DE, Khosla C. Quantitative analysis of loading and extender acyltransferases of modular polyketide synthases. *Biochemistry.* 2003;42:200−207.

162. Kinoshita K, Pfeifer BA, Khosla C, Cane DE. Precursor-directed polyketide biosynthesis in *Escherichia coli. Bioorg Med Chem Lett.* 2003;13:3701−3704.

163. Cane DE, Kudo F, Kinoshita K, Khosla C. Precursor-directed biosynthesis: biochemical basis of the remarkable selectivity of the erythromycin polyketide synthase toward unsaturated triketides. *Chem Biol.* 2002;9:131−142.

164. Regentin R, Kennedy J, Wu N, et al. Precursor-directed biosynthesis of novel triketide lactones. *Biotechnol Progr.* 2004;20:122−127.

165. Ward SL, Desai RP, Hu ZH, Gramajo H, Katz L. Precursor-directed biosynthesis of 6-deoxyerythronolide B analogues is improved by removal of the initial catalytic sites of the polyketide synthase. *J Ind Microbiol Biot.* 2007;34:9−15.

166. Hans M, Hornung A, Dziarnowski A, Cane DE, Khosla C. Mechanistic analysis of acyl transferase domain exchange in polyketide synthase modules. *J Am Chem Soc.* 2003;125:5366−5374.

167. Stassi DL, Kakavas SJ, Reynolds KA, et al. Ethyl-substituted erythromycin derivatives produced by directed metabolic engineering. *Proc Natl Acad Sci USA.* 1998;95:7305−7309.

168. Kato Y, Bai LQ, Xue Q, Revill WP, Yu TW, Floss HG. Functional expression of genes involved in the biosynthesis of the novel polyketide chain extension unit, methoxymalonyl-acyl carrier protein, and engineered biosynthesis of 2-desmethyl-2-methoxy-6-deoxyerythronolide B. *J Am Chem Soc.* 2002;124:5268−5269.

169. Peiru S, Menzella HG, Rodriguez E, Carney J, Gramajo H. Production of the potent antibacterial polyketide erythromycin C in *Escherichia coli. Appl Environ Microb.* 2005;71:2539−2547.

170. Menzella HG, Reid R, Carney JR, et al. Combinatorial polyketide biosynthesis by de novo design and rearrangement of modular polyketide synthase genes. *Nat Biotechnol.* 2005;23:1171−1176.

171. Desai RP, Leaf T, Hu ZH, et al. Combining classical, genetic, and process strategies for improved precursor-directed production of 6-deoxyerythronolide B analogues. *Biotechnol Progr.* 2004;20:38−43.

172. Dayem LC, Carney JR, Santi DV, Pfeifer BA, Khosla C, Kealey JT. Metabolic engineering of a methylmalonyl-CoA mutase-epimerase pathway for complex polyketide biosynthesis in *Escherichia coli. Biochemistry.* 2002;41:5193−5201.

173. Murli S, Kennedy J, Dayem LC, Carney JR, Kealey JT. Metabolic engineering of *Escherichia coli* for improved 6-deoxyerythronolide B production. *J Ind Microbiol Biotechnol.* 2003;30:500−509.

174. Zhang H, Skalina K, Jiang M, Pfeifer BA. Improved *E. coli* erythromycin A production through the application of metabolic and bioprocess engineering. *Biotechnol Prog.* 2011;28:292−296.

175. World Health Organization. *World Malaria Report 2010*; 2010.

176. Mercke P, Bengtsson M, Bouwmeester HJ, Posthumus MA, Brodelius PE. Molecular cloning, expression, and characterization of amorpha-4,11-diene synthase, a key enzyme of artemisinin biosynthesis in *Artemisia annua* L. *Arch Biochem Biophys*. 2000;381:173–180.

177. Martin VJ, Pitera DJ, Withers ST, Newman JD, Keasling JD. Engineering a mevalonate pathway in *Escherichia coli* for production of terpenoids. *Nat Biotechnol*. 2003;21:796–802.

178. Anthony JR, Anthony LC, Nowroozi F, Kwon G, Newman JD, Keasling JD. Optimization of the mevalonate-based isoprenoid biosynthetic pathway in *Escherichia coli* for production of the anti-malarial drug precursor amorpha-4,11-diene. *Metab Eng*. 2009;11:13–19.

179. Pitera DJ, Paddon CJ, Newman JD, Keasling JD. Balancing a heterologous mevalonate pathway for improved isoprenoid production in *Escherichia coli*. *Metab Eng*. 2007;9:193–207.

180. Tsuruta H, Paddon CJ, Eng D, et al. High-level production of amorpha-4,11-diene, a precursor of the antimalarial agent artemisinin, in *Escherichia coli*. *PLoS One*. 2009;4:e4489.

181. Ro DK, Paradise EM, Ouellet M, et al. Production of the antimalarial drug precursor artemisinic acid in engineered yeast. *Nature*. 2006;440:940–943.

182. Westfall PJ, Pitera DJ, Lenihan JR, et al. Production of amorphadiene in yeast, and its conversion to dihydroartemisinic acid, precursor to the antimalarial agent artemisinin. *Proc Natl Acad Sci USA*. 2012;109: E111–E118.

183. Chang MCY, Eachus RA, Trieu W, Ro DK, Keasling JD. Engineering *Escherichia coli* for production of functionalized terpenoids using plant P450s. *Nat Chem Biol*. 2007;3:274–277.

184. Jennewein S, Park H, DeJong JM, Long RM, Bollon AP, Croteau RB. Coexpression in yeast of Taxus cytochrome P450 reductase with cytochrome P450 oxygenases involved in Taxol biosynthesis. *Biotechnol Bioeng*. 2005;89:588–598.

185. Dejong JM, Liu Y, Bollon AP, et al. Genetic engineering of taxol biosynthetic genes in *Saccharomyces cerevisiae*. *Biotechnol Bioeng*. 2006;93:212–224.

186. Engels B, Dahm P, Jennewein S. Metabolic engineering of taxadiene biosynthesis in yeast as a first step towards Taxol (Paclitaxel) production. *Metab Eng*. 2008;10:201–206.

187. Huang Q, Roessner CA, Croteau R, Scott AI. Engineering *Escherichia coli* for the synthesis of taxadiene, a key intermediate in the biosynthesis of taxol. *Bioorg Med Chem*. 2001;9:2237–2242.

188. Ajikumar PK, Xiao WH, Tyo KEJ, et al. Isoprenoid pathway optimization for taxol precursor overproduction in *Escherichia coli*. *Science*. 2010;330:70–74.

CHAPTER 11

Synthetic Biology of Microbial Biofuel Production: From Enzymes to Pathways to Organisms

Gregory Bokinsky[1], Dan Groff[1] and Jay Keasling[1,2,3]
[1]Joint BioEnergy Institute, Emeryville, CA, USA
[2]University of California, CA, USA
[3]Lawrence Berkeley National Laboratory, Berkeley, CA, USA

INTRODUCTION

207

The microbial generation of hydrocarbons, either from inexpensive plant biomass or directly from sunlight, could provide a renewable and carbon-neutral source of liquid transportation fuels, so long as the biomass is grown and harvested sustainably.[1,2] The explosive advances in synthetic biology and metabolic engineering have dramatically accelerated efforts to engineer microbes for fuel production (Table 11.1). Thanks to an ever-decreasing cost of DNA synthesis and new, robust tools for DNA manipulation, our abilities to install whatever DNA sequence we can imagine into a widening range of host microbes are more powerful than ever, even up to the scale of installing whole genomes into cells.[3] The sheer number of enzymes available in gene databases, and the incredible breadth of chemical transformations they catalyze, represents vast biosynthetic versatility that can be harnessed in simple cloning steps. These capabilities are already being applied to engineer microbial production of valuable chemicals, such as pharmaceuticals, fragrances, and vitamins. However, the current low cost of petrochemicals is a tremendous barrier to an economy-wide adoption of microbially produced bulk compounds such as fuels and plastics. Biologically generated commodity compounds will only be competitive with petrochemicals when inexpensive bioprocesses are finally developed. These processes will feature robust microbes capable of high product titers at near-theoretical yields. Achieving this ambitious goal will likely require every trick known to synthetic biology.

Synthetic biologists have already taken many of the steps towards this goal. Microbial production of biofuels with combustion properties similar to existing fuels has been repeatedly demonstrated at proof-of-concept levels.[4–7] However, we still cannot underestimate the challenge of genetically rewiring the physiology and metabolism of microbes to reliably generate chemicals at a commodity scale. Our understanding of how synthetic DNA sequences will behave once they are inside a cell still severely lags behind

Synthetic Biology. DOI: http://dx.doi.org/10.1016/B978-0-12-394430-6.00011-X

TABLE 11.1 Selected Examples of Synthetic Biology Principles for Biofuel and Biomaterial Production

Mechanism of Control	References
DNA	
Copy number	47
	41
	57
	49
Transcription	
Promoter strength	51
	57
	53
Promoter timing	54
	54a
Operon organization	60
mRNA	
RNA half-life	62
Secondary structures	62
Translation	
RBS strength	59
Codon usage	60
Protein	
Protein half-life	15
Protein optimization	7
	16
	34
	15
Metabolites and Products	
Biosensors	54a
	54
Allosteric regulation	29
	32

our DNA synthesis abilities. In addition, the behavior of highly engineered microbes must be robust in the scale-up from shake-flask cultures to thousand-liter tanks. If we are to achieve the high titers required of biofuel production beyond the milligram per liter scales commonly demonstrated in laboratory production, and develop a process that whole economies can rely upon, our capabilities for biological design still need much improvement.

We describe here a selection of the methods available for engineering gene expression, enzyme function, and host cell physiology, many of which have been shown to be effective in improving the yield of a biologically produced compound (see Fig. 11.1 for an overview). This chapter is organized into two parts, beginning with 'Pathway design and optimization.' We address issues and techniques for engineering the biofuel production pathway: the collection of enzymes that catalyze the chemical transformations that turn metabolites into fuels. The second part of the chapter describes issues and techniques for engineering the host organism responsible for expressing, feeding, and sustaining the biofuel production pathway.

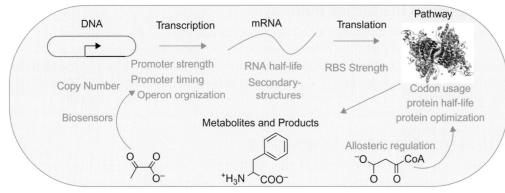

FIGURE 11.1

Synthetic-biology-guided production of biofuels. In black are control points at which one can modify the biochemistry of the host cell. In green are the mechanisms through which each of these control points can be modified as discussed.

PATHWAY DESIGN AND OPTIMIZATION

Initial Pathway Design and Validation

The first step in the design of any biofuel production pathway is deciding which fuel compound will be produced using the pool of metabolic precursors available in the host organism. Many potential fuel compounds or their synthetic intermediates occur naturally, and their biosynthetic pathways have been determined. The suitability of biological compounds for use as fuel replacements is reviewed elsewhere.[9] The next challenge is to assemble a biochemical pathway capable of synthesizing the target molecule using known enzymatic activities.[10] The genes that encode the necessary enzymes can often be drawn from the vast library of publicly available sequences. Pathways to produce potentially any molecule can be assembled from the rich diversity of enzymatic activities that have evolved over billions of years. Entire biosynthetic pathways can be used as they are found in nature, or can be assembled piecemeal by incorporating enzymes from different sources. This diversity can be expanded yet further by engineering existing enzymes to generate nonnatural products.[10] Biosynthetic pathways can also be constructed from catabolic pathways engineered to run in the reverse direction than their usual physiological role. This has been demonstrated using a reversed fatty-acid degradation pathway to produce butanol in high titers in *E. coli*.[5]

Construction of a biofuel pathway typically begins with selection of genes encoding enzymes that comprise the individual pathway components. The full pathway is assembled in one or more plasmids and expressed in the host organism of choice, which is often a well-characterized and genetically tractable organism such as *Escherichia coli* or *Saccharomyces cerevisiae*. Other organisms with different metabolic capabilities and toxicity tolerances that may be better suited to producing the desired compound can certainly be used as hosts, although they often present other challenges, most notably having fewer techniques for genetic manipulations.[11] Production of the desired compound is usually assayed using an appropriate chomatographic method. As it is unlikely that the initial design of the pathway will produce high titer levels, production will be orders-of-magnitude lower than is needed in an industrial process. Nonetheless, an important milestone will have been achieved: proof that biological production of a fuel compound within a new host is possible. The brevity with which the initial stage of pathway design and validation is treated here is not intended to indicate its triviality or simplicity. It is a necessary first step, but an equally important challenge lies in increasing titers and yields closer to commercially viable levels.

209

Optimization of the Biofuel Pathway

The goal of pathway optimization is to maximize the efficiency of the process that synthesizes the desired fuel from existing metabolites in the host, which is determined by concentrations of precursors, intermediates, cofactors, and enzymes, as well as the catalytic activities of the enzymes, which are almost always subject to regulation on multiple levels, from gene expression to enzyme activity. This section will begin by addressing the optimization of the fundamental pathway components: the enzymes themselves.

PROTEIN ENGINEERING TO IMPROVE FLUX

If an enzyme is found to be limiting biofuel production, one straightforward approach to improve production is to screen a library of homologues to find a replacement for the bottleneck enzyme. This approach relies upon the sheer numbers of homologues often available within genetic databases that presumably span a wide range of properties, such as stability and catalytic rates. Often, an improved enzyme can be found, as has recently been demonstrated in work that achieved a high level of production of the biodiesel replacement bisabolene.[7] A library of five bisabolene synthases from four plant species were cloned and screened for production titers, generating yields that varied by 100-fold, and one high-yielding (0.5 g/L) enzyme was discovered. However, homologues may not always be available or numerous (especially for certain classes of terpene cyclases), and high-yielding enzymes may not always be happened upon so fortuitously, especially for enzymes that have not evolved to produce high yields, as is often the case for secondary metabolites.[12] A recent review[13] argues that it is better to optimize production by adjusting the copy number, regulation, and enzymatic properties of the components of a pathway, rather than screening libraries of enzymes. While screening a library of enzymes might provide a coarse-grained search over a productivity landscape, protein and metabolic engineering methods are likely necessary to bring the pathway closer to a productivity maximum.

The lack of biochemical characterization of the vast majority of known enzymes is a natural consequence of the ease with which prospective genes can be found in sequence databases and cloned, compared to the difficulties and uncertainties inherent in protein overexpression and biochemical measurements. This is unfortunate, as knowledge of the properties of each enzyme within a pathway can better focus pathway optimization efforts. For instance, attempting to increase flux by increasing the copy number of a pathway component will be ineffective if the enzyme is nonfunctional (not expressed, or lacking necessary post-translational modifications), poorly functional (not folding properly), or functional but permitting a lower flux than is optimal for production of the biofuel (inhibition by a competitor or allosteric regulator). Of course, the enzymes selected for a pathway must be compatible with the host, a property referred to as 'functional composibility.'[10] An enzyme is more likely to function in the production host if the organism from which it is obtained shares a similar intracellular chemical environment with the new host (such as pH or growth temperature), though the sensitivity of enzymes to these factors likely varies widely, and often compatibility cannot be known until expression is attempted. In particular, genes transplanted between kingdoms may suffer compatibility problems.[14] These issues may be overcome by engineering the protein itself to improve its 'in vivo' properties, a term that refers collectively to the solubility, stability, selectivity, and activity of an enzyme.[15]

All enzymes need to fold into their native three-dimensional structure before they can function. Proteins often misfold when expressed in a foreign host (or even when overexpressed in the original host). When a pathway enzyme does not fold, it can either be replaced or modified to improve its folding. Yoshikuni et al. found that the enzyme γ-humulene synthase (HUM), a sesquiterpene synthase from the gymnosperm *Abies grandis*, demonstrated low activity when transplanted into *E. coli* (15). Investigations found that very little of the expressed HUM (~15%) resided in the soluble fraction of the cell, suggesting that much of the enzyme was misfolding or was insufficiently solublized. An adaptive evolution approach that involved replacing amino acid residues in HUM with glycine and proline residues improved the solubility of the enzyme (~60% in the soluble fraction), resulting in a 220-fold increased production of sesquiterpines. A biochemical analysis of wild-type and mutant HUM enzymes revealed that the catalytic properties (k_{cat} or K_m) of the mutant enzyme remained the same, implying that the production improvement was achieved by increasing concentration of folded, active enzyme. Other methods adapted from the protein overexpression field may help to solubilize an unfolded enzyme,[16,17] and thus help improve final product titer by increasing the flux it catalyzes.

Many enzymes require post-translation modifications for activity, e.g. phosphopantethenylation.[18-20] Often the proteins responsible for catalyzing these modifications need to be transplanted from their host organism, along with the pathway enzyme, if the modification activity is absent from the host or incompatible with the pathway enzyme. Other accessory proteins whose activities are absent in the host cell may also need to be present, and their activity levels optimized, in order to demonstrate activity. For instance, a key intermediate step in the microbial production of 1,3-propanediol using *E. coli* is the dehydration of glycerol to 3-hydroxypropionaldehyde by a glycerol dehydratase gene *dhaB1-3*.[21] The enzyme can also be deactivated by glycerol, requiring reactivation by a glycerol dehydratase reactivase enzyme.[22] An engineered strain capable of converting glucose to 1,3-propanediol on an industrial scale uses glycerol dehydratase and its reactivating factors from *Klebsiella pneumonia*. Other examples of accessory proteins are the cytochrome P450 reductases (CPR) that are required to reduce the heme group in cytochrome P450s.[23] Because commonly used platform organisms for biosynthesis lack CPR enzymes, CPR activity often needs to be imported with the P450 genes, and itself optimized to improve titer.[14]

Flux through a pathway may also be modulated by feedback regulation of the enzymes by metabolites or other small molecules. Feedback regulation can strongly affect enzymatic activity through product or allosteric inhibition. This mode of regulation reacts to metabolite concentration far quicker than either translational or transcriptional control,[24,25] and maintains the concentrations of intermediates within a narrow range to prevent osmotic stress or other toxicity resulting from over-accumulation of small molecules.[26] Negative feedback regulation on an enzyme decreases its activity, effectively lowering the concentration of the enzyme. When production pathways contain allosterically regulated enzymes, these enzymes can become the bottleneck if the regulating metabolite reaches inhibitory levels. Thus, engineering feedback-resistant enzymes often enables higher production. Because our knowledge of allosteric inhibition and similar modes of regulation by metabolites has not kept pace with our vast catalogue of genes,[27] it seems likely that careful biochemical measurements of pathway enzymes would reveal such feedback mechanisms.

Relieving feedback inhibition in the biosynthetic pathway of 2-keto acids,[28] a production intermediate in the biosynthesis of amino acids, enabled workers in the Liao group to greatly increase their titers of alcohols. Work by Shen and Liao[29] found that production of threonine, a precursor to 2-ketobutyrate, was a bottleneck in their alcohol production pathway. The first two steps in threonine production from aspartate are catalyzed by the enzyme aspartate kinase/homoserine dehydrogenase (ThrA), which is inhibited by the downstream product threonine, an effect first discovered in 1976.[30] A feedback-resistant ThrA enzyme was cloned from a threonine-overproducing *E. coli* strain and expressed under control of an inducible promoter. Expressing the feedback-resistant ThrA resulted in a three- to four-fold increase in the titers of both 1-propanol and 1-butanol.[29]

The abrogation of feedback inhibition can be achieved not only by mutations at inhibitor-binding allosteric sites, but also by simply relocating the enzyme to a different compartment within the cell. For instance, removing the leader peptide sequence of an *E. coli* thioesterase TesA prevents its export into the periplasmic space and confines it to the cytoplasm, where it depletes the pool of fatty acid-acyl carrier protein, a repressor of fatty acid biosynthesis. This resulted in accumulation of fatty acids.[31] Overexpression of leaderless TesA in *E. coli* provided high levels of intracellular fatty acids and accelerated the production of fatty acid ethyl esters, a biodiesel.[32]

The catalytic activity of the pathway enzymes may also be a limiting factor, as not all enzymes exhibit a fast turnover rate, particularly when they are members of a pathway that has not evolved to generate large amounts of its product,[12] or if the enzyme is functionally promiscuous (generates multiple products). For instance, enzymes involved in terpene biosynthesis generate several products from a common precursor, a property that tends to correlate with having low catalytic activity.[33] As a result, the flux through the pathway to the terpene products is often low. Leonard et al. screened active site mutants of the enzyme levopimaradiene synthese (LPS) from *Gingko biloba* to improve the product titer of diterpenoids produced by a factor of 10.[34] A colorimetric screen of random mutants of the prenyl transferase enzyme GGPPS, which catalyzes the preceding step and provides LPS with a substrate, generated candidate GGPPS enzymes that produced higher yields of geranylgeranyl diphosphate (GGPP). By coupling the best-performing GPPSS mutant with the LPS mutant, the authors succeeded in increasing the diterpenoid yield by 18-fold, a result obtained using protein engineering alone.

INCREASING FLUX BY TAILORING ENZYME LOCALIZATION OR COMPARTMENTALIZATION

Pathway flux can also be increased without directly engineering higher catalytic activity. Enzymes that catalyze sequential reactions can be localized in close proximity to each other by means of artificial scaffolds,[35] or by fusing enzymes that catalyze sequential reactions. This approach is inspired by examples found in native organisms of multiple enzyme activities being colocalized in one complex or organelle.[36] Because the concentration of the product of an enzyme will be higher near its active site, placing an enzyme that reacts with that product nearby takes advantage of the higher local concentration to increase flux through the following reaction. Colocalization of pathway enzymes may be particularly useful when either the enzymes within a pathway cannot be optimized further, or an intermediate metabolite is toxic to the host cell, or is a substrate of a competing enzyme within the cell. This approach was validated using several enzyme components of the mevalonate pathway. The enzymes AtoB, HMGS, and HMGR were joined on an artificial scaffold coexpressed with the pathway enzymes, and the stoichiometry of the enzymes on the scaffold was adjusted by modifying the numbers of individual binding sites of each enzyme on the scaffold. The best scaffold resulted in a 77-fold increase in titer over the nonscaffolded pathway, while further enabling lower enzyme expression. Even simple protein fusions without a scaffold have demonstrated increased production. Fusing the enzymes FPP synthase and farnesene synthase resulted in a modest increase in farnesene production, compared to free enzymes,[37] and eliminated formation of the side product farnesol.

Colocalization of proteins within a microcompartment, a strategy used by bacteria, is a promising avenue of research into improving pathway efficiency.[38] Carboxysomes, a microcompartment found within cyanobacteria, increase the carbon fixation rate of RubisCo by colocalizing carbonic anhdyrase enzymes, which produces CO_2 from biocarbonate, into a protein shell with RubisCO, which condenses the CO_2 with ribulose 1,5-bisphosphate to produce 3-phosphoglycerate. Because RubisCO has a low affinity for CO_2, the increased concentration of bicarbonate inside the microcompartment enables a higher flux than RubisCO might otherwise achieve. Other compartments, such as the Pdu and Eut (1,2-propanediol and ethanolomine bacterial microcompartments, respectively) are thought to benefit their hosts by blocking the diffusional loss of toxic intermediates produced by the pathway into the cytoplasm. The protein shell of the microcompartment may also act as a regulation point by controlling the entry and efflux of precursors and products, enabling metabolite-level regulation on these pathways. The details of protein targeting into the microcompartments are becoming clear,[39] and promising applications for harnessing microcompartments for biofuel production are as yet unexplored.

CENTRAL DOGMA ENGINEERING: HOW MANY ENZYMES AND AT WHAT TIME?

The ideal starting point in optimizing a biofuel production pathway is with biochemically characterized enzymes that are known to be capable of providing flux at a level needed for high production rates in vivo. The next step is to design a genetic system that expresses the optimal level of each enzyme at the correct time. Unfortunately, just as it is difficult to perform biochemical measurements of pathway enzymes, it is also extremely difficult to know a priori how much enzyme is needed in vivo for an optimum pathway, and to precisely engineer those protein levels. The use of in vitro assays, such as cell-free approaches that enable each pathway component to be titrated individually, can inform optimization.[40]

One goal of synthetic biology is to be able to engineer protein expression systems that robustly produce enzymes at a predetermined level, and at precise times. A simplistic viewpoint that approximates flux as a product of enzyme concentration, substrate concentration, and catalytic rate would assume that more enzymes would produce higher flux. Following this logic, the optimal solution for maximum flux might be to engineer the highest possible enzyme concentrations. Unfortunately, enzyme activities are often feedback-regulated by metabolites, and high concentrations of intermediates can be toxic to the cell.[41] Furthermore, expressing a high level of pathway protein may be detrimental for other reasons, such as squandering transcription and translation machinery generating superfluous levels of protein.[42,43] It has become clear that enzyme levels within a biofuel pathway must be carefully balanced. Metabolic control theory[44] predicts that the best approach to maximizing productivity is to modify each component of a synthetic pathway in a way that increases fluxes but maintains intermediate concentrations near their physiological levels. Considered an extremely difficult undertaking when first articulated, advances in synthetic biology and protein engineering, which enable the careful tuning of enzyme levels and catalytic activities, may make this approach more tractable, if still challenging.

From Gene to mRNA: Modulating Transcription of Pathway Genes

The concentration of a protein can be expected to increase as the concentration of the mRNA transcripts that encode the protein increases, although this is not always a simple relationship (as reviewed in [45]), and there is little to no correlation between mRNA levels and protein levels at the single-cell level.[46] Fortunately for the bioengineer, factors that determine mRNA levels, such as DNA copy number and promoter strength, usually have predictable effects on protein expression, and can be readily modulated using straightforward genetic methods.

Typically when demonstrating biosynthesis of a fuel, production pathways are encoded on plasmids. Plasmid copy number can be easily modulated by using plasmids with various origins of replication that confer low- or high-copy numbers. This can considerably simplify efforts to manipulate gene copy number and resulting protein concentrations, which can identify enzymes that are expressed at insufficient levels. This approach was used to determine the enzyme whose low concentration limited the productivity of the amorphodiene production pathway in *E. coli*. The complete pathway for amorphodiene production was encoded on a low-copy plasmid and transformed into *E. coli*. Higher-copy plasmids that encoded each individual pathway enzyme were subsequently introduced. A plasmid encoding the mevalonate kinase gene (MK) increased amorphodiene yields, suggesting that concentration of the MK enzyme limited production.[47] This approach has also been used to improve production of the amorphodiene precursor mevalonate. Increasing the copy number of rate-limiting enzyme tHMGR relieved the growth inhibition caused by accumulation of a precursor metabolite, relieving toxicity and enabling higher mevalonate levels.[41] Additional copies of a gene may be placed on the same plasmid as the rest of the pathway; however, this may lead to gene loss through recombination and is best avoided.[48]

Plasmid-based expression of pathways may not be well-suited for high-titer production of biofuels. The metabolic requirements of plasmid replication impose on the host cell a metabolic burden, which can scale with plasmid copy number.[49] Plasmids are autonomous genetic elements that may not be distributed evenly upon cell division. If expression of a plasmid-based biofuel pathway decreases cell growth rate or viability, individuals with fewer copies of the plasmid will rapidly out-compete the others, leading to a population with low overall productivity.[50] Techniques that transfer the pathway genes to the chromosome itself avoid this problem; however, as the chromosome is present in very low copy numbers, this may place a very low limit on the pathway copy number that can be introduced. A recently developed innovation, chemically inducible chromosomal evolution (CIChE), enables the introduction of moderate-copy numbers of pathways onto the chromosome (~40). Stability of the repeated genes is ensured via knockout of the native recombinase gene *recA*.

Biofuel production genes are commonly expressed from well-characterized promoters induced by the addition of an exogenous chemical, such as anhydrous tetracycline or IPTG. This provides a simple means with which to modulate the strength of expression, as increasing inducer concentration should result in higher rates of transcription. While the use of promoters driven by exogenously added chemical inducers is very useful in the early stages of pathway demonstration, the price of typical inducers prohibits their use on an industrial scale. Constitutive promoters with a wide range of promoter strengths have been developed, enabling finely tuned enzyme levels,[51] without the need for chemical inducers. Constitutively active promoters can enable an engineer to modulate the transcript levels of either the whole biofuel pathway,[52] or each individual enzyme can be controlled by a specific constitutive promoter.[53] However, constitutive expression of pathway enzymes may present a metabolic burden on the cell, presenting an opportunity for escape mutants to evolve and contribute to pathway instability.[48]

A more sophisticated approach to controlling pathway expression is to use biosensors (see Box 11.1) to couple transcription of pathway enzymes to the concentration of specific metabolites or other cellular or environmental conditions. This takes advantage of control mechanisms evolved to dynamically respond to metabolite levels. An early demonstration of using biosensors to achieve dynamic metabolic control coupled transcription of key pathway genes of a lycopene biosynthesis pathway with the concentration of acetyl phosphate resulted in improved titers of the product in *E. coli*.[54] The expression of the rate-controlling enzymes of lycopene synthesis was placed under control of a promoter native to *E. coli* that responds to acetyl phosphate, a metabolite that increases during periods of excess glycolytic flux.[55,56] Thus, the lycopene pathway was activated when acetyl phosphate levels were high, which indicated an imbalance between carbon influx and carbon consumption. This increased the production of lycopene by 10-fold above what was achieved by

BOX 11.1 BIOSENSORS

One technology that is becoming increasingly important in the synthetic biology toolkit is biosensors. This class of molecules, which can be either naturally occurring or specifically engineered, can include protein or RNA. The output of a biosensor can include increased transcription whose activities change depending on the biochemical state of the cell or the presence of a specific molecule, such as a biofuel or a biofuel precursor. Of particular relevance are transcription factors that respond to precursors, intermediates, or products in the synthetic pathway. This type of biosensor has been used for a variety of applications. The clearest use of this technology is for directed evolution. Biosensors that are responsive to the product or key intermediate in a pathway can be coupled to the transcription of a screenable marker such as GFP,[78] or a selectable marker such as β-lactamase, to isolate the highest producing strains out of a pool of mutants.[79] When a suitable biosensor does not exist, it may be possible to generate a synthetic one using directed evolution. Cirino and coworkers demonstrated that there is some plasticity in the molecular recognition of the arabinose responsive regulator AraC. Placing GFP under the control of P_{BAD} promoter allowed FACS screening of a library of AraC mutants. Using both a positive and negative screen they were able to identify mutants which were activated by D-arabinose, the sugar with the inverse stereochemistry of the natural L-ligand which also exhibited the same tight control and low background associated with the natural protein.[80]

controlling expression of those genes using a chemically inducible promoter. Furthermore, growth retardation associated with overexpression of one of the rate-limiting proteins was not seen, despite a higher protein expression level. These improvements were attributed to a better coordination of lycopene biosynthesis with the metabolic state of the cell, resulting in better diversion of flux away from acetate into lycopene. Finally, because production is coupled to glycolytic excess, the extra carbon is channeled into lycopene rather than the toxic overflow product acetate, reducing the production of this waste by 66%.

A naturally occurring biosensor of an intermediate in the synthesis of the biodiesel fatty acid ethyl esters was recently used to drive expression of genes that further modify the intermediate into the final biofuel product. In *E. coli,* the transcription factor FadR responds to fatty acyl-CoAs and free fatty acids[5] by relieving inhibition of downstream genes. The expression of an ethanol production pathway and a wax ester synthase gene responsible for esterifying the fatty acids with ethanol were placed under control of FadR. These genes were not transcribed until sufficient levels of fatty acyl-CoA had accumulated, preventing not only wasteful expression of downstream proteins before they had substrates to act upon, but also accumulation of the toxin ethanol to high levels. In addition, this type of regulation allows the cell to coordinate the levels of the reactants; the transcriptional output of the biosensor increases with increasing concentrations of acyl-CoA, which in turn drives faster production of ethanol and the wax ester synthase. With this technology it was possible to push FAEE levels up three-fold to 1.6 g/L and to 28% of the theoretical maximum yield.[54a]

A recent study that demonstrated the production of a precursor to the important drug Taxol in *E. coli* is another example of optimization of a pathway by combining copy number modulation with promoter strength. Ajikumar et al.[57] split their pathway into two parts (referred to as the 'upstream module' and the 'downstream module'), and varied copy number and promoter strength of each module independently. By measuring production levels, they mapped out a production landscape as a function of expression level of both the upstream and downstream pathways. Tellingly, production levels did not monotonically increase with increasing expression of either half of the pathway, but rather exhibited localized peaks that indicate improved balancing of the two halves of the pathway.

FROM mRNA TO PROTEIN: TUNING TRANSLATION OF PATHWAY PROTEINS

The overall process of how an mRNA transcript is translated into protein is well-understood. However, making accurate predictions of how much protein will be generated from a given transcript are very difficult. Because fluxes depend strongly on pathway enzyme abundances, especially for rate-limiting enzymes, modulating protein translation from mRNA can help to fine-tune enzyme expression. Translation is also indirectly controlled by mRNA transcript stability. We will focus on efforts to engineer two of the phases of translation (initiation and elongation) in the context of biosynthesis pathways.

It is thought that the rate-limiting step in protein production from a transcript is transcription initiation, a complex process largely determined by binding of the ribosome RNA to the ribosome binding site (RBS), a sequence upstream of the initiation codon.[58] This binding step is controlled by several factors: the presence of secondary structures within the mRNA that occludes the RBS and reduces the availability of RBS to the ribosome; and the equilibrium binding strength of the RBS to the ribosomal RNA. Because these factors are highly dependent on the mRNA sequence upstream and downstream of the RBS, a single RBS will likely result in widely varying translation initiation rates for different surrounding sequences. This may complicate efforts to precisely tune protein levels across a biofuel production pathway.

An elegant method for predicting and controlling RBS-determined translation initiation in *E. coli* has been developed.[59] It estimates protein expression levels using a thermodynamic model for translation initiation. The model uses an mRNA folding algorithm to calculate the free energies involved in disrupting mRNA secondary structure occluding the RBS, as well as the binding energy of the RBS with ribosomal RNA. Using the RBS calculator, the authors were able to predict protein levels that result from a variety of different RBS sequences reasonably well. Furthermore, their model enables forward engineering of desired expression levels by generating RBS sequences with predicted expression levels for specific proteins. The authors confirmed that RBS sequences that result in high expression activity for a given CDS do not universally apply to other proteins. Using a simple genetic switch, the authors demonstrate the utility of their model using RBS sequences generated from their calculator.

The rate of protein elongation can also be engineered. The most common approach to increasing protein yield from a transcript is to optimize the codon usage within each gene for its new host by replacing codons that occur rarely in the host coding sequences with more commonly used codons. This is thought to speed translation by using codons that correspond to tRNA species that are more abundant in the host organism. Codon optimization of pathway genes has been shown to lead directly to increased protein expression, which in turn results in higher yields. Redding-Johanson and coauthors used transcriptomics and proteomics to understand the rate-limiting steps of an amorphadiene production pathway in *E. coli*.[60] They found that two proteins in the pathway, mevalonate kinase (MK) and phosphomevalonate kinase (PMK), were expressed at very low levels. By codon-optimizing the genes encoding MK and PMK, the authors were able to increase their protein abundances by about two-fold each, resulting in a two-fold increase in titer. However, the presence of the codon-optimized MK and PMK sequences resulted in a dramatic decrease in the levels of the other pathway enzymes encoded within the operon. The addition of a transcriptional terminator and a second promoter sequence restored the levels of the proteins, and led to a 2.5-fold further increase in amorphodiene production, a total production increase of five-fold over the original operon design.

The previous study also demonstrates the dangers of treating individual genes transcribed within a common operon as noninteracting entities. Because of poorly-understood and context-dependent phenomena such as folding and degradation of the mRNA transcripts, efforts to optimize the translation level of an individual gene may affect expression of the other pathway genes in ways that are difficult or impossible to predict. This relationship between mRNA secondary structure and its effect upon both mRNA stability and translation has been exploited to create *E. coli* mutants with varying protein concentrations. To demonstrate this principle, a combinatorial library was generated in which different hairpins were introduced into the intergenic regions between the three genes responsible for producing mevalonate *atoB*, *HMGS*, and *tHMGR*. This secondary structure led to expression differences for each gene in different mutants. With this strategy it was possible to identify levels of these proteins that lead to the production of seven-fold more mevalonate than the starting strain.[61] A very recent, more comprehensive model of gene expression takes into account the kinetics of mRNA folding and degradation, and uses RNA devices (aptazymes and ribozymes) to tune these parameters directly. RNA devices were placed within the 5′-untranslated region of mRNA to modulate stability of the transcript. This enabled reliable static and dynamic control over the expression of the downstream gene over a wide range.[62]

HOST ENGINEERING

The organism that hosts the biofuel production pathway will, of course, heavily influence the behavior of the biofuel production pathway. In addition to expressing and maintaining the biofuel synthesis pathway, the host organism supplies the metabolic precursors and enzyme cofactors. However, no organism has evolved to generate chemicals on a commodity scale for our consumption. Organisms have evolved to address goals (such as self-replication) that are often at odds with engineering efforts; thus, the engineer must consider the host physiology and biochemistry if high titers are to be obtained. This section will review approaches to modifying production hosts that result in improved titers of final products. For the sake of brevity, we will consider only rational approaches rather than methods that employ mutagenesis or random, undirected perturbations that are then screened for improved titers. The reader should consider that we are still quite ignorant of many details of cellular physiology and metabolism, so undirected approaches often produce completely unexpected targets for improvements.[63]

KNOCKING OUT COMPETING PATHWAYS

The starting materials of biofuels are typically precursors to building blocks that will ultimately become components of the cell (fatty acids, amino acids), or secondary metabolites involved in other processes (isoprenoids). An engineered biofuel pathway will compete with native pathways for these resources. Removing or attenuating competing pathways through genetic manipulations have been repeatedly demonstrated to improve product titers. For instance, in anaerobic conditions in the absence of an electron acceptor, *E. coli* will use pyruvate as an electron acceptor in mixed acid fermentation, generating lactate, acetate, and succinate. Pathways used to generate higher alcohols, such as butanol, use pyruvate as a carbon source. The deletion of key enzymes required for mixed acid fermentation relieves competition of the biofuel for pyruvate and acetyl-CoA, increasing yield of the desired final product.[28,64] Furthermore, as these pathways compete for the electron carrier NADH, deletion of the pathways also increases the availability of NADH for alcohol production.

Other deletions that have no direct effect on precursor pools may improve titer by eliminating the generation of substrates that might act as a competitive inhibitor of an enzyme along the biofuel production pathway. The product of *ilvD* generates a compound that acts as a competitive inhibitor for one of the enzymes involved in producing butanol from a 2-keto acid. Deletion of the *ilvD* resulted in a three-fold production improvement.[28] Gene deletions may also increase product yield by eliminating pathways that catabolize biofuel precursors that are present in excess. Steen et al. improved the yield of fatty acids in

E. coli by removing the genes responsible for beta-oxidation of free fatty acids that had accumulated as a result of their earlier interventions. This provided a three- to four-fold increase in fatty acid titer.[65]

A caveat for gene deletion attempts is that cellular metabolism can be highly robust to gene deletions, compensating for knockouts by increasing flux to a metabolite through other pathways that may affect the biofuel production pathway.[66] Another caveat is that the deletion of competing pathways may paradoxically decrease the final titer of the desired product, possibly indicating that the competing pathways may in fact generate energy or cofactors that are required for cell growth. Deleting pathways necessary for the generation of an essential metabolite will result in auxotrophy. In this case, rather than deleting the competing pathway, it may be better to decrease the expression of the pathway.[67] One can imagine a compromise between attenuation and complete elimination of a competing but necessary pathway, such as engineering competing pathways to reduce flux or shut off after the growth phase is completed, enabling precursors to flow entirely into biofuel production. This principle was demonstrated by down-regulating the first gene in the biosynthesis pathway for ergosterol in a strain of yeast producing artisemic acid. Ergosterol is required for survival, and decreasing production without completely eliminating it minimized flux into this competing pathway leading to an increase in artisemic acid titers.[68,69]

INCREASING PRECURSOR CONCENTRATION BY UP-REGULATING INPUT PATHWAYS

Along with deleting competing pathways, the availability of precursor metabolites can be increased by up-regulating the native pathways that produce them. Higher levels of 1-butanol were engineered by overexpressing native *E. coli* genes *ilvA* and *leuABCD*, increasing the pool of 2-ketobutyrate and directing it towards the synthesis of 2-ketovalerate, precursors of 1-butanol.[29] This method also uncouples enzyme expression from native transcriptional negative feedback, as highlighted in a production improvement of 2-methyl-1-butanol.[70] However, overexpressing native genes may not increase flux as much as desired due to allosteric regulation of the enzymes themselves, as discussed above. Furthermore, if a pathway enzyme is inhibited by its own substrate, increasing precursor pools may not increase the flux as quickly as without substrate inhibition.[71]

IN SILICO MODELING TO FIND DELETION AND UP-REGULATION CANDIDATES

Because it can be extremely difficult to predict the effect of knockouts on metabolism, and even more difficult to experimentally screen a large library of knockout mutants for production increases, in silico modeling of cellular metabolism can suggest unintuitive candidates for gene deletion that may increase titers. This approach has been used to improve the production of succinate in *E. coli*.[72] In silico methods can also be used to not only pick gene deletion candidates, but also native gene candidates for up-regulation. Choi et al. used an in silico model to predict gene candidates for overexpression that would result in an increased titer of the isoprenoid lycopene.[73] Impressively, most of the gene overexpression candidates selected resulted in increased yield. Combining several of the overexpression candidates into a single strain resulted in a yield increase of approximately five-fold over the wild-type strain. Gene deletion candidates were also chosen from an in silico model and combined with the amplification targets, resulting in further increases in yield. The authors acknowledge that not all suggested targets resulted in increased yield, potentially due to poorly-understood feedback regulation mechanisms. As our knowledge of post-translational regulation of native pathways grows and becomes incorporated into in silico models, their predictions will likely improve.

STRAIN ENGINEERING FOR INCREASED BIOFUEL TOLERANCE

Because many fuel candidates are toxic to microbial strains used in industrial processes, tolerance of a candidate host organism to the fuel, and to biosynthetic intermediates of that fuel, must be considered when selecting an appropriate production host. If the production host cannot tolerate high concentrations of the biofuel it is designed to make, it might be difficult to produce the compound in high titers. For instance, it makes little sense to use *E. coli* to produce ethanol, as its tolerance to this compound is limited to ~4%, whereas *S. cerevesiae* can withstand far higher concentrations. There are many microbial species known to show high tolerance to hydrophobic solvents, but unfortunately, the more genetically tractable organisms typically show low tolerance to many biofuel candidates. Some exceptions include two promising biodiesel compounds, fatty acid ethyl esters[32] and bisabolene,[7] which have been demonstrated to inflict no toxicity to *E. coli*. Interestingly, there are recent reports of *E. coli* producing butanol above the concentrations where toxicity is seen.[4,5] We speculate that biofuel tolerance in late growth phase and early stationary phase, where much of the biofuel is typically produced, might be higher than in the exponential growth phase, where toxicity studies are typically conducted. Furthermore, the most commonly used measure of tolerance (inhibition of growth) may not accurately report on the stability of biochemical pathways within the cell.

Increased tolerance can also be engineered by modifying the host organism, or equipping it with mechanisms to resist toxicity. An excellent review of the mechanisms of biofuel toxicity and techniques for engineering-increased tolerance has been written,[74] but will be briefly sketched out here. Microbial species possess efflux pumps that can expel hydrophobic compounds and enable the cells to thrive in the presence of low levels of the toxic compounds. For instance, in *E. coli*, the AcrAB-TolC efflux pump provides some tolerance to biofuel compounds. More effective efflux pumps from hardier microbial species (e.g. species isolated from oil fields) can be harnessed to increase tolerance in the host organism. This approach was used to demonstrate that expression of solvent-tolerance pumps can result in increased biofuel production.[75] A library of efflux pumps was cloned from a variety of microbial species and expressed in *E. coli* in the presence of several fuel candidates. A competition experiment among strains expressing the various pumps revealed the pumps that bestowed the most tolerance to the fuels. While this study demonstrated the potential of using heterologous efflux pumps for increasing tolerance to biofuels, compatibility of the foreign pumps with the host may need to be adjusted through additional engineering efforts.

A metagenomic library of heterologous genomic DNA was used to increase the tolerance of *E. coli* to various toxins produced in biomass pretreatment.[76] Microbial genomic DNA (gDNA) was isolated from peat bog soil and other environments with high concentrations of toxic inhibitors, and a phage library was made from the gDNA and infected into *E. coli*. Transfected cells that exhibited increased tolerance were obtained by imposing a selection for tolerance to inhibitory concentrations of the toxins. An investigation into the heterologous genes that bestowed higher tolerance suggested that increased tolerance may have been achieved by complementation of an *E. coli* enzyme affected by the toxins with a foreign enzyme that could tolerate the toxin. This study is particularly attractive as it provided a strong clue to the mechanism of toxicity of those particular poisons. While this method was demonstrated to engineer increased tolerance to biomass pretreatment toxins, there seems no obvious reason why it could not also be applied to engineering increased biofuel toxicity as well. It might also provide information as to the specific enzymes that are affected the most in the presence of biofuels, which are often assumed to be systemic.[74]

FUTURE PROSPECTS

This chapter has discussed methods to improve yields of biofuel production pathways. There are many aspects of industrial biofuel production that we lack the space to discuss further. As discussed in the introduction, constraints on economic margins require that the yields of fuel from biomass must be extremely close to the theoretical yield. While impressive yields quite close to theoretical are beginning to appear in the literature, results obtained in laboratory shake flasks, or even small fermentation vessels, rarely scale predictably to large-scale fermentation processes. Researchers often tailor their engineered microbes for laboratory conditions rather than realistic industrial situations. For instance, production on industrial scales will likely occur in anaerobic environments, due to the difficulty in aerating thousand-liter or larger vessels. Also, production results are often (though fortunately not always) reported using rich media, which are far too expensive to use at industrial scales. A biofuel pathway or a host organism optimized to produce biofuel in rich, oxygenated medium may not achieve high yields in minimal medium, in an anaerobic environment. While proof-of-concept results are certainly useful to demonstrate that production in a new host is indeed possible, redesigning a pathway to work at industrial scales often requires efforts that match or exceed those required to achieve the demonstration level. Furthermore, other advantages of laboratory-scale production that are taken for granted, such as lack of contaminants and absence of phage, will probably not apply to industrial scales. These are issues that can by addressed through synthetic biology.

Other complications of biofuel production need to be considered. The sugar stream that the biofuel production host will be converting into biofuels will most likely be the hydrolysate of a high-yielding lignocellulosic feedstock such as switchgrass. Most biofuel produced at the laboratory scale is done with glucose or glycerol; in contrast, plants are composed of five- and six-carbon sugars. Cells growing on purified glucose may have different metabolite pools than cells growing on such complex carbon mixtures, because these sugars enter central carbon metabolism through different routes. It may be necessary to refine the biofuel synthesis pathway to accommodate these metabolic perturbations. In addition, it will probably be necessary to modify the host organism so all sugars are consumed simultaneously rather than sequentially. This is necessary because it leads to more efficient biofuel production and more complete sugar conversion. The host organism can also be engineered to use actual plant biomass, rather than biomass hydrolysate, in order to further reduce processing costs.[8,77] Given these large obstacles, production of economically viable biofuel may seem incredibly daunting. We anticipate that with applications of both existing and as-yet undiscovered synthetic biology tools, advanced biofuels will become an integral part of the energy economy.

References

1. National Research Council (U.S.). Panel on alternative liquid transportation fuels. *Liquid Transportation Fuels From Coal and Biomass: Technological Status, Costs, and Environmental Impacts.* Washington, D.C.: National Academies Press; 2009 pp xvii, 370 p.

2. Tilman D, et al. Beneficial biofuels — the food, energy, and environment trilemma. *Science.* 2009;325:270—271.

3. Gibson DG, et al. Creation of a bacterial cell controlled by a chemically synthesized genome. *Science.* 2010;329:52—56.

4. Bond-Watts BB, Bellerose RJ, Chang MCY. Enzyme mechanism as a kinetic control element for designing synthetic biofuel pathways. *Nat Chem Biol.* 2011;7:222—227.

5. Dellomonaco C, Clomburg JM, Miller EN, Gonzalez R. Engineered reversal of the beta-oxidation cycle for the synthesis of fuels and chemicals. *Nature.* 2011;476:355—U131.

6. Peralta-Yahya PP, Keasling JD. Advanced biofuel production in microbes. *Biotechnol J.* 2010;5:147—162.

7. Peralta-Yahya PP, et al. Identification and microbial production of a terpene-based advanced biofuel. *Nat Commun.* 2011:2.

8. Bokinsky G, et al. Synthesis of three advanced biofuels from ionic liquid-pretreated switchgrass using engineered *Escherichia coli*. *Proc Natl Acad Sci USA*. 2011;108:19949–19954.

9. Fortman JL, et al. Biofuel alternatives to ethanol: pumping the microbial well. *Trends Biotechnol*. 2008;26:375–381.

10. Prather KLJ, Martin CH. De novo biosynthetic pathways: rational design of microbial chemical factories. *Curr Opin Biotechnol*. 2008;19:468–474.

11. Alper H, Stephanopoulos G. Engineering for biofuels: exploiting innate microbial capacity or importing biosynthetic potential? *Nat Rev Microbiol*. 2009;7:715–723.

12. Bar-Even A, et al. The moderately efficient enzyme: evolutionary and physicochemical trends shaping enzyme parameters. *Biochemistry*. 2011;50:4402–4410.

13. Yadav VG, Stephanopoulos G. Reevaluating synthesis by biology. *Curr Opin Microbiol*. 13:371–376.

14. Chang MC, Eachus RA, Trieu W, Ro DK, Keasling JD. Engineering *Escherichia coli* for production of functionalized terpenoids using plant P450s. *Nat Chem Biol*. 2007;3:274–277.

15. Yoshikuni Y, Dietrich JA, Nowroozi FF, Babbitt PC, Keasling JD. Redesigning enzymes based on adaptive evolution for optimal function in synthetic metabolic pathways. *Chem Biol*. 2008;15:607–618.

16. Baneyx F, Mujacic M. Recombinant protein folding and misfolding in *Escherichia coli*. *Nat Biotechnol*. 2004;22:1399–1408.

17. Kolaj O, Spada S, Robin S, Wall JG. Use of folding modulators to improve heterologous protein production in *Escherichia coli*. *Microb Cell Fact*. 2009;8.

18. Elovson J, Vagelos PR. Acyl carrier protein. X. Acyl carrier protein synthetase. *J Biol Chem*. 1968;243:3603–3611.

19. Tsoi CJ, Khosla C. Combinatorial biosynthesis of 'unnatural' natural products: the polyketide example. *Chem Biol*. 1995;2:355–362.

20. Stachelhaus T, Marahiel MA. Modular structure of genes encoding multifunctional peptide synthetases required for non-ribosomal peptide synthesis. *FEMS Microbiol Lett*. 1995;125:3–14.

21. Nakamura CE, Whited GM. Metabolic engineering for the microbial production of 1,3-propanediol. *Curr Opin Biotechnol*. 2003;14:454–459.

22. Kajiura H, Mori K, Tobimatsu T, Toraya T. Characterization and mechanism of action of a reactivating factor for adenosylcobalamin-dependent glycerol dehydratase. *J Biol Chem*. 2001;276:36514–36519.

23. Sevrioukova IF, Li H, Zhang H, Peterson JA, Poulos TL. Structure of a cytochrome P450-redox partner electron-transfer complex. *Proc Natl Acad Sci USA*. 1999;96:1863–1868.

24. Heinemann M, Sauer U. Systems biology of microbial metabolism. *Curr Opin Microbiol*. 2010;13:337–343.

25. Ralser M, et al. Metabolic reconfiguration precedes transcriptional regulation in the antioxidant response. *Nat Biotechnol*. 2009;27:604–605.

26. Voet D, Voet JG. *Biochemistry*. 3rd ed. Hoboken, NJ: J. Wiley & Sons; 2004:Ed pp xv, 1591p.

27. Lindsley JE, Rutter J. Whence cometh the allosterome? *Proc Natl Acad Sci USA*. 2006;103:10533–10535.

28. Atsumi S, Hanai T, Liao JC. Non-fermentative pathways for synthesis of branched-chain higher alcohols as biofuels. *Nature*. 2008;451:86–89.

29. Shen CR, Liao JC. Metabolic engineering of *Escherichia coli* for 1-butanol and 1-propanol production via the keto-acid pathways. *Metab Eng*. 2008;10:312–320.

30. Szczesiul M, Wampler DE. Regulation of a metabolic system in vitro: synthesis of threonine from aspartic acid. *Biochemistry*. 1976;15:2236–2244.

31. Cho HS, Cronan JE. Defective export of a periplasmic enzyme disrupts regulation of fatty-acid synthesis. *J Biol Chem*. 1995;270:4216–4219.

32. Steen EJ, et al. Microbial production of fatty-acid-derived fuels and chemicals from plant biomass. *Nature*. 2010;463:559–562.

33. Nobeli I, Favia AD, Thornton JM. Protein promiscuity and its implications for biotechnology. *Nat Biotechnol*. 2009;27:157–167.

34. Leonard E, et al. Combining metabolic and protein engineering of a terpenoid biosynthetic pathway for overproduction and selectivity control. *Proc Natl Acad Sci USA*. 107: 13654–13659.

35. Dueber JE, et al. Synthetic protein scaffolds provide modular control over metabolic flux. *Nat Biotechnol*. 2009;27:753–759.

36. Huang XY, Holden HM, Raushel FM. Channeling of substrates and intermediates in enzyme-catalyzed reactions. *Annu Rev Biochem*. 2001;70:149–180.

37. Wang C, et al. Metabolic engineering of *Escherichia coli* for alpha-farnesene production. *Metab Eng*. 13: 648–655.

38. Kerfeld CA, Heinhorst S, Cannon GC. Bacterial microcompartments. *Annu Rev Microbiol*. 64:391–408.

39. Fan C, et al. Short N-terminal sequences package proteins into bacterial microcompartments. *Proc Natl Acad Sci USA*. 107: 7509–7514.

40. Liu TG, Vora H, Khosla C. Quantitative analysis and engineering of fatty acid biosynthesis in *E. coli*. *Metab Eng*. 2010;12:378–386.

41. Pitera DJ, Paddon CJ, Newman JD, Keasling JD. Balancing a heterologous mevalonate pathway for improved isoprenoid production in *Escherichia coli*. *Metab Eng*. 2007;9:193–207.

42. Stoebel DM, Dean AM, Dykhuizen DE. The cost of expression of *Escherichia coli* lac operon proteins is in the process, not in the products. *Genetics*. 2008;178:1653–1660.

43. Scott M, Gunderson CW, Mateescu EM, Zhang Z, Hwa T. Interdependence of cell growth and gene expression: origins and consequences. *Science*. 330: 1099–1102.

44. CornishBowden A, Hofmeyr JHS, Cardenas ML. Strategies for manipulating metabolic fluxes in biotechnology. *Bioorg Chem*. 1995;23:439–449.

45. Maier T, Guell M, Serrano L. Correlation of mRNA and protein in complex biological samples. *FEBS Lett*. 2009;583:3966–3973.

46. Taniguchi Y, et al. Quantifying *E. coli* proteome and transcriptome with single-molecule sensitivity in single cells. *Science*. 329: 533–538.

47. Anthony JR, et al. Optimization of the mevalonate-based isoprenoid biosynthetic pathway in *Escherichia coli* for production of the anti-malarial drug precursor amorpha-4,11-diene. *Metab Eng*. 2009;11:13–19.

48. Sleight SC, Bartley BA, Lieviant JA, Sauro HM. Designing and engineering evolutionary robust genetic circuits. *J Biol Eng*. 4:12.

49. Jones KL, Kim SW, Keasling JD. Low-copy plasmids can perform as well as or better than high-copy plasmids for metabolic engineering of bacteria. *Metab Eng*. 2000;2:328–338.

50. Tyo KEJ, Kocharin K, Nielsen J. Toward design-based engineering of industrial microbes. *Curr Opin Microbiol*. 2010;13:255–262.

51. Jensen PR, Hammer K. The sequence of spacers between the consensus sequences modulates the strength of prokaryotic promoters. *Appl Environ Microbiol*. 1998;64:82–87.

52. Jensen PR, Hammer K. Artificial promoters for metabolic optimization. *Biotechnol Bioeng*. 1998;58:191–195.

53. Mijakovic I, Petranovic D, Jensen PR. Tunable promoters in systems biology. *Curr Opin Biotechnol*. 2005;16:329–335.

54. Farmer WR, Liao JC. Improving lycopene production in *Escherichia coli* by engineering metabolic control. *Nat Biotechnol*. 2000;18:533–537.

54a. Zhang F, Carothers JM, Keasling JD. Design of a dynamic sensor-regulator system for production of chemicals and fuels derived from fatty acids. *Nature Biotech*. 2012;30:354–359.

55. Reitzer LJ, Magasanik B. Expression of glnA in *Escherichia coli* is regulated at tandem promoters. *Proc Natl Acad Sci USA*. 1985;82:1979–1983.

56. Liu J, Magasanik B. Activation of the dephosphorylation of nitrogen regulator I-phosphate of *Escherichia coli*. *J Bacteriol*. 1995;177:926–931.

57. Ajikumar PK, et al. Isoprenoid pathway optimization for Taxol precursor overproduction in *Escherichia coli*. *Science*. 330:70–74.

58. Laursen BS, Sorensen HP, Mortensen KK, Sperling-Petersen HU. Initiation of protein synthesis in bacteria. *Microbiol Mol Biol Rev*. 2005;69:101–123.

59. Salis HM, Mirsky EA, Voigt CA. Automated design of synthetic ribosome binding sites to control protein expression. *Nat Biotechnol*. 2009;27:946–950.

60. Redding-Johanson AM, et al. Targeted proteomics for metabolic pathway optimization: application to terpene production. *Metab Eng*. 2011;13(2):194–203.

61. Pfleger BF, Pitera DJ, Smolke CD, Keasling JD. Combinatorial engineering of intergenic regions in operons tunes expression of multiple genes. *Nat Biotechnol*. 2006;24:1027–1032.

62. Carothers JM, Goler JA, Juminaga A, Keasling JD. Design-driven approaches for engineering RNA-regulated pathway controls. *Abstr Pap Am Chem Soc*. 2011:241.

63. Warner JR, Patnaik R, Gill RT. Genomics enabled approaches in strain engineering. *Curr Opin Microbiol*. 2009;12:223–230.

64. Atsumi S, et al. Metabolic engineering of *Escherichia coli* for 1-butanol production. *Metab Eng*. 2008;10:305–311.

65. Steen EJ, et al. Microbial production of fatty-acid-derived fuels and chemicals from plant biomass. *Nature*. 2009;463:559–562.

66. Kim PJ, et al. Metabolite essentiality elucidates robustness of *Escherichia coli* metabolism. *Proc Natl Acad Sci USA*. 2007;104:13638–13642.

67. Park JH, Lee SY, Kim TY, Kim HU. Application of systems biology for bioprocess development. *Trends Biotechnol*. 2008;26:404–412.

68. Paradise EM, Kirby J, Chan R, Keasling JD. Redirection of flux through the FPP branch-point in *Saccharomyces cerevisiae* by down-regulating squalene synthase. *Biotechnol Bioeng*. 2008;100:371–378.

69. Ro DK, et al. Production of the antimalarial drug precursor artemisinic acid in engineered yeast. *Nature*. 2006;440:940–943.

70. Cann AF, Liao JC. Production of 2-methyl-1-butanol in engineered *Escherichia coli*. *Appl Microbiol Biotechnol*. 2008;81:89–98.

71. Reed MC, Lieb A, Nijhout HF. The biological significance of substrate inhibition: a mechanism with diverse functions. *Bioessays*. 2010;32:422–429.

72. Lee SJ, et al. Metabolic engineering of *Escherichia coli* for enhanced production of succinic acid, based on genome comparison and in silico gene knockout simulation. *Appl Environ Microbiol*. 2005;71:7880–7887.

73. Choi HS, Lee SY, Kim TY, Woo HM. In silico identification of gene amplification targets for improvement of lycopene production. *Appl Environ Microbiol*. 76:3097–3105.

74. Dunlop MJ. Engineering microbes for tolerance to next-generation biofuels. *Biotechnol Biofuels*. 2011;4:32.

75. Dunlop MJ, et al. Engineering microbial biofuel tolerance and export using efflux pumps. *Mol Syst Biol*. 2011;7:487.

76. Sommer MOA, Church GM, Dantas G. A functional metagenomic approach for expanding the synthetic biology toolbox for biomass conversion. *Mol Syst Biol*. 2010;6:360.

77. Lynd LR, van Zyl WH, McBride JE, Laser M. Consolidated bioprocessing of cellulosic biomass: an update. *Curr Opin Biotechnol*. 2005;16:577–583.

78. Tang SY, Cirino PC. Design and application of a mevalonate-responsive regulatory protein. *Angew Chem Int Ed*. 2011;50:1084–1086.

79. Aune TEV, et al. Directed evolution of the transcription factor XylS for development of improved expression systems. *Microb Biotechnol*. 2010;3:38–47.

80. Tang SY, Fazelinia H, Cirino PC. AraC regulatory protein mutants with altered effector specificity. *J Am Chem Soc*. 2008;130:5267–5271.

CHAPTER 12

Tools for Genome Synthesis

Mitsuhiro Itaya
Keio University, Tsuruoka, Yamagata, Japan

INTRODUCTION

A cellular system is complex. Present terrestrial lives from unicellular to multicellular have adapted to the present diverse environments on the Earth. The two putative definitions are in favor by the author and sufficient for this chapter: the premise of the biological systems should: (1) constitute biological materials such as proteins, nucleic acids, lipids, and other metabolites; and (2) yield proliferative offspring inherited by genome DNA. Genomes, as information molecules, govern cellular activities in accordance with information flow known as the central dogma: DNA→RNA→protein→metabolites. Fundamental knowledge on biology has been obtained through various genetic, biochemical, and biophysical studies applied on existing lives in nature and long-cultivated domesticated ones in laboratories. The framework of further complicated biological systems will be ultimately proven if man-made ones are available. What are man-made cells deemed to be or recognized by most researchers? Fundamental answers should eventually come when those cells use genetic circuits of interacting genes and proteins to implement diverse cellular functions.[1,2]

A fundamental topic underlying this chapter is how to practically produce genomes that should lead to innovation of newly engineered cells. The genome, the largest unbranched polymer molecule among biological substances, is a warehouse of all genes in a given cell. Hundreds or thousands of genes are embedded in the genome, as illustrated in Figure 12.1. The gene, a major object in modern biology on the other hand, is included in small DNA molecules. Thanks to the genetic engineering technologies developed in the late 1980s and later established as essential fundamental tools, small DNA segments are readily available in all aspects of the current molecular biology field. Genes encoded in small DNA segments are stable in test tubes. Once cloned, they are ready to be amplified by *Escherichia coli* plasmids followed by manipulations aiming at various goals. In contrast, large DNAs become unstable in test tubes, even if they are carefully isolated. The primary reason is that hydrodynamic shearing in solution under regular laboratory manipulations will result in fragmentation to small DNAs, due to intrinsic physicochemical properties associated with the long polymer. Regular DNA implementations, agitation by vortex or precipitation by addition of ethanol, provide fragmented DNA not exceeding 50 kbp. Unnoticed nucleases contaminated during biochemical isolation procedures might be the secondary reason making damage-free genome DNA preparation considerably difficult. A DNA molecule of >50 kbp is designated as a large DNA molecule to discriminate small DNA fragments, unless specified otherwise in this chapter (Fig. 12.1). Protocols working well for small DNA engineering have not simply been expanded to those for large DNA engineering.

225

Synthetic Biology. DOI: http://dx.doi.org/10.1016/B978-0-12-394430-6.00012-1

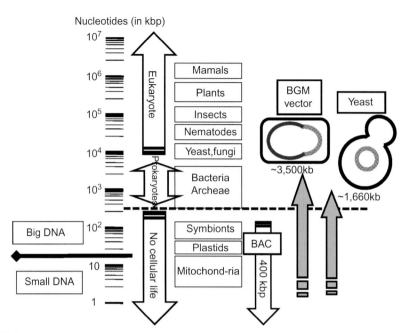

FIGURE 12.1

Variation of cellular genome size. Genome sizes are shown in kbp in log scale with representative species. A horizontal dotted line indicates a 585 kpb, the smallest natural genome of the free-living organism, *M. genitallium*. The bold horizontal line indicates the size to discriminate large and small DNAs used in the text. Two hosts for recombinant genome production, *B. subtilis* and yeast are schematically drawn. The structural difference for guest genomes, circular in yeast and chimera with *B. subtilis* genome, are illustrated. The largest guest genomes reported to date, 3500 kbp from *Synechocystis* PCC6803,[3] and 1660 kbp from *Prochlorococcus marinus* MED4[4] are indicated by vertical arrows to the right. Stable BAC clones are available below 400 kbp only.[5,6]

A technological breakthrough occurred in 2005[3] and 2008,[4] when a large DNA such as a whole bacterial genome was demonstrated clonable and manipulatable by use of novel cloning hosts, *Bacillus subtilis* 168[3] and Bakers' yeast, *Saccharomyces cerevisiae*.[4–6] The impact of entire genome synthesis on modern biology is obvious. All genetic information for cellular life is under our manipulation, and is causing a paradigm shift in all areas of life science. Any working hypothesis in modern cellular sciences and biotechnology development should be examined by experiments where truth lies only in observed reproducibility. 'To know the genome' should ultimately be verified after 'synthesis of genomes.'[7] This chapter will also cover sketches where abundant synthetic genomes are available for potential new organisms in coming years.

DNA SIZE LIMIT BY *E. COLI* PLASMID

DNA cloning in *E. coli* is performed using plasmid vectors. An *E. coli* plasmid vector cloning system covers most small DNA fragments in Figure 12.1. In addition, a Bacterial Artificial Chromosome (BAC) vector is frequently used to clone large DNA fragments.[8] Even DNA fragments stably maintained by a BAC vector of around 100–200 kbp remain far below the smallest existing genome, i.e. the *Mycoplasma genitalium*, as illustrated in Figure 12.1. Recovery of high-quality BAC DNAs in a large quantity normally requires a mild extraction protocol to reduce shearing. The protocol includes time-consuming and labor-intensive cesium chloride gradient ultracentrifugation,[8,9] and therefore quality and quantity are often trade-offs in terms of time and labor in the preparation of large DNA. To address this issue, a novel one-step preparation of DNAs as large as BAC clones was developed very recently,[10,11] as described below.

GENOME CLONING USING A BOTTOM-UP APPROACH

Unlike the straightforward top-down cloning approach, the bottom-up approach stems from the idea that repeated connection of small DNA segments leads to reconstruction of larger ones. As small DNAs are readily prepared in *E. coli*, an appropriate host that allows repeated connection and stable maintenance of large DNAs is needed. To this end, two hosts, *B. subtilis* 168 and *S. cerevisiae* were developed independently.

B. subtilis 168, a Gram-positive endospore-forming bacterium possessing a 4215 kbp genome, shown in Figure 12.2, has been demonstrated to accommodate very large DNAs, as described in the following sections. This remarkable ability was highlighted in 2005 by stable accommodation of a 3.5 Mbp-long genome from a nonpathogenic unicellular photosynthetic bacterium *Synechocystis* PCC6803 (3500 kbp).[3] Meanwhile, the J. Craig Venter Institute (JCVI) produced synthesized genomes designed based on *Mycoplasma* species using yeast as a final host.[4] The *B. subtilis* host case[3,12] is illustrated in Figure 12.2A. Those two groups used different hosts, but shared the bottom-up concept. Target genomes are called 'guest' genomes throughout this chapter in comparison to 'host' for cloning. For both methods, the guest genome was divided into pieces small enough to be readily prepared by an *E. coli* system. Small pieces designed to share overlap sequences between adjacent ones are incorporated into the cytoplasm by virtue of DNA transfer mechanisms shown in Figure 12.2B, and connected together as illustrated in Figure 12.2A.

It should be mentioned that the guest from *Mycoplasma mycoides* synthesized by JCVI in the host yeast was directly implanted to a closely related strain, *Mycoplasma caplicorum*.[6] Via genome replacement in the latter species, the resulting colony was the *Mycoplasma* cell possessing the complete designed and synthesized *M. mycoides* genome. The JCVI's achievement, a spectacular biological and technological feat, opened up a gateway to numerous genome synthesis applications where the functional genomes can be produced even starting from chemically synthesized DNA oligomers.[4,5] Differences in using

227

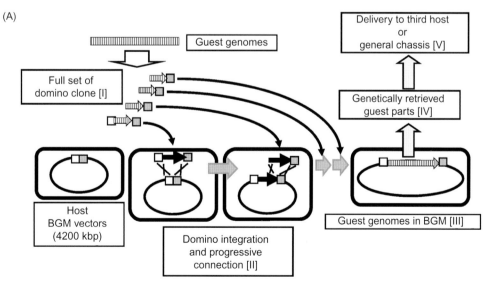

(A)

FIGURE 12.2A

Framework of BGM vector system for production of guest genomes. Host BGM vectors are *B. subtilis* 168 derivative strains possessing accessory sequences. pBR322 or BAC are accessory sequences to designate the cloning locus catalyzed by double homologous recombination with domino clones (I) made on the same plasmid.[3,9,12,16,19] (II) is described in the text and Figure 12.2B. Relevant observations in the case for a largest guest genome (III) are detailed in the text. (IV) is argued in the text and in Figure 12.4. (V) is mentioned throughout this chapter with a keyword called chassis. Antibiotic resistance markers for efficient selection throughout are omitted.

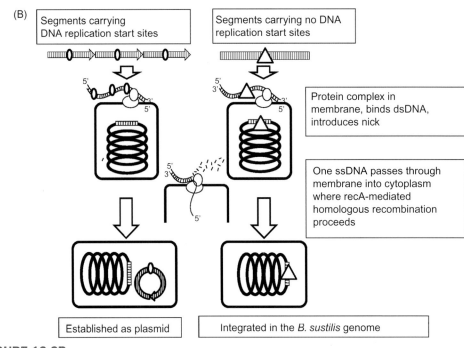

FIGURE 12.2B

B. subtilis transformation provides two DNA forms through natural competent cells. The inherent nature of *B. subtilis* 168 is illustrated and briefly described in the text.

prokaryotic and eukaryotic hosts for genome synthesis applications will also be argued for sporadically in the sections below.

THE KEIO METHOD

How Does *B. subtilis* Carry out DNA Incorporation and Connection?

Why was *B. subtilis* chosen as a host? The strain was known to be able to develop natural competence,[13,14] an inherent remarkable feature, and conduct high and precise natural homologous recombination.[1,12] As illustrated in Figure 12.2B, *B. subtilis* forms protein complexes in membrane, responding to the quorum-sensing molecular mechanism, that are able to trap double-stranded DNA (dsDNA) present outside. One of the proteins introduces a nick in the bound exogenous dsDNA. Single-stranded DNA (ssDNA) starting at the nick is taken up into the cytoplasm. The cleavage site by the nuclease and the strand selection by the transformation complex are basically random.[13,14] The resultant highly recombinogenic ssDNA is immediately recruited by cellular *recA* product and integrated into a *recA*-dependent homologous recombination pathway.[14] The DNA promptly recombines with the counter homologous sequences if present in the genome (Figs 12.2A and 12.2B). The genetically coordinated molecular mechanism to transfer DNA actively through membranes is absent in *E. coli* and yeast. Direct delivery of dsDNA to cytoplasm of both species has to be catalyzed by physicochemical treatment, in sharp contrast to the inherent properties of *B. subtilis*.[15]

A Genome Vector Suited for Giant DNA Handling

Use of *B. subtilis* 168 as a host for stable guest cloning has a long history. BGM vectors are *B. subtilis* hosts specified to giant DNA cloning.[1,3,12] The name BGM, standing for *Bacillus* GenoMe, was given to strains possessing accessory sequences in their genomes.[16] Accessory sequences are used for selection in cloning processes illustrated in Figure 12.2A.

Homologous recombination through transformation shown in Figure 12.2B is a single shared principle of the huge BGM vector systems.[1,17,18] The first technical breakthrough was brought about by accommodation of a complete genome back in 1995.[19] The provisional success came from *E. coli* bacteriophage lambda genome (48.5 kbp), a very large DNA at the time, albeit small when compared with current abundant sequence-known bacterial genomes. All the fundamental ideas and concepts illustrated in Figure 12.2A were recognized and established in the prototype work.[19] Expansion to larger guest DNAs than *E. coli* bacteriophage lambda had to wait until sequence-determined genomes were available for choice as guest DNA.

BGM vectors are derived from a *B. subtilis* Marburg 168 strain, the entire genome sequence of which was reported in 1997.[20] The BGM system does not employ conventional restriction endonucleases and DNA ligase to connect DNA segments that are essential for gene cloning in *E. coli*. On the contrary, connection steps are carried out through *recA*-mediated homologous recombination, an inherent and highly efficient property of the host, as illustrated in Figures 12.2A and 12.2B. Relevant examples of giant DNA production via the BGM system are described below.

A Guest Genome from *Synechocystis* PCC6803

The method to connect small DNA segments in the BGM vector was termed megacloning,[3] which literally came from cloning of DNA with a megabase pair (Mbp) size. The genome from *Synechocystis* PCC6803 was chosen as the largest guest in the BGM host. As sequence-known DNA is a minimal requirement for megacloning, *Synechocystis* PCC6803, a nonpathogenic safe and sound bacterium among a dozen sequence-known species available in 1997, was indeed the only choice when my project started. The genome of this unicellular photosynthetic bacterium belonging to different phylum, and far in the phylogenic distance from *B. subtilis*, was attempted to be reconstructed in BGM. The project took about seven years and resulted in a publication in 2005.[3] Also referring to other reviews,[1,17] stepwise connection of neighboring fragments is illustrated in Figure 12.2A, and the assembly steps of larger fragments are visualized in Figure 12.3A. The guest strain possesses a 3573 kb (3.5 Mbp) genome and seven plasmids with various sizes.[21] The intrinsic property of the multicopy genome, estimated more than 10 per cell,[22] strongly indicated that the PCC6803 strain actually carries total DNA larger than 35 Mbp in a single cell. This notion will be argued in the next section.

Issues Associated with *Synechocystis* Megacloning

Connection of two bacterial genomes in one cell raised a number of issues previously poorly argued. A *Bacillus*—*Synechocystis* chimera, hereafter putatively named as *CyanoBacillus* or CB, was cultivated in *B. subtilis* growth media (Fig. 12.3A). Immediately after the last DNA segment was connected for completion of megacloning, CB was examined to see whether it starts to grow in a *Synechocystis* growth media, complete synthetic media with no carbon sources as mentioned in Figure 12.3B. Logically, a vast number of genes from the guest genome should be expressed properly and coordinately, and dominate cellular gene regulatory networks from that of *B. subtilis*. Accordingly, a number of factors and components, including the cell membrane, cell wall, and metabolic state were expected to be converted from the host (*B. subtilis*) to adapt to the guest (*Synechocystis*) in response to changes in culture medium. No CB culture or colony has yet been obtained in liquid medium, or on solid plate for cyanobacteria.[22]

We have learned many lessons in my project during and after the construction of CB. They are: (1) genes expressed that influence the host; (2) exclusivity of ribosomal RNA genes; (3) balanced genome structure; (4) plasmid-borne genes; (5) copy number of genome per cell; and (6) deleterious mutations possibly incorporated during reconstruction. It is

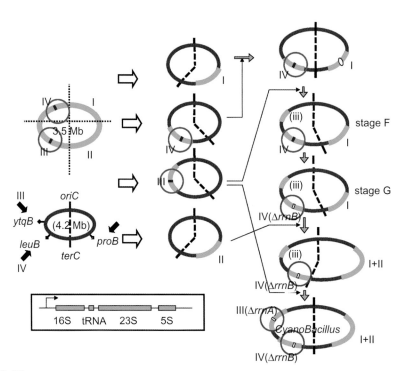

FIGURE 12.3A

Stage-specific interference of the guest *rrn* operon. The *Synechocystis* PCC6803 circular genome in the upper left was megacloned in the BGM vector middle left. Figure 1 of reference[3] was reproduced to emphasize (i) balanced structures around *oriC–terC* axis, indicated by dotted lines and (ii) guest ribosomal operons, presence and absence, of BGM clones. A 5.4 kbp-long ribosomal RNA operon, *rrnA* and *rrnB*, sketched in the bottom left has an identical sequence. Intermediate BGM clone at the stage F was compatible with the presence of blue circled *rrnB*. Attempts to elongate by a next domino elongation accompanied large deletions of the guest part already integrated. Pinpointing deletion of the 5.4-kb *rrnB* indicated at the stage, indicated by a red circle, allowed the rest of megacloning to be completed to give a *CyanoBacillus*. The other *rrnA* operon, red circled, should be also deleted. Reassessment of the *rrn* operon exclusiveness, likely dependent on the amount of guest genes, is described in the text.

noteworthy that a host used by JCVI, *S. cerevisiae* eukaryotic yeast, can provide the guest genome in a circular form sketched in Figure 12.1. As long as guest genomes are originated from eubacteria as is the case for *Mycoplasma* species,[4,6] and even 1.66 Mb *Prochlorococcus marinus* MED4 genome,[5] their gene expression in yeast is not expected and therefore considered harmless for the yeast growth. In sharp contrast, synthesis/megacloning of a bacterial guest in BGM caused various unsuspected issues argued below.

EXPRESSION OF GUEST GENES IN BGM HOST

Chassis is a technical term in synthetic biology for a cell envelope that divides inner solutions from the outer environments. If two complete genomes are forced to be present in a chassis, which set of genes, from host or guest, should be dominant for the chassis? In the case of CB, expression profiles for guest genes of *Synechocystis* (~3000), and host genes of *B. subtilis* (~4000) of the CB strain have been analyzed by microarray and tiling array. Most of the *Synechocystis*-originated genes were suppressed in CB, resulting in about 100 mRNAs at low abundance, in contrast to normal expressions for genes from *B. subtilis*. As to the transcription apparatus, no RNA polymerases and few sigma factors from the guest were present among those expressed (Nishida, Yoshikawa, Itaya, unpublished). A proteome analysis revealed that only dozens of translation products from the guest can clearly be detected but, curiously, they showed little match to the transcripts (Ishihama and Itaya, unpublished). Despite some transcriptions actually expressed from the guest genes, proteins from the guest seemed virtually sterile (Itaya, Yoshikawa, and Nishida, unpublished), as

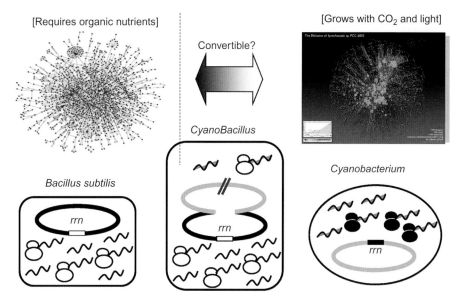

FIGURE 12.3B

CyanoBacillus: Interchangeable gene expression networks? *CynoBacillus* lacking only two *rrnA* and *rrnB* is schematically drawn in the bottom middle. Gene expression and ribosomes in the host *B. subtilis* (left) and guest *Synechocystis* (right) are altered in CB. Expression of reduced transcripts and few translated product from guest genes are described in the text. Robust gene networks for the host and guest (top) might be exchangeable by certain factors.[25]

sketched roughly in Figure 12.3B. These observations did not fulfill any predictions derived from the normal central dogma scenario in which translation tends to be coupled with transcription. The robust host gene network and the sterile guest one should be tightly coupled with presence and absence of ribosomes, as argued below.

EXCLUSIVITY OF RIBOSOMAL RNA IN THE HOST

CB lacked two ribosomal gene operons, *rrnA* and *rrnB* from *Synechocystis*. The guest *rrn* operons displayed complicated roles and have raised a possible scenario on species selection in a previously unsuspected manner. It should be emphasized that the presence of ribosomal gene operons alone, regardless of *rrnA* and *rrnB*, did not cause any detrimental phenotypes on *B. subtilis*. They had to be deleted halfway in elongation before completing the synthesis of the chimera genome as detailed below. The step-by-step step elongation method left many intermediate strains that possess part of guest DNA of different regions and in different lengths (Fig. 12.3A). The *rrn* exclusivity was not found until the accumulation of the *Synechocystis* genome up to 2.6 Mbp in *B. subtilis*, insert stage F of Figure 12.3A. At that stage, another domino elongation always accompanied big deletions randomly from guest genome parts already present in BGM. Eradication of the guest *rrnA* operon only, a 5.4 kbp-long DNA segment shown in Figure 12.3A, solved all the stalling problems and allowed the rest of the megacloning to be completed. Together with the same exclusivity for the other *rrnB*, these observations strongly demonstrated that expression of guest *rrn* species turned out to be toxic if a certain number of *Synechocystis* genes are accumulated and possibly expressed during stage F. Toxicity might be brought in by possible formation of chimera ribosomes that should not stand with pure *B. subtilis* ribosomes. This explanation is one of many scenarios and remains controversial and elusive, which requires more experimental evidence.

Function of chimera ribosomes created by mixing 23 S from *E. coli* with 16 S from different species was reported in *E. coli*, in which 16 S seems to be restricted to species closely related to *E. coli*.[23] No experimental indication has been yet made available for chimera ribosomes between *B. subtilis* and *Synechocystis*. Absence of guest ribosomal RNA is consistent with the

231

data that a small number of transcripts and a few proteins originated from *Synechocystis* genes in CB. These observations, together with the absence of a guest RNA polymerase, imply that ribosomes play central roles for robust cellular gene networks. Results from exchange of ribosomal RNA[23] suggest that chimera ribosomes are allowed only between closely related species. The only way for CB to survive seems to avoid conflict interference with and be in accordance with the host *B. subtilis* gene regulatory networks that are essential for growth in *B. subtilis* medium. Transcriptome and proteome analyses for all the intermediate strains should give remarkable clues, but remain to be done. These scenarios might be examined by adopting guest genomes already synthesized, for example those reported in[4–6] in the BGM host.[3]

Lastly on the ribosomal RNA issues, there is a priori consensus for current molecular phylogenic analyses consistent with the apparent lack of multiple *rrn* species carriers. Exclusivity of distantly related ribosomal RNA species observed in CB gave the first experimental basis on and coordinated with the present molecular phylogenic trees for eubacteria.

STRUCTURAL CONSTRAINTS ON THE HOST GENOME

Bacterial genomes have to be replicated to produce offspring. Many circular bacterial genomes possess particular sequences, called *oriC* and *terC*, where a new round of DNA replication starts and two replication machineries meet.[24] Consensus molecular mechanisms are drawn by a number of studies that two DNA replication machineries initiated at *oriC* divergently traverse half of a circular genome. Consequently, the right and left length around the *oriC–terC* axis have to be similar in length, typically as shown for *E. coli* and *B. subtilis*. Logically, guest genomes megacloned at a single locus of the *B. subtilis* genome increase the one replication length, causing significant imbalance around the original *oriC–terC* axis, as illustrated in Figure 12.3C. The degree of imbalance had not been known

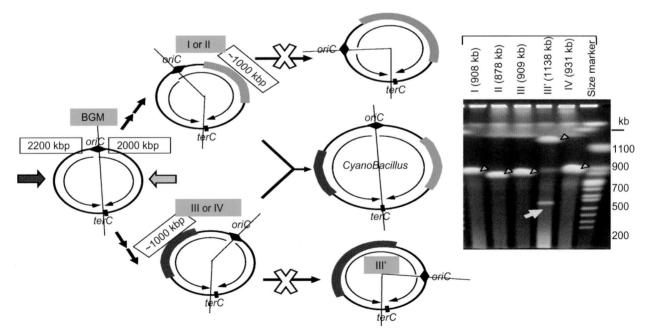

FIGURE 12.3C

Flexible size distribution of *B. subtilis* genome around the *oriC–terC* axis. Symmetry around the *oriC–terC* axis is obstinate to ensure growth. Growth reduction by large DNA insertion (i.e. megacloning: top center) is compensated by DNA insertion into the opposite half to restore symmetry. Right: BGM clones corresponding to I, II, III, and IV in Figure 12.3A were analyzed. Apparent deletion indicated by an arrow is observed from III's, 1138 kbp. It was concluded that guest genome part <1000 kbp displays stable, disregarding the right or left side of BGM. See text.

until our experimental estimations indicated in Figure 12.3C. Unsuspected deletions became obvious when the degree of imbalance exceeded 1000 kbp. Deletions took place in the guest genome part only, with various sizes up to several hundred kbp, disregarding the right or left side of BGM. Accommodation of the whole guest genome, consisting of 3500 kbp, far above the 1000 kbp putative imbalance limit, at one BGM locus was given up. Based on indiscrimination for the right and left side we were encouraged to investigate how the guest genome is divided and distributed at both sides of BGM (Fig. 12.3C). Amazingly and expectedly, megacloning of DNA larger than 1000 kbp was achieved by distributing DNAs into both sides to keep the BGM vector imbalance below 1000-kbp.[3] This empirical and putative rule was effective enough to produce the CB genome of 7.7 Mbp that possessed the *Synechocystis* genome (3.5 Mbp) combined with the *B. subtilis* genome (4.2 Mbp). The CB was the first experimental evidence answering my initial inquiry as to how big the *B. subtilis* genome can become.[25] Can the chimera CB genome function as the *Synechocystis* genome? The present chimera structure for the CB may have disordered the *Synechocystis* genome in terms of *oriC−terC* axis.[26] The origin of DNA replication (*ori*) in this *Synechocystis* PCC6803 genome has not yet been located (personal communications from Watanabe and Yoshikawa).

ARE INDIGENOUS PLASIMD GENES ESSENTIAL FOR CELL GROWTH?

The original *Synechocystis* strain PCC6803 carries seven plasmids, four large plasmids, and three small plasmids.[22] These plasmids, from the smallest (2.4 kbp) to the largest (120 kbp), summed up about 383 kbp which encoded 397 plasmid-borne genes that remain to be added to the CB.[3] Those plasmid-borne genes are presently ignored in CB construction, because of the lack of tools to hold seven plasmids simultaneously in *B. subtilis*.

MULTIPLICITY OF BACTERIAL GENOMES

Cells must perform DNA replication and cell division cycles to ensure faithful inheritance of their genetic materials. The copy number of a bacterial genome per cell in general differs from species to species.[27] One copy genome of familiar strains *E. coli* K − 12 and *B. subtilis* 168 is not a typical state among naturally inhabiting eubacteria. The CB strain is likely a *B. subtilis* derivative possessing one genome per cell (Itaya and Yoshikawa, unpublished). If 10 copies of the genome are essential for *Synechocystis* PCC6803 to grow in a photosynthetic bacterium medium,[22] the increase of CB genome copy number per cell from one to 12 should be an absolute requirement prior to gene expression framework. In contrast to certain genetic regulations on plasmid copy number,[28,29] there is little understanding about how copy number is regulated. Use of origin sequences from a low copy plasmid to produce subgenomes in *B. subtilis* could lead to the discovery of a clue to regulation networks on genome copy number.[30,31] It is worth mentioning here a frequently asked question whether similar megacloning in a reverse order is possible. Is the *B. subtilis* genome able to be accommodated in the *Synechocystis* genome? Probably, it is less plausible because integration of DNA into a multicopy genome (∼10) would be genetically harsh. It seems a major stumbling block, as we experienced,[27] for the host to accommodate and maintain 10 chimera genomes stably.

SEQUENCE FIDELITY OF GUEST GENOMES IN BGM HOST

Recent high-throughput sequencing technologies unveiled the frequent appearance of single nucleotide polymorphisms (SNPs) for even variants from laboratory stock cultures.[32] Beneficial alleles can be selected by adaptation in evolution.[33] One may evade this intrinsic problem by resequencing the DNA, since the time and cost of DNA sequencing became less of a concern recently. Because the sequence of synthesized genome should be confirmed in terms of sequence fidelity, it reminded us of the JCVI report on the first synthetic genome.

As described by Gibson et al.,[6] unsuspected mutations may freely accumulate in proportion to both the increase of the DNA size and the number of cell divisions during prolonged cultivation.

CB has been cultivated only in *B. subtilis* growth conditions. There is no selection pressure on holding the guest genome. One deep concern for the guest genome is fidelity at the nucleotide sequence level of progenies. Because recent sequencing of the CB genome unveiled little traits of insertions, deletions, and SNPS,[34] it was astonishing that large DNA in *B. subtilis* did not suffer mutation incorporations. We are confident that genes required for growth and photosynthesis reactions excluding plasmid-borne genes are not affected.

MITOCHONDRIA AND CHLOROPLAST: ORGANELLE GUEST GENOMES IN BGM

Megacloning of the two whole organelle genomes, mouse mitochondria and rice chloroplast, was implemented.[12] Their sizes, 16.5-kb for the mitochondria DNA that carries only 13 genes and 135.5-kbp for the rice plastid carrying 128 genes, were dramatically smaller if compared with the *Synechocystis* genome as sketched in Figure 12.1. Both organelle genomes were good systems to establish slightly different assembly methods in BGM. The domino method is a one-way cloning and the cloned DNA stays in the *B. subtilis* genome. A retrieval method out of the *B. subtilis* genome is needed to isolate megacloned DNA for further use. More importantly, both organelle genome DNAs were retrieved out of the BGM genome and isolated as a circular form.[12] As illustrated in Figure 12.4, a Bacillus Recombinational Transfer (BReT) method we developed[35–37] was applied. Complete circular organelle genomes freely designed would be valuable to examine systems for direct plastid genome delivery in plant cells. In particular, chloroplast genomes available in the BGM vector[12] should be suitable to compare gene expression profiles with those of *Synechocystis* in CB.

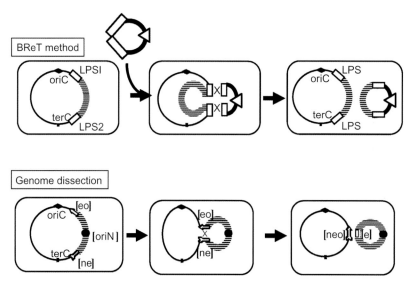

FIGURE 12.4

Genetic methods to retrieve cloned DNA out of the BGM vector. Two methods were developed for the BGM vector. Top: BReT, Bacillus Recombinational Transfer. BReT plasmid possessing an antibiotic resistance marker (open triangle) and LPS1 and LPS2 sequences identical to those in BGM is linearized prior to transformation. Circular BReT plasmid forms only by copy and paste of the intervening DNA region between genomic LPS1 and LPS2 by homologous recombinatons indicated by X. Bottom: The target part to be retrieved is designated by two DNA repeats *ne* and *eo* installed at both ends and a new origin of replication *oriN*. Intrachromosomal homologous recombination between the two repeats is indicated by X producing *neo* and *e*. The creation of *neo*, a complete neomycin resistance gene, allows positive selection for intervening region DNA circularized under *oriN*-dependent replication.

BAC-Mediated Large DNA Shuttling between Plasmid and Genome in BGM

The source of guest DNAs for the BGM vector is not limited to those from eubacteria. In addition to organelle genomes, existing plant and animal genomes are promising as DNA resources and have been shown to be widely available. However, whole plant or animal genomes are too huge to handle as viewed in Figure 12.1. Instead, the BGM system specific for relatively small guest DNAs from plant and animal genomes was developed by use of a prominent *E. coli* cloning vector, the BAC vector.[38] Indeed, DNA libraries constructed in BAC from plant and mouse genomes have greatly contributed to sequence determination and molecular engineering.[39,40] Cloned DNA in BAC that covers long-range genomic regions, normally 100–200 kb, can be modified by genetic recombination systems in the original host *E. coli*.[41] It is already proven that once BAC is transferred into the BGM system, any type of sequence alterations of the BAC insert is possible by virtue of intrinsic natural transformation and associated *recA*-dependent homologous recombination, as shown in Figure 12.2B. The BAC/BGM manipulation system[9,16,42–44] is a novel pipeline to produce transgenic animals and promote a shift for potential BAC users to the BGM system.[44] Existing BAC libraries if transferred to BGM would produce another DNA resource to generate remarkably engineered DNA.[8] The more the BGM is used, the more opportunities and fruitful results are expected. This virtuous cycle should invoke important contemporary issues in a wider range of genome-related biotechnology.[8]

GC Contents of Guests in BGM

Among factors that influence the accommodation of guest genomes into BGM, the content of guanine and cytosine of guest DNA was inspected.[45] As an extremely high G + C content DNA, a megaplasmid pTT27, 250 kbp with GC content (68%) from *Thermus thermophilus* HB8 was attempted to accommodate at one locus of BGM by the domino method. During domino elongation of the pTT27 DNA in the BGM vector, stability of the cloned guest as small as 200 kbp was apparently reduced, which was unexpected based on the case of *Synechocystis* as a guest which allows 1000 kbp 44% G + C content on average.[3,21] The reduced stability was reflected by frequent random deletions within the cloned pTT27 insert,[45] and high GC content seems likely a sole reason. Replication of high GC DNA might be ineffective by *B. subtilis* molecular machinery for DNA replication, supposedly adapted to its own genome (43% GC on average). The cellular dNTP pool might be too small to adapt for a new rounds of high GC DNA synthesis. Molecular mechanisms of the adverse effect of high GC content in the BGM remain to be examined. Alternatively, megacloning of DNA with high AT content DNA could give some clues. Any limits on the BGM cloning system should be tested for broader and more versatile applications. Several guest genomes possessing various GC contents have been chosen to investigate possible restrictions (Itaya, unpublished).

BOTTOM-UP APPROACHES FOR DE NOVO GENOME PRODUCTION

Chemically synthesized oligomers are connected to produce longer DNA in test tubes by use of promising methods, such as Gibson assembly.[46] As chemical synthesis does not require a template DNA, guest genome synthesis can be expanded to nonexisting ones designed even from scratch. A primary obstacle seems to be a lack of the ability to draw complete nucleotide sequence information, and the second obstacle is that of the delivery method of the synthesized DNA to the appropriate chassis. This oversimplified scenario was ascertained by a fascinating epoch-making report from JCVI,[6] where a guest, the modified *Mycoplasma micoides* genome prepared in the yeast host, was delivered to a chassis, a closely related species *Mycoplasma capricorum*. Employment of

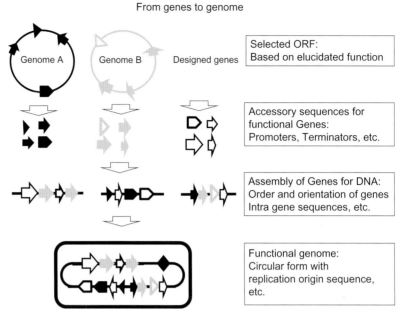

FIGURE 12.5

Flows from genes to genome. Design of a new genome needs a number of factors in addition to the list of genes. Genes are selected from various existing genome resources or *de novo* designed ones on top. Variety of accessory sequence information is needed for these genes to function appropriately in the second step. Relevant genes assembled in one DNA molecule separately in the third step. Finally, combining all the DNA molecules into one circular form with nucleotide sequences in need. Our model is shown in Figure 12.6.

general 'chassis' should be far more valuable and BGM cell might be likely a plausible one (Itaya, unpublished).

Complete Nucleotide Sequence for Designed Genome

There are numerous approaches to reorganize genomes starting with a small number of genes.[47,48] These approaches unveil how to modify certain genes in primary metabolic pathways of interest, and are not expanded to large-scale alteration of host genomes as illustrated in Figure 12.5. Considering that genes with unknown functions still amount to up to 40% of all genes even for most well-studied *E. coli* K-12, genes pertinent to a de novo genome blueprint ought to be limited to genes with known functions. Function unknown genes are conversely in favor of reducing the number of genes and promoters towards a minimal genome construction. For example, approximately 270−350 essential genes for single bacterial life have been estimated.[47,48] The list of essential genes possesses ORF information only. Their promoters, terminators, differences in codon usages, etc. are all neglected, and remain options to draw complete nucleotide sequences of newly designed genomes as illustrated in Figure 12.5. Complete nucleotide sequences in addition to promoters and termination should be properly provided. Additional factors, gene orders, gene orientations, sequences between genes, origin of chromosome replication (*oriC*), termination sequence (*terC*), and location of genes in the genome should be taken into account to draw the complete sequence, as summarized in Figure 12.5. Less attention at present is paid to additional factors such as ribosome RNA genes, GC content, copy number per genome, and multiplicity of genomes per cell.

Smallest Genome versus Minimal Set of Genes

The increasing number and variety of complete genome sequences raised a fundamental question of how many genes are essential for cellular life, and if a genome carrying those essential genes can be obtained. A date-driven concept of minimal genome[49] and minimal set of genes[50] appeared for the first time. It should be emphasized here that a minimal set of genes (MSG) is primarily determined by growth environments. Suffice it to mention that genes for amino acid biosynthesis are regarded as dispensable if abundant amino acids are present in the culture medium. Therefore, in terminology, MSG is not equivalent to a minimal genome that is a real DNA molecule including all information at the sequence level described above. In an extreme and ideal environment where all nutrients are supplied in growth medium at proper concentrations and at constant pH, and temperature, the core set of genes for hypothetical minimal bacterial cell are highlighted. The most straightforward manner for construction of a minimal genome seems to be starting from any existing small genome. Eradication of all the nonessential genes may be the starting point to obtain reduced *Mycoplasma* species genomes.[51] Alternatively, *B. subtilis* is shown to require approximately 300 genes under fixed growth conditions.[52]

COSTS TO SYNTHESIZE GENOMES

Genome synthesis is offering versatile and broad applications for cell engineering never discussed previously. Meanwhile, the state-of-the-art technology still remains at a nascent stage. Genome synthesis is surprisingly expensive in both time and cost. If KEIO and/or JCVI would repeat the guest genome synthesis already published by the same highly skillful staff and materials, costs are roughly 0.5 million US$ per 1000 kbp at a synthesis rate of 1000 kbp/staff/year on average. Synthesis of the familiar *E. coli* K-12 genome (4650 kbp > 1000 kbp × 4) might require 3 years and 3 million US$ for completion. Higher-performance oligomer synthesis and reduced cost should allow applications to various guest genomes.

RELEVANT METHODS TO SUPPORT GENOME SYNTHESIS
Production of Domino DNA by Connection of Small DNA in *B. subtilis*

The bottom-up assembly approach requires a rapid, precise, and efficient method to connect small DNA segments. Connection of many small DNAs in fewer reaction steps ensures a rapid and prerequisite method for bottom-up assembly. A special method to produce through *B. subtilis* was innovated and developed by our group.[17,18] Details of the method, called OGAB, standing for Ordered Gene Assembly in *B. subtilis*, can be separately referred to in these reviews.[53,54]

Rapid preparation of a domino fragment is one of the key factors to reduce the labor cost, as well as assembly time. If domino is made in a BAC vector, the total time to transform *B. subtilis* ought to decrease as indicated earlier. Use of dominos prepared in the BAC of *E. coli* have already been reported.[45] However, inefficiency lies in the step of preparation of high-quality BAC DNA suited for the domino method. The need for a careful purification procedure may be a bottleneck for its general use.[9]

Rapid and Simple Preparation of Dominos in *E. coli* Lysate

Recently, we invented a protocol for rapid preparation of plasmid DNA carried by special *E. coli* by which plasmids are purified at high yield and have no damage at all.[10,11] An *E. coli* strain lysogenic with lambda prophage carrying cI857ts mutation can grow normally at 30°C, but is unable to grow at restricted temperature above 37°C. The strain starts inducing lambda phage and lyses if exposed at a high temperature, for example, at 42°C. In the resulting clear *E. coli* lysate, what are expected include cell debris and degraded DNA, but not the genome and plasmids. To our great surprise, plasmid DNA was found undamaged and stayed stable for certain periods.[10] The amount and quality of plasmids in *E. coli* lysate remained high enough to be used for transformation of *B. subtilis*. This finding led to the simplest recipe for plasmid preparation from *E. coli* ever made; just raise the temperature of growing *E. coli* culture and wait for about 2 hours for *E. coli* to lyse completely.[10] The lysate was directly mixed with a *B. subtilis* competent cell. Because no biochemical reagents are used to lyse *E. coli*, no additional cost was paid other than regular incubation facilities. More valuable, BAC with inserts as large as 100 kbp showed similar high yields and quality to those purified through traditional biochemical methods.[11] We believe the method should be a useful tool for the BAC domino method in the near future.[11]

It is noteworthy to mention a brief history of lambda lysogen use. After completion of cloning of the first guest genome, a 48.5-kbp lambda genome in a BGM prototype in 1995,[19,55] my next attempt was to modify the guest lambda DNA directly by lysing *E. coli* cells infected by lambda phages. It led to a discovery not of lambda DNA, but of stable plasmid DNA in lambda-induced *E. coli* lysate. The finding of stable plasmids in *E. coli* lysate contradicted the prevailing common sense that degraded *E. coli* cell hold no complete *E. coli* genome and plasmid. These DNAs should be rapidly degraded because of no protection mechanism from nucleases abundant in cells. The major nuclease-deficient mutants of *E. coli* that prolong plasmid stability were investigated to widen available *E. coli* strains as practical tools.[56] A related protocol where prompt lysis by virulent lambda phage infection allows almost all unmodified *E. coli* transformants to be used, is promising (Kaneko et al., manuscript in preparation).

Dissection of the Guest Genomes out of the Host Genome

Separation of the guest genome out of the BGM vector is vital to deliver the DNA into the cell. A method to copy-out the target sequences in BGM and move to *B. subtilis* plasmids was briefly mentioned earlier and in Figure 12.4. Plasmid parts are being improved to carry larger DNAs.[57] Another method to separate target sequences of the *B. subtilis* genome was examined.[30,31] As mentioned briefly above and illustrated in Figure 12.4, two short identical sequences are inserted at both ends of the target genome sequence. The *recA*-dependent intrachromosomal homologous recombination between the short sequences precedes circularization and physical separation of the intervening genome segment. The separated segment had to possess an origin of DNA replication (*ori*) different from the chromosomal *oriC* to replicate independently in *B. subtilis*. Although only one case is reported to date, where the wild-type circular *B. subtilis* genome (4200 kbp) dissected into two circular parts, 300- and 3900-kbp without losing cell viability,[30] future generation of BGM vectors is being planned to employ the gadget intrinsically. Any parts of guest genomes would be prepared as dissected DNA in a circular form appropriate to be directly posed into the chassis. In particular, choice of *ori* from various plasmids with different multiplicity would provide clues to unveil genome multiplicity.[31]

Long-Term Storage of Valuable Genome Resources in BGM

Another BGM-related technology less argued is long-term storage of parts or complete forms of guest genomes. Small DNA can be readily synthesized, and therefore there is no need to store them. In contrast, reconstructed guest genomes are indispensable because they are not readily reproduced. Not only complete guest genomes, but also their intermediate DNA during reconstructions in BGM vectors, are also valuable DNA resources for future innovations. A BGM vector literally derived from *B. subtilis* is capable of forming spores. Spores survive for a long time under many hazardous environments including dryness. *B. subtilis* spores start to germinate instantly when exposed in nutrition-rich environments. Stability of certain guest DNAs inside spores for years is being confirmed by us ([9], and our unpublished observations). BGM spores should be inevitable as a low-cost, long-term reservoir of genomes demanding no special equipment and facilities.

FUTURE PERSPECTIVES OF BGM SYSTEMS ACCRUED FROM THE PRESENT ACHIEVEMENTS

Current technological barriers in handling between genes and genomes will be lowered or removed in 10 years. Pipelines of certain modifications on existing genomes and innovation of general chassis will be expected. In 20 years, guest genomes would be gradually expanded from: (1) existing beneficial microbial genomes for further modification to those of (2) VNC (Viable Not Culturable) microbes in natural environments; (3) large-sized modified DNA to be delivered to unicellular eukaryotes, algae, or fungi, and higher multicellular eukaryotes such as mammals or plants; and (4) designed ones in silico for basic researches and future applications.

(1) Designed microbes based on existing ones should be of great use for basic research and industrial applications. Purpose-oriented genomes can be organized and produced without the need of template DNA if whole nucleotide sequence information is available.

(2) VNC microbes are considered major inhabitants in natural environments and therefore tremendous untouched guest genome resources.[58] Once their complete genome sequences are determined by, for example, metagenomic procedures,[59,60] they are prospectively constructed starting from chemical oligomers which would lead to unprecedented challenges. By seeking the reason why they do not grow in laboratory conditions, and investigating how to convert them to be culturable in the laboratory, those huge natural resources presently unnoticed are desired for future wider and unpredicted applications in various fields of life sciences.

(3) Genome delivery technology should not be limited to the case of bacteria microbes. Handling protocols for microbial guest genomes described in this chapter are readily applied to far-larger parts of mammal and plant genomes than ever. The large DNA would dramatically require improved protocols in the delivery process and create target regions previously thought to be very hard. Although we do not draw concrete pipelines and time schedules on this issue, once the DNA size limit is cleared, applications should be open to any cell types.

(4) Genome is the ultimate molecule manipulation which reorganizes and regulates global cellular activities. De novo genome design should be a combined task of a number of on-going cutting-edge technologies in various areas. Our strategy starting from bottoming-up gene clusters illustrated in Figure 12.6 is one of many approaches. Guest genomes synthesized with perfect match to the blueprint made in silico should provide initial kick-off strain only. Evolutionary ideas and selection methods should be included at certain stages to produce desired ones more rapidly.[32,33]

SUMMARY

Recent next generation genome-sequencing technologies are being applied to an unlimited number of species on the Earth, which provides various directions of research and applications in synthetic biology.[2,7,61–64] In contrast, full engineering of genomes even from bacteria, far smaller than those of higher eukaryotes, are very limited. The state-of-the-art technology remains immature for a wide range of applications due to the difficulty of daily handling and high cost of producing correct genomes. Genome synthesis/cloning technologies discussed in this chapter have proven that entire bacterial genomes and even the smallest eukaryote genome, *S. cerevisiae*, can be synthesized from DNA oligomers.[65] *De novo* synthesis of functional genomes from scratch, frequently mentioned as the ultimate motivation of genome synthesis, remains the greatest challenge for the future. The rational design of bacterial genomes requires rational sequence-writing disciplines to draw complete nucleotide sequences. Practical and solid bacterial genome design for applications and

239

Category of function	Number	Clustered DNA with single promoter
DNA metabolism	27	
RNA metabolism	14	
Protein synthesis	95	
Cell envelope	44	
Cell shape and division	8	
Glycolysis	8	
Respiratory pathways	22	
Nucleotides	10	
Cofactors	15	
Other	15	
Unknown	11	
Total	271	

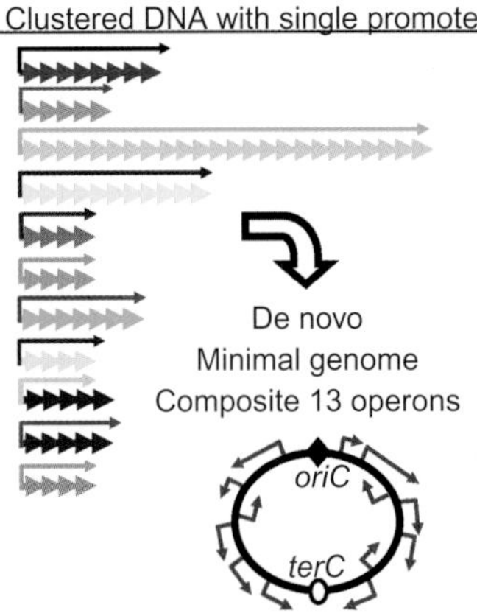

FIGURE 12.6

A scenario to produce a novel genome with a few promoters. A minimum set of genes elucidated on *B. subtilis*[52] is listed on the left by function category. Relevant genes in each category if assembled under a promoter shown in the center might function as an operon (Tsuge and Itaya, unpublished). Assembly of these operons in a circular form should result in a genome with fewer promoters and accordingly less complicated regulation.

practical use becomes more complex when an increased number of genes have to be handled.[47,48]

Here is a good opportunity to address my motivation to advance the entire genome cloning. As soon as I noticed plausible rough borders on genome size between eukaryotes and prokaryotes, as well as prokaryotes and noncellular DNA as indicated in Figure 12.1, an experiment was designed to prove how big the *B. subtilis* genome, at present 4.2 Mbp,[20] can become. About 10 years later, the genome size increased to 7.7 Mbp, still in a circular form.[3] I've learned many fundamental rules underlying the circular genome structure described in this chapter through DNA assembly in the BGM system. In order for the size of the *B. subtilis* genome to approach and exceed the yeast genome size, certain large DNA, neutral in terms of gene expression influencing host *B. subtilis*, should be megacloned. Use of the yeast genome might be a plausible neutral DNA option.[25]

Aside from the personal ideas for innovation, the multipurpose BGM system will play significant roles in versatile gene/genome production and delivery steps. The huge and solid promising framework is able to aim at not only microbial breeding, but also cell engineering and genome engineering in all disciplines of life sciences.

References

1. Itaya M. In: Pengcheng F, Latterich M, Panke. S, eds. *Systems Biology and Synthetic Biology.* John Wiley & Sons, Inc.; 2009:155.

2. Cambray G, Mutalik VK, Arkin AP. Toward rational design of bacterial genomes. *Curr Opin Microbiol.* 2011;14:624—630.

3. Itaya M, Tsuge K, Koizumi M, Fujita K. Combining two genomes in one cell: stable cloning of the Synechocystis PCC6803 genome in the *Bacillus subtilis* 168 genome. *Proc Natl Acad Sci USA.* 2005;102:15971—15976.

4. Gibson DG, Benders GA, Andrews-Pfannkoch C, Denisova EA, Baden-Tillson H. Complete chemical synthesis, assembly, and cloning of a *Mycoplasma genitalium* genome *Science.* 2008;319:1215—1220.

5. Tagwerker C, Dupont CL, Karas BJ, Ma L, Chuang RY. Sequence analysis of a complete 1.66 Mb *Prochlorococcus marinus* MED4 genome cloned in yeast. *Nucleic Acids Res.* 2012;40(20):10375—10383.

6. Gibson DG, Glass JI, Lartigue C, Noskov VN, Chuang RY, Algire MA. Creation of a bacterial cell controlled by a chemically synthesized genome. *Science.* 2010;329:52—56.

7. Nandagopal N, Elowitz MB. Synthetic biology: integrated gene circuits. *Science.* 2011;333:1244—1248.

8. Kaneko S, Itaya M. Production of multi-purpose BAC clones in the novel *Bacillus subtilis* based host systems. In: Chatterjee P, ed. *Bacterial Artificial Chromosomes.* InTech open. 2011:119—136.

9. Kaneko S, Akioka M, Tsuge K, Itaya M. DNA shuttling between plasmid vectors and a genome vector: Systematic conversion and preservation of DNA libraries using the *Bacillus subtilis* genome (BGM) vector. *J Mol Biol.* 2005;349:1036−1044.

10. Kaneko S, Itaya M. Designed horizontal transfer of stable giant DNA released from *Escherichia coli. J Biochem.* 2010;147:819−822.

11. Itaya M, Kaneko S. Integration of stable extra-cellular DNA released from *Escherichia coli* into the *Bacillus subtilis* genome vector by culture mix method. *Nucleic Acids Res.* 2010;38:2551−2557.

12. Itaya M, Fujita K, Kuroki A, Tsuge K. Bottom-up genome assembly using the *Bacillus subtilis* genome vector. *Nat Methods.* 2008;5:41−43.

13. Chen I, Christie PJ, Dubnau D. The ins and outs of DNA transfer in bacteria. *Science.* 2005;310 (5753):1456−1460.

14. Kidane D, Graumann PL. Intracellular protein and DNA dynamics in competent *Bacillus subtilis* cells. *Cell.* 2005;122:73−84.

15. Maniatis T, Fritsch EF, Sambrook J. *Molecular Cloning: A Laboratory Manual.* Cold Spring Harbor, NY: Cold Spring Harbor Laboratory; 1982.

16. Itaya M, Shiroishi T, Nagata T, Fujita K, Tsuge K. Efficient cloning and engineering of giant DNAs in a novel *Bacillus subtilis* genome vector. *J Biochem.* 2000;128:869−875.

17. Itaya M, Tsuge K. Construction and manipulation of giant DNA by a genome vector. *Methods Enzymol.* 2011;498:427−447.

18. Itaya M, Tsuge K. Chapter 9: towards bacterial genome reconstruction: comparison of expression profiles of genes from different cellular DNA vehicles, plasmids and genomes. In: Sadaie, Matsumoto, eds. Escherichia coli and Bacillus subtilis: *The Frontiers of Molecular Microbiology Revisited.* Research Signpost 2012.

19. Itaya M. Toward a bacterial genome technology: integration of the *Escherichia coli* prophage lambda genome into the *Bacillus subtilis* 168 chromosome. *Mol Gen Genet.* 1995;248:9−16.

20. Kunst F, Ogasawara N, Dunchin A, et al. The complete genome sequence of the Gram-positive bacterium *Bacillus subtilis. Nature* 1997;390:249−256.

21. Kaneko T, Sato S, Kotani H, et al. Sequence analysis of the genome of the unicellular cyanobacterium Synechocystis sp. strain PCC6803. II. Sequence determination of the entire genome and assignment of potential protein-coding regions. *DNA Res.* 1996;3:109−136.

22. Labarre J, Chauvat F, Thuriaux P. Insertional mutagenesis by random cloning of antibiotic resistance genes into the genome of the cyanobacterium *Synechocystis* strain PCC6803. *J Bacteriol.* 1989;171:3449−3457.

23. Kitahara K, Miyazaki K. Specific inhibition of bacterial RNase T2 by helix 41 of 16 S ribosomal RNA. *Nat Commun.* 2011;2:549.

24. Katayama T, Ozaki S, Keyamura K, Fujimitsu K. Regulation of the replication cycle: conserved and diverse regulatory systems for DnaA and oriC. *Nat Rev Microbiol.* 2010;8:163−170.

25. Itaya M. News & views. *Nat Biotechnol.* 2010;28:687−689.

26. Kono N, Arakawa K, Tomita M. Validation of bacterial replication termination models using simulation of genomic mutations. *PLoS One.* 2012;7(4):e34526.

27. Ohtani N, Tomita M, Itaya M. An extreme thermophile, *Thermus thermophilus,* is a polyploid bacterium. *J Bacteriol.* 2010;192:5499−5505.

28. Cervantes-Rivera R, Pedraza-López F, Pérez-Segura G, Cevallos MA. The replication origin of a repABC plasmid. *BMC Microbiol.* 2011;11:158−171.

29. Llosa M, Schröder G, Dehio C. New perspectives into bacterial DNA transfer to human cells. *Trends Microbiol.* 2012;20(8):355−359.

30. Itaya M, Tanaka. T. Experimental surgery to create subgenomes of *Bacillus subtilis* 168. *Proc Natl Acad Sci USA.* 1997;94:5378−5382.

31. Itaya M, Tanaka T. Fate of unstable *Bacillus subtilis* subgenome: reintegration and amplification in the main genome. *FEBS Lett.* 1999;448:235−238.

32. Hindre T, Knibbe C, Beslon G, Schneider D. New insights into bacterial adaptation through in vivo and in silico experimental evolution. *Nat Rev Microbiol.* 2012;10:352−365.

33. Woods RJ, Barrick JE, Cooper TF, Shrestha U, Kauth MR, Lenski RE. Second-order selection for evolvability in a large *Escherichia coli* population. *Science.* 2011;331:1433−1436.

34. Watanabe S, Shiwa Y, Itaya M, Yoshikawa H. Complete sequence of the first chimera genome constructed via cloning of a whole genome from Synechocystis PCC6803 into the *Bacillus subtilis* 168 genome. *J Bacteriol.* 2012;194.

35. Tsuge K, Itaya M. Recombinational transfer of 100-kb genomic DNA to plasmid in *Bacillus subtilis* 168. *J Bacteriol.* 2001;183:5453−5456.

36. Tomita S, Tsuge K, Kikuchi Y, Itaya M. Targeted isolation of a designated region of the *Bacillus subtilis* 168 genome by recombinational transfer. *Appl Environ Microbiol.* 2004;70:2508−2513.

37. Tomita S, Tsuge K, Kikuchi Y, Itaya M. Regional dependent efficiency for recombinational transfer of the *Bacillus subtilis* 168 genome. *Biosci Biotech Biochem.* 2004;68:1382−1384.

38. Shizuya H, Birren B, Kim UJ, et al. Cloning and stable maintenance of 300-kilobase-pair fragments of human DNA in *Escherichia coli* using an F- factor-based vector. *Proc Natl Acad Sci USA.* 1992;89:8794−8797.

39. Osoegawa K, Mammoser AG, Wu C, Frengen E, Zeng C, Catanese JJ. A bacterial artificial chromosome library for sequencing the complete human genome. *Genome Res.* 2001;11:483−496.

40. Frengen E, Weichenhan D, Zhao B, Osoegawa K, van Geel M. A modular, positive selection bacterial artificial chromosome vector with multiple cloning sites. *Genomics.* 1999;58:250−253.

41. Hardy S, Legagneux V, Audic Y, Paillard L. Reverse genetics in eukaryotes. *Biol Cell.* 2010;102:561−580.

42. Kaneko S, Tsuge K, Takeuchi T, Itaya M. Conversion of submegasized DNA to desired structures using a novel *Bacillus subtilis* genome vector. *Nucleic Acids Res.* 2003;31(18):e112.

43. Kaneko S, Takeuchi T, Itaya M. Genetic connection of two contiguous bacterial artificial chromosomes using homologous recombination in *Bacillus subtilis* genome vector. *J Biotechnol.* 2009;139:211−213.

44. Tomita S, Tsuge K, Kikuchi Y, Itaya, M. Regional dependent efficiency for recombinational transfer of the *Bacillus subtilis* 168 genome. *Biosci Biotech Biochem.* 2004;68:1382−1384.

45. Ohtani, N., Hasegawa, M., Sato, M., Tomita, M., Kaneko, S. and Itaya, M. Serial assembly of thermus megaplasmid DNA in the genome of *Bacillus subtilis* 168: A BAC-based domino method applied to DNA with high GC content. *Biotechonol J.* 2012;7(7):867−876.

46. Gibson DG. Enzymatic assembly of overlapping DNA fragments. *Methods Enzymol.* 2011;498:349−361.

47. Gerdes S, Edwards R, Kubal M, Fonstein M, Stevens R, Osterman A. Essential genes on metabolic maps. *Curr Opin Biotechnol.* 2006;17:448−456.

48. Juhas M, Eberl L, Glass JI. Essence of life: essential genes of minimal genomes. *Trends Cell Biol.* 2011;21:562−568.

49. Itaya M. An estimation of minimal genome size required for life. *FEBS Lett.* 1995;362:257−260.

50. Mushegian AR, Koonin EV. A minimal gene set for cellular life derived by comparison of complete bacterial genomes. *Proc Natl Acad Sci USA.* 1996;93:10268−10273.

51. Glass JI, Hutchison III CA, Smith HO, Venter JC. A systems biology tour de force for a near-minimal bacterium. *Mol Syst Biol.* 2009;5:330.

52. Kobayashi K, Ehrlich SD, Albertini A, et al. Essential *Bacillus* subtilis genes. *Proc Natl Acad Sci USA.* 2003;100:4678−4683.

53. Tsuge K, Matsui K, Itaya M. One step assembly of multiple DNA fragments with designed order and orientation in *Bacillus subtilis* plasmid. *Nucleic Acids Res.* 2003;31(21):e133.

54. Tsuge K, Matsui K, Itaya M. Production of the non-ribosomal peptide plipastatin in *Bacillus subtilis* regulated by 3 relevant gene blocks assembled in a single movable DNA segment. *J Biotechnol.* 2007;129:592−603.

55. Itaya M. Genetic transfer of large DNA inserts to designated loci of the *Bacillus subtilis* 168 genome. *J Bacteriol.* 1999;181:1045−1048.

56. Itaya M, Kawata Y, Sato M, et al. A simple method to provide a shuttling plasmid for delivery to other host ascertained by prolonged stability of extracellular plasmid DNA (excpDNA) released from *Escherichia coli* K12 endA mutant, deficient in major endonuclease. *J Biochem.* 2012;152(6):501−504.

57. Kuroki A, Ohtani N, Tsuge K, Tomita M, Itaya M. Conjugational transfer system to shuttle giant DNA cloned by the *Bacillus subtilis* genome (BGM) vector. *Gene.* 2007;399:72−80.

58. Keller M, Zengler K. Tapping into microbial diversity. *Nat Rev Microbiol.* 2004;2:141−150.

59. http://dels-old.nas.edu/metagenomics/overview.shtml

60. Wooley JC, Godzik A, Friedberg I. A primer on metagenomics. *PLoS Comput Biol.* 2010;6:e1000667.

61. Zhu L, Zhu Y, Zhang Y, Li Y. Engineering the robustness of industrial microbes through synthetic biology. *TIMS.* 2012;20:94−101.

62. Ray JC, Tabor JJ, Igoshin OA. Non-transcriptional regulatory processes shape transcriptional network dynamics. *Nat Rev Microbiol.* 2011;9:817−828.

63. Isaacs FJ, Carr PA, Wang HH, et al. Precise manipulation of chromosomes in vivo enables genome-wide codon replacement. *Science.* 2011;333(6040):348−353.

64. Blount ZD, Barrick JE, Davidson CJ, Lenski RE. Genomic analysis of a key innovation in an experimental *Escherichia coli* population. *Nature.* 2012;489(7417):513−518:27

65. Dymond JS, Richardson SM, Coombes CE, et al. Synthetic chromosome arms function in yeast and generate phenotypic diversity by design. *Nature.* 2011;477(7365):471−476.

Synthetic Microbial Consortia and their Applications

Robert P. Smith, Yu Tanouchi and Lingchong You
Duke University, Durham, NC, USA

INTRODUCTION

A major goal of synthetic biology is to predictably engineer novel behaviors in cells. To date, most efforts have been focused on engineering single populations of *Escherichia coli*, yeast, and mammalian cells. Behaviors engineered in single populations of cells include, but are not limited to, switches,[1−3] oscillators,[4−8] logic gates (e.g.[9]) and transcriptional cascades (e.g.[10]).

It has become increasingly recognized that programming interactions between multiple populations represents a new frontier in synthetic biology, in terms of technical challenge and the potential implications of the resulting systems. Such multipopulation systems, often termed consortia, have some potential advantages over systems consisting of one population. First, a well-established consortium, such as a biofilm, is likely to be more resistant to invasion by other microbial species.[11] Second, a properly defined mixed population can survive transient nutrient deprivation better when different subpopulations share essential metabolites.[12] Third, by implementing cooperating circuits in two populations, a task that has a large metabolic burden can be shared, thus dividing the labor between populations.[13] As such, unique behaviors can be realized in a consortium that would be difficult to engineer in a single population. Finally, the study of synthetic consortia may lead to novel insights into ecological and evolutionary relationships that are difficult to study in a natural setting.[14]

In this chapter, we review recent examples of synthetic microbial consortia. We describe the unique gene circuits that allow engineered consortia to either cooperate to perform a task or to compete against one another. Furthermore, we discuss how the spatial structure of a consortium can lead to a novel behavior. We touch upon recent examples of how such consortia may have relevance in industrial applications and in medicine. Finally we outline current and future challenges in engineering consortia.

COMMUNICATION IN SYNTHETIC MULTICELLULAR SYSTEMS

A critical requirement for engineering microbial consortia is programmed communication between populations. Communication can be implemented using small diffusible

243

Synthetic Biology. DOI: http://dx.doi.org/10.1016/B978-0-12-394430-6.00013-3

molecules, including quorum sensing (QS) signals or essential metabolites, or using secreted enzymes. In a natural setting, QS is used by bacteria to determine the density of their population.[15] Each cell in the population synthesizes and secretes a small diffusible molecule (i.e. the QS signal), the concentration of which increases with cell density. At a sufficiently high concentration, the QS signal can activate expression of target genes that allow the population to respond to increasing cell density. Synthetic biology takes advantage of this property inherent to QS; activation of a gene circuit under the regulation of QS-activated promoter only occurs at a sufficiently high bacterial density (Fig. 13.1a).

Communication can also be realized by using diffusible metabolites.[13,16] Here, each cell population cannot synthesize a required metabolite. These auxotrophic strains can be engineered by replacing a gene required for the synthesis of the metabolite with an antibiotic-resistant marker (e.g.[17]). An auxotroph cannot grow in a minimal medium lacking the corresponding essential metabolite. As such, these strains require additional supplementation of the metabolite to survive, which may be supplied by a cocultured population of engineered bacteria. This dependence on a partner provides a means to enable obligate cooperation, without which a consortium is not viable.

Finally, communication can also be realized using secreted enzymes. In contrast to small molecules, enzymes often require additional circuitry for them to be secreted and shared

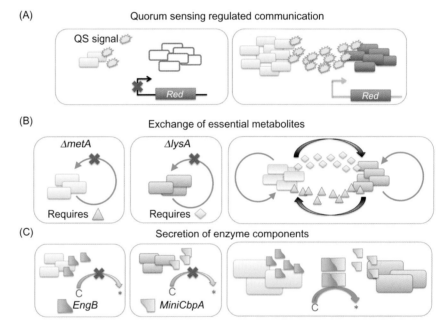

FIGURE 13.1

Communication between populations in a synthetic consortium. (A) Quorum sensing (QS) systems. One population (green) secretes a density-dependent QS signal (yellow shapes). A second population contains QS responsive promoter that drives expression of gene that turns the cells red. When the green population is at low density (left panel) insufficient QS signal is produced, and thus the 'red' gene in the second population is not expressed. In contrast, when a high density of the green population is present, sufficient concentration of QS signal is synthesized, driving expression of the red gene (right panel). (B) Exchange of essential metabolites. Two auxotrophic populations,[22] one that cannot synthesize methionine (purple triangles, green cells), and one that cannot synthesize lysine (blue diamonds, red cells) cannot grow (green arrow) in monoculture in minimal medium. However, when cocultured as a consortium, both populations exchange these essential metabolites and thus grow. (C) Exchange of enzymatic components. Two populations are engineered to synthesize and secrete components of an enzymatic pathway.[23] The green population synthesizes and secretes EngB (blue shapes) while the red population synthesizes and secretes MiniCbpA (orange shapes). When grown in monoculture, the hydrolysis (*) of cellulose (C) does not occur. However, when grown in coculture, the secreted EngB and MiniCbpA form an enzymatic complex, which allows the hydrolysis of cellulose.

between populations. Once secreted into the culture, the enzymes from each population can act in concert to allow a cooperative task to be completed. The advantage of an enzyme-based system is that it can combine communication and production of a useful molecule via the secreted enzymes.

ENGINEERING UNIDIRECTIONAL COMMUNICATION

Early examples of gene circuits that employ QS elements have focused on communication between cells within a single population,[18] or unidirectional communication between two populations. In 2004, Basu et al. implemented a QS-based circuit to control the spatiotemporal expression of genes.[19] Their system consisted of two *E. coli* populations: a sender population and a receiver population. In the sender population, the LuxI protein, which synthesizes 3-oxohexanoyl-homoserine lactone (3OC6HSL), is under the control of a TetR-repressible promoter ($P_{LtetO-1}$). Upon induction of $P_{LtetO-1}$ by anhydrotetracycline (aTc), 3OC6HSL is synthesized and accumulates in the culture. The receiver population constitutively expresses a 3OC6HSL receptor protein, LuxR. At a sufficiently high concentration, 3OC6HSL activates LuxR, which in turn activates expression of the CI repressor protein from λ phage, which is under control of the cognate promoter ($luxP_R$) of LuxR. Finally, the circuit readout, a green fluorescent protein (GFP), is under the regulation of a dual promoter ($luxP_RcI-O_R1$), which is repressed by the CI repressor protein, but activated by active LuxR.

The logic of this circuit is as follows: addition of aTc leads to the synthesis of 3OC6HSL from the sender population. As the concentration of 3OC6HSL builds up in the culture, it binds to LuxR and activates expression of both CI and GFP in the receiver population. However, once CI reaches a sufficiently high concentration in the cell, it represses GFP expression. Overall, this circuit is expected to generate a transient, pulsatile response in GFP expression in the receiver population.

To test the receiver circuit function, the authors inoculated the receiver cells in a liquid culture, applied various concentrations of exogenously added 3OC6HSL, and examined GFP expression using cell sorting. Indeed, the receiver cells generated a transient pulse of GFP. The amplitude and length of the GFP expression could be modulated by the amount of exogenously added 3OC6HSL or the rate increase of 3OC6HSL. In the latter property, increasing the rate of 3OC6HSL concentration led to higher peak expression of GFP. As such, the authors surmised that, when the sender and receiver cells were coinoculated on solid medium (i.e. the spatial domain), the distance between the sender and receiver populations would lead to a differential response in GFP expression. Here, the accumulation rate of 3OC6HSL in the culture would depend upon the distance from the sender population. Indeed, when both populations were seeded on an agar plate, the authors demonstrated that receiver cells that were closer to the sender population had a stronger pulse of GFP expression than the receiver cells farther away from the sender population.

In 2005, Basu et al. used the same sender population and a redesigned receiver population to generate a consortium able to generate robust spatial patterns.[20] Again, the receiver cells constitutively express LuxR. They also express the CI repressor and a modified LacI repressor ($LacI_{M1}$), both driven by 3OC6HSL responsive P_{luxR} promoter, as well as the wild-type LacI repressor under control of a CI-repressible promoter (λP(R-O12)). As an output, the expression of GFP is driven by a P_{lac} promoter, which is repressed by either $LacI_{M1}$ or LacI.

Using a mathematical model, the authors predicted that the circuit could generate ring patterns of fluorescent protein expression in receiver cells in response to an AHL gradient generated by sender cells inoculated at a spot. Their model predicted that receiver cells sufficiently far from the sender cells would not have the LuxR regulated circuit activated due to a low concentration of 3OC6HSL. Here, the wild-type LacI repressor would repress GFP expression. In contrast, at a sufficiently close distance to the sender cells, the high concentration of 3OC6HSL in the medium would result in the expression of both the $LacI_{M1}$ and CI repressors. Here, $LacI_{M1}$ will directly repress GFP, while CI would repress expression of the wild-type LacI repressor. However, at an intermediate distance from the sender cells, the concentration of 3OC6HSL would be sufficient to trigger the repression of wild-type LacI repressor by the CI repressor, but not the repression of GFP by the $LacI_{M1}$ repressor, as the CI repressor is a more potent repressor of expression when compared to $LacI_{M1}$. Overall, this would result in suppression of the wild-type LacI repressor only, which in turn would allow expression of GFP. On the whole, the circuit would result in high expression of GFP from the receiver cells at an intermediate distance from the sender cells.

To control the upper 3OC6HSL threshold at which GFP expression was repressed, the authors engineered three variants of the 'high detect' circuit, which determines the 3OC6HSL concentration at which GFP expression was repressed, inside the receiver cells. Each different 'high detect' circuit contains a LuxR protein with a different sensitivity threshold to the 3OC6HSL signal. When combined with the 'low detect' circuit, which determines the lowest concentration of 3OC6HSL that drives GFP expression, the authors created three receiver cell strains: BD1 (high sensitivity to 3OC6HSL); BD2 (wild-type sensitivity to 3OC6HSL, which was labeled with a red fluorescent protein (RFP), instead of GFP); and BD3 (low sensitivity to 3OC6HSL).

Using simulations, the authors predicted that by combining various BD receiver strains, and the sender population, they could produce different patterns. Guided by modeling, the authors experimentally created a bulls-eye pattern by plating a mixture of BD3 and BD2 cells on an agar plate and by placing a disk containing sender cells in the middle. The BD3 cells formed a GFP ring near the sender population, whereas the BD2 cells produced a red ring farther away. Similarly, by coculturing BD1 and BD2 cells on an agar plate, and in the presence of sender cells, a GFP ring formed outside of the RFP ring. By varying the positioning of the sender cell population within a lawn of receiver cells, the authors generated a wide variety of patterns including a four-leaf clover, an ellipse, and a heart.

ENGINEERED COOPERATION IN SYNTHETIC MICROBIAL CONSORTIA

Auxotrophic strains have often been used to create cooperative synthetic consortia (e.g.[21] which will be discussed later). Recently, Wintermute and Silver examined the properties that led to efficient metabolite exchange, and thus cooperation, between pairs of auxotrophic strains.[22] The authors tested 46 engineered, conditional lethal *E. coli* auxotrophs to determine if coculturing pairs of auxotrophs could lead to cooperation. Each auxotrophic strain carries a mutation that abolishes its ability to synthesize an essential metabolite. The metabolites range from amino acids to enzymes involved in cellular respiration. When pairs of auxotrophs were cocultured, a subset of the cocultured strains resulted in varying degrees of cooperation. For example, when an auxotrophic population lacking *panD* (required for biosynthesis of pantothenate) and an auxotrophic population lacking *proC* (required for proline biosynthesis) were grown in coculture, the *panD* auxotroph grew considerably, whereas the *proC* auxotroph did not. In contrast, when a *metA* auxotroph (required for methionine biosynthesis) population and a *lysA* auxotroph (required for lysine biosynthesis) population were cocultured, both populations grew markedly (Fig. 13.1b). A single 'universal cooperator' could not be identified.

That is, not one specific auxotroph was able to form a cooperative relationship with all other auxotrophs. Interestingly, pairings between auxotrophs that contain mutations in the same biosynthetic pathway grew significantly less than pairings between auxotrophs that contain mutations in distinct metabolic pathways. Using mathematical analysis, the authors determined that pairs of auxotrophs grow the most when their growth requirement is fulfilled using small quantities of a metabolite that are inexpensive for their growth partner to produce and share.

Another approach to engineering communication is to use secretion of the individual components of an enzymatic system. Arai et al. engineered three populations of *Bacillus subtilis* to secrete the enzymatic components of an endoglucanase or a xylanase from *Clostridium cellulovorans*.[23] One strain was constructed to secrete a HIS-tagged MiniCbpA, which serves as a scaffold protein that binds to and localizes enzymes that contain dockerin domains. Two additional strains were created to express and secrete two cellulosomal enzymes: EngB, which has endoglucanase activity (i.e. hydrolysis of cellulose); or XynB, which has xylanase activity (i.e. hydrolysis of xylose). Both of these proteins contain dockerin domains, and as such could readily bind to the MiniCbpA protein. Initially, the authors determined that each enzyme component could be readily expressed and secreted into the growth medium. Next, they cocultured two *B. subtilis* populations, the MiniCbpA and XynB-expressing populations, or the MiniCbpA and EngB-expressing populations (Fig. 13.1c), to determine if functional enzyme complexes could be formed in the medium. By isolating HIS-tagged proteins from the medium, the authors observed that coculture of these populations led to the formation of protein complexes between MiniCbpA and XynB or MiniCbpA and EngB, both of which retained xylanase or endoglucanase activity, respectively.

While many synthetic consortia have used either small molecules or enzymes to enable communication, a recent study has combined both methods to program novel dynamics in a consortium. Goldberg et al. engineered a synthetic consortium where two populations can detect, communicate, and chemotax in concert towards two substrates.[24] The authors engineered two *E. coli* strains that could respond to either asparagine or phenylacetyl glycine (PAG), both of which are nonnatural chemotaxic agents in *E. coli*. To construct an asparagine sensing strain, the authors engineered one strain to constitutively express and export the enzyme asparaginase II (*ansB* gene), which converts asparagine into aspartate, into the periplasm. Furthermore, they engineered this strain to express an aspartate receptor, Tar, that, when complexed with aspartate, results in chemotaxis towards asparagine. To construct a PAG-sensing strain, the authors placed penicillin lyase (*pac*), which converts PAG to produce PAA, under the regulation of a constitutive promoter. To sense the presence of PAG, the authors created a Tar receptor variant (TarPA) that could bind to PAA and activate chemotaxis towards PAG.

When the authors experimentally tested each strain individually, both strains readily chemotaxed toward their respective substrates. However, to create a consortium, the authors swapped the receptors (i.e. Tar and TarPA) between the cell strains, creating two novel strains. Here, one strain contains Tar and *pac* (i.e. chemotaxes towards asparagine but hydrolyzes PAG to PAA), and the other contains TarPA and *ansB* (i.e. chemotaxes towards PAA but hydrolyzes asparagine into aspartate). As such, when the authors grew either strain in monoculture, neither strain performed chemotaxis towards an attractant, as each strain lacked the ability to both 'sense' (i.e. hydrolyze a product) and respond to an attractant. However, when the two strains were grown together on soft agar, the consortium readily chemotaxed towards asparagine and PAG. As such, the enzyme products produced by one strain activated chemotaxis in the other. The authors noted that both attractants were required in order to result in chemotaxis as the consortium would not chemotax when only one attractant was present.

247

PROGRAMMING ANTAGONISTIC INTERACTIONS BETWEEN POPULATIONS

In addition to cooperation between populations, microbial consortia have been designed based on antagonistic interactions. Engineering of such consortia has led to the study of ecologically and evolutionary relevant phenomena, while also serving as a platform to advance the ability to engineer complex behaviors. Balagaddé et al. constructed a synthetic predator—prey system consisting of two *E. coli* strains that communicate bidirectionally using two QS systems.[25] Each strain harbors a QS signal synthase gene (*lasI* for predator and *luxI* for prey) and a cognate receptor protein for the QS signal produced by the other strain (LuxR for predator and LasR for prey). As such, the receptor protein in one strain is activated when the other strain is present at sufficiently high density. The logic of the predation, where the predator kills the prey and the prey helps the predator's survival and growth, is realized by utilizing a toxin—antitoxin pair. The predator's QS signal activates a toxin gene, *ccdB*, in the prey while the prey's QS signal induces an antitoxin gene, *ccdA*, which neutralizes the effect of the constitutively expressed *ccdB* in the predator (Fig. 13.2A).

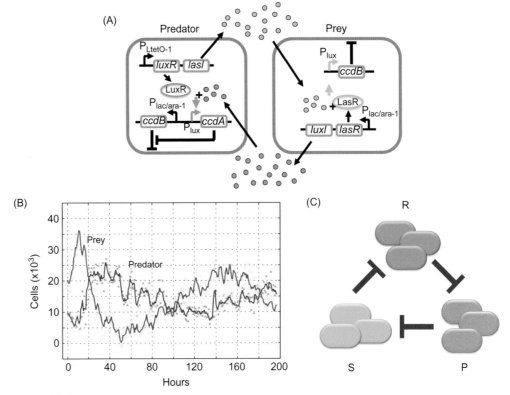

FIGURE 13.2

Engineering competition within a synthetic consortium. (A) A synthetic predator—prey system. The predator and prey populations control each other's survival and death via two different QS signals. The predator and prey synthesize 3OC12HSL (green circles) and 3OC6HSL (orange circles) via LasI and LuxI, respectively. The signal from the predator population is sensed by the prey population via LasR, while that from prey population is sensed by the predator population via LuxR. The QS genes in the predator are constitutively expressed from $P_{LtetO-1}$, and those in the prey are under IPTG-inducible promoter, $P_{lac/ara-1}$. The *ccdB* toxin gene is controlled by $P_{lac/ara-1}$ in the predator and P_{lux} in the prey. P_{lux} also regulates *ccdA* antitoxin gene in the predator. Redrawn from Balagaddé et al.[25] (B) Typical oscillatory predator—prey dynamics observed using a custom-made microfluidic device. Adapted from Balagaddé et al.[25] (C) Schematic of a rock—paper—scissors interaction. The colicin-producing population (P) kills the colicin-sensitive population (S). The S population outcompetes the colicin-resistant population (R), which can outcompete the P population.

A simple mathematical model predicted that with sufficiently high expression of CcdB in the predator and the QS module in the prey (i.e. LuxI and LasR), the system could exhibit oscillatory dynamics. By experimentally modulating these two factors, and observing the population dynamics using a custom-made microfluidic device, the authors confirmed this prediction (Fig. 13.2B).

This consortium was further extended to examine the role of cellular motility and spatial structure and their effects on biodiversity (i.e. coexistence of the predator and prey strains).[26] While previous studies had shown that reducing motility or dispersal promoted biodiversity, it was unclear how cellular motility could impact biodiversity if the interactions occurred over a long distance. As the synthetic predator−prey interaction is mediated by QS signals, which can diffuse over a long distance, the consortium was suitable to address this question. Unlike the previous study, in which the cells were cultured in liquid medium,[25] Song et al. inoculated the predator and prey populations at two separate spots on an agar plate.[26] Cellular motility was modulated by changing the agar density, while seeding the two strains at various segregation distances modulated the spatial configuration. Through modeling and experimentation, the authors found that motility had a negligible impact on biodiversity when the two strains were well mixed in liquid medium or if they were inoculated at the same spot on an agar plate. In contrast, when the two strains were seeded at a relatively long distance from one another, increasing cellular motility was found to significantly decrease biodiversity. Using a mathematical model, the authors determined that motility plays a critical role in biodiversity only when the length scale of the predator−prey interaction matched the segregation distance.

While the predator−prey interaction represents a canonical example of antagonistic relationships in ecology, such interactions in nature often involve more than two populations. Along this line, Kerr et al. constructed a synthetic rock−paper−scissors system consisting of three *E. coli* populations.[27] The system revolves around the clever use of colicin, a toxic protein that is secreted into the environment. Three strains, a colicin-producing strain (P), a colicin-sensitive strain (S), and a colicin resistant strain (R) (Fig. 13.2C), realize the rock−paper−scissors interaction as follows: P kills S through colicin secretion; S outgrows R due to the cost associated with colicin resistance; and R outgrows P due to the metabolic cost associated with the plasmid housing the colicin gene, as well as the genes encoding colicin immunity proteins, which protect P from colicin. Using this consortium, the authors examined previous theoretical predictions that spatial scales of dispersal, movement, and interaction are important factors in determining whether multiple competitive species can coexist. When the three strains were cocultured in liquid medium, where dispersal and movement are high and the interactions among the three populations occurred globally, the community was unstable and only R persisted in the community. In contrast, when the populations were grown on an agar plate in patches, where dispersal and movement are limited and the interactions occur only locally, the three strains stably coexisted for at least 70 generations. These observations thus provided a direct demonstration of the theoretical prediction that local dispersal and interaction promote biodiversity.

A similar rock−paper−scissors system was employed to examine the evolution of restrained growth.[28] Restraint in the use of common resources (i.e. restrained growth) can be seen as an altruistic trait. Prudent use of common resources likely prolongs the persistence of a community, but such a trait has a competitive disadvantage against less-restrained individuals who enjoy a higher fitness than restrained individuals.[29,30] As such, how restrained growth can evolve stands as a fundamental question in ecology and evolutionary biology. Previous computational studies had suggested that a rock−paper−scissors interaction in a spatially structured community could lead to evolution of restrained growth.[31,32] For example, if the rock population increased its growth rate by mutation,

249

it would drive the scissors population to extinction. The decrease in the scissors population would benefit the paper population, which would cause extinction of the rock population. In other words, the rock—paper—scissors interaction acted as a negative feedback. Although the mutated rock population had a competitive advantage over the wild-type rock population, when the community was spatially structured and the interaction occurred locally, the negative feedback was predicted to favor restrained growth.

To test this hypothesis, Nahum et al. propagated the synthetic rock—paper—scissors community described above as metapopulations using 96-well microplates, where each well represented a distinct subpopulation. The degree of population structure was manipulated by having restricted migration, where migration occurred only between adjacent wells (i.e. a local interaction), or unrestricted migration, where migration occurred between any wells (i.e. a global interaction). As a control, the authors also propagated the R strain alone. After 36 rounds of transfer, the R isolates were randomly sampled and competed against a labeled variant of their common resistant ancestor to measure their relative fitness. All three strains persisted in the community for the duration of the experiment regardless of their migration protocol. However, the authors found that the R strain propagated in the community with restricted migration evolved to show a significantly lower growth rate than when the R strain was propagated in the community with unrestricted migration, or when the R strain was propagated alone.

SPATIAL ORGANIZATION IN SYNTHETIC CONSORTIA

While most synthetic consortia have been engineered to exist as well-mixed cultures, where the division of labor is separated only through the genetic components intrinsic to each population, several recent publications have demonstrated that the spatial orientation of each subpopulation can enhance controllability and the ability of the consortium to complete a novel task. A spatially organized consortium can arise through two fashions; it can be organized into biofilms or it can be rationally distributed using technology. Indeed, the spatial organization of microbial consortia in natural settings has long been perceived to assist in the establishment of a cooperative behavior that benefits both populations (e.g.[33]).

SYNTHETIC BIOFILMS LEAD TO STABLE CONSORTIA BEHAVIOR OVER A LONG PERIOD OF TIME

Biofilms have been observed to confer unique benefits in natural systems.[34] Biofilms are more robust against cellular stress, and thus may allow the persistence of novel behaviors over greater periods of time as compared to their nonbiofilm-forming counterparts.[13] As such, the construction of synthetic multipopulation biofilms is of particular interest. In 2007, Brenner et al. created a synthetic consortium that consisted of two *E. coli* strains designed to communicate bidirectionally via QS signals.[35] To engineer this system, the authors used QS systems originally found in *Pseudomonas aeruginosa*. Two circuits (A and B) were implemented in separate *E. coli* populations. Both circuits synthesize a small QS molecule; LasI in circuit A synthesizes 3-oxododecanoyl-HSL (3OC12HSL), while RhlI in circuit B synthesizes butanoyl-HSL (C4HSL). Genes encoding these proteins are under the control of a promoter that responds to the opposite population's QS molecule. That is, expression of LasI, and thus synthesis of 3OC12HSL, is driven by a C4HSL responsive promoter. Conversely, expression of RhlI, and thus synthesis of C4HSL, is driven by a 3OC12HSL responsive promoter. To allow identification of both populations, the QS responsive promoters also drive expression of GFP and RFP, in circuit A and B, respectively. As such, only when the consortium effectively communicates bidirectionally will both GFP and RFP be expressed.

When grown independently, each population was unable to produce a significant amount of fluorescent protein (either RFP or GFP). However, when both populations were grown in separate chambers, which allowed passage of the QS signals but not cells, both GFP and RFP were observed in the mixed population. That is, the passage of a QS signal between populations was sufficient to trigger the expression of both fluorescent proteins. The authors then sought to determine if both RFP and GFP could be expressed when both E. coli populations formed a biofilm. When either population was grown independently as a biofilm, the biofilms were viable for up to two weeks. However, neither population expressed its respective fluorescent protein. However, when both populations were grown together in one biofilm, both GFP and RFP were detected within 24–36 hours of inoculation. This behavior was stable for ~6 days, whereupon the biofilm grew to exceed its experimental limit (i.e. where oxygen diffusion began to affect fluorophore expression). As such, this study demonstrates that a synthetic consortium can be engineered to exist as a biofilm, and that the behavior of such biofilms is consistent over extended periods of time.

Recently, Brenner and Arnold created a synthetic consortium that could self-organize into a structurally layered biofilm.[36] Their synthetic system consisted of two populations, each labeled with a different fluorescent protein for ease of identification (Fig. 13.3). The first population consisted of an E. coli dapD auxotroph (i.e. cannot synthesize the essential metabolites diaminopimelate and lysine) labeled with a cyan fluorescent protein (CFP). In this strain, the authors implemented a simple gene circuit that expresses dapD only in the presence of the QS signal, C4HSL. When grown in the absence of C4HSL, this population initially forms a healthy biofilm, but then quickly dies due to a lack of the aforementioned metabolites. A second population of E. coli, labeled with a yellow fluorescent protein (YFP), was engineered to lack the ability to form healthy biofilms by deleting genes required for the expression of curli, type I pili, colanic acid, and capsular polysaccharides. Furthermore, the authors engineered this strain to constitutively express C4HSL. When equal amounts of these populations were grown in biofilm flow chambers, the two populations grew as a highly intertwined biofilm. Here, the YFP population rescued the CFP population by supplying C4HSL, thus activating dapD expression. Conversely, the CFP population allowed the YFP population to form a biofilm, which consisted of an entangled biomass of both YFP and CFP cells.

After 80 hours of growth, the authors observed that the biofilm would form a well-defined structure where the YFP cells formed aggregates around CFP cells. As the biofilm matured, the biofilm became stratified. Here, the CFP cells formed a uniform layer over the substrate, while the YFP cells formed an uneven layer attached to the layer of CFP cells. Note that YFP

251

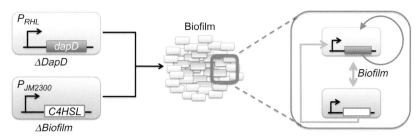

FIGURE 13.3

Engineering synthetic biofilms. To engineer a synthetic biofilm, Brenner and Arnold[36] created two strains that communicate via a QS signal. The yellow population contains a C4HSL responsive promoter (P$_{RHL}$), that when activated, drives expression of dapD thus rescuing the population from death (left panel). The blue population constitutively expresses C4HSL via a P$_{JM2300}$ promoter, but cannot make biofilms. When the populations are grown together, a biofilm forms where a population of yellow cells surrounds a population of blue cells (center panel), which eventually leads to the formation of a structurally layered biofilm (not shown). Within the biofilm, the C4HSL from the blue population rescues the yellow population, allowing growth (green arrow) by driving expression of dapD while the yellow population allows the blue population to form a biofilm.

cells were not detected in the CFP cell biomass. Interestingly, when the authors removed effluent from the biofilm before and after the formation of this structured biofilm, a difference in the ability to colonize a downstream location was observed. Specifically, cells removed from the biofilm prior to the formation of the structured biofilm formed a downstream biofilm that accumulated less biomass as compared to when the CFP strain was grown alone and did not form a structured biofilm. Conversely, when effluent from the biofilm was taken after the structured biofilm was formed, the cells quickly initiated structurally layered biofilms that produced more biomass than their predecessor biofilm. The authors demonstrated that an essential requirement for the formation of these robust downstream biofilms was the YFP/CFP population aggregates. If these aggregates were disrupted during the transfer process, the populations still formed biofilms, but did not generate a layered structure. While the authors did not pinpoint the exact benefit that these aggregates have on the formation of downstream biofilms, this interesting study nonetheless demonstrates that one can design a synthetic consortium that can readily self-organize into a spatially structured biofilm.

The controllability of synthetic biofilms was significantly enhanced in a 2012 publication by Hong et al.[37] In this study, a synthetic consortium was engineered to form a biofilm and disperse in response to externally applied signals. The authors engineered two *E. coli* cell types: a 'disperser' cell type and a 'colonizer' cell type. The disperser cells harbor two gene circuits. One circuit consists of a constitutively expressed GFP (for strain identification) and LasI, which synthesizes the QS signal 3OC12HSL. The other consists of an IPTG-inducible (P_{lac}) promoter driving the expression of Hha13D6, a protein that causes dispersal of biofilms. The colonizer cells also contain two circuits. One circuit constitutively expresses RFP and LasR, the receptor protein for 3OC12HSL. The other contains a 3OC12HSL-inducible promoter (P_{lasI}) driving the expression of BdcAE50Q, another protein that causes biofilm dispersal. The circuit logic is as follows: when the disperser cells reach a critical cell density, sufficient 3OC12HSL is produced, causing the colonizer cells to express BdcAE50Q and disperse. The disperser cells themselves are dispersed through the use of exogenously supplemented IPTG.

The authors first verified that the circuit functioned when the two strains were grown separately as biofilms: the addition of IPTG dispersed a biofilm consisting solely of disperser cells; whereas the addition of 3OC12HSL dispersed a biofilm consisting solely of colonizer cells. The authors then sought to determine if the two strains could function together when seeded into a microfluidic device. To this end, the authors used colonizer cells to form a biofilm upon which they added disperser cells. Indeed, as the disperser cells grew, the colonizer cells began to disperse from the biofilm. After ~ 40 hours of coculture, nearly 80% of the colonizer biofilm had been dispersed. Upon addition of IPTG to the microfluidic device, 92% of the disperser biofilm was successfully dispersed. Furthermore, by replacing the 3OC12HSL responsive promoter with an arabinose responsive promoter, the authors demonstrated that they could control the amount of colonizer cells dispersed from colonizer/disperser biofilms.

While biofilms can self-organize, they can also be rationally organized. Stubblefield et al. developed a flow system to modulate the formation of biofilms.[38] The flow system uses a peristaltic pump that passes a bacterial culture over a flow cell, the contents of which can be monitored using a microscope.[39] The authors first created a biofilm consisting of two *E. coli* strains, one of which expresses GFP. Independent passage of either population over the flow cell resulted in a near linear relationship between the initial inoculum density and the amount of cells deposited on the flow cell. However, when the two bacterial populations were passed over the flow cell as a well-mixed culture, the linearity observed during the single population analysis was abolished. Nonetheless, the authors predicted that sequential passage of each population could reintroduce the previously observed linear relationship, which was indeed confirmed experimentally. Furthermore, the authors demonstrated that, using this technology, they could construct multispecies biofilms consisting of *E. coli* with either *B. subtilis* or *P. aeruginosa*. This study illustrates a powerful technique that may be very useful in engineering complex biofilms.

ADVANCES IN TECHNOLOGY ALLOW THE PRECISE SPATIAL ARRANGEMENT OF SYNTHETIC CONSORTIA

While the engineering of synthetic biofilms represents a tremendous feat in synthetic biology, equally important are the technological advances that allow the rational design of spatially structured, nonbiofilm-forming consortia. In 2008, Kim et al. created a synthetic consortium that consisted of three wild-type strains of soil bacteria, *Azotobacter vinelandii* (Av), *Bacillus licheniformis* (Bl), and *Paenibacillus curdlanolyticus* (Pc).[40] Each strain performs a different task that is vital for the survival of the entire population when grown in a selective (contains penicillin) and nutrient-poor medium: Av fixes gaseous nitrogen into amino acids using a molybdenum-requiring nitrogenase; Bl degrades penicillin G using β-lactamase enzymes; and Pc generates a carbon source using cellulases to cleave carboxymethyl-cellulose (Fig. 13.4a). Initially, the authors determined that, when cultured independently, none of the strains grew in a nutrient-limited medium. When all three populations were grown together as a well-mixed population, the consortium was not sustainable regardless of the nutrient availability.

However, the authors predicted that spatial structure could allow all three populations to grow together. To spatially separate the populations, the authors created a microfluidic chamber that consisted of three physically separated culture wells that are joined through a chemical communication chamber. When each population was seeded into a separate

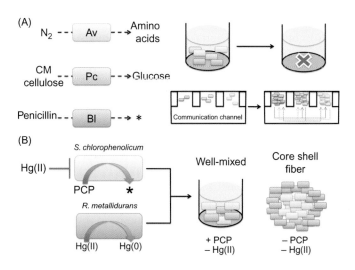

FIGURE 13.4

The structure of a synthetic consortium modulates a novel behavior. (A) Spatial separation leads to a novel behavior. Three bacterial populations were spatially arranged to allow them to cooperate in a selective (contains penicillin) and nutrient-poor minimal medium.[40] The Av population converts gaseous nitrogen into amino acids. The Pc population converts CM-cellulose into glucose while the Bl population degrades penicillin (*). When the three populations were well-mixed, the consortium did not grow (top right panel). However, when the three populations are physically separated, but are allowed to communicate, all three populations grow (bottom right panel). (B) Spatial orientation creates a novel behavior.[42] The bacterium *S. chlorophenolicum* (green cells) can degrade PCP, but is inhibited by Hg(II). The bacterium *R. metallidurans* (red cells) can convert Hg(II) into Hg(0). When these strains are grown in the presence of Hg(II) and PCP as a well-mixed consortium, Hg(II) is converted to Hg(0), but PCP is not degraded (center panel). When spatially arranged in a core (*S. chlorophenolicum*)-shell (*R. metallidurans*) fiber, Hg(II) is converted to Hg(0) and PCP is degraded.

culture well, all three subpopulations grew to establish a stable consortium. By varying the distance between the culture wells, and thus the subpopulations, the stability of the consortium was altered. When the culture wells were seeded with a mixture of all three populations (i.e. a separation distance of 0 μm), a marked decrease in the consortium's viability was observed. Similarly, when the distance of the culture wells was increased too much (~1800 μm), the consortium also declined in viability. Only when the culture wells were separated at an intermediate distance was the consortium viable. From modeling, the authors determined that spatially separating the subpopulations allowed for the balance between consumption of nutrients (competition) and the performance of each population's vital task (cooperation) within the consortium.

Recently, it has been demonstrated that coculturing populations within close proximity to each other can also generate a functional consortium. Park et al. created a microfluidic device to encapsulate distinct populations of bacteria in aqueous droplets.[41] Using this system, they coencapsulated three auxotrophic populations, where the distribution of each population within a droplet could be predicted using a Poisson distribution. When tyrosine (Y-) and tryptophan (W-) auxotrophs were encapsulated in separate droplets, neither could grow in a minimal medium. However, when coencapsulated within the same droplet, both auxotrophic populations grew in minimal medium by exchanging essential metabolites. Interestingly, amino acids secreted and exchanged within the same droplet could not diffuse across the droplet membrane. As such, while a population of droplets consisted of a distribution of Y- only, W- only, and Y- and W- strains, only those droplets containing Y- and W- were observed to grow. The authors then extended this concept and coencapsulated three auxotrophic populations, the aforementioned auxotrophs and a serine auxotroph (S-), which could not exchange metabolites with either Y- or W-. Only when the droplets contained a combination of S-, Y-, and W-, was S- able to grow. By varying the initial ratios of the populations, the authors were able to predict and experimentally verify the distribution of the coencapsulated populations.

The spatial organization of distinct bacterial populations can also affect their ability to cooperate (Fig. 13.4b). In 2011, Kim et al. created a synthetic consortium consisting of two bacteria strains, *Sphingobium chlorophenolicum*, which can degrade the environmental pollutant pentachlorophenol (PCP), and *Ralstonia metallidurans*, which can convert Hg(II) into Hg(0).[42] The authors sought to develop a consortium that could metabolize both environmental pollutants simultaneously. When both populations were well-mixed, the consortium could not metabolize both pollutants, as the PCP concentration remained nearly unchanged. The authors determined that Hg(II) was inherently toxic to *S. chlorophenolicum*, and thus prevented it from degrading PCP. As such, they hypothesized that a spatially structured community, consisting of a core-shell spatial structure, would improve the ability of the consortium to metabolize both pollutants. They hypothesized that they could shelter *S. chlorophenolicum* from Hg(II) by placing it on the inside of the structure, while placing *R. metallidurans* on the outside of the structure, thus forming a protective shell around *S. chlorophenolicum*. Using a mathematical model, they rationally determined the spatial scale of the structure. They predicted that an outer layer of *R. metallidurans* that was too thin would be insufficient to convert Hg(II) to Hg(0) before Hg(II) diffused into the *S. chlorophenolicum* layer killing the population. Alternatively, if the *R. metallidurans* outer layer was too thick, the diffusion rate of PCP would be insufficient to allow sufficiently fast PCP degradation by *S. chlorophenolicum*. After predicting the appropriate scale of the outer layer, they used a microfluidic laminar flow system to create core-shell fibers consisting of the two bacterial populations. In fibers consisting of an *R. metallidurans* shell and an *S. chlorophenolicum* core, both PCP and Hg(II) were metabolized. Alternatively, when the orientations of the populations were switched, a significant reduction in PCP degradation was observed. As such, the authors developed a novel method by which a consortium can be created and arranged to perform novel tasks.

THE EVOLUTION OF COOPERATION CAN YIELD NOVEL BEHAVIORS IN SYNTHETIC CONSORTIA

Studies of engineered consortia have demonstrated that two distinct populations can evolve a novel, cooperative behavior in a relatively short period of time. While such evolvability has implications in the construction of consortia for medical and industrial applications, it can also address questions regarding the evolution of symbiosis and cooperation.[14]

Shou et al.[21] examined evolvability of a cooperative consortium by creating two auxotophic yeast strains: one which requires adenine to grow and the other which requires lysine to grow. When the two strains were cocultured in a minimal medium, they were unable to exchange sufficient amounts of essential metabolites to sustain growth of either strain. To create a cooperative consortium, the authors engineered each strain to overproduce the amino acids required by the other strain. That is, the adenine auxotroph overproduces lysine and the lysine auxotroph overproduces adenine. When cocultured, each overproducing strain was able to supply sufficient metabolites to the opposing population, and thus the consortium cooperated and grew. Using a mathematical model, the authors predicted that a critical factor in determining whether the consortium would grow depended upon the initial densities of each population. When each population was inoculated at a sufficiently low density, the consortium was not viable. However, if both populations were inoculated at a sufficiently high density, the consortium grew. Interestingly, the authors observed that when the consortium was cultured and sequentially diluted (i.e. bottlenecked) over extended periods of time (~70 generations), the initial population density required for viability of the consortium was reduced. Specifically, prior to long-term culturing, 10^3-10^4 cells/mL were required for a viable consortium, whereas after long-term culturing, 10^1-10^2 cells/mL could initiate a viable consortium.

Recently, Hosoda et al. determined a potential mechanism by which auxotrophic cooperation could emerge in a synthetic consortium.[43] The authors engineered two auxotrophic *E. coli* strains: an auxotrophic isoleucine strain (I-) and an auxotrophic leucine strain (L-). When grown independently, neither strain grew in a minimal medium, nor could either strain supply sufficient metabolites to support the other population's growth. That is, the L- strain could not produce and secrete sufficient isoleucine to rescue the growth of the I- strain, and vice versa. Surprisingly, however, when cocultured in a well-mixed minimal medium, the consortium could cooperate and grow, thus suggesting sufficient production and metabolite exchange. Growth measurements revealed that the I- strain grew first. This suggested that the L- population supplied isoleucine initially to rescue I-, whereupon growth of I- rescued the L- strain. Interestingly, the isoleucine supply from the L- population was sufficiently high to allow the I- population to grow two-fold higher than the L- population. Furthermore, the concentration of isoleucine in the medium was significantly greater than when the L- population was grown in monoculture. This suggested that, upon interaction with the I- population, the L- population began to oversupply isoleucine.

To understand how the L- population became an apparent oversupplier of isoleucine upon coculture with the I- population, the authors examined the transcriptional profiles of both populations after coculture. Several categories of genes were up-regulated in the L- population (amino acid biosynthesis of tryptophan, proline, methionine, phenylalanine, leucine, cysteine, and chorismate, the precursor to the aromatic amino acids), while a few were down-regulated (pathways for catabolism including tricarboxylic acid cycle, fatty acid oxidation). Expression of genes regulating isoleucine biosynthesis and transport remained unchanged. As such, while it appeared that the direct up-regulation of isoleucine production did not contribute to the mechanism of enhanced cooperation, an overall increase in amino acid anabolism was detected, suggesting its role in the oversupplier phenotype. The authors suggested two plausible explanations for their results. First, a small population of isoleucine oversuppliers existed in the initial inoculum and was enriched for during growth.

Second, the changes in the amino acid supply occurred via the stringent response. While the authors did not rule out either explanation, this study demonstrates that two populations can readily and quickly evolve to form a successful consortium.

APPLICATIONS OF SYNTHETIC CONSORTIA IN INDUSTRIAL PROCESSES AND MEDICINE

Increasing attention is being paid to the industrial and medical applications that microbial consortia may offer. To date, synthetic circuits with medical applications have been engineered in single populations, and include bacteria that target cancer cells (e.g.[44]), viruses that destroy biofilms (e.g.[45]), bacteria that prevent infection,[46] and circuits designed to prevent the spread of vector-borne pathogens (e.g.[47]). While consortia with direct medical applications have yet to be created, such consortia may offer a more robust mechanism that is tolerant to a wide variety of environments and is stable for longer periods of time as compared to their single population counterparts. As such, it is conceivable that consortia have future roles in the production of pharmaceuticals, drug delivery, and the control of infectious diseases.

As compared to medical applications, more progress has been made in engineering consortia for industrial processes. The benefit of dividing tasks with a high metabolic burden between two populations has been readily utilized in the fermentation of sugars towards the creation of biofuels.[13,48] While the majority of biofuel production to date uses expensive crops such as sugar cane, there is a growing desire to use lignocellulosic materials from less-expensive crops for the production of biofuels. Hydrolysis of lignocellulosic materials yields two sugars, glucose and xylose, both of which require efficient fermentation in the biofuels process.[49] However, *E. coli* preferentially ferments glucose, followed by xylose, thus making concurrent degradation of both sugars impossible. However, Eiteman et al.[48] engineered two strains of *E. coli* which preferentially degrade glucose or xylose. In one strain, the authors deleted *glk* (a cytoplasmic glucokinase), *ptsG* (a component of glucose import), and *manZ* (required for mannose import) genes, thus inhibiting the strain from degrading glucose. In a second strain, the authors deleted *xylA* (catalyzes the first step in xylose catabolism), thus inhibiting the strain from degrading xylose. Coculture of these populations resulted in efficient and simultaneous fermentation of both sugars.

FUTURE CHALLENGES

The design and engineering of synthetic consortia face many of the same challenges that synthetic biology using single populations does. While current research is forming a solid foundation describing how robust and predictable behaviors may be rationally engineered, the field must now move towards more application-based circuits. However, there are several challenges that need to be addressed. While it is often cited that synthetic systems are predictable, interactions with the host physiology may have unintended and surprising effects on a desired behavior (e.g.[8]). While consortia are often more stable than single populations, they are still subject to 'circuit blind' mutations, which may lead to loss of circuit function and destabilization of the consortium. As such, a consortium may perform a desired behavior for a discrete length of time before the consortium must be replaced. Finally, there are currently many ethical issues surrounding the use of synthetic biology,[50] which must be addressed by both scientists and lawmakers alike.

References

1. Gardner TS, Cantor CR, Collins JJ. Construction of a genetic toggle switch in *Escherichia coli*. *Nature*. 2000;403 (6767):339−342.

2. Kobayashi H, Kærn M, Araki M, et al. Programmable cells: interfacing natural and engineered gene networks. *Proc Natl Acad Sci USA*. 2004;101(22):8414−8419.

3. Kramer BP, Viretta AU, Baba MD-E, Aubel D, Weber W, Fussenegger M. An engineered epigenetic transgene switch in mammalian cells. *Nat Biotech*. 2004;22(7):867−870.

4. Elowitz MB, Leibler S. A synthetic oscillatory network of transcriptional regulators. *Nature*. 2000;403 (6767):335−338 [10.1038/35002125].

5. Tigges M, Marquez-Lago TT, Stelling J, Fussenegger M. A tunable synthetic mammalian oscillator. *Nature*. 2009;457(7227):309−312.

6. Tigges M, Denervaud N, Greber D, Stelling J, Fussenegger M. A synthetic low-frequency mammalian oscillator. *Nucl Acids Res*. 2010;38(8):2702−2711.

7. Stricker J, Cookson S, Bennett MR, Mather WH, Tsimring LS, Hasty J. A fast, robust and tunable synthetic gene oscillator. *Nature*. 2008;456(7221):516−519.

8. Marguet P, Tanouchi Y, Spitz E, Smith C, You L. Oscillations by minimal bacterial suicide circuits reveal hidden facets of host-circuit physiology. *PLoS ONE*. 2010;5(7):e11909.

9. Lou C, Liu X, Ni M, et al. Synthesizing a novel genetic sequential logic circuit: a push-on push-off switch. *Mol Syst Biol*. 2010;6:350.

10. Hooshangi S, Thiberge S, Weiss R. Ultrasensitivity and noise propagation in a synthetic transcriptional cascade. *Proc Natl Acad Sci USA*. 2005;102(10):3581−3586.

11. Burmolle M, Webb JS, Rao D, Hansen LH, Sorensen SJ, Kjelleberg S. Enhanced biofilm formation and increased resistance to antimicrobial agents and bacterial invasion are caused by synergistic interactions in multispecies biofilms. *Appl Environ Microbiol*. 2006;72(6):3916−3923.

12. LaPara TM, Zakharova T, Nakatsu CH, Konopka A. Functional and structural adaptations of bacterial communities growing on particulate substrates under stringent nutrient limitation. *Microb Ecol*. 2002;44(4):317−326.

13. Brenner K, You L, Arnold FH. Engineering microbial consortia: a new frontier in synthetic biology. *Trends Biotechnol*. 2008;26(9):483−489.

14. Tanouchi Y, Smith R, You L. Engineering microbial systems to explore ecological and evolutionary dynamics. *Curr Opin Biotechnol*. 2012;23(5):791−797.

15. Waters CM, Bassler BL. Quorum sensing: cell-to-cell communication in bacteria. *Annu Rev Cell Dev Biol*. 2005;21(1):319−346.

16. Smith R. Design principles and applications of engineered microbial consortia. *Acta Hortic*. 2011;905:63−69.

17. Baba T, Ara T, Hasegawa M, et al. Construction of *Escherichia coli* K-12 in-frame, single-gene knockout mutants: the Keio collection. *Mol Syst Biol*. 2006;2:2006.0008.

18. You L, Cox RS, Weiss R, Arnold FH. Programmed population control by cell−cell communication and regulated killing. *Nature*. 2004;428(6985):868−871.

19. Basu S, Mehreja R, Thiberge S, Chen M-T, Weiss R. Spatiotemporal control of gene expression with pulse-generating networks. *Proc Natl Acad Sci USA*. 2004;101(17):6355−6360.

20. Basu S, Gerchman Y, Collins CH, Arnold FH, Weiss R. A synthetic multicellular system for programmed pattern formation. *Nature*. 2005;434(7037):1130−1134.

21. Shou W, Ram S, Vilar JMG. Synthetic cooperation in engineered yeast populations. *Proc Natl Acad Sci*. 2007;104(6):1877−1882.

22. Wintermute EH, Silver PA. Emergent cooperation in microbial metabolism. *Mol Syst Biol*. 2010;6:407.

23. Arai T, Matsuoka S, Cho H-Y, et al. Synthesis of *Clostridium cellulovorans* minicellulosomes by intercellular complementation. *Proc Natl Acad Sci*. 2007;104(5):1456−1460.

24. Goldberg SD, Derr P, DeGrado WF, Goulian M. Engineered single- and multi-cell chemotaxis pathways in *E. coli*. *Mol Syst Biol*. 2009;5.

25. Balagaddé FK, Song H, Ozaki J, et al. A synthetic *Escherichia coli* predator-prey ecosystem. *Mol Syst Biol*. 2008;4:187.

26. Song H, Payne S, Gray M, You L. Spatiotemporal modulation of biodiversity in a synthetic chemical-mediated ecosystem. *Nat Chem Biol*. 2009;5(12):929−935.

27. Kerr B, Riley MA, Feldman MW, Bohannan BJM. Local dispersal promotes biodiversity in a real-life game of rock-paper-scissors. *Nature*. 2002;418(6894):171−174.

28. Nahum JR, Harding BN, Kerr B. Evolution of restraint in a structured rock-paper-scissors community. *Proc Natl Acad Sci USA*. 2011;108:10831−10838.

29. MacLean RC. The tragedy of the commons in microbial populations: insights from theoretical, comparative and experimental studies. *Heredity*. 2008;100(5):471−477.

30. Rankin DJ, Bargum K, Kokko H. The tragedy of the commons in evolutionary biology. *Trends Ecol Evol (Amst)*. 2007;22(12):643−651.

31. Johnson CR, Seinen I. Selection for restraint in competitive ability in spatial competition systems. *Proc Biol Sci/R Soc*. 2002;269(1492):655−663.

257

32. Prado F, Kerr B. The evolution of restraint in bacterial biofilms under nontransitive competition. *Evolution.* 2008;62(3):538–548.

33. Fux CA, Costerton JW, Stewart PS, Stoodley P. Survival strategies of infectious biofilms. *Trends Microbiol.* 2005;13(1):34–40.

34. Hansen SK, Rainey PB, Haagensen JAJ, Molin S. Evolution of species interactions in a biofilm community. *Nature.* 2007;445(7127):533–536.

35. Brenner K, Karig DK, Weiss R, Arnold FH. Engineered bidirectional communication mediates a consensus in a microbial biofilm consortium. *Proc Natl Acad Sci.* 2007;104(44):17300–17304.

36. Brenner K, Arnold FH. Layered structure, and aggregation enhance persistence of a synthetic biofilm consortium. *PLoS ONE.* 2011;6(2):e16791.

37. Hong SH, Hegde M, Kim J, Wang X, Jayaraman A, Wood TK. Synthetic quorum-sensing circuit to control consortial biofilm formation and dispersal in a microfluidic device. *Nat Commun.* 2012;3:613.

38. Stubblefield B, Howery K, Islam B, Santiago A, Cardenas W, Gilbert E. Constructing multispecies biofilms with defined compositions by sequential deposition of bacteria. *Appl Microbiol Biotechnol.* 2010;86(6):1941–1946.

39. Gilbert ES, Keasling JD. Bench scale flow cell for nondestructive imaging of biofilms. In: Spencer J, Ragout de Spencer A, eds. *Environmental Microbiology Methods and Protocols.* Humana Press; 2004:109–118.

40. Kim HJ, Boedicker JQ, Choi JW, Ismagilov RF. Defined spatial structure stabilizes a synthetic multispecies bacterial community. *Proc Natl Acad Sci.* 2008;105(47):18188–18193.

41. Park J, Kerner A, Burns MA, Lin XN. Microdroplet-enabled highly parallel co-cultivation of microbial communities. *PLoS ONE.* 2011;6(2):e17019.

42. Kim HJ, Du W, Ismagilov RF. Complex function by design using spatially pre-structured synthetic microbial communities: degradation of pentachlorophenol in the presence of Hg(II). *Integr Biol.* 2011;3:2.

43. Hosoda K, Suzuki S, Yamauchi Y, et al. Cooperative adaptation to establishment of a synthetic bacterial mutualism. *PLoS ONE.* 2011;6(2):e17105.

44. Anderson JC, Clarke EJ, Arkin AP, Voigt CA. Environmentally controlled invasion of cancer cells by engineered bacteria. *J Mol Biol.* 2006;355(4):619–627.

45. Lu TK, Collins JJ. Dispersing biofilms with engineered enzymatic bacteriophage. *Proc Natl Acad Sci.* 2007;104(27):11197–11202.

46. Duan F, March JC. Engineered bacterial communication prevents *Vibrio cholerae* virulence in an infant mouse model. *Proc Natl Acad Sci.* 2010;107(25):11260–11264.

47. Fu G, Lees RS, Nimmo D, et al. Female-specific flightless phenotype for mosquito control. *Proc Natl Acad Sci.* 2010;107(10):4550–4554.

48. Eiteman M, Lee S, Altman E. A co-fermentation strategy to consume sugar mixtures effectively. *J Biol Eng.* 2008;2(1):3.

49. Zaldivar J, Nielsen J, Olsson L. Fuel ethanol production from lignocellulose: a challenge for metabolic engineering and process integration. *Appl Microbiol Biotechnol.* 2001;56(1):17–34.

50. Schmidt M, Ganguli-Mitra A, Torgersen H, Kelle A, Deplazes A, Biller-Andorno N. A priority paper for the societal and ethical aspects of synthetic biology. *Syst Synth Biol.* 2009;3(1):3–7.

Future Prospects

CHAPTER 14

Semi-Synthetic Minimal Cells: Biochemical, Physical, and Technological Aspects

Pasquale Stano[1], Tereza Pereira de Souza[2], Yutetsu Kuruma[3], Paolo Carrara[1] and Pier Luigi Luisi[1]
[1]University of Roma Tre, Rome, Italy
[2]Universidade Estadual Paulista Júlio de Mesquita Filho, Sao Jose do Rio Preto, Brazil
[3]The Tokyo University, Tokyo, Japan

INTRODUCTION

Synthetic biology (SB) is an offspring of bioengineering that has focused on a very ambitious aim, that of synthesizing in the laboratory forms of life, or at least biological structures, which are alternative to natural ones. Thus, the most well-known examples of SB are in genetic engineering, in the attempt to genetically modify extant bacteria in order to produce novel bacterial species which are capable of producing fuels, specialized drugs, or particular biopolymers.[1] Typical in this field is the work by Venter and his group, who reported the 'creation of a bacterial cell controlled by a chemically synthesized genome.'[2] And there is much expectation that SB can offer new tools and products, both for therapy and for bioengineering utilities.

Genetic manipulation is not the only way to SB. In fact, the chemical approach can lead to biopolymeric structures and to complex supramolecular complexes which are not present in nature, and are in principle alternative to the extant ones. Examples are the pyranose DNAs developed by Eschenmoser and his group at the ETH Zurich,[3] or the proteins with a reduced alphabet of amino acids.[4] These kinds of SBs are referred to as 'chemical synthetic biology,' and our group has been influential in emphasizing the importance of this approach to SB.[5,6] Genetic engineering and chemical SB also differ in their basic conceptual framework, and the epistemology of SB has been discussed in such a context.[7]

Our group has been active in chemical SB along two lines of research. One is the project on the so-called 'never born proteins' (NBPs). Recognizing that the proteins existing in nature are only an extremely small fraction of the possible protein structures, we have developed methods to synthesize in the laboratory libraries of proteins which have never been produced by nature, or at least which are not present today in our life structures. Here the questions asked are whether and to what extent these NBPs display novel structures, and possibly catalysis features, with respect to the natural proteins.[8] Linked to this research are also basic epistemology questions, for example the controversy between determinism and contingency in the biogenesis of the fundamental life structure — like the proteins. The

261

Synthetic Biology. DOI: http://dx.doi.org/10.1016/B978-0-12-394430-6.00014-5

other line of research is the one which will be the subject of this article. It concerns the 'minimal cells.'[9–11] A minimal cell is broadly defined as the cell containing the minimal and sufficient number of macromolecular components to be recognized as living — or at least displaying some of the basic functions of a living cell (e.g. protein synthesis, metabolism, or self-reproduction). This is a program of chemical SB, based on vesicles, as supramolecular surfactant aggregates, and the basic operation is chemical manipulation. This kind of investigation is important for determining the minimal structural conditions for cellular life, and is fundamental for the origin of life — in particular for the origin of cellular life. The relevance of this study is also apparent if one considers that the simplest bacterial form of our present life contains 400–500 genes, and therefore each compartment has many thousands of molecular components: conceivably, life cannot have started from the very beginning with this high complexity, and therefore the study of a minimal cell may shed light on those protocellular structures preceding the full-fledged genome cells.

In this article we will briefly highlight the main results obtained in this field, we will also consider some novel developments in the filed, such as the 'cell colonies' (aggregates of minimal cells), and the recent surprising finding of the spontaneous 'overcrowding' of biological macromolecules in vesicles.

THE CONCEPTUAL FRAMEWORK OF SEMISYNTHETIC MINIMAL CELLS

The concept of minimal cells was developed in the context of origin of life research. For understanding how primitive living cells could be originated from the association and the dynamic organization of nonliving components, our research group initiated a research field focused on the construction of cell-like compartments that could display living-like properties such as growth, internal metabolism, self-reproduction, etc. by using chemical and biochemical approaches. These investigations, which have been recognized as a branch of SB, are currently carried out by several groups and encompass different aspects of cellular organization (an overview of these aspects can be found in a recently published book).[10]

The starting point for our discussion is the theory of autopoiesis (self-production), developed in the 1970s by Humberto Maturana and Francisco J. Varela.[12] Autopoiesis is a distinctive feature of all living systems. It describes how a living system is organized in terms of the internal transformation of its components. In particular, by referring to unicellular organisms (which are the best example to describe what is an autopoietic mechanism), the genetic/metabolic network of a cell is a set of biochemical processes that produce all its own molecular components, which in turn interact between each other giving rise to the biochemical processes that produce (Fig. 14.1A).

Moreover, the autopoietic network needs to be limited by a semipermeable boundary (also created autopoietically) for self-containment, and that distinguishes the autopoietic unit (the cell, in this case) from the environment. Note that this definition focuses on the dynamic organization of the molecules that constitute the autopoietic unit (e.g. the cell), rather than their chemical identity or their lifetime. Autopoiesis provides a framework for the construction of synthetic cells in the laboratory and can be considered as a blueprint of cellular life. This gives rise to two considerations. First, 'minimal autopoietic cells' can be thought of and designed as molecular systems composed of the minimal number of molecules that, by interacting with each other, give rise to an autopoietic dynamic and an autopoietic self-confinement. Second, considering that early living cells were certainly simpler than modern (evolved) cells, the construction of minimal autopoietic cells in the laboratory might reveal what are the key steps, the criticalities, the emergent properties, and

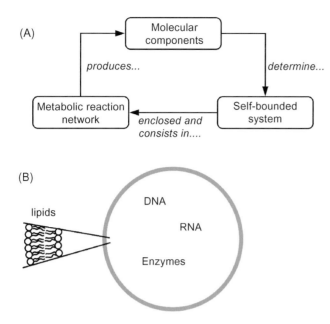

FIGURE 14.1

(A) The concept of autopoiesis. Autopoietic systems are composed of molecular components that self-assemble (self-organize) into self-bounded systems. The systems consist of a metabolic reaction network that is organized in order to produce the molecular components of the autopoietic unit. As specified by Maturana and Varela, the autopoietic organization is a network of productions of components 'which participate recursively in the same network of productions of components which produced these components, and realize the network of productions as a unity in the space in which the components exist.'[12] This may also be called 'operational closure'. (B) The concept of semisynthetic minimal cells (SSMCs). The minimal number of functional molecules, like DNA, RNA, and enzymes, are inserted in a lipid vesicle with the aim of constructing cell-like structures endowed with some functions.[9] For instance, if the SSMC can produce all its components by internalized reactions, it will be an autopoietic minimal cell. SSMCs are used as models of primitive cells and as tools in synthetic biology.

the unexpected features that could have had a role in the route from non-living matter to living organisms.

Is it then possible to build minimal autopoietic cells in the laboratory? Early studies date back to about 20 years ago when the first attempts were done by using fatty acid micelles[13] and fatty acid reverse micelles[14] as minimal chemical autopoietic systems. In these very simple systems, fatty acids were produced autopoietically by feeding the micelles with suitable reactive precursors (generally fatty acid esters or anhydrides), which were converted to fatty acid molecules inside or at the boundary of the micelles. These transformations were consistently defined as autopoietic self-reproduction, because the micelles firstly grow in size, thanks to an autopoietic production of their components, and eventually divide into two or more smaller micelles that could start again the growth-division cycle. Next, the autopoietic self-reproduction of fatty acid vesicles was achieved,[15] which are cell-like shell structures consisting of a semipermeable fatty acid bilayered spherical membrane that encloses an aqueous core. Also in this case, the production of fatty acids was obtained by the external addition of a reactive precursor that was transformed after being taken up by the vesicles. The whole subject of micelles and vesicle self-reproduction, which also includes the discovery of the 'matrix' effect and additional studies by other researchers, has been recently reviewed.[16] The results on micelles, reverse micelles, and vesicle self-reproduction collectively demonstrate that simple lipid compartments, plausibly existing in primitive times (and being precursors of primitive cells), already show a set of dynamic properties that might have played a role in the origin of cellular life.

Stimulated by the success in the autopoietic self-reproduction of vesicles, the next step was to create more complex systems in order to achieve living-like functions in vesicle-based systems. Due to the structural similarity between lipid vesicles and natural cells, the goal became the laboratory construction of synthetic minimal autopoietic cells (as a model of primitive minimal autopoietic cells), in order to understand what could happen when lipid vesicles hosted simple and complex biochemical reactions. Clearly, in order to build a model of primitive minimal autopoietic cells primitive molecules are needed. Unfortunately, our knowledge of which are the primitive molecules that give rise to living cells is still very poor, and this approach cannot be followed experimentally. What can be done, however, consists in the use of currently available molecules, such as DNA, RNA, ribosomes, enzymes, and lipids to try to construct a 'semisynthetic' minimal autopoietic cell (SSMC) (Fig. 14.1b). As we have highlighted above, autopoiesis resides in the organization of a network of transformations, not in the chemical nature of its components. This means that in order to demonstrate experimentally the transition from nonliving separated molecules to a living autopoietic system it is not important which molecules are used. In other words, SSMCs can serve as a model of primitive cells only from the viewpoint of functional organization, not from the viewpoint of molecular structures. Despite the loss of some 'primitiveness' aspects, SSMCs are certainly an interesting tool for origin of life research. Moreover, it is already possible to glimpse some possible future applications in biotechnology, as we will discuss in the concluding remarks.

RECONSTRUCTION OF GENETIC/METABOLIC PROCESSES IN SEMISYNTHETIC MINIMAL CELLS

In order to construct an autopoietic minimal cell in the laboratory using the semisynthetic approach, as we have seen in the previous section, one has to design and realize a genetic/metabolic network that gives as output all its components (including the boundary). In turn, these components will again give rise to the same set of transformation to produce copies of themselves, and so on. As a result, the minimal cell grows in size, reaches an unstable physical state, and produces offspring by a fragmentation-like process. What is the minimal number of molecules needed for accomplishing this task?

A starting point for this discussion is given by the studies on 'minimal genomes.' It is well known that the genome of free-living prokaryotes, like *Escherichia coli*, contains regions that encode for all proteins and RNAs for its self-maintenance, namely for producing all its components from simple nutrients that are present in its environment. There are, however, microorganisms that are symbionts or parasites, and live inside other cells. Evolution has shaped their genome by eliminating unnecessary genes, so that they have a minimal genome where only essential genes are kept. This is possible because their 'environment' is another cell that provides for it the missing biochemicals. By comparing the natural minimal genomes of several microorganisms, several researchers have tried to answer the question of what is the universal minimal set of genes for sustaining life in very permissive conditions (reviewed in [9]). In a recent study, Andres Moya and coworkers[17] have suggested that the minimal genome should contain 206 genes, most of which (121 genes) are devoted to the transcription/translation and protein processing processes. Additional genes are needed for DNA processing (16 genes), basal metabolism (56 genes), and for other cellular processes (13 genes). In principle, therefore, a minimal autopoietic cell could be built by constructing a cell-like compartment that contains about 200 genes (those of the minimal genome) and all molecules required for their expression. Despite the recent advances, however, the experimental approaches that we will describe below are still too rudimentary to achieve this goal. They have been developed firstly for understanding each single process separately, and possibly reaching the final goal of synthesizing a living cell stepwise.

The first step in SSMC research was done in 1991, when extension of the initial studies on vesicle self-reproduction[15] to phospholipid vesicles was attempted. The biochemical machinery for phosphatidylcholine production, composed of four enzymes working sequentially, was inserted inside phosphatidylcholine vesicles, aiming at producing lipids from internal reactions at the expense of lipid building blocks.[18] The low chemical yield did not allow the observation of the desired behavior. Next, it was demonstrated by our group that RNA and DNA could be replicated inside lipid vesicles, by means of an RNA template and Qβ-replicase, and a DNA template and DNA polymerase, respectively. Moreover, the intravesicle production of poly(adenylic acid) was obtained from ADP and polynucleotide phosphorylase. Finally, in 1999 we reported for the first time the ribosomal production of a polypeptide (poly(phenylalanine)) inside lipid vesicles. Details and references to these and other studies on biopolymerization reactions inside lipid vesicles can be found in recently published reviews.[9,11]

Time was ripe for the first experimental report on the synthesis of a functional protein (the green fluorescent protein, GFP) inside lipid vesicles, which was carried out in 2001.[19] After about 10 years, several advancements have been recorded in this field, but the synthesis of proteins remains the key ingredient in SSMC research. This can be understood from two points of view. The first is related to the complexity of this multistep reaction that models well the complexity of a minimal genetic/metabolic system. The approximately 80 different macromolecules needed to perform such a reaction (DNA, RNA polymerase, ribosome and translation factors, tRNAs, aminoacyl-tRNA synthases) already represent an important part of an autopoietic SSMC. As a consequence, physical and biochemical considerations obtained by studying the synthesis of proteins inside compartments have a sort of general implication for the emergence of functional compartments from separated molecules. Second, the control of protein synthesis allows the development of the next generation of SSMC, where more complex functions are developed thanks to the in situ production of proteins (e.g. enzymes, transcription factors, etc.).

The construction of SSMCs relies on cell-free technology. Initial reports on protein synthesis were carried out by encapsulating cell extracts (e.g. typically from *E. coli*) inside lipid vesicles. Cell extracts, however, although they contain all required molecules for transcription and translation, are not well characterized in terms of molecular composition, and are not very attractive from an SB point of view. An important breakthrough in the field of protein synthesis was achieved by the group of Takuya Ueda, who invented in 2001 the PURE system (Protein synthesis Using Recombinant Elements), a fully reconstituted molecular system containing the minimal number of macromolecules required to carry out protein biosynthesis in vitro, starting from the corresponding coding DNA (or RNA).[20] The PURE system consists of four subsystems, namely: (1) transcription; (2) translation; (3) amino acid charging onto tRNA (aminoacylation); and (4) energy regeneration. The composition of the PURE system is shown in Table 14.1. As can be seen, the PURE system is composed of 36 individually purified *E. coli* proteins, highly purified 70 S ribosome, tRNAs (for a total of about 80 macromolecules), and low-molecular-weight compounds such as NTPs, 20 amino acids, DTT, spermidine, formyl-tetrahydrofolate, salts, and creatine phosphate as ultimate phosphate donor. In contrast to cell-extract-based systems, the PURE system realizes a totally defined condition concerning the number and concentration of its components. This system can be considered as a standard chassis for synthetic biology (http://partsregistry.org/Main_Page).

There are six significant examples of 'advanced' systems created by the semisynthetic approach. In all these examples, the desired function has been achieved by expressing one or more proteins inside lipid vesicles, and therefore realizing the first steps toward the construction of an SSMC. The first one[21] deals with the control of genetic expression and consists of the construction of a two-step genetic cascade inside a lipid vesicle. In particular,

TABLE 14.1 Composition of the PURE System, adapted from[20]

No.	Factor	No.	Factor
1	Ala RS	20	Val RS
2	Arg RS	21	MTF
3	Asn RS	22	IF1
4	Asp RS	23	IF2
5	Cys RS	24	IF3
6	Gln RS	25	EF-G
7	Gly RS	26	EF-Tu
8	His RS	27	ET-Ts
9	Ile RS	28	RF1
10	Leu RS	29	RF2
11	Lys RS	30	RF3
12	Met RS	31	RRF
13	Phe RS	32	Myokinase
14	Pro RS	33	Creatine kinase
15	Ser RS	34	NDK
16	Ser RS	35	PPiase
17	Thr RS	36	T7 RNA polymerase
18	Trp RS	37	Ribosome
19	Tyr RS	38	tRNAs mix

RS: tRNA synthetase; MTF: methionyl-tRNA transformylase; IF: initiation factor; EF: elongation factor; RF: release factor; NDK: nucleoside-diphosphate kinase; PPiase: pyrophosphatase.

a first gene (under SP6 promoter), was transcribed by SP6 RNA polymerase. The gene encoded for T7 RNA polymerase. Therefore, after translation, functional T7 RNA polymerase was produced inside liposomes. In turn, T7 RNA polymerase started the transcription of the second gene (under T7 promoter). The second gene encoded for a reporter protein, the GFP, which was revealed fluorimetrically.

The second[22] deals with the problem of low permeability of the lipid vesicle membrane, and consists of the expression of a pore-forming protein (the α-hemolysin) inside liposomes. Thanks to the formation of the pore, the synthetic cell could be fed by low-molecular-weight compounds added externally, so that the synthesis of a reporter protein could be prolonged up to 100 hours, whereas in the absence of the pore, it generally stops after a couple of hours. The size of the pore (molecular weight cut-off: 3 kDa), although permitting the diffusive flow of small molecules inward and outward, did not allow the escape of large internal macromolecules from the synthetic cell.

The third[23] deals with the replication of genetic information, and consists in the replication of genetic material operated by a replicase produced in situ by a cell-free system. In particular, RNA was taken as a genetic polymer. It encoded the sequence for Qβ-replicase. RNA was translated so that Qβ-replicase was effectively produced inside liposomes. Once formed, Qβ-replicase — acting on RNA — produced the RNA complementary strand. The latter was so designed that it could be translated into an enzyme (β-galactosidase) that was revealed fluorimetrically. The RNA complementary strand could also act as a template for Qβ-replicase for producing the first RNA strand. In other words, the RNA and its complementary sequences were both produced by Qβ-replicase, which is an RNA-dependent RNA-polymerase. This report demonstrates that a genetic polymer can produce an enzyme that is able to replicate the genetic polymer.

The fourth example[24] deals with the synthesis of phospholipids inside lipid vesicles, aimed at obtaining an autopoietic lipid production. In this work, the target was the synthesis of phosphatidic acid, which occurs in two steps from simple precursors like glycerol-3-phosphate and acyl-coenzyme As. To this end, it was necessary to synthesize the two enzymes that catalyze this conversion. The first enzyme (the glycerol-3-phosphate acyltransferase) is an integral membrane protein that could be functionally synthesized only by a careful design of the liposome membrane (prepared from 1-palmitoyl-2-oleoyl-sn-glycero-3-phosphatidylcholine (POPC), 1-palmitoyl-2-oleoyl-sn-glycero-3-phosphatidylethanolamine (POPE), 1-palmitoyl-2-oleoyl-sn-glycero-3-phosphatidylglycerol (POPG), and cardiolipin). The second enzyme (the lysophosphatidic acid acyltransferase) is a peripheral membrane protein and was synthesized. Both results were functional (catalytically active), but required different redox conditions for assuming the proper conformation so that their simultaneous synthesis inside the same lipid vesicle was not possible. The stepwise activation of the two enzymes (in two different populations of vesicles), however, brought about the production of phosphatidic acid from the precursors. Due to the low yield, however, the growth of vesicles was not observed.

The fifth case[25] deals with the production of DNA inside self-reproducing giant vesicles. First, DNA was amplified inside giant vesicles by entrapping all components of the polymerase chain reaction. After DNA synthesis, lipid precursors were added externally. They were taken up by the vesicle and transformed into the membrane-forming compound thanks to a catalyst dissolved in the membrane. As a consequence of the increased number of membrane molecules, the giant vesicle could grow and divide into several vesicles, each containing part of the synthesized DNA. Contrary to the first four examples, in this study lipid vesicles were not composed from phospholipids, but from an ad hoc designed membrane-forming compound, that could be easily obtained after a chemical reaction catalyzed by a simple imidazole-based membrane-soluble compound (without needing an enzyme).

The sixth case[26] focuses on the production of cytoskeletal elements inside giant vesicles. In this study, MreB (an *E. coli* protein that is homologous of actin) and MreC (an anchoring protein) have been synthesized inside vesicles by transcription/translation processes. It was observed that flexible filamentous structures were formed in liposomes (smaller than 15 μm). This study paves the way to the reconstruction of synthetic divisome in SSMCs.

From this short survey on the state-of-the-art of SSMC studies it is evident that there have been great efforts for developing synthetic cells capable of performing some nontrivial functions, such as the control of genetic expression, the opening of a pore on the vesicle membrane, the replication of genetic material, and lipid synthesis. When compared to a hypothetical minimal cell based on a minimal genome, the current constructs might appear too simple, and indeed they are. But they have to be considered as the pioneer steps into a new technology, i.e. the laboratory construction of synthetic cells, that has just been born. In a certain sense, this way of progressing fully reflects one facet of SB (perhaps not fully recognized), namely the possibility of learning about living systems by attempting their construction, in order to understand what the physicochemical constraints are underlying the observed biological behavior. The ever-increasing involvement of several new research groups in this field will certainly impact on the advances in this field, so that more progress is foreseeable in the next few years (Table 14.2).

PHYSICAL ASPECTS OF SSMC CONSTRUCTION: IMPLICATIONS FOR THE ORIGIN OF LIFE

In the previous section we have discussed in detail one of the two ingredients of the current experimental approaches: the cell-free system. Here we will discuss further the second ingredient, namely the lipid vesicles and the physical aspects related to their formation and encapsulation of the solutes.

From the practical viewpoint, SSMCs (or, more generally, solute-filled lipid vesicles) are generally obtained by forming lipid vesicles in a solution that contains all solutes to be

TABLE 14.2	Experimental Advances in SSMC Studies		
Entry	Year	Description	References
1	2004	Two-step genetic cascade (T7 RNA polymerase and later GFP)	21
2	2004	Expression of a pore-forming protein (α-hemolysin) and prolongation of protein synthesis up to 100 h	22
3	2008	Self-replication of RNA template by Qβ-replicase, which is expressed in situ from the RNA template itself. Production of β-galactosidase after translation of the complementary RNA strand	23
4	2009	Production of two functional membrane enzymes for the two-step synthesis of phosphatidic acid (a membrane component)	24
5	2011	DNA amplification (by PCR reaction) and vesicle self-reproduction by the external addition of a precursor	25
6	2012	Synthesis of cytoskeleton proteins (MreB and MreC)	26

encapsulated. Several lipid vesicle preparation methods shared this strategy. Most of the studies reported above have been carried out by forming lipid vesicles by hydrating thin lipid films obtained directly from lipids or by freeze-drying preformed liposomes. The injection method has also been occasionally used (this method consists of the injection of lipids, as ethanol solution, in an aqueous solution). These methods are not only classical methods from the viewpoint of liposome technology, but are also good models of how lipid vesicles originated in primitive times. One typical scenario, in fact, foresees that lipid vesicles formed in aqueous solution where a sort of rudimentary metabolism was already established. Thanks to the encapsulation of these molecules, a primitive cell emerged from the primitive 'soup.' It is thought that the confinement into a semipermeable membranous compartment not only protected the primitive metabolic network from dilution, parasitic reactions, and interfering reactions, but also allowed the establishment of gradients that could be ultimately used for generating chemical energy (e.g. ATP), and, fundamentally, allowed the formation of a unit that could behave autopoietically.

But what is really known about the physics of solute encapsulation, especially in the case of multicomponent systems, and in the case of diluted solutions (as it is expected to be the primitive 'soup')? Until a couple of years ago, it was taken for granted that encapsulation of solutes did not represent a critical aspect of the emergence of primitive cells — the argument was simply not discussed in the literature. Our investigation on the physics of solute entrapment stemmed from a study on the minimal physical size of primitive cells.[27] Previous studies on protein synthesis — like those reported above (see also [11] for a more complete review) — were carried out by constructing SSMCs by means of large lipid vesicles, namely, with diameters above 800 nm. However, together with a minimal biochemical complexity, a minimal cell could also feature a minimal physical size, which would perform better than a large one because of the limited volume that would facilitate reactions between a few components. Moreover, smaller vesicles could be physically more stable than larger ones, and need less time (and less material) to self-reproduce.

We approached the issue of minimal cellular size by attempting to synthesize a protein inside conventional lipid vesicles with diameters of 200 nm. We reasoned that the complexity of the transcription/translation machinery well represents the complexity of primitive cells. The size of 100 nm was chosen on the basis of theoretical considerations, elaborated by biophysicists, chemists, and biologists.[28] POPC vesicles were prepared — by injection methods — in a solution containing all the molecular components (i.e. the PURE system) needed to synthesize a reporter protein (enhanced GFP, eGFP). As we have remarked above, the PURE system contains about 80 macromolecules (actually 83) and several small molecules (nucleotides triphosphate, amino acids, salts, etc.). When needed, the dimension of vesicles was regulated by extrusion, a mechanical procedure that converts large vesicles into small ones with a desired size. After incubation, the synthesis of eGFP was revealed fluorimetrically, demonstrating that the process of protein synthesis can occur inside conventional lipid vesicles with diameters of 200 nm. It is noteworthy to say that the eGFP yield was estimated as 0.01−0.05 molecules/vesicle (indicating that not all vesicles were able to produce eGFP, this observation will be important for the discussion below).

This apparently trivial result, however, at close inspection, reveals a conundrum. The fact that eGFP was successfully produced inside small vesicles implies that at least one copy of the 83 different macromolecules (DNA, RNA polymerase, ribosome, etc. see Table 14.1) contained in the PURE system were simultaneously encapsulated within the same lipid vesicle in the moment of vesicle formation. It should be remarked that the concentration of each PURE system component is between 0.1 and 1 μM. It is interesting to calculate the probability of co-entrapping 82 different molecules on the basis of the expected number of solutes/vesicles and the Poisson distribution. For each solute, this probability is dependent on the vesicle size (as the average number of entrapped molecules is given by CV, where C is the solute concentration in bulk and V the vesicle volume), and the overall probability is calculated by considering the independent entrapment of 82 species. The final curve is shown in Figure 14.2A. As intuitively expected, the overall probability of co-entrapping at least one copy of the 82 macromolecular species (each at concentrations below between 0.1 and 1 μM) decreases when the vesicle volume decreases, reaching extremely low values (ca. 10^{-26}!) for vesicles with diameters of 200 nm. In order to have physically significant probabilities, the only reasonable hypothesis is that the *local* concentration of the entrapped solutes was not the same as the bulk values, but at least 10−20 times higher. This hypothesis, together with the observation made above (not all vesicles were capable of synthesizing the protein) brought us to the conclusion that the encapsulation of solutes was not following the Poisson distribution. Rather, few 'special' vesicles were formed by following a mechanism that is not related to the expected capture of solutes (proportional to the vesicle volume). In these vesicles, the entrapment of solutes should follow a mechanism that facilitates solute capture during vesicle formation, so that the corresponding probability of entrapment results were underestimated by the Poisson distribution. Our next question was on how to test this hypothesis.

The current available knowledge related to entrapment of solutes inside lipid vesicles has been generally obtained via averaging techniques (batch absorbance or fluorescence, etc.), and little attention has been devoted to studying individual vesicle encapsulation. But only the direct measurement of the content of each vesicle can determine the true solute occupancy distribution (i.e. the distribution of the number of entrapped solute molecules). We therefore based our investigations on the direct visualization of the entrapped solutes (inside liposomes) by means of cryogenic transmission electron microscopy (cryo-TEM).[29,30] As solutes we employed: (1) ferritin (from 4 to 32 μM), a protein that contains in its core about 4 500 iron atoms, and it is therefore clearly visible in cryo-TEM images as a black round spot; (2) ribosomes (from 0.4 to 8 μM), that are also electron dense enough to be distinguishable in cryo-TEM pictures; (3) *E. coli* extracts and the PURE system, which also contain ribosomes (1.2 μM in the PURE system). The main idea was to study the statistics of entrapment of these macromolecules in liposomes by varying the solute concentration, the vesicle preparation method, and the kind of lipids. We focused on unilamellar vesicles with diameters between 50 to 300 nm, formed by thin-film hydration (with or without extrusion), and by the injection method, starting from POPC, POPC/cholesterol 8/2, or POPC/oleate 8/2 or 6/4.

If the solute occupancy distribution was a Poisson one, we would expect to have a bell-shaped curve centered around the average number N_0 of entrapped solute, calculated as stated above from the product of solute concentration C and the vesicle volume V

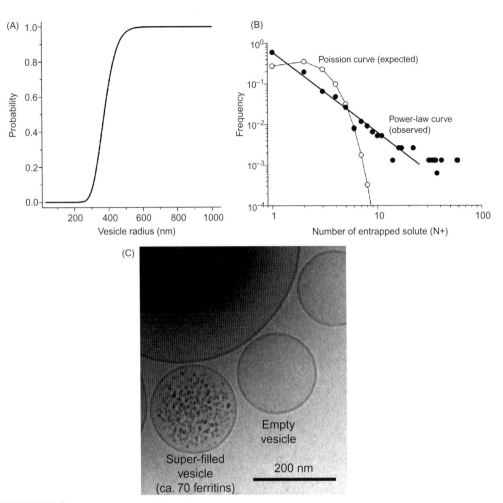

FIGURE 14.2

Probability of coentrapping the about 80 different components of the PURE system (as single copy) inside a lipid vesicle of a certain radius.[27] The probability becomes 1 for large lipid vesicles (radius >600 nm), whereas it rapidly decreases to zero for small vesicles (radius <250 nm). In particular, for vesicles with a radius of 100 nm, the calculated probability is around 10^{-26}. (B) Comparison between Poisson and power-law solute occupancy distribution.[29,30] The Poisson curve refers to vesicles with radius of 100 nm, in the case of the entrapment of 8 µM solutes (one species). In contrast to the expectations, the analysis of about 8000 vesicles reveals that the distribution of ferritin- and ribosome-containing POPC vesicles does not follow the Poisson distribution, being shaped instead as a power-law. Note that the power-law probability of finding a lipid vesicle containing more than 10 solutes lies in the 0.001 probability range, whereas the Poisson curve rapidly decreases to very low values. (C) Cryo-TEM reveals the existence of super-filled vesicles (in this case, filled with ferritin[29]) together with empty vesicles. Empty and filled vesicles, when having approximately the same size as shown here, should entrap approximately the same number of solutes, and the fluctuations should be explained by the Poisson law. But experiments controvert the expectations.

(a shortcut approximate formula gives $N_0 = 2.52\ 10^{-6} C_{\mu M} R^3_{nm}$, where R is the vesicle radius). For example, if vesicles with radius $R = 50$ nm are formed in a $C = 3.2$ µM ferritin solution, it is expected then, on average, each vesicle contains 1 ferritin ($N_0 = 1$). The Poisson probability $p(N)$ of finding a vesicle with N ferritin molecules is calculated accordingly (Fig. 14.2B). For example, given the stochastic nature of the entrapment process, the probability of finding an empty vesicle is not so low ($p(0) = p(1) = 37\%$), whereas the probability of finding a vesicle with two ferritins is 18%. The Poisson curve decreases rapidly, so that $p(10) = 10^{-5}$ %, and $p(20) = 10^{-17}$ % (these values represent, respectively, the probabilities of finding a vesicle with 10 or 20 ferritins, when on average 1 ferritin is expected to be entrapped).

We analyzed individually 7700 vesicles in the case of ferritin entrapment, and 400 vesicles in the case of ribosome entrapment (the third case, namely the entrapment of the PURE system, was not analyzed quantitatively but only qualitatively). As shown in Figure 14.2C, samples consisted of many empty vesicles and very few solute 'super-filled' vesicles. The classical Poisson model cannot explain the existence of such vesicles. The experimentally determined solute occupancy distribution, in all tested cases, does not follow the Poisson distribution, but it is rather shaped like *power-law* distributions. In particular, a long tail of decreasing probabilities (0.1−1%) was typically found in all the experimentally determined distributions, indicating that the probability of finding vesicles containing a very high number of solutes (ferritins, or ribosomes) was low, but not so unrealistic as estimated by the Poisson curve.

Considering the experimentally determined number of macromolecules inside liposomes of a certain diameter, it is possible to determine the internal concentration of these macromolecules in each liposome. The results indicate that for 'superfilled' vesicles, the internal concentration can be up to ca. 20 times higher (but typical values are around 12 for ferritin and around 6 for ribosomes) than the external bulk solute concentration (exceptionally high internal concentrations have been also recorded in rare cases, up to about 50 times the expected concentration). This effect is size-dependent: smaller vesicles (diameter <100 nm) reach higher enhancement factors when compared to larger vesicles.

Having demonstrated that the spontaneous formation of vesicles (in a solute-containing solution) brings about not only empty and regularly filled vesicles, but also a small but measurable fraction of 'super-filled' vesicles, we have speculated about the possible generative mechanisms. Our working hypothesis, at the moment, is based on the interplay between solute/bilayer interaction and the mechanism of vesicle formation.[30] Let us discuss the vesicle closure mechanism that is thought to be involved in several vesicle formation processes such as the injection method. Open bilayer fragments, formed by self-association of lipids, are the intermediates that give rise to closed vesicles. Their half-life has been estimated in the order of milliseconds. We suppose that the solutes stochastically adsorb on the bilayer fragments and slow down the closure rate so that additional solutes can be recruited by the still-open bilayer. The expulsion of bound water would be the driving force of the process.

A possible biological relevance of these results and their importance in the origin-of-life scenarios can be discussed as follows. The spontaneous formation of first membrane-enclosed compartments might have been fundamental to the origins of primitive cells. These results suggest that the compartment is no longer a simple physical container, but it becomes an agent that induces reactivity under conditions where reactions are not possible (diluted solutions). If we assume the possible development of macromolecules and some metabolic network outside compartments, the fact that a large number of solutes can indeed be unexpectedly entrapped inside lipid vesicles during their formation would provide good evidence to explain the emergence of functional primitive cells from free components (separated solutes and lipids). In fact, one of the major theoretical and practical obstacles to explain the formation of functional primitive cells is the very limited solute permeability of most membranes. The other interesting consequence of the 'super-concentration' of solutes inside vesicles is that the resulting high concentration of enzymes or macromolecules allows the overcoming of critical concentration for effective reactivity (e.g. overcoming the dissociation constant of molecular complexes, or enhancing reaction rates of enzymatic reaction when the substrate concentration is higher than the Michaelis constant). This effect, therefore, means that the intravesicle water pool can be considered a privileged environment for biochemical transformation. Clearly, this is also linked to the concept of cellular crowding. The results suggest that a kind of internal crowding can spontaneously emerge from unexpected physical events, probably triggered by the interplay of stochastic fluctuations and the mechanism of vesicle formation.

SSMCS AS A BIOTECHNOLOGICAL TOOL

As we have described in the previous sections, great emphasis has been given to the self-assembly events underlying the formation of minimal cells from separated compounds, freely available in solution. This somehow reflects the spirit of the pioneers in this research field, born within the origin-of-life community. However, also considering the observations described earlier, in order to fully develop an 'SSMC technology' as a more efficient (controllable and reproducible) way of producing solute-filled vesicles would be advantageous. In other words, the fact that only few vesicles are able to encapsulate a complex reaction mixture is seen as an intriguing phenomenon to be studied by scientists interested in emergent behaviors and self-organization, but can be seen as a problem by bioengineers trying to build SSMCs for biotechnology.

Inspired by these considerations, we started a thorough investigation on a new and very interesting vesicle formation method, which was firstly described by the group of Weitz in 2003.[31] The method, that can be referred to as the 'droplet transfer' method, is particularly useful for preparing lipid vesicles (giant lipid vesicles, or GVs, with typical diameters from 1 to 100 μm) filled with a desired content, from simple 'one-solute' to complex 'multisolute' solutions. The produced GVs can be transformed in submicron vesicles by extrusion. The droplet transfer method starts with the formation of lipid-stabilized water-in-oil (w/o) droplets. These are easily obtained by dispersion of a small volume (e.g. 10 μL) of a water solution into hydrocarbons (e.g. 500 μL). Lipids, which are dissolved in the hydrocarbon (the 'oil'), will stabilize the microscopic w/o droplets by forming an oriented layer (ideally, a monolayer) between the water and the hydrocarbon. In the dispersion process, a macrodroplet of the aqueous solution is transformed into millions of microdroplets by a mechanical shearing force (e.g. pipetting, vortexing, etc.). Typically, several thousands of w/o droplets/μL of oil are produced from this fragmentation mechanism, each containing a portion of the solution of interest. Due to the large volume of each droplet (of the order of pL), stochastic factors are partially smoothed out (however they are not eliminated, as evidenced by the fact that the content of each w/o droplet is unique: there is indeed a distribution of the solute concentration among the droplets, whereas in the ideal case the solutes should have the same concentration in every droplet).

After their formation, w/o droplets can be transformed into GVs by the transfer process. This consists of letting the w/o droplets cross an oil–water interface where a lipid monolayer is stratified. In this way, the w/o droplet, covered by a lipid monolayer, becomes coated by another lipid monolayer and acquires in this way the lipid bilayer that characterizes a vesicle. In practical terms, the w/o emulsion is stratified over an aqueous phase. W/o droplets, present in the upper phase tend to sediment toward the bottom of the tube, and after traveling in the oil, reach and cross the oil–water interface, and become transferred into the bottom aqueous phase as GVs. Centrifugation is often used to speed up the process. Not all w/o droplets are successfully transformed in GVs. Actually, most of them break during the oil–water interface crossing, and release their content in the bottom aqueous phase. However, when a GV forms, it contains all the solutes initially present in the w/o droplet. This makes the droplet transfer method so attractive for SSMC technology. We have recently optimized this method for producing oleate/POPC-containing GVs and characterized the process from several viewpoints. The droplet transfer method has been used for producing GVs that contain the complex transcription/translation kit.[22,26] Due to its superior entrapment efficiency, we believe that the droplet transfer method will become the method of choice for all those applications where SSMCs are used as a biotechnological tool.

Thinking of future developments, together with the cell-free and liposome technologies, which dominate the current approaches to SSMCs, we think that microfluidic technology will also impact very much on the use of SSMCs for biotechnology. Several microfluidic devices have been applied, in recent years, to the production of w/o droplets (which are easy to make), in order to exploit the 'in vitro compartmentalization' (IVC) strategy (à la Tawfik/Griffiths[32]) for high-throughput screening the activity of proteins, genes, etc. for directed evolution, and assay miniaturization, as well as for several physicochemical investigations.[33] In contrast to the use of microfluidic devices for creating and manipulating the w/o droplets, very few efforts have been reported on the application of microfluidic technology to lipid vesicles (in particular GVs). In this respect, the final goal would be the high-throughput production of GVs in a very reproducible manner, as already happens with w/o droplets, and with the possibility of controlling their internal solution. Ideally, a microfluidic device could produce GVs for biotechnological applications thanks to its reliability in terms of the features described above. In other words, this would correspond to a controllable and fully tunable 'lipid-vesicle-producing machine.' To the best of our knowledge, only five reports have been published dealing with the issue of producing GVs by microfluidic devices. These have been reviewed in[11], with a recent addition.[34] If these methodologies, which are still young, are further developed in a robust way, they certainly will improve our technological ability for constructing SSMCs, not necessarily living, for a variety of applications.

One of the most interesting applications refers to the construction of 'bioreactors,' 'nanofactories,' or 'wet-soft bionanorobots' for intelligent drug delivery applications. In their vision, Le Duc and coworkers[35] foresee that a kind of SSMC is built by encapsulating a minimal metabolic network capable of producing a drug. The drug production should occur only when the SSMC has reached the target tissue inside the human body, for instance, thanks to the specific addressing mechanism (e.g. by functionalizing the lipid vesicle with antibodies that recognize specific antigens presented on the surface of the target tissue). Then, the SSMC, by 'sensing' its environment, should trigger the intravesicle drug production, and its exportation in the external environment. From there, the drug should diffuse into the target cell and exert a pharmacological effect. It is interesting to note that several of the functions required for the construction of such nanofactories are typical functions that are also important in basic research on SSMCs. Possibly, therefore, the knowledge acquired in basic research will be applied to design and construct advanced SSMCs for specific biotechnological applications. Remarkably, the sensing function and the corresponding activation of an internal response bring us directly to the concepts of control and communications.

Quite recently, we argued that the technology of SSMCs can be useful for developing new tools in bio-inspired information and communication technology (bio-ICT). In particular, we have proposed an experimental program for letting synthetic cells communicate with natural cells, via chemical signals.[36] The experimental model to start with would be the development of a chemical language that is common with bacteria, for instance based on acyl homoserine lactones (AHLs), signaling peptides, or furanones. We plan to develop SSMCs step-wise that can send signals to living cells, like in the first pioneer work, published in 2009 by the group of Davis.[37] With this aim, SSMCs should produce AHLs by expressing the corresponding synthase (e.g. via the PURE system). AHLs, once synthesized, can freely diffuse away and reach the receiving (living) cell, activating the expression of a reporter gene. Conversely, SSMCs can be designed in order to express a receptor for sensing a signal molecule sent by the natural partner. After being activated by the chemical signal (also in this case it is convenient to use AHLs), the receptor allows the transcription of a reporter gene. Both the receptor and the reporter gene will be expressed by transcription/translation kits, like the PURE system. Next, we plan to combine these mechanisms for establishing bidirectional synthetic cell–natural cell communication, and after this, synthetic communication between synthetic cells. It is remarkable that this research program implies that SSMCs can be used as computing entities that could lead to new forms of bio-ICTs.

273

SOME OPEN QUESTIONS AND FUTURE PERSPECTIVES

In this chapter we have shown a possible (and realistic) experimental approach to the construction of minimal cells in the laboratory. We also gave a broad overview of the current state of the art in this field, from three different viewpoints: (1) the reconstruction of genetic/metabolic networks inside lipid vesicles for achieving minimal living entities; (2) the importance of physical effects for the material construction of minimal cells, in particular the physics of solute entrapment; and (3) the technological viewpoint, that will certainly affect the future developments of SSMCs in the biotechnology field.

Within the viewpoint (1), we have seen that genes can be expressed inside lipid vesicles. In the case of water-soluble proteins, their intravesicle synthesis should be considered a viable route for developing biochemical functions, except for the case of proteins that require post-translational modifications. In the case of lipid-associated proteins, instead, results indicate that careful design of the membrane composition is a prerequisite for their successful production. Note that by varying the chemical nature of lipids, side problems might arise with respect to encapsulation efficiency and chemical interference with the transcription/translation kits.

One of the most relevant and ambitious open questions in SSMCs research is the achievement of complete 'core-and-shell' self-reproduction. In particular, in order to self-reproduce all its components, as required by the autopoietic theory, minimal cells should be able to synthesize lipids (for membrane growth), and also synthesize all macromolecular components that are enclosed in the shell. These two production processes should be synchronized, in order to let minimal cells grow harmoniously. An unbalanced growth would be deleterious for self-reproduction. The achievement of synchronized core-and-shell self-reproduction will help to understand what the physicochemical constraints that regulate that process are, and possibly, how primitive cells could have faced this problem with a metabolic network of minimal complexity.

From the viewpoint of solute entrapment (2), we have described a novel and unexpected behavior that brings about the spontaneous formation of solute-filled compartments, even starting from diluted solution. We have already discussed the relevance of such observation for understanding the origin of cellular life. An open question in this respect is certainly linked to the investigation of the mechanism that originates these special vesicles. This could not only explain the origin of cellularity, but also could allow the formation of 'super-concentrated' vesicles in a controlled manner (the production of enzyme-filled vesicles could find applications in enzyme therapy). Additional questions are related to the extension of the results obtained to other proteins and other lipids, to verify the generality of the mechanism. An interesting scenario, which is currently under study in our laboratory, focuses on the exploitation of the spontaneous concentration of solutes inside lipid vesicles to activate an otherwise sluggish reaction (e.g. a diluted mixture of compounds that does not react in bulk solution – due to the low concentration – becomes concentrated, and reacts, only inside lipid vesicles).

Several open questions and perspectives are instead related to the biotechnology issues (3), because this field is still in its infancy. The first one is about the goal of standardizing SSMC production. Only when this target is reached will synthetic cells (non-'living' cells) enter the applicative phase. In addition to the already presented example of intelligent and programmable drug-producing vectors, SSMCs could be used for in vitro assays on cell-like systems, as biosensors, as microrepairing devices, for multienzyme immobilization, or to study cellular function without the interference of background processes. It is remarkable that the employment of SSMCs in human health and medicine is advantageous because they should not present biosafety issues, in contrast to engineered living cells.

CONCLUDING REMARKS

The construction of synthetic cells is one of the most ambitious goals of SB, and more generally of biological science and technology. Firstly, it allows the study of structural and dynamic (biochemical) self-organization of separated molecules into systems, shedding light on the deepest scientific questions: how living systems emerged from nonliving molecules. Probably, the most fascinating facet of this question is the reciprocal influence between the parts (i.e. the molecules, at a lower hierarchical level) and the system (i.e. the cell, at a higher hierarchical level), as we have shown by illustrating the theory of autopoiesis, the various biochemical transformations inside lipid vesicles, and the spontaneous formation of super-filled vesicles. Intriguingly, the complexity scale can be expanded upward by introducing the concept of communities of systems (e.g. community of cells). The establishment of synthetic cell–natural cell communication or synthetic cell–synthetic cell communication is a possible future experimental target. Currently, however, we are investigating the association of GVs in colonies, to model the origin of primitive cell communities.[38] This study is interesting because for the first time a scenario is proposed where groups of cells rather than isolated ones are considered relevant to the origin of life. This happens thanks to properties that emerge at their hierarchical level (vesicle fusion, solute capture, enhanced permeability through the membrane of colony vesicles), whereas the separated components (isolated vesicles, bridging agents), which are at a lower hierarchical level, do not share these properties. Finally, we have seen that the efforts for constructing cell models, mainly developed to answer questions on the origin of life, have also produced a new kind of biotechnological approach, namely SSMC technology. We believe that, when properly developed, this technology can become a central subject in synthetic biology.

Acknowledgments

This work derived from our recent involvement in studies on the construction of semisynthetic minimal cells, funded by the Sixth Framework EU Program (SYNTHCELLS: Approaches to the Bioengineering of Synthetic Minimal Cells 043359), HFSP (RGP0033/2007–C), ASI (I/015/07/0), PRIN2008 (2008FY7RJ4); and further expanded thanks to networking initiatives such as SynBioNT(UK), and the COST Systems Chemistry action (CM0703).

References

1. Andy D. Foundations for engineering biology. *Nature*. 2005;438:449–453.

2. Gibson DG, Glass JI, Lartigue C, et al. Creation of a bacterial cell controlled by a chemically synthesized genome. *Science*. 2010;329:52–56.

3. Bolli M, Micura R, Eschenmoser A. Pyranosyl-RNA: chiroselective self-assembly of base sequences by ligative oligomerization of tetranucleotide-2′,3′-cyclophosphates (with a commentary concerning the origin of biomolecular homochirality). *Chem Biol*. 1997;4:309–320.

4. Doi N, Kakukawa K, Oishi Y, Yanagawa H. High solubility of random-sequence proteins consisting of five kinds of primitive amino acids. *Protein Eng Des Sel*. 2005;18:279–284.

5. Luisi PL. Chemical aspects of synthetic biology. *Chem & Biodiv*. 2007;4:603–621.

6. Luisi PL, Chiarabelli C, eds. *Chemical Synthetic Biology*. Chichester: Wiley; 2011.

7. Luisi PL. The synthetic approach in biology: epistemological notes for synthetic biology. In: Luisi PL, Chiarabelli C, eds. *Chemical Synthetic Biology*. Chichester: Wiley; 2011:343–362.

8. Chiarabelli C, Vrijbloed JW, De Lucrezia D, et al. Investigation of de novo totally random biosequences. part ii. on the folding frequency in a totally random library of de novo proteins obtained by phage display. *Chem & Biodiv*. 2006;3:840–859.

9. Luisi PL, Ferri F, Stano P. Approaches to semi-synthetic minimal cells: a review. *Naturwissenschaften*. 2006;93:1–13.

10. Luisi PL, Stano P, eds. *The Minimal Cell. The Biophysics of Cell Compartment and the Origin of Cell Functionality*. Dordrecht: Springer; 2011.

11. Stano P, Carrara P, Kuruma Y, Souza T, Luisi PL. Compartmentalized reactions as a case of soft-matter biotechnology: synthesis of proteins and nucleic acids inside lipid vesicles. *J Mat Chem*. 2011;21:18887–18902.

12. Varela F, Maturana H, Uribe R. Autopoiesis: the organization of living systems, its characterization and a model. *BioSystems*. 1974;5:187–195.

13. Bachmann PA, Luisi PL, Lang J. Autocatalytic self-replicating micelles as models for prebiotic structures. *Nature*. 1992;357:57–59.

14. Bachmann PA, Walde P, Luisi PL, Lang J. Self-replicating micelles: aqueous micelles and enzymatically driven reactions in reverse micelles. *J Am Chem Soc*. 1991;113:8204–8209.

15. Walde P, Wick R, Fresta M, Mangone A, Luisi PL. Autopoietic self-reproduction of fatty acid vesicles. *J Am Chem Soc*. 1994;116:11649–11654.

16. Stano P, Luisi PL. Achievements and open questions in the self-reproduction of vesicles and synthetic minimal cells. *ChemComm*. 2010;46:3639–3653.

17. Gil R, Silva FJ, Peretó J, Moya A. Determination of the core of a minimal bacteria gene set. *Microbiol Mol Biol Rev*. 2004;68:518–537.

18. Schmidli PK, Schurtenberger P, Luisi PL. Liposome mediated enzymatic synthesis of phosphatidylcholine as an approach to self-replicating liposomes. *J Am Chem Soc*. 1991;113:8127–8130.

19. Yu W, Sato K, Wakabayashi M, et al. Synthesis of functional protein in liposome. *J Biosci Bioeng*. 2001;92:590–593.

20. Shimizu Y, Inoue A, Tomari Y, et al. Cell free translation reconstituted with purified components. *Nat Biotechnol*. 2001;19:751–755.

21. Ishikawa K, Sato K, Shima Y, Urabe I, Yomo T. Expression of a cascading genetic network within liposomes. *FEBS Lett*. 2004;576:387–390.

22. Noireaux V, Libchaber A. A vesicle bioreactor as a step toward an artificial cell assembly. *PNAS*. 2004;101:17669–17674.

23. Kita H, Matsuura T, Sunami T, et al. Replication of genetic information with self-encoded replicase in liposomes. *ChemBiochem*. 2008;9:2403–2410.

24. Kuruma Y, Stano P, Ueda T, Luisi PL. A synthetic biology approach to the construction of membrane proteins in semisynthetic minimal cells. *Biochim Biophys Acta*. 2009;1788:567–574.

25. Kurihara K, Tamura M, Shohda K, Toyota T, Suzuki K, Sugawara T. Self-reproduction of supramolecular giant vesicles combined with the amplificationof encapsulated DNA. *Nat Chem*. 2011;3:775–781.

26. Maeda YT, Nakadai T, Shin J, Uryu K, Noireaux V, Libchaber A. Assembly of MreB filaments on liposome membranes: a synthetic biology approach. *ACS Synth Bio*. 2012;1:53–59.

27. Souza T, Stano P, Luisi PL. The minimal size of liposome-based model cells brings about a remarkably enhanced entrapment and protein synthesis. *ChemBioChem*. 2009;10:1056–1063.

28. Knoll A. *Size Limits of Very Small Microorganisms, Proceedings of a Workshop*. Washington DC: National Academic Press; 1999:pp. 1–3.

29. Luisi PL, Allegretti M, Souza T, Steineger F, Fahr A, Stano P. Spontaneous protein crowding in liposomes: a new vista for the origin of cellular metabolism. *ChemBioChem*. 2010;11:1989–1992.

30. Souza T, Steiniger F, Stano P, Fahr A, Luisi PL. Spontaneous crowding of ribosomes and proteins inside vesicles: a possible mechanism for the origin of cell metabolism. *ChemBioChem*. 2011;12:2325–2330.

31. Pautot S, Frisken BJ, Weitz DA. Production of unilamellar vesicles using an inverted emulsion. *Langmuir*. 2003;19:2870–2879.

32. Tawfik DS, Griffiths AD. Man-made cell-like compartments for molecular evolution. *Nat Biotechnol*. 1998;16:652–656.

33. Theberge AB, Courtois F, Schaerli Y, et al. Microdroplets in microfluidics: an evolving platform for discoveries in chemistry and biology. *Angew Chem Int Ed Engl*. 2010;49:5846–5868.

34. The SY, Khnouf R, Fan H, Lee AP. Stable biocompatible lipid vesicle generation by solvent extraction-based droplet microfluidics. *Biomicrofluidics*. 2011;5:044113.

35. Leduc PR, Wong MS, Ferreira PM, et al. Towards an in vivo biologically inspired nanofactory. *Nat Nanotechnol*. 2007;2:3–7.

36. Stano P, Rampioni G, Damiano L, Carrara P, Leoni L, Luisi PL. Semi-synthetic minimal cells as a tool for biochemical ICT. *Biosystems*. 2012; http://dx.doi.org/10.1016/j.biosystems.2012.01.002.

37. Gardner PM, Winzer K, Davis BG. Sugar synthesis in a protocellular model leads to a cell signalling response in bacteria. *Nat Chem*. 2009;1:377–383.

38. Carrara P, Stano P, Luisi PL. Giant vesicles 'colonies:' a model for primitive cell communities. *ChemBioChem*. 2012;13:1497–1502.

Transforming Synthetic Biology with Cell-Free Systems

Arnaz Ranji[1], Jeffrey C. Wu[2], Bradley C. Bundy[2] and Michael C. Jewett[1]
[1]Northwestern University, Evanston, IL, USA
[2]Brigham Young University, Provo, UT, USA

INTRODUCTION

Synthetic biology is a new approach for engineering biology by design. At its heart, this emerging discipline seeks to make biology easier to engineer and to harness biology to serve society. Already, this new paradigm for engineering biology is enabling a deeper understanding of living systems and opening the way to sustainable and renewable energy production, cost-effective and widely accessible programmable medicines, and new solutions for environmental stewardship. While most synthetic biology efforts have focused on engineering living organisms, the cell's operating system and endogenous pathways pose an obstacle to synthetic biology efforts. This is because the cell's evolutionarily optimized agenda for growth and adaptation are often at odds with the engineer's objectives. Additionally, the complexity of living cells is still beyond our comprehension, and serves as a disadvantage to using living cells to develop synthetic biology applications. Cell-free systems circumvent these obstacles, and are emerging as a powerful approach for harnessing and expanding the capabilities of natural biological systems.

Over the last several years there has been tremendous development and innovation in the growing, yet still small, field of cell-free synthetic biology. This chapter focuses on recent developments and current applications in this field. We begin by defining cell-free biology and its advantages. We then highlight existing applications in cell-free synthetic biology from the nucleic acids to proteins to metabolites (Fig. 15.1). Finally, we consider challenges and opportunities necessary to propel the field forward.

CELL-FREE BIOLOGY

Cell-free biology is the activation of complex biological processes without using intact, living cells. Bypassing cell walls, this approach opens up a whole new biological world, one without the need or ability to preserve DNA heritage.[1] Moreover, this approach enables one to access and manipulate biology directly. In contrast to in vivo engineering efforts, this direct relationship with biocatalytic enzymes provides the ability to focus metabolism on the production of a single compound and removes physical barriers (allowing easy substrate addition, product removal, and rapid sampling). Furthermore, avoiding competition

Synthetic Biology. DOI: http://dx.doi.org/10.1016/B978-0-12-394430-6.00015-7

277

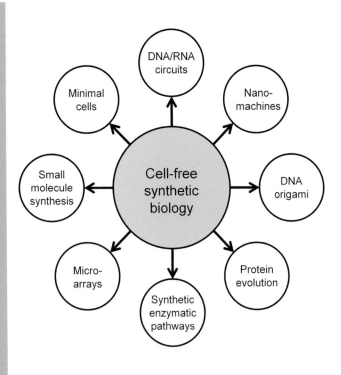

FIGURE 15.1

Cell-free synthetic biology applications: The various applications of cell-free synthetic biology covered in this chapter are indicated.

between cell growth (catalyst synthesis) and production (catalyst utilization) through the use of cell-free systems provides efficiency benefits for how we use synthetic biology for synthetic chemistry. In short, cell-free systems provide an unprecedented freedom of design to modify and control biological systems.

Cell-free systems can be generated from crude extracts or purified components. Crude extracts are prepared by lysing cells (typically harvested in the exponential growth phase) and are cheaper, as well as simpler to produce than purified systems. While crude extract-based systems are prominent in the field and have now been commercialized,[2] purified systems also have their merits. Specifically, the fact that every component of a purified system is well-characterized and can be tuned provides advantages for engineering design. Furthermore, purified systems lack unwanted activities (such as proteases and nucleases) that might be deleterious to crude extract systems. An example of a well-utilized purified system is the PURE system for cell-free translation, pioneered by the Ueda group.[3]

Scale has long been considered a technological limitation of cell-free systems. Due to recent advances in cost-effective high-level protein synthesis,[4] *E. coli*-based cell-free protein synthesis systems have been successfully scaled-up to the manufacturing scale.[2] Production rates, production yields, and product activity were maintained over a volumetric expansion of 6 orders of magnitude.[2,5] This pioneering advance has transformed cell-free systems from a foundational bench-top tool into an enabling technology with promise for commercial scale applications.

ADVANTAGES OF CELL-FREE BIOLOGY

Cell-free technologies simplify biological systems and remove unnecessary overheads for engineering efforts. In this section, we highlight some of the many advantages gained by removing physical barriers and decoupling cellular catalysts from the genetic architecture of the cell.

Direct Environment Control of an Open System

Cell-free biology enables direct access to and control of the biological environment. In living cells, the organism must maintain an environment conducive to every biocatalytic mechanism necessary for survival. In contrast, the cell-free environment can be directly optimized for a user-defined objective. The indispensable microbiology workhorse of polymerase chain reaction (PCR) is a simple but powerful cell-free synthetic biology example where the DNA polymerase, template, primer, and salt concentrations are directly controlled and optimized in a nuclease-free environment.[6] In another useful but simple cell-free example, RNA transcription can also be directly optimized while maintaining an RNase-free environment.[7] In cell-free protein synthesis, the capability to directly control variables such as ionic strength, pH, redox potential, hydrophobicity, and enzyme and reactant concentrations has provided extraordinary advantages.

Direct Influence of Reaction Networks

Open cell-free biology enables the addition or removal of catalysts and/or reagents to directly influence reaction networks. In one example, Zhang and coworkers have demonstrated the optimized combination of 12 enzymes for 99.6% energy-efficient biohydrogenation of xylose and cellobiose to xylitol.[8] This example, along with others like it,[9-11] showcases the freedom of design for building metabolic networks from the ground up using cell-free systems. In another example, cell-free circuits can be built and controlled by the addition of enzymes to the nucleic-acid-based circuits to control their rates of synthesis and degradation.[12] In the case of cell-free protein synthesis, the open environment enables simplified preparations of PCR products for protein expression. Expression templates can be directly added and the concentration controlled for optimization. In one example, linear expression templates (LETs) prepared using PCR have been effectively used in cell-free translation systems by multiple research groups.[13-15] LETs obviate the need for time-consuming gene-cloning steps, accelerating process and product development pipelines. The ability to use LETs in the cell-free system is particularly valuable when the goal is the production of a large array of gene products, such as in genomic studies.[14]

Direct Product Access

In cell-free systems, the lack of cell walls or membranes facilitates online product monitoring, one-step purification and recovery of DNA, RNA, or protein.[16,17] In the case of cell-free protein synthesis, several single-step purification processes using affinity chromatography have been demonstrated.[16,18] These include strep- and his-tag purification and in situ product purification using magnetic affinity beads.[19] In another illustrative model, Bujara et al. coupled cell-free metabolic engineering to mass spectrometry analysis for direct profiling of rate-limiting steps to optimize multienzyme catalysis of dihydroxyacetone phosphate (DHAP) from glucose.[20]

Focusing Biological Machinery on a Single User-Defined Objective

Without the need, or ability, to support ancillary processes required for adaptation and growth, cell-free systems offer the ability to focus metabolism towards the exclusive objective of the engineer, and not the cell. This affords efficiency advantages necessary for producing complex biochemical products.

Accelerating Design—Build—Test Cycles

Engineering biology today is a time- and money-intensive process. Therefore, one of the key objectives of synthetic biology is to accelerate the design—build—test loops required for engineering biology. Right now, in vivo approaches take, on average, 3—4 months to complete this cycle (Dr. Alicia Jackson, DARPA 'Living Foundries' Industry Day 2011, personal communication). Much of this time is taken to identify and modify potential genes, and assemble and synthesize the potential pathways in living organisms. However, the design cycle limit of in vivo synthetic biology projects could at best potentially scale (in the future) as the growth rate of the organism and time for transformation. In cell-free systems, the design cycle is not limited by how fast cells reproduce. Rather, it can be faster, potentially approaching the limit of synthesizing the components (DNA, RNA, proteins). Ultimately, cell-free systems may therefore speed up the design cycle for engineering by more than 10-fold relative to in vivo approaches, and could be used as their own biomanufacturing platforms, or as feedback in the design of in vivo platforms.

Decreased Effects of Toxicity

Due to cell death and hindered cell growth, in vivo production of both cytotoxic products and products synthesized from cytotoxic substrates is a challenge. Unhindered by cell viability constraints, cell-free systems address this challenge. Over the last decade, researchers have demonstrated the use of cell-free systems to produce a growing number of cytotoxic products.[21] These include proteins with cytotoxic amino acids,[22,23] cytolethal disease toxins,[24] cytotoxic proteins,[25] and others.[26,27] In one exemplary approach, hepatitis researchers studying a cytolethal distending toxin from *Helicobacter hepaticus* had been unable to produce sufficient levels of protein in vivo to observe its mechanism of action due to its high cytotoxicity. Using a cell-free approach, researchers were able to produce sufficient quantities to examine the toxin's effect on the liver.[24] In another illustrative case, cell-free translation of the cytotoxic A2 protein led to yields almost 1000 times higher than previously reported with in vivo production.[25]

Expanding the Chemistry of Life

The products of biological systems are governed by the chemistry of life, which is limited to natural building blocks. Cell-free systems offer advantages for using synthetic biology for nonstandard chemistry.[28] One of the most prominent examples is using cell-free protein synthesis for site-specific incorporation of nonnatural amino acids. In contrast to in vivo systems, there are no transport limitations for getting nonnatural amino acids into the cell, and there is flexibility for reprogramming the genetic code because cellular viability need not be maintained. By employing cell-free protein synthesis, the putatively transport-limited p-propargyloxyphenylalanine was incorporated site-specifically into proteins at significantly higher production yields than in vivo systems.[29] The high-yielding site-specific incorporation of azidophenylalanine and the global replacement of methionine by an unnatural analogue have also been demonstrated with cell-free systems.[30,31]

EXISTING TECHNOLOGIES AND APPLICATIONS IN CELL-FREE SYNTHETIC BIOLOGY

Cell-free synthetic biology projects focus on exploiting and harnessing biopolymer synthesis, replication, and evolution. To this end, we organize our discussion of existing research frontiers by levels of the biological hierarchy: nucleic acids; proteins; metabolites; and minimal cells. Focus is given towards recent efforts in designing and constructing programmable circuits, advances in synthesizing biopolymers and metabolites, and steps toward enabling self-replication.

Nucleic Acids

As the field of synthetic biology has emerged, we are now at a position where both writing (i.e. synthesis) and reading (i.e. sequencing) DNA are not limiting. Indeed, nucleic acid synthesis forms the basis for synthetic biology.[32,33] DNA synthesis can be achieved in vitro by PCR, recombinant DNA cloning, or by chemical synthesis. PCR has extensive applications in medical diagnosis, cloning, forensics, phylogenic analysis, etc. Like DNA, RNA can be synthesized in vitro by chemical synthesis. A good review detailing the history and advances in RNA synthesis is available.[34] RNA can also be transcribed in vitro using purified T7 polymerase, NTPs, and optimized buffers. A timeline of developments in this area of oligonucleotide-based cell-free synthetic biology is illustrated in Figure 15.2A, and some of the main developments in the field are listed below.

NUCLEIC ACID CIRCUITS

Nucleic acid circuits involve programming biological circuits, which connect the fields of molecular biology and computational science. Over the last five years, there has been a dramatic growth in reports describing synthetic in vitro circuits.[34a] This growth underscores the importance of cell-free systems as a testing ground for our ability to engineer and analyze even more complex biological circuits; a testing ground where at least we can clearly see when and how we fail. Rather than encoding ones and zeros into high and low voltages, nucleic acid computing involves choosing specific base sequences on synthetic strands of DNA to process information. Nucleic acid circuits are modular,

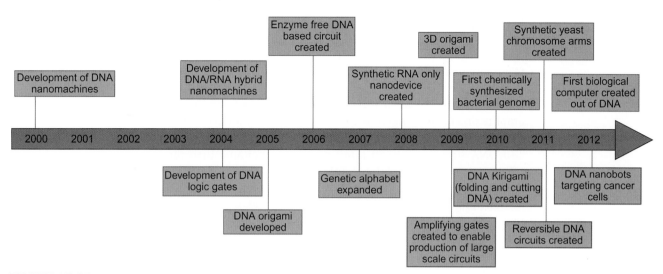

FIGURE 15.2A

Significant discoveries in nucleic-acid-based synthetic biology: Breakthrough discoveries in DNA and RNA synthetic biology that propelled the field forward are shown along a timeline.

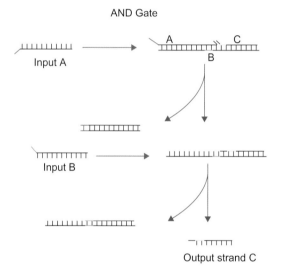

FIGURE 15.2B

DNA logic gate: An example of a DNA logic gate: the AND gate, is illustrated. The gate is composed of three strands (A, B, and C) along with Toeholds indicated in color. Two input strands (strand A and strand B) are required to displace the A and B strands resulting in the output C strand. The released output strand can be further used as input for a downstream gate or may fluoresce upon release.

scalable, and rely on Watson-Crick base-pairing between strands. Both DNA and RNA have been used.

DNA logic circuits initiated the field[35] and have remained a prominent fixture of development. Indeed, by modulating DNA species (i.e. bit information), concentration, and quantity, researchers have constructed a variety of enzyme-free in vitro DNA-based logic gates capable of Boolean logic functions (AND, OR, NOT). In one exemplary report, Seelig et al. described the construction of a circuit composed of 11 gates.[36] An example of a DNA logic gate is illustrated in Figure 15.2B.

Recent developments have seen orders of magnitude increases in complexity (i.e. more nucleic acid components or components with multiple functionalities) for more predictable and complicated artificial biochemical circuits. In one approach, Erik Winfree's team designed entropy-driven catalytic gates to provide an amplifying circuit element, increasing the speed and performance of cell-free circuits.[37] Later, the same group made additional advances in creating scalable DNA architecture by generating gates that could amplify downstream signal with minimal noise and errors. Specifically, they designed a DNA computer capable of calculating the square root of numbers up to 15 with up to 130 DNA strands, albeit in a much slower timeframe than digital computers. This technology could eventually enable the production of large-scale circuits involving thousands of gates.[38,39] However, with the scaling-up of circuits comes the added burden of designing the DNA. To address this challenge, a programming language to facilitate the design of DNA circuits based on DNA strand displacement has now been created.[40] This tool will help generate sequences for DNA circuit construction after simulation of nucleic acid strand displacement behavior. In other advances, researchers in the Tuberfield group demonstrated the creation of a reversible DNA circuit that continuously changed output depending on changing input (unlike in an irreversible circuit).[41] Further, modular DNA circuits have been created to detect different analytes and give off various output signals (electrochemical, colorimetric, and fluorescent outputs).[42]

While DNA computation enables us to model and build sophisticated biochemical circuits, RNA transcriptional logic is used more frequently for cellular regulation. Additionally, RNA has unique properties (relative to DNA) that add a new dimension when engineering circuits. RNA can act as a sensor, be catalytic, and exhibit different conformations to elicit a response.[43] Further, the ability of RNA to base pair enables the creation of synthetic circuits when placed in *cis* and *trans* or in layers.[44] The development of hybrid nucleic acid systems, comprising both DNA and RNA, is enabling us to understand new principles of biological circuitry from a constructive biology approach.

Erik Winfree's group at Caltech has contributed pioneering efforts to the development of hybrid systems. Their first foray into programmable biochemical circuitry was made using simplified circuit construction that utilized nucleic acids as regulatory molecules. As a stripped-down version of a genetic circuit, they created an artificial transcriptional network with modular and scalable molecular switches. They also attempted to control nucleic acid production and degradation by including T7 polymerase and RNAse H.[12]

Highlighting the utility of cell-free systems for accelerating our understanding of the principles of biological modules that could be used to obtain the same functional output, several studies in 2011 have reported synthetic in vitro oscillators with different molecular implementations. Montagne et al.[45] demonstrated a remarkably elegant (and simple) oscillator that uses just three nucleic acid strands and can be quantitatively modeled. At the exact same time, Kim et al. reported in vitro oscillators using transcriptional circuits from seven nucleic acid strands that can be modularly wired into artificial networks by changing the regulatory and coding domains of DNA.[46] In this system, RNA (not DNA) was the only molecule synthesized and degraded, thereby preventing accumulation of mutations in the DNA. An interesting observation is that even simple systems such as this seven-component oscillator challenge our engineering and analysis abilities. Franco et al. later expanded on this oscillator work of Kim et al. by combining the synthetic in vitro oscillator with DNA nanomachines (discussed below), as well as the synthesis of a functional RNA structure.[47] The key aspect of this research was the ability to use an insulator circuit to isolate the DNA nanomachine operation from the oscillator dynamics, allowing for much larger loads in the circuit. In the future the oscillator can be coupled to more complex downstream processes, and could even be incorporated into artificial cells (see below). However, waste product accumulation, limitations in energy sources, and enzyme function over time are obstacles that need to be addressed.

Beyond proving valuable for exploration of biochemical circuit design principles, continued developments in cell-free circuits promise to have a variety of applications. Key potential areas for use include: in situ detection of expression patterns;[36] disease detection based on mRNA levels;[48] controlling nanoscale devices; and organizing biological processes in minimal cells.[45,46]

NANOMACHINES

As with nucleic acid circuits, the simplicity, specificity, and predictability of DNA make it an ideal building block for synthetic nanomachines.[49] Nucleic acid nanomachines are synthetic devices that switch between different molecular conformations based on interactions with signaling molecules or changes in the environment.[50] They have received considerable attention for their possible applications in molecular-scale electronics and as biomedicines.

The earliest nanomachines were controlled by environmental factors such as pH and temperature, but never by a sequence-specific code.[50] A breakthrough paper by Yurke et al. outlined the construction of DNA 'tweezers' using DNA hybridization and strand displacement technology to create a DNA nanomachine that could repeatedly be opened and closed.[51] More recently, a nucleic acid nanomachine comprising DNA and RNA was created by researchers using DNA tweezers, as well as another gene that encodes for an mRNA that closes/changes the conformation of the tweezers. A single-stranded DNA removal strand can then hybridize with the mRNA and result in opening of the tweezers. Using this simplified system researchers can create a more complicated and autonomous nucleobase nanomachine.[52]

Researchers also designed DNA nanomachines with integrated instructions into a gene to establish control of the DNA nanodevice in vitro using gene regulation switches. By incorporating DNA regulated by regulatory proteins from *E. coli* operons, the researchers were able to create an independently functioning nanomachine that responds to environmental cues, such as changes in lactose concentrations. This work paved the way for development of more complicated nanomachines that are able to respond to environmental changes.[53] Beyond DNA, RNA devices are also being developed.[54]

DNA ORIGAMI

The field of structural DNA nanotechnology was born in the 1980s with initial work by Seeman, who proposed building geometric structures with multiple strands of DNA at a nanoscale level.[1,55] However, complications in synthesis and stoichiometric control left the field lacking in more promising developments.[1] In 2004 the Joyce group demonstrated that a long single-stranded DNA could be folded into a geometric shape (octahedron) with the help of synthetic shorter oligomers.[56] Taking this further, Paul Rothemund, in a landmark discovery, demonstrated that long single-stranded DNA could be folded at nanoscale levels with the help of smaller staple strands to create various shapes.[57] Specifically, Rothermund demonstrated the creation of complex single-layered nanostructures that were 100 nm in diameter and had a spatial resolution of 6 nm.[57] Termed DNA origami, this approach creates custom nanoscale shapes that are atomically precise by taking advantage of the specificity of Watson Crick base-pairing.

In recent efforts, there has been a thrust to make 3D DNA origami structures and expand the dimensionality and functionality of molecularly engineered objects. Douglas et al., for example, demonstrated the ability to make a variety of 3D shapes from linear DNA molecules.[58] However, proper assembly of the 3D structure required week-long folding

283

times and was much more complicated than 2D origami.[58] New technology by the Yan group has created a more sophisticated version of DNA origami termed DNA kirigami that involves folding and cutting DNA into topological objects.[59] Additionally, the same group created a strategy to engineer 3D DNA structures with complex curvature.[60] As the fundamental design principles of complicated architectures become more and more elucidated, including computer-aided design caDNAno,[61] this field is now turning to applications.[1] Key among these is the creation of DNA nanochips that can be used to observe single molecule behavior of DNA-binding enzymes,[62] and DNA nanorobots for medical therapeutics and medical diagnostics.[63] In particular, the recent demonstration that DNA nanorobots can target cancer cells and deliver an antibody payload[63] is expected to usher in a new era of DNA device utilization.

Proteins

In contrast to in vitro efforts in DNA and RNA that mainly center on building and understanding nucleic acid circuitry, major in vitro efforts in proteins have been focused on synthesis and evolution strategies. Indeed, a technological renaissance has reinvigorated cell-free protein synthesis (CFPS) technologies over the past decade.[4] This progress has realized protein yields exceeding grams of protein produced per liter reaction volume, cost reductions of multiple orders of magnitude, and microscale to manufacturing scale production. Here we discuss both bottom-up and top-down approaches to CFPS, along with frontier applications enabled by recent advances.

BOTTOM-UP SYSTEMS

The bottom-up approach to CFPS centers on using purified components. Pioneered by Ueda and colleagues, the most prevalent bottom-up system is the PURE system.[3,64] In the PURE system, cellular machinery necessary for translation is independently overexpressed, purified, and combined in a test tube.[65] An advantage of the PURE system is that the researcher is afforded the greatest management of every aspect of the protein synthesis process. Indeed, the ability to mix and match nearly any component of translation has proven remarkably useful for efforts to fold proteins[64] and to incorporate nonnatural amino acids.[28] The main disadvantage to this system is its cost. The necessity of expressing, purifying, and adding each component greatly increases the reagent cost and time required compared to top-down systems.[66] A cost comparison of the CFPS systems discussed herein is provided as Table 15.1.

TOP-DOWN SYSTEMS

The top-down approach involves the modification and engineering of crude cell extract for protein production. Although any organism can potentially provide a source of crude lysate, the most frequently utilized crude extract systems are made from *Escherichia coli*, wheat germ, rabbit reticulocytes, and insect cells.[4]

The *E. coli*-based CFPS systems are currently the most widely used due to lower labor requirements for cell growth and extract preparation, lower-cost energy sources, higher yields, greater rate of protein synthesis, and commercial scalability.[2,67] Given the significant number of proteins previously expressed in this system and its high linear scalability (from 15 μL to 100 liters), the *E. coli*-based CFPS currently demonstrates the greatest versatility.[4] However, this cell-free system still faces post-translational modification challenges, although a new report has described glycosylation in these systems for the first time.[68] Other bacteria-based cell-free systems have been developed, notably from thermophiles such as *Thermus thermophilus*.[69]

Eukaryote-based CFPS systems are mainly derived from wheat germ, rabbit reticulocytes, and insect cells. Wheat germ extract-based cell-free systems provide the highest yields of the

TABLE 15.1 Common Energy Sources and the Relative Cost of Cell-Free Protein Production Using Commercial Kits

CFPS Type	Commonly used Energy Sources	Extract Preparation Time	Relative Cost (per ng Protein Yield)*
Escherichia coli lysate	Phosphoenolpyruvate[146] Pyruvate[146] Acetyl kinase/acetyl phosphate[147] Glucose[148] Creatine kinase/creatine phosphate[15] Creatine kinase/creatine phosphate/glucose[149] Creatine kinase/creatine phosphate/cAMP[150]	1–2 days	~$0.70–1**[75]
wheat germ lysate	Creatine kinase/creatine phosphate[70]	4–5 days	~$1.50–3***[4]
Insect cell lysate	Creatine kinase/creatine phosphate[151]	1–2 days	~$4–9.25****
Purified component	Creatine kinase/creatine phosphate[152]	Extract not used, 32 individually purified components	~$11******
Rabbit reticulocyte cell lysate	Creatine kinase/creatine phosphate[153]	4 days for rabbit treatment, 1 day for extract prep	~$30–100*****[75]

*Cost assessed using maximum yields of respective commercial kits.
**Expressway Maxi Cell-free E. coli Expression System from Invitrogen, S30 T7 High-yield Expression Kit from Promega.
***Wheat Germ Extract Cell Free Protein Synthesis Kits from Genecopoeia.
****TnT® T7 Insect Cell Extract Protein Expression System from Promega.
*****TnT-coupled Reticulocyte Lysate System from Promega.
******PURExpress In vitro Protein Synthesis Kit from NEB.

major eukaryotic cell-free systems.[70,71] Using a cell-free protein-producing gel, one group reported yields as high as 5 mg/ml, approaching the highest yields reported for E. coli-based cell-free synthesis, but at much lower volumes.[71] Unfortunately, wheat germ extract preparation is more costly and more labor-intensive than E. coli-based systems.

Rabbit reticulocyte-based cell-free systems require a highly technical and difficult extract preparation method, but are used because of their ability to enable some post-translational modifications.[72–74] Low expression yields linked to the significant amount of ribonuclease M in the lysate, however, limit their utility.[75] Additionally, the collection of sufficient amounts of reticulocytes requires harmful chemical treatments of rabbits.[73]

Commonly produced from Spodoptera frugiperda cells, insect-based CFPS provide an effective combination of ample protein yield and post-translational modifications.[76] Commercial insect cell-based systems have reached yields of up to 50 µg/ml.[76] They also have been used successfully for several forms of post-translational modifications including glycosylation, isoprenylation, acetylation, N-myristoylation, and others.[77–79] Due to the cost of insect cell cultivation and extract preparation, insect cell-based cell-free systems are more costly than E. coli-based or wheat germ-based systems. While each of the CFPS systems developed to date has their merits and trade-offs, we turn our attention to several case studies that highlight the advantages of having direct access and control to the reaction conditions, as well as emerging applications made possible by recent technical advances.

ADVANTAGES ACHIEVED BY DIRECTLY INFLUENCING REACTION CONDITIONS

As mentioned above, cell-free biology provides the ability to directly manipulate and modulate reaction conditions. We now highlight several illustrative examples where ionic strength, pH, redox potential, hydrophobicity, and translation components were tuned. The overall ionic strength and the relative concentration of a specific ion can significantly impact the structure and activity of many enzymes and proteins.[80−83] For example, macromolecular protein structures such as virus-like particles have been shown to assemble at higher efficiencies as the ionic strength is increased.[84] Viral RNA polymerases are commonly inhibited by high concentrations of zinc.[82] Cell-free biology enables direct optimization of the ionic strengths for both the performance of the enzyme machinery and the activity of the product protein. By utilizing cell-free translation and assembly reactions, the highly efficient translation and assembly of macromolecular virus capsids has been demonstrated.[25,85,86] Also, the cell-free assembly of the human hepatitis B virus and human papillomavirus capsids were observed to be the highest at ionic strengths that would be cytotoxic to most cells.[84,87]

Cell-free biotechnology also enables the ability to directly control the redox potential, which is not possible in cells. In an exemplary illustration, Zawada et al. carried out a combinatorial optimization of reaction conditions for both protein expression and folding.[2] By modulating the redox potential and disulfide bond isomerase concentrations, they achieved greater than 95% solubility for a protein containing multiple disulfide bonds. This is in contrast to in vivo studies where correctly folding such complex proteins is very difficult. Indeed, cell-free production of many active disulfide-bonded proteins and disulfide-bonded macromolecular complexes has been achieved, including *E. coli* alkaline phosphatase, human granulocyte-macrophage colony-stimulating factor, *Candida antarctica* lipase B, human lysozyme, *Gaussia princeps* luciferase, and the Qβ virus-like particle which contains up to 180 disulfide bonds.[2,86,88−90] Although a subtle point, the less-crowded environment of the cell-free reaction (relative to in vivo concentrations) helps provide for improved protein folding.

The open nature of cell-free translation also enables the use of several synthetic approaches to increase solubility of highly insoluble proteins such as membrane proteins.[91] In many instances, detergents have been used to alleviate protein solubility issues.[92] In one example, detergents and chaperone proteins were successfully used in the cell-free system to produce an active form of an insoluble cancer-linked protein that had never been produced in vivo.[93] More recently, others have focused on the use of nondetergent surfactants, amphipols, and fluorinated surfactants to increase solubility without damaging protein structural integrity.[94] Other recent breakthroughs have included insertion of insoluble proteins into synthetic liposomes with verified postinsertion functionality.[95]

Recently, advances in protein-producing scaffold technology have produced synthetic hydrogels capable of cell-free production yields approaching those of any solution-based cell-free systems.[71,96,97] Hydrogels are scaffolds made up of linked networks of strongly hydrophilic polymers and mimic hydrophilic physiological environments.[97] These hydrogels have been reported to be compatible with various cell-free systems, and capable of generating protein yields up to 5 mg/ml.[71]

APPLICATIONS

Improvements in productivity, scale, and complexity of recombinant protein synthesized have rapidly expanded the utility, and now industrialization, of CFPS systems. In this section, we highlight several emerging applications, focusing on protein microarrays, protein evolution, and synthetic proteins.

Currently, the synthesis of proteins for functional analysis, structural genomics projects, or the identification of novel characteristics is a multi-step task. These tasks involve gene/vector construction, protein expression, and functional analysis of the expressed protein. Unfortunately, this process is time-consuming and costly (both in terms of funds and human capital), since it is primarily a serial process where only a single or a few proteins are produced simultaneously. High-throughput production of proteins using CFPS systems is now addressing this challenge. Valuable for studying the synthesis of many proteins simultaneously, protein microarrays are being developed at a rapid pace for improved large-scale protein expression and purification.[98] The basic principle of protein microarrays is that DNA encoding a protein target of interest is printed onto a glass slide in a physically isolated location, then CFPS synthesizes the target of interest.[99] To improve protein isolation and stability, proteins are commonly engineered to include epitope tags (e.g. C-terminal glutathione S-transferase tags) which bind the protein products to antibodies immobilized on the slides.[99] In one exemplary report, parallel large-scale expression of more than 13 000 human genes using a wheat germ extract has been achieved.[100] Because the cell-free approach obviates the need to synthesize, purify, and immobilize proteins separately, it allows for proteins with novel characteristics to be quickly generated and analyzed. Further efforts to integrate gene synthesis on chips for protein analysis promise even greater capability.

CFPS offers a rich and versatile platform for protein engineering. Specifically, the cell-free system enables direct control of the environment to efficiently drive protein evolution to contain the desired traits.[101] Two major protein evolution strategies are in vitro compartmentalization and ribosome display. In vitro compartmentalization (IVC) involves the generation of an emulsion to create cell-like compartments.[102] Ideally, each compartment contains a single gene and all reaction components necessary for protein transcription/translation. These separate compartments trap all gene products within an individual compartment, preventing cross-contamination. The reaction components may either be present all at once or added in several phases.[103] The formation of compartments down to the femtoliter scale has made IVC by emulsion particularly attractive for high-throughput protein evolution (Fig. 15.3).[102] In contrast, the low-tech physical method of IVC by microtiter plate has high reagent costs and the screening of one million genes could

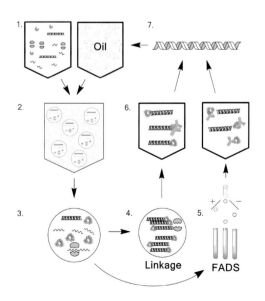

FIGURE 15.3

In vitro compartmentalization by emulsion for protein evolution: (1—2) Cell-free transcription/translation machinery and mineral oil form water in oil emulsions. (3) In each emulsion, a single gene is transcribed and translated. (4) Protein—gene linkage may be performed prior to selection. (5) Proteins are selected based on binding, catalytic, or regulatory capability. (6) Alternately, proteins may be selected through a fluorescence-assisted droplet sorting system (FADS). (7) The genes for selected proteins undergo a new cycle of IVC.

cost over $2.4 million.[104] The implementation of new ultrahigh-throughput microfluidic systems have dramatically reduced these costs.[105] In one such system, the use of picoliter-sized droplets, low polydispersity emulsions, and effective fluorescence sorting allowed the screening of one million genes in less than an hour at a cost of less than $31.[104]

As a complement to IVC, ribosome display provides a simple mechanism for evolution of proteins that will bind to a specific ligand.[106,107] The ribosome display process begins with a DNA sequence or library of sequences containing a spacer sequence lacking a stop codon. Due to the spacer sequence, the ribosome remains bound to the mRNA–protein complex post-translation and the nascent polypeptide is able to successfully fold.[101] The mRNA–protein–ribosome complex can then be exposed to a surface-immobilized ligand for binding. Weak-binding complexes are washed away, leaving only high-affinity complexes. The mRNA of the highest affinity complexes are recovered, reverse transcribed, PCR amplified, and then collected for future selection rounds.[107] In one example, ribosome display enabled the rapid selection of a designed ankyrin repeat protein which binds the cancer-relevant epidermal growth factor 2 (Her2) at high selectivity and nanomolar affinity.[108]

Beyond protein microarrays and protein evolution, the synthesis of synthetic proteins is another frontier application of CFPS. While Hecht and coworkers have elegantly shown the ability to synthesize de novo proteins with unique functionality from combinatorial libraries,[109] we focus our discussion here on efforts to expand the chemistry of life by the introduction of unnatural amino acids. Efforts to use CFPS for unnatural amino acid incorporation are beginning to grow. This is because of recent advances enabling cost-effective, high-level CFPS systems, and advantages over in vivo approaches. Namely, there are no transport issues for unnatural amino acids, and there is greater flexibility for reprogramming the translation system.[110]

288

The incorporation of unnatural amino acids to create novel proteins has been performed by globally replacing a natural amino acid with an unnatural analogue, or by site-specifically incorporating the unnatural amino acid while maintaining the natural amino acid cannon. The global replacement method can be performed in the cell-free system by simply not adding an amino acid such as methionine, and adding an unnatural amino analogue in its place. In a recent report, methionine was globally replaced with azidohomoalanine to enable the efficient attachment of proteins to virus-like particles for the development of a B-cell lymphoma vaccine.[31] While simple in implementation, the global replacement method is more likely to adversely affect protein function due to the loss of methionine from the amino acid canon.[111]

A number of site-specific incorporation methods have been developed. However, one of the highest yielding and transferable methods involves utilizing an orthogonal tRNA/tRNA synthetase pair specific to the amber stop codon, as developed by Schultz and coworkers.[112] The expression or addition of the tRNA/tRNA synthetase pair is relatively straightforward in both in vivo and cell-free systems. However, due to the insoluble nature of some unnatural amino acids and transport limitations of the unnatural amino acid into the cell, the efficient incorporation of some unnatural amino acids in vivo can be challenging.[29] Because of the direct access provided by cell-free systems, these limitations can be overcome.[29,30,111] By optimizing the tRNA/tRNA synthetase pair and using a continuous exchange cell-free system with linear templates, Ozawa et al. recently produced protein yields of over 2 mg/ml for a protein with a single unnatural amino acid mutation.[113]

For site-specific incorporation, the presence of release factor 1 (RF1) in cell-free extracts can compete with the exogenous tRNA for the amber stop codon.[114] A significant improvement in site-specific multisite incorporation of unnatural amino acid protein yields has been made by producing an RF1 knockout strain of E. coli. With this strain, protein yields reached

57 μg/ml for a protein with one unnatural amino acid site mutation, and 68 μg/ml for a protein with three site mutations.[114] Although not attempted, the use of this strain in a cell-free system could further improve the production of proteins containing unnatural amino acids.

Small Molecule Metabolites

Similar to CFPS, cell-free metabolic engineering for overproducing small molecule metabolites occurs via either a bottom-up or top-down approach. Since the original discovery that cell extracts could convert sugar into ethanol,[115] there have been relatively few top-down examples of metabolic engineering. Efforts primarily by the Swartz and Panke laboratories have developed the field; the former for activating energy pathways for fueling cell-free protein synthesis reactions,[116] and the latter for making desired small molecule metabolites. In one example of this approach, the Panke group constructed an enzyme catalytic system by fine-tuning a catalytic pathway with information from real-time analysis of concentrations of the pathway intermediates.[117]

In contrast to extract-based systems, the bottom-up approach exploits the organization of synthetic enzymatic pathways from purified components, sometimes to facilitate a process or reaction that does not occur in nature.[9,118] In one instance, a 13-step synthetic enzymatic pathway using starch and water produced hydrogen at yields far higher than the theoretical yields of biological hydrogen fermentations (illustrated in Fig. 15.4).[9] Indeed, because carbon flux is not directed towards cell growth, cell-free systems can achieve higher theoretical yields than natural biological processes found in living organisms.[116] Unfortunately, the high cost of purifying stable, standalone enzymes, and cofactor regeneration costs, currently limit synthetic enzymatic pathways to laboratory scale research.

289

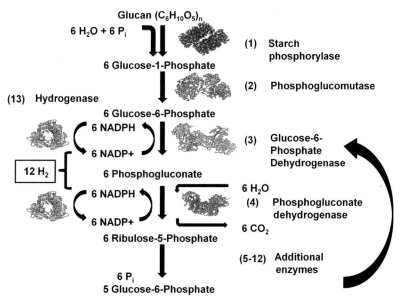

FIGURE 15.4

A cell-free synthetic enzymatic pathway for converting starch and water into hydrogen and carbon dioxide: this artificial pathway has been reported to produce hydrogen at yields greater than the theoretical limits for industrial anaerobic fermentations (12 H_2 per glucose compared to 4 H_2 per glucose).[9] Molecular graphics images were produced using the UCSF Chimera package from the Resource for Biocomputing, Visualization, and Informatics at the University of California, San Francisco (supported by NIH P41 RR-01081).

Minimal Cells

Minimal cell construction seeks to assemble the minimal number of cellular components such as DNA, RNA, and protein encapsulated in a lipid membrane necessary for life.[119] The goals are fundamental knowledge, testing our understanding of life and its origins, and facilitating engineering for biotechnology applications, including evolutionary optimization of natural and unnatural biopolymers.[119−121] Minimal cell construction is proceeding via two approaches: (1) the top-down approach and (2) the bottom-up approach. The top-down approach attempts to identify the minimal number of genes required for the organism to live by genome reduction. Pioneering efforts from the J Craig Venter Institute have made tremendous progress towards experimentally determining essential genes. Specifically, they identified that 387 out of 482 protein-coding genes of *Mycoplasma genitalium* and 43 RNA-coding genes were essential for life.[122] Further efforts to minimize this genome have led to the creation of the first synthetic cell controlled by a genome synthesized from scratch.[123]

However, this book chapter is focused on 'cell-free' synthetic biology, and therefore we will not delve further into the top-down approach, which was reviewed recently by Jewett and Forster.[121] Instead we will focus on the bottom-up approach, which involves creating a minimal cell by bringing together essential purified biological macromolecules, their genes, and their small molecule substrates (Fig. 15.5A). The main goal of an artificial cell built from the bottom-up is its ability to self-replicate with efficient removal of byproducts being exchanged with the outside environment.[124] We will focus on attempts to construct a DNA−RNA−protein-based system derived from current biological systems, although it should be noted that there are efforts and recent advances centered on modeling an RNA world. In a breakthrough study last year, for example, in vitro evolution was used to develop an RNA polymerase ribozyme that was capable of synthesizing a hammerhead ribozyme from an RNA template, demonstrating that RNA can self-replicate and create functional RNAs in the process.[125]

Key developments in the construction of minimal cells have centered on integrating the subsystems of biology necessary for self-replication. This includes creating the compartment (or membrane vesicles), enabling substrates and waste products to flow in and out of compartments, and activating complex biochemical reactions and networks inside compartments.

The membrane of the minimal cell must be capable of growth and division, must be permeable to allow for entry and exit of substrates and waste products, and must be stable under the conditions of replication and gene expression.[126] Conditions and timing of membrane replication must be ultimately similar to conditions for replication of the enclosed nucleic acid for the cell to successfully replicate and survive. Core and shell reproduction (i.e. reproduction of all the components needed to sustain a minimal cell along with the surrounding lipid vesicle) is currently not possible, but several important developments have been made, three of which are detailed below. For a more detailed review of this process refer to Stano et al.[119,127]. In one example, Luisi's group was able to construct a synthetic cell that consisted of the reconstituted translation system (PURE) encapsulated in a liposome that expressed two enzymes/membrane proteins involved in phospholipid biosynthesis.[128] This was the first successful demonstration that membrane proteins could be synthesized in lipid vesicles of proper lipid composition. Although the authors initially aimed at observing cell growth and division, low product yield and issues in the reaction biochemistries did not allow for an observation of morphological changes. In a second example, researchers were able to chemically link amplification of DNA and self-reproduction of a giant vesicle, thereby creating a self-reproducing supramolecular giant (Fig. 15.5B).[129] While a huge advance for the field, this system has two limitations: (1) the self-reproduction is limited as the percentage of the phospholipid in the membrane

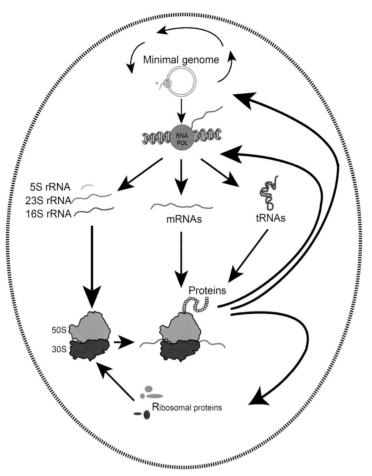

FIGURE 15.5A

Theoretical protein-based minimal cell: in the illustration the minimal genome is incorporated into a lipid bilayer along with a purified transcription/translation system. For the creation of a self-replicating/self-sustaining minimal cell, the cell must be able to efficiently transcribe RNA and translate the genome into proteins that are then able to sustain the cell and additionally aid in replication of all the components of the cell.

gradually decreases; and (2) the PCR conditions that are used to replicate the DNA in the giant vesicle would cause denaturation of any proteins that were synthesized or introduced into the vesicle to create a fully competent minimal cell. In a third example, controlled vesicle fusion of oppositely charged vesicles demonstrated for the first time the ability to 'regulate' and maintain metabolism in minimal cell models. Specifically, vesicles containing either DNA or RNA polymerase could trigger gene expression when fused. This method provides a means to introduce larger biomolecules into the system that are unable to diffuse through the pore.[130]

Outside of efforts to create membranes that can grow, divide, and integrate cellular processes, another key area of growth has centered on relieving limitations imposed on nutritional and energy requirements in a minimal cell. One landmark study was the creation of pores (the α-hemolysin protein from *Streptomyces aureus*) in a phospolipid membrane that encapsulated a cell-free protein synthesis reaction (Fig. 15.5C).[131] This pore enabled selective permeabilization to realize long-lived cell-free transcription and translation.

In addition to developments in membranous minimal cellular compartments, activating complex biochemical processes inside vesicles is required for self-replication. Recent

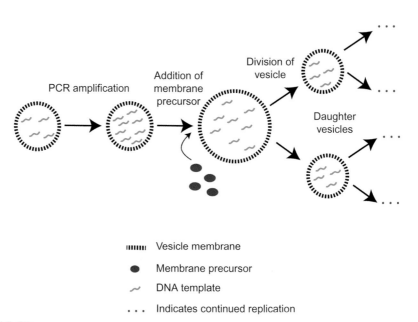

|||||| Vesicle membrane

● Membrane precursor

～ DNA template

. . . Indicates continued replication

FIGURE 15.5B

Combined replication of DNA and giant vesicles: vesicles are created that contain DNA and other components to amplify the DNA such as primers, dNTPs, and DNA polymerase. PCR results in amplification of DNA. Addition of a membrane precursor induces swelling of the vesicle that then leads to division of the vesicle into two daughter vesicles.

advances towards this goal include: (1) the PURE system,[3,64] which reconstitutes translation from its individual component parts; (2) transcription initiation carried out in the PURE system;[132] (3) the PURE system utilized for protein synthesis in liposomes;[133] and (4) the PURE system utilized to translate and translocate presecretory and integral membrane proteins into liposomes, which is necessary for making an artificial cell that can grow and maintain a bilayer membrane.[134]

Other studies have attempted to enable self-copying systems. A study in 2008 created a replicating system that involved containment of the RNA of the β subunit of the Qβ replicase along with purified translation factors (Fig. 15.5D).[135] A functional Qβ replicase was assembled when the RNA was translated into the β subunit protein, which assembled with other subunits already present in the translation system. This replicase could then replicate the RNA template, leading to simplified replication cycles. Unfortunately, this RNA–protein system does not connect directly to the DNA–RNA–protein world. Researchers have also shown that DNA replication (via PCR) was possible to a limited extent inside liposomes.[136] However, DNA replication using PCR is not possible in minimal cells, as the elevated temperatures would denature the proteins required for transcription and translation.[124] While major challenges remain for building minimal cells, tremendous progress has been made in the last decade, and there are plans now reported for making a self-replicating system dependent only on small molecule substrates,[120] paving the way for outstanding scientific discovery in the years to come.

CHALLENGES AND OPPORTUNITIES

In the last 5 years, efforts in cell-free synthetic biology have grown at a remarkable rate, producing synthetic gene circuits, enabling commercial production of proteins, and setting the stage for minimal cells. Immediate challenges and opportunities for this emerging field include reducing costs and increasing scale. Further challenges and opportunities include the development of methods for multienzyme synthesis and purification of multienzyme ensembles.

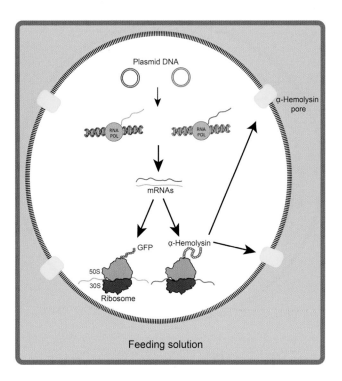

FIGURE 15.5C

Creation of pores in the membrane relieves limitation on nutritional and energy requirements: a cell-free transcription and translation system is encapsulated in a lipid vesicle along with DNA encoding GFP and α-hemolysin. Translated α-hemolysin protein is then able to form a pore through the lipid membrane, thereby allowing exchange of nutrients and toxic byproducts with the surrounding feeding solution. Creation of the pore allows for longer expression of synthesized protein as measured by GFP production. The lipid bylayer is indicated by the black dashed line.

Cost Reduction

Cell-free biology offers remarkable opportunities for engineering biology. However, until commercial products (e.g. proteins and metabolites) are made using cell-free systems, these systems will mainly serve as test-beds for understanding biology and accelerating the design of synthetic programs. To enable commercial applications, costs must continue to be reduced. Cells have evolved many pathways to reliably produce energy from inexpensive raw materials. In contrast, more expensive energy sources are typically used in cell-free systems. An overarching goal for cell-free biology is to develop strategies for using the same energy substrate molecules and waste feedstreams that are used to fuel in vivo synthetic biology efforts. Bottom-up approaches will benefit from more efficient methods for rapid and cost-effective protein production and purification, while more efficient and cost-effective extract preparation methods are desired for top-down cell-free systems. With the *E. coli*-based cell-free protein synthesis system, engineering the extract preparation method decreased the extract preparation time and reagent cost by over 50%,[137–139] and similar advances are expected to improve other top-down cell-free systems. Further advances will likely couple high-throughput screens to characterize and assess the functions of proteins found in the extract. In one illustrative example, Swartz and colleagues recently surveyed endogenous proteins in *E. coli* for their influence on CFPS. Using this information, they were able to improve CFPS yields up to ~4 g/L in an 8-hour batch reaction.[14,140]

Scale

With the exquisite control the open cell-free system enables, researchers have demonstrated the scalability of the cell extract-based cell-free protein synthesis reaction from the microliter

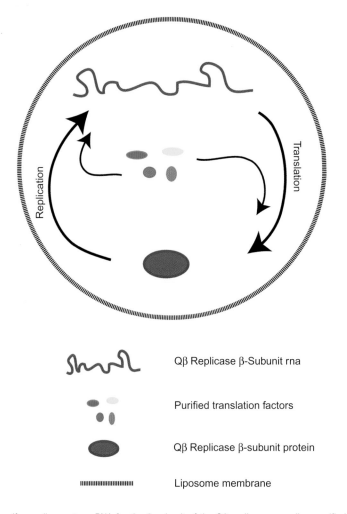

QΒ Replicase β-Subunit rna

Purified translation factors

QΒ Replicase β-subunit protein

Liposome membrane

FIGURE 15.5D

Illustration of a basic self-encoding system: RNA for the β-subunit of the Qβ replicase as well as purified translation factors are compartmentalized into a liposome. The β-subunit of the replicase is then translated into protein by the purified translation system, following which it forms the hetero-tetramer Qβ replicase composed of the β-subunit as well as EF-Tu, EF-Ts, and ribosomal protein S1 which are present in the purified translation system. The reconstituted Qβ replicase can then replicate the RNA template, leading to a basic self-encoding system whereby the template RNA is replicated by the protein it encodes.

level to the 100-liter scale.[2,5] In addition to analogous product protein yields, the production rates at the μL, mL, and L scale were virtually identical.[2] Although scalability over six orders of magnitude to the 100 L scale has been demonstrated, increasing to the 1000 L or even 10 000 L scale would enhance the attractiveness of cell-free biology for commercial applications. Moreover, the development of a scalable eukaryotic system is needed.

Development of Complex Biocatalytic Networks

The cell-free system provides an environment where the expression rate of multiple products can be controlled in parallel to assess the impact of coproduction on an individual product's activity. This is particularly important when assembling complex macromolecular assemblies such as the ribosome or virus-like particles. As recently demonstrated, small changes in expression levels of one protein of multiprotein macromolecular complex can significantly impact the efficiency and activity of the assembled macromolecular complex.[25] In addition, researchers have noted the engineering advantages of employing in vitro enzyme-catalyzed biochemical reaction cascades to produce the desired natural product.[141] The easily controlled and engineered open cell-free environment will be exploited for optimizing multienzyme synthesis and activity for next-generation biocatalysts. Efforts such as those by Forster and colleagues could provide directions.[142,143]

Purification and Localization of Multienzyme Ensembles

To produce a desired product from an inexpensive feedstock, multiple biocatalytic steps are often necessary. A key challenge, therefore, for bottom-up cell-free systems is cost-effective purification of multienzyme ensembles. Offering one possible solution, a recent report from Wang et al. leverages advanced genome engineering for simultaneous modification of catalysts to be used in copurification strategies.[144] This type of strategy, once further developed, could lead to commercially relevant in vitro catalytic systems. Beyond strategies for simultaneously modifying and copurifying large protein complexes and pathways, localization may also be important. Nature has evolved highly efficient mechanisms to facilitate the transfer of reactants through complex enzymatic pathways. For the production of secondary metabolites, for example, type I polyketide synthetases actively transfer the reactant through a number of catalytic domains that are locally constrained as part of the same protein. Biomimetic engineering, the ability to control the proximity of multiple enzymes, could facilitate the rapid transfer of reactants through the correct sequence of biocatalytic events. Innovations in the rapidly growing enzyme immobilization field could help facilitate the development of such biomimetic multienzyme ensembles.[145]

SUMMARY

Cell-free synthetic biology is a powerful tool for understanding and expanding the basic biology of living systems. Recent advances in nucleic acid circuitry, protein synthesis, and small molecule production, amongst others, have heralded a new age of in vitro bioengineering. The general simplicity and flexibility of in vitro systems serves as an excellent complement to in vivo studies, and enables a more complete understanding of the inner workings of a cell. Cell-free synthetic biology has versatile applications in protein synthesis and evolution, as well as the development of nucleic acid circuits, nanomachines, and therapeutics.

Although cell-free systems have been used for over 50 years, more work is needed before they can be used as commercial platforms for therapeutic and metabolite production. However, recent advances have demonstrated commercial-scale protein synthesis with crude extract cell-free systems. In addition, continued engineering has rapidly driven down the cost and simplified the preparation of many cell-free systems. With the remarkable progress made over the last 10 years, it is but a matter of time before several of the significant challenges facing the field evolve into transformative synthetic biology opportunities.

Acknowledgments

MCJ and AR gratefully acknowledge funding from the National Institutes of Health (Grant Number R00GM081450), the National Academies Keck Futures Initiative (Grant Number NAFKI-SB5), the National Science Foundation (Grant Number MCB-0943393), the Office of Naval Research (Grant Number N00014-11-1-0363), and the DARPA YFA Program (Grant Number N66001-11-1-4137). BCB and JCW thankfully acknowledge funding from the National Science Foundation (Award Number 1115229), the Rocky Mountain NASA Space Grant Consortium, and Utah NASA EPSCoR.

References

1. Sacca B, Niemeyer CM. DNA origami: the art of folding DNA. *Angew Chem Int Ed Engl*. 2012;51(1):58–66.

2. Zawada JF, Yin G, Steiner AR, Yang J, Naresh A, Roy SM, et al. Microscale to manufacturing scale-up of cell-free cytokine production – a new approach for shortening protein production development timelines. *BiotechBioeng*. 2011;108(7):1570–1578.

3. Shimizu Y, Inoue A, Tomari Y, Suzuki T, Yokogawa T, Nishikawa K, et al. Cell-free translation reconstituted with purified components. *Nat Biotechnol.* 2001;19(8):751–755.

4. Carlson ED, Gan R, Hodgman CE, Jewett MC. Cell-free protein synthesis: applications come of age. *Biotechnol Adv.* 2012;30(5):1185–1194. http://dx.doi.org/10.1016/j.biotechadv.2011.09.016.

5. Voloshin AM, Swartz JR. Efficient and scalable method for scaling up cell free protein synthesis in batch mode. *Biotech Bioeng.* 2005;91(4):516–521.

6. Saiki RK, Gelfand DH, Stoffel S, Scharf SJ, Higuchi R, Horn GT, et al. Primer-directed enzymatic amplification of DNA with a thermostable DNA polymerase. *Science.* 1988;239(4839):487–491.

7. Sousa R, Mukherjee S. T7 RNA polymerase. *Prog Nucleic Acid Res Mol Biol.* 2003:1–41.

8. Wang Y, Huang W, Sathitsuksanoh N, Zhu Z, Zhang YHP. Biohydrogenation from biomass sugar mediated by in vitro synthetic enzymatic pathways. *Chem Biol.* 2011;18(3):372–380.

9. Zhang YHP, Evans BR, Mielenz JR, Hopkins RC, Adams MWW. High-yield hydrogen production from starch and water by a synthetic enzymatic pathway. *PLoS One.* 2007;2(5):e456.

10. Zhang YHP, Myung S, You C, Zhu ZG, Rollin JA. Toward low-cost biomanufacturing through in vitro synthetic biology: bottom-up design. *J Mater Chem.* 2011;21(47):18877–18886.

11. Jewett MC, Calhoun KA, Voloshin A, Wuu JJ, Swartz JR. An integrated cell-free metabolic platform for protein production and synthetic biology. *Mol Syst Biol.* 2008;4:220.

12. Kim J, White KS, Winfree E. Construction of an in vitro bistable circuit from synthetic transcriptional switches. *Mol Syst Biol.* 2006;2:68.

13. Lesley SA, Brow MA, Burgess RR. Use of in vitro protein synthesis from polymerase chain reaction-generated templates to study interaction of *Escherichia coli* transcription factors with core RNA polymerase and for epitope mapping of monoclonal antibodies. *J Biol Chem.* 1991;266(4):2632–2638.

14. Woodrow KA, Airen IO, Swartz JR. Rapid expression of functional genomic libraries. *J Proteome Res.* 2006;5(12):3288–3300.

15. Park CG, Kwon MA, Song JK, Kim DM. Cell-free synthesis and multifold screening of *Candida antarctica* lipase B (CalB) variants after combinatorial mutagenesis of hot spots. *Biotechnol Prog.* 2011;27(1):47–53.

16. Alimov AP, Khmelnitsky A, Simonenko PN, Spirin AS, Chetverin AB. Cell-free synthesis and affinity isolation of proteins on a nanomole scale. *Biotechniques.* 2000;28(2):338–344.

17. Ahn J-H, Keum J-W, Kim D-M. High-throughput, combinatorial engineering of initial codons for tunable expression of recombinant proteins. *J Proteome Res.* 2008;7(5):2107–2113.

18. Jewett MC, Swartz JR. Rapid expression and purification of 100 nmol quantities of active protein using cell-free protein synthesis. *Biotechnol Prog.* 2004;20(1):102–109.

19. Kim T-W, Oh I-S, Ahn J-H, Choi C-Y, Kim D-M. Cell-free synthesis and in situ isolation of recombinant proteins. *Protein Expression Purif.* 2006;45(2):249–254.

20. Bujara M, Schumperli M, Pellaux R, Heinemann M, Panke S. Optimization of a blueprint for in vitro glycolysis by metabolic real-time analysis. *Nat Chem Biol.* 2011;7(5):271–277.

21. Katzen F, Chang G, Kudlicki W. The past, present and future of cell-free protein synthesis. *Trends Biotechnol.* 2005;23(3):150–156.

22. Usón I, Sheldrick GM. Advances in direct methods for protein crystallography. *Curr Opin Struct Biol.* 1999;9(5):643–648.

23. Kigawa T, Yamaguchi-Nunokawa E, Kodama K, Matsuda T, Yabuki T, Matsuda N, et al. Selenomethionine incorporation into a protein by cell-free synthesis. *J Struct Funct Genomics.* 2002;2(1):29–35.

24. Avenaud P, Castroviejo M, Claret S, Rosenbaum J, Mégraud F, Ménard A. Expression and activity of the cytolethal distending toxin of *Helicobacter hepaticus. Biochem Biophys Res Commun.* 2004;318(3):739–745.

25. Smith MT, Varner CT, Bush DB, Bundy BC. The incorporation of the A2 protein to produce novel Qβ virus-like particles using cell-free protein synthesis. *Biotechnol Prog.* 2012;28(2):549–555.

26. Wuu JJ, Swartz JR. High yield cell-free production of integral membrane proteins without refolding or detergents. *Biochim Biophys Acta.* 2008;1778(5):1237–1250.

27. Lian J, Ma Y, Cai J, Wu M, Wang J, Wang X, et al. High-level expression of soluble subunit b of F1F0 ATP synthase in *Escherichia coli* cell-free system. *Appl Microbiol Biotechnol.* 2009;85(2):303–311.

28. Harris DC, Jewett MC. Cell-free biology: exploiting the interface between synthetic biology and synthetic chemistry. *Curr Opin Biotechnol.* 2012;23(5):672–678. http://dx.doi.org/10.1016/j.copbio.2012.02.002.

29. Bundy BC, Swartz JR. Site-specific incorporation of p-propargyloxyphenylalanine in a cell-free environment for direct protein–protein click conjugation. *Bioconjug Chem.* 2010;21(2):255–263.

30. Goerke AR, Swartz JR. High-level cell-free synthesis yields of proteins containing site-specific non-natural amino acids. *BiotechBioeng.* 2009;102(2):400–416.

31. Patel KG, Swartz JR. Surface functionalization of virus-like particles by direct conjugation using azide-alkyne click chemistry. *Bioconjug Chem.* 2011;22(3):376−387.

32. Forster AC, Church GM. Synthetic biology projects in vitro. *Genome Res.* 2007;17(1):1−6.

33. Carr PA, Church GM. Genome engineering. *Nat Biotechnol.* 2009;27(12):1151−1162.

34. Marshall WS, Kaiser RJ. Recent advances in the high-speed solid phase synthesis of RNA. *Curr Opin Chem Biol.* 2004;8(3):222−229. http://dx.doi.org/10.1016/j.cbpa.2012.05.179.

34a. Hockenberry AJ, Jewett MC. Synthetic in vitro circuits. *Curr Opin Chem Biol.* 2012;16(3−4):253−259.

35. Adleman LM. Molecular computation of solutions to combinatorial problems. *Science.* 1994;266 (5187):1021−1024.

36. Seelig G, Soloveichik D, Zhang DY, Winfree E. Enzyme-free nucleic acid logic circuits. *Science.* 2006;314 (5805):1585−1588.

37. Zhang DY, Turberfield AJ, Yurke B, Winfree E. Engineering entropy-driven reactions and networks catalyzed by DNA. *Science.* 2007;318(5853):1121−1125.

38. Qian L, Winfree E. A simple DNA gate motif for synthesizing large-scale circuits. *J R Soc Interface.* 2011;8 (62):1281−1297.

39. Qian L, Winfree E. Scaling up digital circuit computation with DNA strand displacement cascades. *Science.* 2011;332(6034):1196−1201.

40. Phillips A, Cardelli L. A programming language for composable DNA circuits. *J R Soc Interface.* 2009;6 (Suppl 4):S419−S436.

41. Genot AJ, Bath J, Turberfield AJ. Reversible logic circuits made of DNA. *J Am Chem Soc.* 2011;133 (50):20080−20083.

42. Li B, Ellington AD, Chen X. Rational, modular adaptation of enzyme-free DNA circuits to multiple detection methods. *Nucleic Acids Res.* 2011;39(16):e110.

43. Weigand JE, Wittmann A, Suess B. RNA-based networks: using RNA aptamers and ribozymes as synthetic genetic devices. *Methods Mol Biol.* 2012;813:157−168.

44. Davidson EA, Ellington AD. Synthetic RNA circuits. *Nat Chem Biol.* 2007;3(1):23−28.

45. Montagne K, Plasson R, Sakai Y, Fujii T, Rondelez Y. Programming an in vitro DNA oscillator using a molecular networking strategy. *Mol Syst Biol.* 2011;7:466.

46. Kim J, Winfree E. Synthetic in vitro transcriptional oscillators. *Mol Syst Biol.* 2011;7:465.

47. Franco E, Friedrichs E, Kim J, Jungmann R, Murray R, Winfree E, et al. Timing molecular motion and production with a synthetic transcriptional clock. *Proc Natl Acad Sci USA.* 2011;108(40):E784−E793.

48. Benenson Y, Gil B, Ben-Dor U, Adar R, Shapiro E. An autonomous molecular computer for logical control of gene expression. *Nature.* 2004;429(6990):423−429.

49. Zhang DY, Seelig G. Dynamic DNA nanotechnology using strand-displacement reactions. *Nat Chem.* 2011;3 (2):103−113.

50. Bath J, Turberfield AJ. DNA nanomachines. *Nat Nano.* 2007;2(5):275−284.

51. Yurke B, Turberfield AJ, Mills Jr. AP, Simmel FC, Neumann JL. A DNA-fuelled molecular machine made of DNA. *Nature.* 2000;406(6796):605−608.

52. Dittmer WU, Simmel FC. Transcriptional control of DNA-based nanomachines. *Nano Lett.* 2004;4(4):689−691.

53. Dittmer WU, Kempter S, Rädler JO, Simmel FC. Using gene regulation to program DNA-based molecular devices. *Small.* 2005;1(7):709−712.

54. Win MN, Smolke CD. Higher-order cellular information processing with synthetic RNA devices. *Science.* 2008;322(5900):456−460.

55. Seeman NC. Nucleic acid junctions and lattices. *J Theor Biol.* 1982;99(2):237−247.

56. Shih WM, Quispe JD, Joyce GF. A 1.7-kilobase single-stranded DNA that folds into a nanoscale octahedron. *Nature.* 2004;427(6975):618−621.

57. Rothemund PW. Folding DNA to create nanoscale shapes and patterns. *Nature.* 2006;440(7082):297−302.

58. Douglas SM, Dietz H, Liedl T, Hogberg B, Graf F, Shih WM. Self-assembly of DNA into nanoscale three-dimensional shapes. *Nature.* 2009;459(7245):414−418.

59. Han D, Pal S, Liu Y, Yan H. Folding and cutting DNA into reconfigurable topological nanostructures. *Nature Nano.* 2010;5(10):712−717.

60. Han D, Pal S, Nangreave J, Deng Z, Liu Y, Yan H. DNA origami with complex curvatures in three-dimensional space. *Science.* 2011;332(6027):342−346.

61. Douglas SM, Marblestone AH, Teerapittayanon S, Vazquez A, Church GM, Shih WM. Rapid prototyping of 3D DNA-origami shapes with caDNAno. *Nucleic Acids Res.* 2009;37(15):5001−5006.

62. Endo M, Katsuda Y, Hidaka K, Sugiyama H. A versatile DNA nanochip for direct analysis of DNA base-excision repair. *Angew Chem Int Ed Engl*. 2010;49(49):9412–9416.

63. Douglas SM, Bachelet I, Church GM. A logic-gated nanorobot for targeted transport of molecular payloads. *Science*. 2012;335(6070):831–834.

64. Shimizu Y, Kanamori T, Ueda T. Protein synthesis by pure translation systems. *Methods*. 2005;36(3):299–304.

65. Ohashi H, Kanamori T, Shimizu Y, Ueda T. A highly controllable reconstituted cell-free system – a breakthrough in protein synthesis research. *Curr Pharm Biotechnol*. 2010;11(3):267–271.

66. Hodgman CE, Jewett MC. Cell-free synthetic biology: thinking outside the cell. *Metab Eng*. 2012;14(3): 261–269. http://dx.doi.org/10.1016/j.ymben.2011.09.002.

67. Swartz J. Developing cell-free biology for industrial applications. *J Ind Microbiol Biot*. 2006;33(7):476–485.

68. Guarino C, Delisa MP. A prokaryote-based cell-free translation system that efficiently synthesizes glycoproteins. *Glycobiology*. 2011;22(5):596–601.

69. Uzawa T, Yamagishi A, Oshima T. Polypeptide synthesis directed by DNA as a messenger in cell-free polypeptide synthesis by extreme thermophiles, *Thermus thermophilus* HB27 and *Sulfolobus tokodaii* strain 7. *J Biochem*. 2002;131(6):849–853.

70. Madin K, Sawasaki T, Ogasawara T, Endo Y. A highly efficient and robust cell-free protein synthesis system prepared from wheat embryos: plants apparently contain a suicide system directed at ribosomes. *Proc Natl Acad Sci USA*. 2000;97(2):559–564.

71. Park N, Um SH, Funabashi H, Xu J, Luo D. A cell-free protein-producing gel. *Nat Mater*. 2009;8(5):432–437.

72. Pelham HRB, Jackson RJ. An efficient mRNA-dependent translation system from reticulocyte lysates. *Eur J Biochem*. 1976;67(1):247–256.

73. Ryabova LA, Ortlepp SA, Baranovp VI. Preparative synthesis of globin in a continuous cell-free translation system from rabbit reticulocytes. *Nucleic Acids Res*. 1989;17(11):4412.

74. Hancock JF. Reticulocyte lysate assay for in vitro translation and posttranslational modification of ras proteins. *Methods Enzymol*. 1995;244:60–65.

75. Endo Y, Sawasaki T. Cell-free expression systems for eukaryotic protein production. *Curr Opin Biotechnol*. 2006;17(4):373–380.

76. Ezure T, Suzuki T, Shikata M, Ito M, Ando EA. Cell-free protein synthesis system from insect cells. In: Endo Y, Ueda T, Takai K, eds. *Cell-Free Protein Production*. Humana Press; 2010:31–42.

77. Tarui H, Murata M, Tani I, Imanishi S, Nishikawa S, Hara T. Establishment and characterization of cell-free translation/glycosylation in insect cell extract prepared with high pressure treatment. *Appl Microbiol Biotechnol*. 2001;55(4):446–453.

78. Suzuki T, Ito M, Ezure T, Shikata M, Ando E, Utsumi T, et al. Protein prenylation in an insect cell-free protein synthesis system and identification of products by mass spectrometry. *Proteomics*. 2007;7 (12):1942–1950.

79. Suzuki T, Ito M, Ezure T, Shikata M, Ando E, Utsumi T, et al. N-terminal protein modifications in an insect cell-free protein synthesis system and their identification by mass spectrometry. *Proteomics*. 2006;6 (16):4486–4495.

80. Sedov SA, Belogurova NG, Shipovskov S, Levashov AV, Levashov PA. Lysis of *Escherichia coli* cells by lysozyme: discrimination between adsorption and enzyme action. *Colloids Surf B Biointerfaces*. 2011;88(1):131–133.

81. Hu T, Li HY, Zhang XZ, Luo HJ, Fu JM. Toxic effect of NaCl on ion metabolism, antioxidative enzymes and gene expression of perennial ryegrass. *Ecotoxicol Environ Saf*. 2011;74(7):2050–2056.

82. te Velthuis AJ, van den Worm SH, Sims AC, Baric RS, Snijder EJ, van Hemert MJ. Zn(2 +) inhibits coronavirus and arterivirus RNA polymerase activity in vitro and zinc ionophores block the replication of these viruses in cell culture. *PLoS Pathog*. 2010;6(11):e1001176.

83. Jaenicke R. Protein folding: local structures, domains, subunits, and assemblies. *Biochemistry*. 1991;30 (13):3147–3161.

84. Zlotnick A, Ceres P, Singh S, Johnson JM. A small molecule inhibits and misdirects assembly of hepatitis B virus capsids. *J Virol*. 2002;76(10):4848–4854.

85. Bundy BC, Franciszkowicz MJ, Swartz JR. *Escherichia coli*-based cell-free synthesis of virus-like particles. *BiotechBioeng*. 2008;100(1):28–37.

86. Bundy BC, Swartz JR. Efficient disulfide bond formation in virus-like particles. *J Biotechnol*. 2011;154 (4):230–239.

87. Ceres P, Zlotnick A. Weak protein–protein interactions are sufficient to drive assembly of hepatitis B virus capsids. *Biochemistry*. 2002;41(39):11525–11531.

88. Ezure T, Suzuki T, Shikata M, Ito M, Ando E, Nishimura O, et al. Expression of proteins containing disulfide bonds in an insect cell-free system and confirmation of their arrangements by MALDI-TOF MS. *Proteomics*. 2007;7(24):4424–4434.

89. Goerke AR, Loening AM, Gambhir SS, Swartz JR. Cell-free metabolic engineering promotes high-level production of bioactive gaussia princeps luciferase. *Metab Eng.* 2008;10(3-4):187−200.

90. Park CG, Kim TW, Oh IS, Song JK, Kim DM. Expression of functional *Candida antarctica* lipase B in a cell-free protein synthesis system derived from *Escherichia coli. Biotechnol Prog.* 2009;25(2):589−593.

91. Betton J-M, Miot M. *Cell-free production of membrane proteins in the presence of detergents. Cell-Free Protein Synth.* Wiley-VCH Verlag GmbH & Co. KGaA; 2008:165−178.

92. Gilbert GP. Detergents for the stabilization and crystallization of membrane proteins. *Methods.* 2007;41 (4):388−397.

93. Pedersen A, Hellberg K, Enberg J, Karlsson BG. Rational improvement of cell-free protein synthesis. *New Biotechnol.* 2011;28(3):218−224.

94. Park K-H, Billon-Denis E, Dahmane T, Lebaupain F, Pucci B, Breyton C, et al. In the cauldron of cell-free synthesis of membrane proteins: playing with new surfactants. *New Biotechnol.* 2011;28(3):255−261.

95. Hovijitra NT, Wuu JJ, Peaker B, Swartz JR. Cell-free synthesis of functional aquaporin Z in synthetic liposomes. *Biotech Bioeng.* 2009;104(1):40−49.

96. Um SH, Lee JB, Park N, Kwon SY, Umbach CC, Luo D. Enzyme-catalysed assembly of DNA hydrogel. *Nat Mater.* 2006;5(10):797−801.

97. Park N, Kahn JS, Rice EJ, Hartman MR, Funabashi H, Xu J, et al. High-yield cell-free protein production from P-gel. *Nat Protocols.* 2009;4(12):1759−1770.

98. Chandra H, Srivastava S. Cell-free synthesis-based protein microarrays and their applications. *Proteomics.* 2010;10(4):717−730.

99. Ramachandran N, Hainsworth E, Bhullar B, Eisenstein S, Rosen B, Lau AY, et al. Self-assembling protein microarrays. *Science.* 2004;305(5680):86−90.

100. Goshima N, Kawamura Y, Fukumoto A, Miura A, Honma R, Satoh R, et al. Human protein factory for converting the transcriptome into an in vitro-expressed proteome. *Nat Methods.* 2008;5(12):1011−1017.

101. Zahnd C, Amstutz P, Pluckthun A. Ribosome display: selecting and evolving proteins in vitro that specifically bind to a target. *Nat Meth.* 2007;4(3):269−279.

102. Zhu Y, Power B. Lab-on-a-chip in vitro compartmentalization technologies for protein studies. In: Werther M, Seitz H, eds. *Protein−Protein Interaction.* Berlin/Heidelberg: Springer; 2008:81−114.

103. Sepp A, Choo Y. Cell-free selection of zinc finger DNA-binding proteins using in vitro compartmentalization. *J Mol Biol.* 2005;354(2):212−219.

104. Fallah-Araghi A, Baret JC, Ryckelynck M, Griffiths AD. A completely in vitro ultrahigh-throughput droplet-based microfluidic screening system for protein engineering and directed evolution. *Lab Chip.* 2012;12 (5):882−891.

105. Agresti JJ, Antipov E, Abate AR, Ahn K, Rowat AC, Baret J-C, et al. Ultrahigh-throughput screening in drop-based microfluidics for directed evolution. *Proc Natl Acad Sci USA.* 2010;107(9):4004−4009.

106. Wada A, Sawata SY, Ito Y. Ribosome display selection of a metal-binding motif from an artificial peptide library. *Biotech Bioeng.* 2008;101(5):1102−1107.

107. Yan X, Xu Z. Ribosome-display technology: applications for directed evolution of functional proteins. *Drug Discov Today.* 2006;11(19-20):911−916.

108. Zahnd C, Pecorari F, Straumann N, Wyler E, Pluckthun A. Selection and characterization of Her2 binding-designed ankyrin repeat proteins. *J Biol Chem.* 2006;281(46):35167−35175.

109. Hecht MH, Das A, Go A, Bradley LH, Wei Y. De novo proteins from designed combinatorial libraries. *Protein Sci.* 2004;13(7):1711−1723.

110. Ohta A, Yamagishi Y, Suga H. Synthesis of biopolymers using genetic code reprogramming. *CurrOpinChem Biol.* 2008;12(2):159−167.

111. Wang L, Xie J, Schultz PG. Expanding the genetic code. *Annu Rev Biophys Biomol Struct.* 2006;35(1):225−249.

112. Xie J, Schultz PG. A chemical toolkit for proteins − an expanded genetic code. *Nat Rev Mol Cell Biol.* 2006;7 (10):775−782.

113. Ozawa K, Loscha KV, Kuppan KV, Loh CT, Dixon NE, Otting G. High-yield cell-free protein synthesis for site-specific incorporation of unnatural amino acids at multiple sites. *Biochem Biophys Res Commun.* 2012;418 (4):652−656.

114. Johnson DBF, Xu J, Shen Z, Takimoto JK, Schultz MD, Schmitz RJ, et al. RF1 knockout allows ribosomal incorporation of unnatural amino acids at multiple sites. *Nat Chem Biol.* 2011;7(11):779−786.

115. Buchner E. Alkoholischegahrung ohnehefezellen. *BerChemGes.* 1897;30:117−124.

116. Swartz JR. Transforming biochemical engineering with cell-free biology. *AIChE Journal.* 2012;58(1):5−13.

117. Bujara M, Schümperli M, Billerbeck S, Heinemann M, Panke S. Exploiting cell-free systems: implementation and debugging of a system of biotransformations. *BiotechBioeng.* 2010;106(3):376−389.

118. Khalil AS, Collins JJ. Synthetic biology: applications come of age. *Nat Rev Genet*. 2010;11(5):367−379.

119. Stano P. Minimal cells: relevance and interplay of physical and biochemical factors. *Biotechnol J*. 2011;6 (7):850−859.

120. Forster AC, Church GM. Towards synthesis of a minimal cell. *Mol Syst Biol*. 2006;2:45.

121. Jewett MC, Forster AC. Update on designing and building minimal cells. *Curr Opin Biotechnol*. 2010;21 (5):697−703.

122. Glass JI, Assad-Garcia N, Alperovich N, Yooseph S, Lewis MR, Maruf M, et al. Essential genes of a minimal bacterium. *Proc Natl Acad Sci USA*. 2006;103(2):425−430.

123. Gibson DG, Glass JI, Lartigue C, Noskov VN, Chuang RY, Algire MA, et al. Creation of a bacterial cell controlled by a chemically synthesized genome. *Science*. 2010;329(5987):52−56.

124. Noireaux V, Maeda YT, Libchaber A. Development of an artificial cell, from self-organization to computation and self-reproduction. *Proc Natl Acad Sci USA*. 2011;108(9):3473−3480.

125. Wochner A, Attwater J, Coulson A, Holliger P. Ribozyme-catalyzed transcription of an active ribozyme. *Science*. 2011;332(6026):209−212.

126. Szostak JW, Bartel DP, Luisi PL. Synthesizing life. *Nature*. 2001;409(6818):387−390.

127. Stano P, Luisi PL. Achievements and open questions in the self-reproduction of vesicles and synthetic minimal cells. *Chem Commun (Camb)*. 2010;46(21):3639−3653.

128. Kuruma Y, Stano P, Ueda T, Luisi PL. A synthetic biology approach to the construction of membrane proteins in semi-synthetic minimal cells. *Biochim Biophys Acta*. 2009;1788(2):567−574.

129. Kurihara K, Tamura M, Shohda K, Toyota T, Suzuki K, Sugawara T. Self-reproduction of supramolecular giant vesicles combined with the amplification of encapsulated DNA. *Nat Chem*. 2011;3(10):775−781.

130. Caschera F, Sunami T, Matsuura T, Suzuki H, Hanczyc MM, Yomo T. Programmed vesicle fusion triggers gene expression. *Langmuir*. 2011;27(21):13082−13090.

131. Noireaux V, Libchaber A. A vesicle bioreactor as a step toward an artificial cell assembly. *Proc Natl Acad Sci USA*. 2004;101(51):17669−17674.

132. Asahara H, Chong S. In vitro genetic reconstruction of bacterial transcription initiation by coupled synthesis and detection of RNA polymerase holoenzyme. *Nucleic Acids Res*. 2010;38(13):e141.

133. Murtas G, Kuruma Y, Bianchini P, Diaspro A, Luisi PL. Protein synthesis in liposomes with a minimal set of enzymes. *Biochem Biophys Res Commun*. 2007;363(1):12−17.

134. Kuruma Y, Nishiyama K, Shimizu Y, Muller M, Ueda T. Development of a minimal cell-free translation system for the synthesis of presecretory and integral membrane proteins. *Biotechnol Prog*. 2005;21 (4):1243−1251.

135. Kita H, Matsuura T, Sunami T, Hosoda K, Ichihashi N, Tsukada K, et al. Replication of genetic information with self-encoded replicase in liposomes. *Chembiochem*. 2008;9(15):2403−2410.

136. Oberholzer T, Wick R, Luisi PL, Biebricher CK. Enzymatic RNA replication in self-reproducing vesicles: an approach to a minimal cell. *Biochem Biophys Res Commun*. 1995;207(1):250−257.

137. Liu DV, Zawada JF, Swartz JR. Streamlining *Escherichia coli* S30 extract preparation for economical cell-free protein synthesis. *Biotechnol Prog*. 2005;21(2):460−465.

138. Kim TW, Keum JW, Oh IS, Choi CY, Park CG, Kim DM. Simple procedures for the construction of a robust and cost-effective cell-free protein synthesis system. *J Biotechnol*. 2006;126(4):554−561.

139. Shrestha P, Holland T, Bundy BC. Streamlined extract preparation for *Escherichia coli*-based cell-free protein synthesis by sonication or bead vortex mixing. *BioTechniques*. 2012;53(3):163−174.

140. Airen IO, Swartz JR, eds. *Functional Genomic Analysis of* Escherichia Coli *Using Sequenctial Cell-Free Protein Synthesis*. Salt Lake City, UT; 2010: AIChE Annual Meeting.

141. Santacoloma PA, Sin G, Gernaey KV, Woodley JM. Multienzyme-catalyzed processes: next-generation biocatalysis. *Org Process Res Dev*. 2011;15(1):203−212.

142. Du L, Gao R, Forster AC. Engineering multigene expression in vitro and in vivo with small terminators for T7 RNA polymerase. *Biotech Bioeng*. 2009;104(6):1189−1196.

143. Du L, Villarreal S, Forster AC. Multigene expression in vivo: supremacy of large versus small terminators for T7 RNA polymerase. *BiotechBioeng*. 2012;109(4):1043−1050.

144. Wang HH, Huang P, Xu G, Haas W, Marblestone A, Li J, et al. Multiplexed in vivo his-tagging of enzyme pathways for in vitro single-pot multieznyme catalysis. *ACS Synth Biol*. 2012;1(2):43−52.

145. Sheldon RA. Enzyme immobilization: the quest for optimum performance. *Adv Synth Catal*. 2007;349 (8-9):1289−1307.

146. Jewett MC, Swartz JR. Mimicking the *Escherichia coli* cytoplasmic environment activates long-lived and efficient cell-free protein synthesis. *Biotech Bioeng*. 2004;86(1):19−26.

147. Ryabova LA, Vinokurov LM, Shekhovtsova EA, Alakhov YB, Spirin AS. Acetyl phosphate as an energy source for bacterial cell-free translation systems. *Anal Biochem.* 1995;226(1):184−186.

148. Calhoun KA, Swartz JR. Energizing cell-free protein synthesis with glucose metabolism. *Biotech Bioeng.* 2005;90(5):606−613.

149. Kim TW, Oh IS, Keum JW, Kwon YC, Byun JY, Lee KH, et al. Prolonged cell-free protein synthesis using dual energy sources: combined use of creatine phosphate and glucose for the efficient supply of ATP and retarded accumulation of phosphate. *Biotech Bioeng.* 2007;97(6):1510−1515.

150. Ma R, Yang Z, Huang L, Zhu X, Kai L, Cai J, et al. Construction of an efficient *Escherichia coli* cell-free system for in vitro expression of several kinds of proteins. *Eng Life Sci.* 2010;10(4):333−338.

151. Ezure T, Suzuki T, Higashide S, Shintani E, Endo K, Kobayashi S-I, et al. Cell-free protein synthesis system prepared from insect cells by freeze-thawing. *Biotechnol Prog.* 2006;22(6):1570−1577.

152. Ueda T, Kanamori T, Ohashi H. Ribosome display with the PURE technology. In: Endo Y, Ueda T, Takai K, eds. *Cell-Free Protein Production.* Humana Press; 2010:219−225.

153. Hino M, Kataoka M, Kajimoto K, Yamamoto T, Kido J-I, Shinohara Y, et al. Efficiency of cell-free protein synthesis based on a crude cell extract from *Escherichia coli*, wheat germ, and rabbit reticulocytes. *J Biotechnol.* 2008;133(2):183−189.

Towards Engineered Light—Energy Conversion in Nonphotosynthetic Microorganisms

Ilya Tikh and Claudia Schmidt-Dannert
University of Minnesota, St. Paul, MN, USA

INTRODUCTION

Photosynthesis has been essential for the development of complex life on earth and is generally defined as the ability to convert light into chemical energy coupled to carbon fixation. Depending on their needs, organisms have developed different strategies for capturing and utilizing energy from the sun. The diversity of light-capturing machinery ranges from single protein-driven proton pumping via simple proteorhodopsin, to the incredibly complex multiprotein assemblies of photosystems I and II found in plants and cyanobacteria.[1]

The way that light energy is transferred and stored as chemical energy during photosynthesis varies greatly between organisms. The primary mechanism by which chemical energy is stored is through the generation of a proton gradient across a membrane. In most bacteria, such a proton gradient is created across the inner membrane. In more complex photosynthetic systems, a special organelle, such as a thylakoid in cyanobacteria, houses all of the phototsynthetic machinery and is used to generate a proton gradient.[2] After a proton gradient is created, its energy can be converted to chemical energy in the form of ATP or NAD(P)H which are utilized by the organism for growth. Ultimately, some of the captured energy ends up as C—C bonds during CO_2 fixation. Unlike truly photosynthetic organisms, phototrophic organisms generally use a simpler method for light capture and are not able to utilize CO_2 as their sole carbon source. Instead of CO_2 fixation, these organisms use the additional energy from light capture to help drive other metabolic pathways (reviewed in [1]).

It is important to note that the core photosystem components responsible for light capture and conversion do not exist in a vacuum. Light-energy conversion in a photosynthetic organism requires a number of accessory proteins.[1] Complex pigments like chlorophylls and carotenoids are also required for the proper function of bacterial photosynthetic reaction centers (RCs) and photosystems I and II (PSI, PSII) from plants and algae. The enzymes responsible for the production of these pigments are encoded in large operons, and their importance will be discussed later.[3]

303

Synthetic Biology. DOI: http://dx.doi.org/10.1016/B978-0-12-394430-6.00016-9

In the past several decades major advances have been made in our knowledge about the diverse metabolic networks, and our ability to manipulate and engineer metabolic pathways of microorganisms for the production of complex molecules previously impossible or expensive to synthesize chemically. Metabolically engineered microorganisms are now used for the synthesis of a large number of chemicals ranging from biofuels to complex polymers and secondary metabolites such as drugs and anticancer compounds.[4] However, production of many of these compounds poses a heavy metabolic burden for such engineered producer organisms by depleting their ATP and NAD(P)H pools. Augmenting their energy pools by introducing a functional light-energy capture and conversion mechanism has the potential to reduce the energy strain on production hosts, especially during the synthesis of products with high energy requirements such as fatty acids. Carbon fixation is another very desirable process in an engineered microbe, especially during biofuel production, but it is also an energetically expensive process. With the help of recent developments in synthetic biology, genome-scale modifications of microbes are now becoming more practical. Consequently, introduction of light-energy converting machinery into a nonphotosynthetic host does not seem as impossible as it was even five years ago. In the following sections we will summarize current progress and challenges in implementing light-energy capture and conversion into microbial hosts.

INCORPORATION OF SIMPLE LIGHT-DRIVEN PROTON PUMPS INTO ENGINEERED MICROORGANISMS

The simplest and best characterized of the currently known mechanisms for light-energy conversion involves the rhodopsin protein family which is found in all kingdoms of life. Rhodopsin is a *trans*-membrane protein containing a single, light-responsive retinal cofactor. Depending on the host, rhodopsins function as light sensors, proton pumps, or ion pumps (reviewed in [5–7]). Bacteriorhodopsin from the archaea *Halobacterium salinarium* was the first light-driven proton pump discovered in the early 1970s.[8] To date, many archaeal bacteriorhodopsins have been identified in a wide range of environments using metagenomic sequencing techniques. Bacteriorhodopsin homologues appear to be especially prevalent in the oceans, where resources are sparse and any boost in energy generation can give an organism a significant competitive advantage. The first rhodopsin homologue from a proteobacteria, proteorhodopsin, was isolated from a metagenomic sample from the Sargasso Sea.[9] Recent work on light-driven proton pumping has centered around proteorhodopsins as opposed to bacteriorhodopsin, as the proteobacterial membrane proteins proved to be much easier to express and isolate from a wide range of heterologous hosts.[9] Both bacteriorhodopsin and proteorhodopsin work by translocating a proton from the cytoplasm into the periplasm during the light-induced isomerization of a retinol cofactor.[7] Overall, this creates a very simple mechanism of establishing and maintaining a proton gradient.

Being such a simple system, proteorhodopsin is an obvious candidate to add into a recombinant host as a first step in engineering a heterologous light-energy conversion system. Successful addition of proteorhodopsin to a new host requires not only proper folding and transmembrane localization, but also the availability of retinal as cofactor. Retinal can either be supplemented in the growth media or alternatively, a heterologous retinal biosynthetic pathway can be engineered for endogenous cofactor synthesis. For example, in *E. coli*, four additional gene products are needed to convert the isoprenoid precursor farnesyl diphosphate into β-carotene, which is then cleaved by another heterologous enzyme at the central 15,15′-double-bond to yield retinal.[10]

Since the discovery and characterization of proteorhodopsin, several groups have attempted to use it to supplement a heterologous host's energy pool, though with somewhat limited success.[11,12] There is evidence that expression of proteorhodopsin can aid in the regeneration of ATP in a cell, especially under starvation or stress conditions. Studies have shown that starved *E. coli* cells that express proteorhodopsin and are illuminated by light have more ATP than similar cells that are held in the dark.[11] Furthermore, *E. coli* cells expressing proteorhodopsin under illumination conditions can help negate the effects of respiration inhibition by azide.[12]

Even a small boost to a nonphotosynthetic organism's maximum metabolic capacity could prove to be very important. This was recently tested by coexpressing both a hydrogenase and proteorhodopsin in *E. coli*.[13] In theory, the presence of proteorhodopsin coupled with a hydrogenase should offset some of the metabolic burden on the cell during hydrogen production, and increase final H_2 yields by *E. coli*.[13] Addition of proteorhodopsin in that system did improve the final H_2 yields, though the improvement was fairly small and it is still not completely clear if the higher yield came from more protons being available for the hydrogenase, or from a proteorhodopsin-generated boost to the available energy levels inside the cell. If the increased hydrogen production came solely from more protons being available for the hydrogenase, then the applications for this strategy are somewhat limited. On the other hand, if proteorhodopsin is indeed able to provide a significant increase to the energy available for metabolic expenditures in *E. coli*, many potential applications in metabolically engineered *E. coli* can be envisioned.

An intriguing application is the coupling of proteorhodopsin to carbon fixation in a nonphotosynthetic host producing biofuels. Fixing CO_2 would not only reduce feedstock requirements of the host, but it would also help create a more carbon-neutral fuel product. But is proteorhodopsin able to generate enough energy to facilitate carbon fixation? A recent report shows that the presence of proteorhodopsin increases CO_2 fixation in the nonphotosynthetic marine bacterium *Polaribacter* sp. MED152, though the bacterium does require organic material in the medium for growth.[14] It is still unclear if expression of proteorhodopsin in a nonphotosynthetic host will generate sufficient energy to drive central metabolism, carbon fixation, or other potentially useful metabolic reactions. Several studies have shown that bacteria benefit the most from the presence of proteorhodopsin if they are under starvation, resource-limiting conditions, or alternatively, if the proton gradient across the inner membrane has been disrupted.[15,16] There are several reasons why one could expect limited help from the addition of proteorhodopsin to an energy-intensive metabolic process. One reason is the fairly low membrane potential generated by proteorhodopsin compared to the membrane potential maintained by *E. coli* under normal growth conditions.[12] If proteorhodopsin could be engineered to generate a higher membrane potential, it would allow for an increased generation of ATP under all conditions, and not primarily when the host's respiration is otherwise compromised. Another potential caveat to using proteorhodopsin, as well as other rhodopsins, for engineered light-energy conversion, is the narrow range of their light absorption spectra because of the single pigment molecule involved. This could pose a problem if a system reliant on proteorhodopsin is scaled for industrial applications as it may be difficult to obtain sufficient light penetration into the culture medium, a problem that is currently faced when using cyanobacteria or algae for industrial applications. One potential remedy for the narrow absorption wavelengths is the use of several different rhodopsin molecules tuned to different wavelengths in the same cell. It has been shown possible to tune the peak absorption wavelength by as much as 40 nm.[17] While this could broaden the useful light spectrum, the total spectrum used and the efficiency of protons translocated across a membrane in this system still pales in comparison with bacterial RC complexes.

Traditionally, metabolic engineering in microbes has been used in order to produce small molecules that are difficult or expensive to synthesize through chemical methods. In recent years, a major focus has been the production of biofuels or biopolymers to replace petroleum-based plastics, as well as complex drugs that are economically impractical to make using standard chemical synthesis. The majority of this section so far discussed how light-energy conversion may increase the metabolic output of an engineered organism. In contrast to biofuels, a new field of research recently became prominent, looking at how we can use microbes to bypass production of chemical fuels and instead generate electricity directly. Two organisms, *Geobacter sulfurreducens* and *Shewanella oneidensis* MR-1, have become model organisms for bacterial current-generating systems.[18,19] Both bacteria use a set of cytochromes to move excess reducing equivalents, generated during cell growth, from the quinone pool to the outside of the cell where the terminal electron acceptor is often an insoluble metal, or in a laboratory setup, an electrode. *S. oneidensis*, especially, has proven to be a good organism for engineering as it is similar enough to *E. coli* that many of the same genetic techniques are still applicable. Through genetic manipulation, the complete extracellular-respiratory metal reduction (Mtr) pathway of *Shewanella oneidensis* strain MR-1, which consists of three cytochromes and a porin-like outer membrane protein, is now well characterized (see Fig. 16.2).[20,21] Efforts by multiple groups are underway to engineer *S. oneidensis* and other bacteria's abilities to reduce insoluble metals for industrial applications. One of the more interesting of these potential applications is the conversion of water-soluble uranium (VI) to mostly insoluble uranium (IV).[20] Another obvious use for such bacteria is in microbial fuel cells, where they would be fed with waste streams from other applications, utilize any of the leftover carbon compounds, and produce electricity directly. Presently, the current production is still relatively low, though researchers are trying to optimize this system.[22]

Intrigued by the possibility of converting light-energy directly into electricity, we sought to investigate the possibility of enhancing current production by *S. oneidensis* under anaerobic conditions by expressing a functional proteorhodopsin in this bacterium. Indeed, illumination of proteorhodopsin expressing *S. oneidensis* led to a significant increase in current production.[16] Under anaerobic conditions, *S. oneidensis* can utilize lactate as both the carbon source and electron donor. Lactate is believed to be transported into the cells using a transporter reliant on a proton gradient. Illumination of cells containing proteorhodopsin therefore increases proton gradient coupled lactate uptake, and consequently electron transfer from lactate through the Mtr pathway to the electrode.[16]

These experiments also demonstrated that *S. oneidensis* may be an ideal host for engineered light-energy conversion, because it appears to bypass a respiratory control mechanism found in other bacteria and even mitochondria. For example, in *E. coli*, as the proton gradient increases across the membrane, translocation of more protons coupled to NADH oxidation becomes energetically less favorable, and thus respiration slows down.[23] It is estimated that that up to 10 protons are translocated during the oxidation of NADH to NAD under certain conditions.[24] During anaerobic growth of *S. oneidensis* on lactate, approximately one proton is pumped across the membrane for every electron transferred, greatly reducing the potential negative feedback of proton gradients on redox reactions. Furthermore, transport of lactate into the cell is coupled to proton uptake, further reducing the proton gradient.[16]

INCREASING THE COMPLEXITY AND EFFICIENCY OF LIGHT-ENERGY CAPTURE AND CONVERSION

The greater efficiency of light-energy conversion by the bacterial reaction centers (RCs) compared to proteorhodopsin comes at the price of increased complexity. Bacterial RCs are multiprotein complexes containing a large number of pigments, including bacteriochlophylls (Bchl) and carotenoids. Depending on the organism, either type I or type II RCs are present (Fig. 16.1A). In the case of algae and cyanobacteria, both type I and II RCs make up the photosynthetic apparatus.[1] Type II RCs, best studied in the purple bacteria, are cyclic, nonoxygen-evolving photosystems. The high-energy electrons generated by type II RCs are shuttled through the quinone pool, the cytochrome bc_1 complex, and are

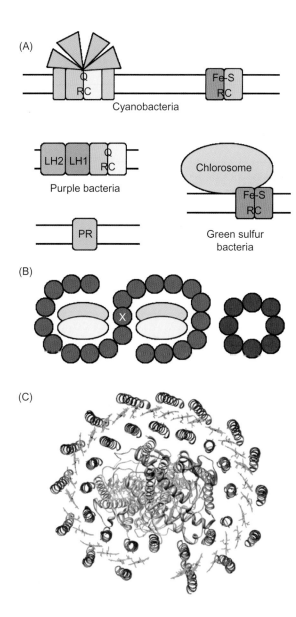

FIGURE 16.1

Overview of composition of light-harvesting machinery across different organisms. (A) Overview of the basic organization of the core reaction center (RC) complexes and the light-harvesting subunits (LH). Colors of the core complexes indicate whether they are composed of homo-dimers or hetero-dimers. Q or Fe—S indicate quinones or iron—sulfur clusters (respectively) as the pathway for the transfer of excited electrons from the special pair to the final acceptor (modified from Bryant et al.[1]). (B) Schematic representation of the organization of the RC—LH1—LH2 complex in *R. sphaeroides* as is present in the plane of the membrane. The core L, M, and H subunits (green and yellow) are surrounded by the LH1 complex (purple). The LH2 complex (red) is localized to the periphery of the RC—LH1 complex. (C) Top view of a crystal structure of the core RC—LH1 complex from *R. capsulatus* (PDB# 1PYH). The core L, M, and H subunits are located in the center surrounded by the $\alpha\beta$-peptides composing the LH1 (purple). Bacteriochlorophyll molecules are shown in green. The structure illustrates the complexity and precise assembly of the light-capturing machinery.

eventually returned to the special Bchl pair of the RC via cytochrome c_2.[25] In contrast, type I RCs found in green-sulfur bacteria can utilize ferrodoxin (Fd) as the final electron acceptor.[2] Since the electrons do not return to the special Bchl pair of the RC, they need to be regenerated from a different source, frequently from oxidation of sulfur compounds. The majority of bacterial RCs are surrounded by light-harvesting complexes containing additional pigments that increase the efficiency of light capture (Fig. 16.1).[1]

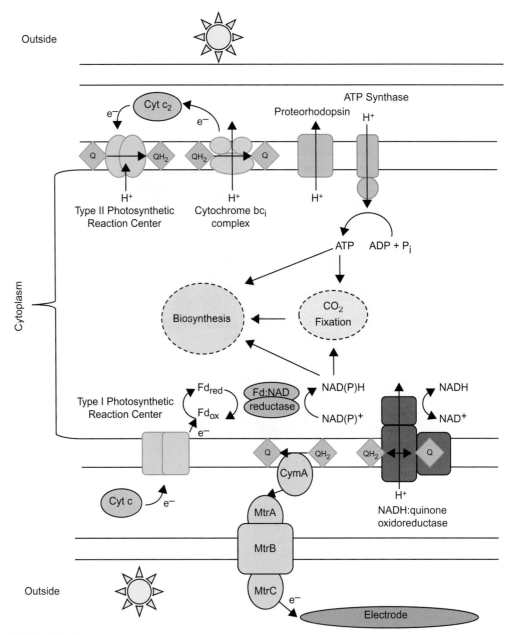

FIGURE 16.2

Integration of light-harvesting components into cell metabolism. Top portion of the figure demonstrates how a proton-translocating light-capturing system such as proteorhodopsin (PR) or a type II reaction center would integrate with the host's ATP synthase, providing a boost to cell energy stores. Upon light excitation, PR is able to transfer a proton from the cytoplasm to the periplasm using the isomerization of a retinol cofactor. Type II RC does not translocate protons directly. Two protons bind to the reduced quinone on the cytoplasmic side of the membrane and are released on the periplasmic side during the oxidation of the quinone. The cytochrome bc_1 complex, which couples the oxidation of quinones to the reduction of cytochrome c2, also translocates protons. The bottom section illustrates how a noncyclic type I RC can be used by the cell to regenerate reduced equivalents such as NADPH. After light excitation, electrons are transferred onto a ferrodoxin (Fd) molecule, which can be subsequently oxidized by Fd:NAD reductase in order to regenerate NADH. The transfer of electrons from the quinone pool through the *S. oneidensis* MR1 metal reduction (Mtr) pathway to an external acceptor, such as an electrode, is also shown in the bottom section of the figure.

Over the past several decades, the photosynthetic RCs from different *Rhodobacter* species have been used as models for the study of light-capturing systems due to their relatively simple purification and stability after purification, thus making them most amenable for introduction into a recombinant host.[26] The core RCs, as well as the surrounding light-harvesting complexes, have been thoroughly characterized. The RC in *R. sphaeroides* consists of three core proteins, referred to as the M, L, and H subunits, which are encoded by *pufM*, *pufL*, and *puhH*, respectively. This core complex, with its bound Bchl and carotenoid pigments, is surrounded by a light-harvesting (LH) superstructure formed by the LH1 and LH2 protein complexes that contain accessory Bchl pigments (Fig. 16.1B,C).[27]

After light strikes the reaction center, it excites a low-energy electron from one of the two Bchla molecules (referred to as the 'special pair') located at the center of the type II RC complex, the excited electron is transferred in the RC via two additional chlorophyll pigments, and a quinone to a final quinone electron acceptor that exits the RC core upon reduction by two electrons. Thus, two excitation reactions are necessary before a quinone molecule is released from the RC complex (reviewed in [28]). Proton translocation across the inner membrane is coupled to the decrease in the potential energy of the electrons during each step of the cyclic electron. The established proton gradient can then be utilized by proton-gradient coupled enzymes such as ATP synthase to regenerate the metabolic needs of the cell.[12] Alternatively, reduced quinones can be used directly by enzymes such as NADH: quinoneoxidoreductase to regenerate NADH reducing equivalents needed for CO_2 fixation or other metabolically taxing reactions (Fig. 16.2).[2] In the noncyclic, type I RC mentioned above, the high-energy electrons are placed on a ferrodoxin molecule instead of a quinone. The ferrodoxin is subsequently released, and is used as an energy source to convert NAD(P)$^+$ to NAD(P)H by the ferrodoxin-NAD reductase (Fig. 16.2).[2] Similarly, the electrons in the special pair are regenerated from a c-type cytochrome, though this time the electrons originate from a different donor, frequently a sulfur compound.[2]

The extra energy boost generated by either the type I or type II RC would be ideally coupled to an energy-expensive metabolic reaction such as the synthesis of long-chain fatty acids as liquid transportation biofuels. Fatty acid synthesis utilizes a malonyl-CoA molecule as a basic building block, and requires hydrolysis of ATP to create C–C bonds (reviewed in [4,29]). If a host is engineered for overproduction of fatty acids, any potential boost to its total ATP pool should prove beneficial to the final product yields. Not only are ATP molecules needed during the production of long-chain fatty acids, but a number of NADPH molecules are also used up in the process of making a saturated fatty acid. As discussed previously, NADPH could be regenerated by the use of reduced quinones, or from ferrodoxin molecules generated during electron transfer from the reaction center (Fig. 16.2).[2,25]

The addition of a more efficient system for light-energy conversion in an electricity-generating bacterium like *S. oneidensis* should translate into a greatly increased current production. Because the RC complex is able to directly generate reduced quinones, it should provide *S. oneidensis* with the ability to send more electrons to the electrode. One issue that would need to be circumvented in recombinant *S. oneidensis* is the cyclic transfer of electrons generated by the RCs from purple bacteria. If electrons are transferred from the quinone pool to the electrode, a system will need to be engineered to reduce the special pair in the RC. A potential solution to compensate for the loss of the special pair electrons would be to engineer cells with a noncyclic, type I, RC complex, and the protein machinery needed to regenerate special pair electrons through the oxidation of inorganic sulfur or nitrogen.

Functional expression of bacterial RCs in a nonnative host presents a number of challenges. First, while the core RC from *R. sphaeroides* consists of only three proteins, two of the core subunits, L and M, are transmembrane proteins. It has been shown that in *R. sphaeroides* both of those subunits are needed for the stability of the RC complex.[30] Furthermore, overexpression of heterologous membrane proteins is problematic, often leading to toxicity issues. Successful overexpression of just the core RC complex in a heterologous host such as *E. coli* will likely require engineering of a strain that is able to better tolerate high levels of membrane proteins. *R. sphaeroides* solves this problem by creating a large amount of membrane space by using

309

membrane invaginations. A similar tactic could be applied to a recombinant host. Unfortunately, the mechanism by which *R. sphaeroides* creates membrane invaginations is poorly-understood. There are *E. coli* strains, such as C43(DE3),[31] that have more membrane space and are used for the overexpression of some membrane proteins, but it does not solve toxicity problems associated with overexpression for all membrane proteins. In addition to the reaction center core complex, the peripheral light-harvesting complexes should ideally also be added to an engineered host for maximum light-capture efficiency. In *R. sphaeroides*, these are encoded by at least four more genes that express the alpha and beta peptides of the LH1 and LH2 complexes. It is likely that other genes, such as *pufX*, will need to be coexpressed as well to facilitate proper super complex formation and function.[32]

Lastly, bacteriochlorophyll and carotenoid pigments need to be introduced into a nonphotosynthetic host either via exogenous addition or by engineering of their biosynthetic pathways. Biosynthesis of porphyrin-derived Bchl a pigments in *E. coli* requires the assembly of at least 10 different enzymatic steps. In previous research, we have shown that overexpression of heme biosynthetic genes leads to accumulation of significant quantities of porphyrins in *E. coli*.[33,34] In subsequent work, we demonstrated that expression of the three-subunit Mg−chelatase complex in *E. coli* introduces magnesium into protoporphyrin IX as the first committed step of (bacterio)chlorophyll biosynthesis. Extension with the next enzyme in the Bchl pathway produced Mg-protoporphyrin IX monomethylester in *E. coli*.[35] The remaining enzymes required for Bchla synthesis in *Rhodobacter* species have been elucidated using gene deletions; however, not all enzyme activities have been confirmed via biochemical methods.[3]

One key enzyme that has yet to be biochemically characterized is the anaerobic Mg-protoporphyrin IX monomethylestercyclase (BchE). This enzyme is responsible for the introduction of the characteristic fifth ring of (bacterio)chlorophylls.[36] The same reaction has also been proposed to be carried out under aerobic conditions by AcsF. Very little is known about either of the two enzymes, except that strains with genomic deletions of those genes accumulate Mg-protoporphyrin IX monomethylester.[36] BchE is predicted to have a Fe−S cluster and is expressed under anaerobic conditions, so any host cell must have the machinery for proper Fe−S cluster maturation. If the host is not inherently competent at Fe−S protein maturation, a suitable maturation pathway would have to be engineered into the host. Even less is known about AcsF, except that it likely requires the presence of at least one more protein in order to be active.[36]

Carotenoid pigments serve as accessory pigments in light-harvesting complexes, and in the RC complex. Many different carotenoid biosynthetic pathways have been successfully engineered into *E. coli*, including pathways for the synthesis of the acyclic carotenoid present in *Rhodobacter* reaction centers.[37,38] Although these pigments are not immediately needed to show function of an engineered RC complex, they can be easily produced in a recombinant host to provide photoprotection to the photosystem.

The primary focus has so far been on possibilities and challenges of taking a light-capturing system and reconstituting it in a nonphotosynthetic host. The primary assumption has been that the nonphotosynthetic host is well-studied, with a good understanding of its metabolic processes and, most importantly, is amenable to genetic engineering. A lack of a good set of molecular biology tools for the majority of photosynthetic organisms prevents the introduction of complex metabolic features required for the development of industrial production processes for chemicals or biofuels. Well-studied photosynthetic organisms, such as *R. sphaeroides* or some cyanobacteria, do have some genetic tools presently available, but they pale in comparison to what is possible in the well-studied biotechnology workhorse *E. coli*. Once we generate a comprehensive genetic toolkit, as well as thorough understanding of metabolic flux and signaling pathways in *R. sphaeroides*, it could become a very attractive host for industrial use. Being naturally photosynthetic, only water, CO_2, and sunlight would be needed for growth of *R. sphaeroides,* and given the large membrane space generated during photosynthetic growth, it could be an ideal host for production of hydrophobic compounds or membrane proteins.

COMBINING LIGHT-ENERGY CONVERSION AND CO_2 FIXATION

Carbon fixation is an integral part of photosynthesis, and something that must be taken into account when engineering photosynthesis into a new host. The addition of carbon fixation into a nonnative host can present numerous advantages for an engineered system. Carbon fixation can be used to reduce the host's dependence on organic material as a carbon source and allow for a wider range of growth conditions. The prospect of utilizing atmospheric CO_2 during growth is especially desirable for biofuel production processes, as this directly generates carbon-neutral fuels. Recent advances in our understanding and engineering of protein-based bacterial microcompartments, which serve to increase the efficiency of metabolic pathways — including CO_2 fixation in carboxysomes — enable new approaches for the introduction of efficient carbon fixation in heterologous hosts.[39−42] In fact, the functional expression of *H. neapolitanus* RuBisCO, inside of carboxysome, has very recently been demonstrated in *E. coli.*[43] Efforts to increase the efficiency of recombinant RuBisCO, coupled to optimized carboxysome expression in *E. coli,* could produce cells capable of efficiently coupling carbon fixation to a biosynthetic process.[44] Reconstructing the carboxysome along with carbon fixation using RuBisCO is appealing, because its catalytic activity is inherently poor (reviewed in [45]). Native RuBisCo has turnover rates on the order of several reactions per second, depending on environmental conditions; RuBisCO can also incorporate O_2 instead of CO_2 during the reaction cycle to generate a molecule of 2-phosphoglycolate. 2-Phosphoglycolate is of little use to the host and must be enzymatically converted to a useful metabolite.[45] However, by localizing RuBisCo together with a carbonic anhydrase to carboxysomes, cells are able to substantially increase local CO_2 concentrations by transporting inorganic bicarbonate from the cytoplasm into compartments where it is converted into CO_2, which helps to overcome some of RuBisCo's inefficiencies. Any attempt to engineer efficient carbon fixation into a novel host must take into account these limitations.

Not all organisms use RuBisCo to fix CO_2, other strategies for carbon fixation include the reductive TCA cycle or the 3-hydroxypropionate cycle.[46,47] Both of those cycles have interesting biotechnology implications if they were engineered into a heterologous host. Driving the regular TCA cycle in reverse yields the reductive TCA cycle, and the majority of reactions in the TCA cycle are reversible. Only three key enzymes need to be replaced to operate the TCA cycle in reverse and yield the reductive TCA cycle.[46] Carbon fixed in this manner would enter cell metabolism as acetyl-CoA, or with the addition of another CO_2 molecule, pyruvate, and would be easily integrated into cellular metabolism. The 3-hydroxypropionate cycle fixes CO_2 by combining it with an acetyl-CoA molecule, making malonyl-CoA, followed by reduction and elongation using a second molecule of CO_2.[46] This form of carbon fixation is especially useful in hosts engineered to produce biofuels or biopolymers derived from fatty acids, because malonyl-CoA is a basic building block for those compounds. The first step, addition of CO_2 to acetyl-CoA, is catalyzed by acetyl-CoA carboxylase in a reaction that requires hydrolysis of ATP.[46] As we have previously discussed, ATP can be regenerated using ATP-synthase and the proton gradient established by either proteorhodopsin or a bacterial RC. If the final goal is to simply boost the availability of malonyl-CoA, while consuming CO_2, then the addition of acetyl-CoA carboxylase to the host should be sufficient. Subsequent enzymes of the 3-hydroxypropionate cycle transform malonyl-CoA into an assortment of small, branched biosynthetic precursors including 4-hydroxybutyrate, a feedstock for biopolymer production.[47]

We mentioned previously that organisms such as *S. oneidensis* and *G. sulfurreducens* are able to balance any excess-reducing equivalents by shuttling those electrons onto an electrode, thus generating a current, and it has been shown that the same process can be driven in reverse.[48] Driving electrons back into *S. oneidensis* through the Mtr pathway generates reduced quinones that can 'in theory' be oxidized to regenerate energy carrying molecules such as NAD(P)H. While this system needs further testing, it is an attractive prospect to engineer a bacterium such as *S. oneidensis* to fix CO_2 and produce high value-added commodity chemicals by using electricity as an energy source for the cell.

FROM BIOLOGICAL TO ARTIFICIAL PHOTOSYNTHETIC SYSTEMS

Nature has had millions of years to perfect photosynthetic machinery; however, for specific applications, a smaller, artificial mimic might be more desirable. For example, it is difficult to utilize a photosynthetic organism in a photovoltaic cell, or conditions in a particular application may be too harsh, damaging the cells. In such instances, a chemically synthesized mimic of a naturally occurring process may offer a solution. The robust nature and functionality of photosynthesis is a marvel of biological mechanics, and a great deal of research has focused on artificially replicating different parts of the photosynthetic process (reviewed in [49,50]). Investigations have yielded new catalysts such as TiO_2 that are able to split water into hydrogen and oxygen when excited by the UV light.[51] Synthetic compounds have also been developed that mimic chlorophylls and are used to transfer electrons to a conductive surface once they have been excited by light (Fig. 16.3). A majority of the dyes utilized in engineered photosynthetic systems are derived from a chemically modified porphyrin backbone, which has an absorption wavelength in the 400 nm range.[50] Porphyrin-derived dyes are excited by light and donate a high-energy electron, similar to what the special pair can do in the bacterial RC. Once the electron is donated from a synthetic dye, it is replaced by an electron from an electrolyte solution, creating a noncyclic system unlike that of the type II RCs where the electron must return to its source.

Recent research into artificial photosynthesis has attempted to couple light excitation to the reduction of $NAD(P)^+$, essentially mimicking what occurs during photosynthesis in nature.[52,53] These techniques involve photosensitizing a nanotube made from a conductive material such as TiO_2 or diphenylalanine (F-F). Photosensitizing compounds can be artificial, such as quantum dots, or of biological origin, such as porphyrin mimics.[52] A mediator compound is used to transfer the electrons from the nanotubes onto $NAD(P)^+$ molecules. This is exciting research, as it would allow for the regeneration of reducing equivalents in vitro, using only sunlight, replacing the multienzyme systems currently used.

The next step could be combining an artificial light-harvesting nanotube system with a bacterium such as *S. oneidensis*, which is able to utilize external electrons for biosynthesis

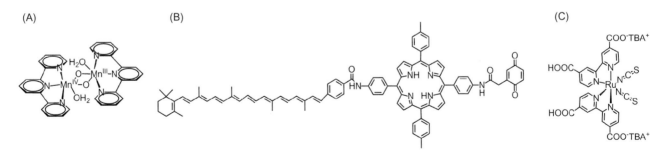

FIGURE 16.3

Examples of photoactive dyes typically used for artificially engineered systems. (A) A dinuclear, Mn-based water-oxidation catalyst that mimics the water splitting by the oxygen-evolving photosystem II. (B) A biological mimic dye containing elements from a carotenoid, porphyrin, and quinone molecules joined together. (C) An example of a Ru-based dye used in dye-sensitized solar cells.[49,50]

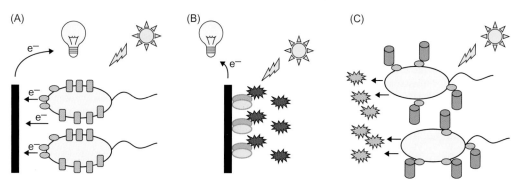

FIGURE 16.4

Combining artificial and biological light-capture mechanisms for energy generation. (A) *S. oneidensis* cells in a bioreactor expressing proteorhodopsin (orange). Upon light excitation, proteorhodopsin (PR) generates a proton gradient which can be used by the cell for ATP synthesis or lactate import. Excess reducing equivalents generated during cell growth are transferred through the *S. oneidensis* metal reduction pathway (red) via either direct contact or an intermediate to the electrode surface generating a current. (B) Purified bacterial reaction centers (yellow and green) attached to an electrode. Upon light excitation, electrons are driven from the RCs to the electrode generating a current. The low-energy electrons at the special pair are regenerated from synthetic electron donors (red) present in the solution. (C) *S. oneidensis* cells coupled to an artificial light-harvesting system (blue) either directly through the metal reduction pathway (red) or through a mediator. The additional energy from the light-energy conversion can be used by the cell during the production of complex, value-added products (orange).

(Fig. 16.4). If *S. oneidensis* is able to interface with such a system, it would eliminate the need to engineer a biological light-harvesting system into the cell, or the need to drive electrons into *S. oneidensis* using an electrode, simplifying the final system. A *S. oneidensis*–nanotube interface could be set in a standard bioreactor, or perhaps the cells could be immobilized in a solid matrix.

The reverse scenario is also intriguing. Creating photovoltaic cells which couple isolated core RC proteins, with a sensitized nanotube or an electrode, has the potential to dramatically increase the efficiency of the light-capturing reactions, thus boosting the final electrical output of the photovoltaic cell (Fig. 16.4). Efforts to do this are currently underway, and it had been demonstrated to be a viable system.[54] The studies have also shown that including other cofactors, such as a cytochromes, increases the efficiency of electron transfer from the RC to the electrode.[54,55] Ensuring the presence of proper accessory proteins has been a theme throughout this chapter, and it remains true when working in vitro on an engineered surface, such as an electrode.

FUTURE DIRECTIONS

There is little doubt about the numerous benefits and potential biotechnological application that can be realized by combining photosynthetic machinery with the biosynthesis of industrially relevant compounds. Initial steps in this direction have been taken, though there are still many hurdles to overcome before a seamless incorporation of light-capturing complexes in a nonphotosynthetic host is possible. Synthetic biology offers many of the tools required to successfully integrate light-capture into novel hosts and novel applications. Applications may include introducing light-energy conversion machinery into a host natively able to generate electricity, effectively creating biological photovoltaic cells, or modifying current photovoltaic cells by adding purified light-harvesting components to increase light-capture efficiency. Another exciting possibility is combining artificial light-capture mechanisms, like photosensitized nanotubes, with a biological system producing value-added products. However, some hurdles like inability to express BchE recombinantly, or optimizing the interface between a biological and artificial system, still remain. The combination of biology and engineering allows for the development of creative solutions to overcome current hurdles which neither field alone could overcome.

References

1. Bryant DA, Frigaard NU. Prokaryotic photosynthesis and phototrophy illuminated. *Trends Microbiol.* 2006;14:488–496.

2. Frigaard NU, Chew AGM, Li H, Maresca JA, Bryant DA. *Chlorobium tepidum*: insights into the structure, physiology, and metabolism of a green sulfur bacterium derived from the complete genome sequence. *Photosyn Res.* 2003;78:93–117.

3. Bollivar DW. Recent advances in chlorophyll biosynthesis. *Photosyn Res.* 2006;90:173–194.

4. Zhang F, Rodriguez S, Keasling JD. Metabolic engineering of microbial pathways for advanced biofuels production. *Curr Opin Biotechnol.* 2011;22:775–783.

5. DeLong EF, Béjà O. The light-driven proton pump proteorhodopsin enhances bacterial survival during tough times. *PLoS Biol.* 2010;8:e1000359.

6. Smith SO. Structure and activation of the visual pigment rhodopsin. *Annu Rev Biophys.* 2010;39:309–328.

7. Lanyi JK. Bacteriorhodopsin. *Annu Rev Physiol.* 2004;66:665–688.

8. Oesterhelt D, Stoeckenius W. Functions of a new photoreceptor membrane. *Proc Natl Acad Sci USA.* 1973;70:2853–2857.

9. Beja O, Aravind L, Eugene K, et al. Bacterial rhodopsin: evidence for a new type of phototrophy in the sea. *Science.* 2000;289:1902–1906:80.

10. Marasco EK, Vay K, Schmidt-Dannert C. Identification of carotenoid cleavage dioxygenases from *Nostoc* sp. PCC 7120 with different cleavage activities. *J Biol Chem.* 2006;281:31583–31593.

11. Martinez A, Bradley AS, Waldbauer JR, Summons RE, DeLong EF. Proteorhodopsin photosystem gene expression enables photophosphorylation in a heterologous host. *Proc Nat. Acad Sci USA.* 2007;104:5590–5595.

12. Walter JM, Greenfield D, Bustamante C, Liphardt J. Light-powering *Escherichia coli* with proteorhodopsin. *Proc Natl Acad Sci USA.* 2007;104:2408–2412.

13. Kim JYH, Jo BH, Jo Y, Cha HJ. Improved production of biohydrogen in light-powered *Escherichia coli* by co-expression of proteorhodopsin and heterologous hydrogenase. *Microb Cell Fact.* 2012;11:2.

14. González JM, Fernández-Gómez B, Fernàndez-Guerra A, et al. Genome analysis of the proteorhodopsin-containing marine bacterium Polaribacter sp. MED152 (Flavobacteria). *Proc Natl Acad Sci USA.* 2008;105:8724–8729.

15. Gómez-Consarnau L, González JM, Coll-Lladó M, et al. Light stimulates growth of proteorhodopsin-containing marine Flavobacteria. *Nature.* 2007;445:210–213.

16. Johnson ET, Baron DB, Naranjo B, Bond DR, Schmidt-Dannert C, Gralnick JA. Enhancement of survival and electricity production in an engineered bacterium by light-driven proton pumping. *Appl Environ Microbiol.* 2010;76:4123–4129.

17. Man-Aharonovich D, Sabehi G, Sineshchekov OA, Spudich EN, Spudich JL, Béjà O. Characterization of RS29, a blue-green proteorhodopsin variant from the Red Sea. *Photochem Photobiol Sci.* 2004;3:459–462.

18. Logan BE. Exoelectrogenic bacteria that power microbial fuel cells. *Nat Rev Microbiol.* 2009;7:375–381.

19. Shi L, Squier TC, Zachara JM, Fredrickson JK. Respiration of metal (hydr)oxides by Shewanella and Geobacter: a key role for multihaem c-type cytochromes. *Mol Microbiol.* 2007;65:12–20.

20. Fredrickson JK, Romine MF, Beliaev AS, et al. Towards environmental systems biology of Shewanella. *Nat Rev Microbiol.* 2008;6:592–603.

21. Coursolle D, Gralnick JA. Modularity of the Mtr respiratory pathway of *Shewanella oneidensis* strain MR-1. *Mol Microbiol.* 2010;77:995–1008.

22. Fornero JJ, Rosenbaum M, Angenent LT. Electric power generation from municipal, food, and animal wastewaters using microbial fuel cells. *Electroanalysis.* 2010;22:832–843.

23. Burstein C, Tiankova L, Kepes A. Respiratory control in *Escherichia coli* K 12. *Eur J Biochem.* 1979;94:387–392.

24. Simon J, van Spanning RJM, Richardson DJ. The organisation of proton motive and non-proton motive redox loops in prokaryotic respiratory systems. *Biochim Biophys Acta.* 2008;1777:1480–1490.

25. Lavergne J, Vermeglio A. Functional coupling between reaction centers and cytochrome bc 1 complexes. *The Purple Phototrophic Bacteria.* 2008:509–536.

26. Goldsmith JO, Boxer SG. Rapid isolation of bacterial photosynthetic reaction centerss with an engineered poly-his tag. *Biochim Biophys Acta.* 1996;1276:171–175.

27. Roszak AW, Howard TD, Southall J, et al. Crystal structure of the RC-LH1 core complex from *Rhodopseudomonas palustris*. *Science*. 2003;302:1969—1972.

28. Wraight C, Gunner M. The acceptor quinones of purple photosynthetic bacteria — structure and spectroscopy. *The Purple Phototrophic Bacteria*. 2008:379—405.

29. Handke P, Lynch SA, Gill RT. Application and engineering of fatty acid biosynthesis in *Escherichia coli* for advanced fuels and chemicals. *Metab. Eng.* 2011;13:28—37.

30. Tehrani A, Beatty JT. Effects of precise deletions in rhodobacter sphaeroides reaction center genes on steady-state levels of reaction center proteins: a revised model for reaction center assembly. *Photosyn Res.* 2004;79:101—108.

31. Miroux B, Walker JE. Over-production of proteins in *Escherichia coli*: mutant hosts that allow synthesis of some membrane proteins and globular proteins at high levels. *J Mol Biol.* 1996;260:289—298.

32. Farchaus JW, Gruenberg H, Oesterhelt D. Complementation of a reaction center-deficient *Rhodobacter sphaeroides* pufLMX deletion strain in trans with pufBALM does not restore the photosynthesis-positive phenotype. *J Bacteriol.* 1990;172:977—985.

33. Kwon SSJ, de Boer AL, Petri R, Schmidt-Dannert C. High-level production of porphyrins in metabolically engineered *Escherichia coli*: systematic extension of a pathway assembled from overexpressed genes involved in heme biosynthesis. *Appl Environ Microbiol.* 2003;69:4875—4883.

34. Kwon SJ, Petri R, DeBoer AL, Schmidt-Dannert C. A high-throughput screen for porphyrin metal chelatases: application to the directed evolution of ferrochelatases for metalloporphyrin biosynthesis. *Chem Bio Chem.* 2004;5:1069—1074.

35. Johnson ET, Schmidt-Dannert C. Characterization of three homologs of the large subunit of the magnesium chelatase from *Chlorobaculum tepidum* and interaction with the magnesium protoporphyrin IX methyltransferase. *J Biol Chem.* 2008;283:27776—27784.

36. Ouchane S, Steunou A-S, Picaud M, Astier C. Aerobic and anaerobic Mg-protoporphyrin monomethyl ester cyclases in purple bacteria: a strategy adopted to bypass the repressive oxygen control system. *J Biol Chem.* 2004;279:6385—6394.

37. Lee PC, Momen AZ, Mijts BN, Schmidt-Dannert C. Biosynthesis of structurally novel carotenoids in *Escherichia coli*. *Chem Biol.* 2003;10:453—462.

38. Mijts BN, Lee PC, Schmidt-Dannert C. Identification of a carotenoid oxygenase synthesizing acyclic xanthophylls: combinatorial biosynthesis and directed evolution. *Chem Biol.* 2005;12:453—460.

39. Cheng S, Liu Y, Crowley CS, Yeates TO, Bobik TA. Bacterial microcompartments: their properties and paradoxes. *Bioessays.* 2008;30:1084—1095.

40. Kerfeld CA, Heinhorst S, Cannon GC. Bacterial microcompartments. *Annu Rev Microbiol.* 2010;64:391—408.

41. Yeates TO, Kerfeld CA, Heinhorst S, Cannon GC, Shively JM. Protein-based organelles in bacteria: carboxysomes and related microcompartments. *Nat Rev Microbiol.* 2008;6:681—691.

42. Choudhary S, Quin MB, Sanders MA, Johnson ET, Schmidt-Dannert C. Engineered protein nano-compartments for targeted enzyme localization. *PLoS ONE.* 2012;7:e33342.

43. Bonacci W, Teng PK, Afonso B, Niederholtmeyer H, Grob P, Silver PA. Modularity of a carbon-fixing protein organelle. *Proc Natl Acad Sci USA.* 2011;109:478—483.

44. Parikh MR, Greene DN, Woods KK, Matsumura I. Directed evolution of RuBisCO hypermorphs through genetic selection in engineered *E. coli*. *Protein Eng. Des Sel.* 2006;19:113—119.

45. Badger MR, Bek EJ. Multiple Rubisco forms in proteobacteria: their functional significance in relation to CO_2 acquisition by the CBB cycle. *J Exp Bot.* 2008;59:1525—1541.

46. Atomi H. Microbial enzymes involved in carbon dioxide fixation. *J Biosci Bioeng.* 2002;94:497—505.

47. Berg IA, Kockelkorn D, Buckel W, Fuchs GA. 3-Hydroxypropionate/4-hydroxybutyrate autotrophic carbon dioxide assimilation pathway in Archaea. *Science.* 2007;318:1782—1786.

48. Ross DE, Flynn JM, Baron DB, Gralnick JA, Bond DR. Towards electrosynthesis in shewanella: energetics of reversing the mtr pathway for reductive metabolism. *PLoS ONE.* 2011;6:e16649.

49. McConnell I, Li G, Brudvig GW. Energy conversion in natural and artificial photosynthesis. *Chem Biol.* 2010;17:434—447.

50. Kalyanasundaram K, Graetzel M. Artificial photosynthesis: biomimetic approaches to solar energy conversion and storage. *Curr Opin Biotechnol.* 2010;21:298—310.

51. Linsebigler AL, Lu G, Yates Jr JT. Photocatalysis on TiO_2 surfaces: principles, mechanisms, and selected results. *Chem Rev.* 1995;95:735—758.

315

52. Ryu J, Lee SH, Nam DH, Park CB. Rational design and engineering of quantum-dot-sensitized TiO_2 nanotube arrays for artificial photosynthesis. *Adv Mater (Weinheim, Ger.)*. 2011;23:1883–1888.

53. Kim JH, Lee M, Lee JS, Park CB. Self-assembled light-harvesting peptide nanotubes for mimicking natural photosynthesis. *Angew Chem*. 2012;124:532–535.

54. Lebedev N, Trammell SA, Spano A, Lukashev E, Griva I, Schnur J. Conductive wiring of immobilized photosynthetic reaction center to electrode by cytochrome C. *J Am Chem Soc*. 2006;128:12044–12045.

55. Das R, Kiley PJ, Segal M, et al. Integration of photosynthetic protein molecular complexes in solid-state electronic devices. *Nano Lett*. 2004;4:1079–1083.

Applications of Engineered Synthetic Ecosystems

Harris H. Wang[1],*, **Michael T. Mee[3,4]*** **and George M. Church[2,3]**
[1]Columbia University, New York, NY, USA
[2]Harvard University, Boston, MA, USA
[3]Harvard Medical School, Boston, MA,USA
[4]Boston University, Boston, MA, USA

INTRODUCTION

A major goal in synthetic biology is to construct biological systems with robust and controllable behavior and functionality.[1] Over recent decades, biologists have tried to deconvolve the complexity of life by elucidating key genetic and regulatory determinants, with the hope of eventually engineering biological systems in more predictable ways. Genetic circuits have now been designed and rewired with relative ease to produce interesting and useful cellular phenotypes.[2] A natural extension of this bioengineering framework is the combination of different cells into groups and artificial consortia of increasing complexity.[3] This approach is important, since heterogeneous populations can often outperform homogeneous populations of genetically identical individuals in many tasks that require more sophisticated divisions of labor.[4,5] Natural microbial consortia, for example, are able to degrade complex substrates more efficiently than any single member can.[6] Furthermore, mixed populations are more robust to environmental variations,[7] and can potentially be reprogrammed in modular ways. However, building higher-order biological systems relies on improving our understanding of the ecology of dynamic multicomponent communities, both natural and synthetic. The area of synthetic ecology is poised to grow in this endeavor.

Just like the age-old tradition of brewing using yeast, we have a history of successes utilizing microbial consortium, albeit crudely, for applications such as composting and waste treatment.[8,9] Coupling biodegradation of complex feedstocks to bioproduction of useful products is also achievable.[10,11] However, attempts to optimize these processes rely on treating natural communities naïvely as black-box operations, because we do not understand much of the underlying design principles and constraints needed to engineer microbial communities. Furthermore, individual cells have their own growth objectives, subject to Darwinian selection, that often do not align with human-designed synthetic objectives such as overproduction of metabolically expensive compounds.[12]

*These authors contributed equally to this work.

Synthetic Biology. DOI: http://dx.doi.org/10.1016/B978-0-12-394430-6.00017-0

Methods to engineer microbial communities have thus far relied on natural selection of desirable traits, which is very limiting.[13] Engineering at this next level will require new advances in synthetic biology. Integration of coupled metabolism, directed cell—cell communication, and programmable community structure for real-world applications will require tools in in silico design, large-scale genome engineering, and high-throughput DNA synthesis, among other cutting edge techniques. Here, we highlight important recent developments that have advanced the field of synthetic ecosystems, and outline crucial areas for future innovations.

TARGETING MICROBIAL COMMUNITIES FOR FORWARD ENGINEERING

Genomics has significantly advanced our knowledge of microbial communities, enabling us to have the potential to engineer and control microbes at the genetic level. DNA and RNA sequencing have allowed us to determine genomic diversity and transcriptomic profiles of microbial populations in areas such as bioremediation or bioproduction.[14,15] In silico models of cellular metabolism can be used to assess flow of metabolites through individual cells, and are now being scaled across communities of cells.[16—20] New methods in recombineering,[21] oligo-directed genomic modifications,[22—25] and gene synthesis[26] have revolutionized how we perturb, understand, and improve the interactions between cellular components through strain engineering. Applying these tools to mixed consortia will bring further elucidation of population-level phenomena, such as interspecies metabolic exchange, community stability, and adaptive evolution.[7,27,28] Emerging advances in synthetic ecology require us to control three important features outlined below (Fig. 17.1).

First, engineering ecosystems requires precise understanding and control of metabolism and metabolic exchange. Engineering metabolism has thus far mainly focused on the biosynthetic capabilities of individual cells.[29,30] A key challenge is the elucidation of metabolism at the population level,[31] and the development of techniques to optimally combine different metabolic pathways together in useful ways.[32] One avenue of pursuit may be to physiologically link cells together through metabolic exchange using different metabolite transport systems. By mining metagenomic libraries for transporters,[33] combined with cytosolic exchange strategies,[34] interactions across a metabolic network of cells can be exploited.[35,36] Engineering these metabolic interactions provides a means to control the degree of antagonistic or beneficial relationships between population members. Modifying metabolic exchange will improve our understanding of how metabolism can be partitioned across a heterogeneous mixture of cells, and its effects on the dynamics, functionality, and efficiency of the system.

Second, engineering ecosystems requires directed cell—cell communication. Quorum sensing (QS) molecules, such as acyl-homoserine lactones, enable microbes to communicate with one another by diffusion through the extracellular space.[37] Intercellular signaling may also occur via direct cell-to-cell contact through membrane-bound protein complexes such as the Notch-Delta system[38] and various bacterial intercytosolic transfer systems.[39] Both types of signal exact transcriptional responses that affect various cooperative processes.[40,41] While synthetic gene circuits have been constructed to exploit QS modules,[3,42,43] more sophisticated circuits need to be developed to produce complex population behaviors.

Third, engineering ecosystems requires formation of defined spatial structure. Natural biofilms are a source of inspiration for spatially organizing microbial populations as an engineering objective.[44] These structures can define intercellular interactions and improve robustness of the consortia to environmental insults such as antibiotics and predation.[45]

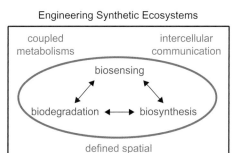

Engineering Synthetic Ecosystems

FIGURE 17.1

Engineering ecology requires the synthesis of metabolism, communication, and spatial architecture to generate synthetic ecosystems that can productively sense, degrade, or produce a myriad of biomolecules of industrial, medical or commercial value.

Spatial organization importantly determines the scale of interactions that can occur between given cells as a result of local proximity. Locality has a critical influence over metabolic exchange and signaling in synthetic communities.[46–48] New microfluidic and microdroplet devices have recently been developed to reproduce interactions between spatially defined communities,[47,49,50] formation of structured biofilms,[51] and cell–cell aggregations,[52] and will continue to play an important experimental role.

TOWARDS SYNTHETIC COMMUNITY ENGINEERING

Advances in synthetic biology are beginning to be applied to multispecies systems with direct real-world applications. While most examples of synthetic ecosystems involve only two or three distinct strains, they represent substantial improvement in our capability to engineer complex microbial interactions.[28] We outline different applications of synthetic microbial communities to highlight their potential in improving areas of biosensing, biosynthesis, and biodegradation, where the capabilities of homogeneous populations of genetically identical cells are insufficient (Fig. 17.2).

Biosensing

Abilities to sense diverse environmental signals and actuate appropriate responses are necessary and key features of engineered microbial communities. For example, autodetection of changes among networks of gut microbes in the intestine would allow for real-time monitoring and pinpointed responses to alarming events such as infections or toxins. Such capabilities would present a marked improvement over current monitoring strategies, where symptoms are only recognized once an infection has fully developed and treatment requires indiscriminant depletion of the native community using antibiotics. These population-level behaviors are only now being demonstrated using synthetic communication circuits with quorum sensing modules. Nonpathogenic *Escherichia coli* has been engineered to recognize specific QS molecules diffusing from virulent strains of *Vibrio*[53] or *Pseudomonas*.[54] Upon detection, pathogen-specific antimicrobial proteins or compounds are released, resulting in 99% reduction in the pathogen load.[54]

Synthetic consortia can also be designed to detect and respond to other compounds to regulate programmed behaviors. Consortia growth rate and relative abundances of different members can be tuned in response to the environment.[36,55] These approaches can be used to engineer biofilms to alter its physical architecture and membership composition, to optimize bioprocesses that rely on spatially associated communities.[56] Engineered communities can also be used not only to microscopically sense low-level metabolites, but also to amplify the signal for macroscopic detection. Building on an oscillatory fluorescence-generating

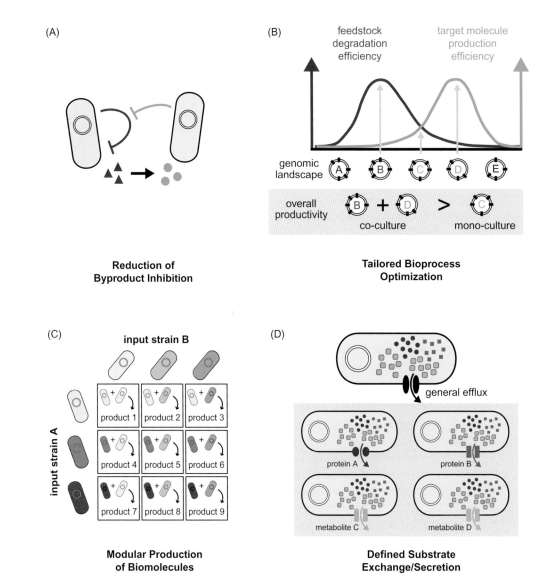

(A)

(B)

feedstock degradation efficiency

target molecule production efficiency

genomic landscape Ⓐ Ⓑ Ⓒ Ⓓ Ⓔ

overall productivity Ⓑ + Ⓓ > Ⓒ

co-culture   mono-culture

Reduction of Byproduct Inhibition

Tailored Bioprocess Optimization

(C)

input strain B

input strain A

product 1 | product 2 | product 3

product 4 | product 5 | product 6

product 7 | product 8 | product 9

(D)

general efflux

protein A   protein B

metabolite C   metabolite D

Modular Production of Biomolecules

Defined Substrate Exchange/Secretion

FIGURE 17.2

Engineering improvements for synthetic consortia. (A) A synthetic consortium can be designed to reduce byproduct inhibition that accompanies overaccumulation of a toxic intermediate. (B) A coculture bioprocessing strategy can be used where multiple steps are separately optimized in different cells to maximize overall productivity. (C) Modular assembly of synthetic consortia using common intermediate metabolites enables reprogrammable bioprospecting. (D) Production and secretion of multiple metabolites may saturate general cellular machinery. Specific metabolite export across different cell-types may increase productivity.

circuit, Prindle et al. synchronized local and global sensing mechanisms to generate periodically synchronized signals that changed in response to arsenic concentration, thereby generating a macroscopic biosensor using populations of cells.[57] Further demonstrations of synthetic consortia for biosensing applications are needed.

Biodegradation

Microbial communities naturally degrade various compounds into nutrients to sustain metabolism. Synthetic communities are increasingly being used to degrade xenobiotic and recalcitrant compounds. Similarly to the process of synthesis, degradation can be improved through careful engineering of organisms with desired functionalities that may be modular and complementary in physiology, resulting in overall improvement in performance of the community. The sequestration of undesirable compounds or pollutants can be augmented by reducing inhibition of cell growth that results from accumulation of inhibitory intermediates. For example, Li et al. demonstrated that an engineered *E. coli* and *Ochrobactrum* consortia can enhance the degradation of methyl parathion (MP), an insecticide and toxin, through removal of the growth-inhibitory intermediate p-nitrophenol (PNP), resulting in 98% MP removal.[58] Degradation profiles can further be improved using engineered microbes that supplement a consortium with limiting metabolites, such as biotin, thiamine, cobalamine, and siderophores that help facilitate growth and bioconversion. Examples of these communities of cyanobacteria or microalgae with bacteria have been documented to greatly improve degradation of hydrocarbons in oil spills.[59]

Microbial communities play a significant role in digestion and metabolism of foods in the mammalian gut, and its dysfunction may lead to diseases of maldigestion.[60] Perturbation of synthetic communities of gut microbes in gnotobiotic mice using altered diets demonstrated that digestive capabilities may be a viable avenue of forward engineering through synthetic biology.[61] For example, gut communities that additionally carry methanotrophic Archaea can lead to an overall increase in microbial metabolism though removal of inhibitory levels of hydrogen that are otherwise generated. This has the direct effect of increased degradation of nutrient into absorbable nutrition, leading to elevated nutritional uptake by the host which when in excess can cause obesity.[62] Engineering and altering the degradation capacity of gut-associated microbial communities will likely be an important avenue to develop for synthetic microbial ecosystems.

Biosynthesis

Synthesis of new compounds, or existing ones using safer and better approaches, is critically needed — a task well-suited for engineered consortia. More efficient use of otherwise waste feedstocks as input materials into fermentation bioreactors is a highly desirable objective. Many of these materials, such as cellulosic biomass, are complex feedstocks that are not well-suited for current bioproduction pipelines. Use of engineered communities[63] presents a better solution than current monoculture production approaches, as excretion of different cellulases from different strains can improve degradation of complex cellulose polymers into smaller monomers. Additionally, cells that optimally excrete these cellulases may not be well-suited for bioproduction due to inherent metabolic costs. Shin et al. demonstrated the advantage of using a synthetic consortium with a divided labor structure for ethanol production from hemicellulosic feedstock.[64] Two *E. coli* strains were cocultured; one genetically optimized for cellulase production and excretion, and the other for utilization of the digested substrate for conversion to ethanol. Ethanol production reached 70% of theoretical yield in the coculture compared to 26−28% with single strains. The use of orthologous secretion systems can further improve specificity of secretion, and improve efficiency of secretion by limiting saturation through dividing different processes across multiple strains.[65]

Medical applications of engineered microbes include in situ biosynthesis and excretion of therapeutic compounds such as cytokines and immunomodulating proteins at the site of injury.[66] Introduction of nonpathogenic engineered *Lactococcus lactis* that can produce interleukin 10 in the mouse gut ameliorated autoinflammatory diseases such as colitis, Crohn's, and inflammatory bowel disease.[67] Improving the engineering of complex microbial ecosystems to stably maintain desired strains in challenging environments such as the human gastrointestinal tract will increase the longevity and effectiveness of these therapies.

Microbial consortia with modular architecture may enable more programmable reconfiguration of biosynthesis objectives and optimization conditions. Degradation strains and production strains can be combined modularly using shared common intermediate metabolites to generate useful products such as biofuels or biomaterials. Metabolic interactions, inhibitory or beneficial, across the microbial networks must be carefully engineered.[6,27,68,69] For example, byproduct inhibition occurs when growth or productivity of a strain is impaired by the compound it produces.[70] For example, *Actinotalea fermentan* can efficiently process cellobiose feedstocks (switchgrass, corn stover, bagasse, etc.) into acetate, but its growth rate is significantly impaired by even moderate levels of acetate. Bayer et al. demonstrate that acetate byproduct inhibition of *A. fermentan* can be removed by addition of an engineered *Saccharomyces cerevisiae* strain which utilizes acetate for growth.[71] The yeast is then engineered to produce methyl halides, which is a useful biofuel precursor. A 12 000-fold improvement was achieved using this approach compared to levels from single-culture bioreactors.

Synthetic consortia additionally enable membrane-bound enzyme complexes, such as those for engineering H_2 production in *E. coli*, to be maximally utilized.[72] Integration of strain into an engineered consortium through metabolic cross-feeding is a more modular approach that allows optimization of partitioned functions such as protein engineering of membrane-bound complexes. Similarly, membrane-associated extracellular mini-cellulosomes that spatially colocalize can improve reaction rate and efficiency to improve performance of synthetic consortia.[73] A four-member cellulosome-generating yeast consortia was recently demonstrated for ethanol production, reaching 87% of theoretical yield – a three-fold increase over a monoculture strain that expressed all four enzymes.[74] Thus, utilizing synthetic consortia for modular and programmable biosynthesis of useful compounds remains very promising.

FUTURE PROSPECTS FOR SYNTHETIC ECOSYSTEMS

Here we have discussed many of the interesting applications that have recently emerged from the area of synthetic microbial ecosystems. In most cases, these represent the low hanging fruits of this area – systems involving only a handful of strains with relatively simple interactions. However, these advances can be further extended using new tools in synthetic biology, mathematical modeling, and molecular biotechnology. Future developments in this area have the potential to transform fields of medicine, bioproduction, bioprocessing, and environmental engineering. Precise manipulation and multiplexed control of community composition, capabilities, and dynamics will generate a suite of reconfigurable cellular modules that can meet the myriad of health and environmental challenges that we face in the near future. Additional insights that reveal how these engineered systems respond to evolutionary pressures in natural environments will guide their proper use in socially responsible ways.

Acknowledgments

H.H.W. acknowledges funding from the National Institutes of Health Director's Early Independence Award (Grant 1DP5OD009172-01). M.T.M. is supported through funding from the Department of Energy Genomes to Life Center (Grant DE-FG02-02ER63445).

References

1. Khalil AS, Collins JJ. Synthetic biology: applications come of age. *Nat Rev Genet.* 2010;11(5):367–379.

2. Lu TK, Khalil AS, Collins JJ. Next-generation synthetic gene networks. *Nat Biotechnol.* 2009;27(12):1139–1150.

3. Basu S, et al. A synthetic multicellular system for programmed pattern formation. *Nature.* 2005;434 (7037):1130–1134.

4. Schink B. Synergistic interactions in the microbial world. *Antonie Van Leeuwenhoek.* 2002;81(1–4):257–261.

5. Shong J, Jimenez Diaz MR, Collins CH. Towards synthetic microbial consortia for bioprocessing. *Curr Opin Biotechnol.* 2012;23(5):798–802.

6. Kato S, et al. Effective cellulose degradation by a mixed-culture system composed of a cellulolytic Clostridium and aerobic non-cellulolytic bacteria. *FEMS Microbiol Ecol.* 2004;51(1):133–142.

7. Eisenhauer N, Scheu S, Jousset A. Bacterial diversity stabilizes community productivity. *PLoS One.* 2012;7(3): e34517.

8. Wagner M, Loy A. Bacterial community composition and function in sewage treatment systems. *Curr Opin Biotechnol.* 2002;13(3):218−227.

9. Ishii K, Takii S. Comparison of microbial communities in four different composting processes as evaluated by denaturing gradient gel electrophoresis analysis. *J Appl Microbiol.* 2003;95(1):109−119.

10. Kan J, et al. Current production by bacterial communities in microbial fuel cells enriched from wastewater sludge with different electron donors. *Environ Sci Technol.* 2011;45(3):1139−1146.

11. Sekiguchi Y, Kamagata Y, Harada H. Recent advances in methane fermentation technology. *Curr Opin Biotechnol.* 2001;12(3):277−282.

12. Sleight SC, et al. Designing and engineering evolutionary robust genetic circuits. *J Biol Eng.* 2010;4:12.

13. Swenson W, Wilson DS, Elias R. Artificial ecosystem selection. *Proc Natl Acad Sci USA.* 2000;97(16):9110−9114.

14. Gihring TM, et al. A limited microbial consortium is responsible for extended bioreduction of uranium in a contaminated aquifer. *Appl Environ Microbiol.* 2011;77(17):5955−5965.

15. Hollister EB, et al. Structure and dynamics of the microbial communities underlying the carboxylate platform for biofuel production. *Appl Microbiol Biotechnol.* 2010;88(1):389−399.

16. Klitgord N, Segre D. Ecosystems biology of microbial metabolism. *Curr Opin Biotechnol.* 2011;22 (4):541−546.

17. Stolyar S, et al. Metabolic modeling of a mutualistic microbial community. *Mol Syst Biol.* 2007;3:92.

18. Taffs R, et al. In silico approaches to study mass and energy flows in microbial consortia: a syntrophic case study. *BMC Syst Biol.* 2009;3:114.

19. Zomorrodi AR, Maranas CD. OptCom: a multi-level optimization framework for the metabolic modeling and analysis of microbial communities. *PLoS Comput Biol.* 2012;8(2):e1002363.

20. Zhuang K, et al. Genome-scale dynamic modeling of the competition between Rhodoferax and Geobacter in anoxic subsurface environments. *ISME J.* 2011;5(2):305−316.

21. Sharan SK, et al. Recombineering: a homologous recombination-based method of genetic engineering. *Nat Protoc.* 2009;4(2):206−223.

22. Wang HH, et al. Programming cells by multiplex genome engineering and accelerated evolution. *Nature.* 2009;460(7257):894−898.

23. Wang HH, Church GM. Multiplexed genome engineering and genotyping methods applications for synthetic biology and metabolic engineering. *Methods Enzymol.* 2011;498:409−426.

24. Wang H, Kim K, Cong L, Jeong J, Bang D, Church GM. Genome-scale promoter engineering by coselection MAGE. *Nat Methods.* 2012;9(6):591−593.

25. Isaacs FJ, et al. Precise manipulation of chromosomes in vivo enables genome-wide codon replacement. *Science.* 2011;333(6040):348−353.

26. Kosuri S, et al. Scalable gene synthesis by selective amplification of DNA pools from high-fidelity microchips. *Nat Biotechnol.* 2010;28(12):1295−1299.

27. Kato S, et al. Network relationships of bacteria in a stable mixed culture. *Microb Ecol.* 2008;56 (3):403−411.

28. Jousset A, et al. Intraspecific genotypic richness and relatedness predict the invasibility of microbial communities. *ISME J.* 2011;5(7):1108−1114.

29. Keasling JD. Manufacturing molecules through metabolic engineering. *Science.* 2010;330(6009):1355−1358.

30. Na D, Kim TY, Lee SY. Construction and optimization of synthetic pathways in metabolic engineering. *Curr Opin Microbiol.* 2010;13(3):363−370.

31. Pernthaler A, et al. Diverse syntrophic partnerships from deep-sea methane vents revealed by direct cell capture and metagenomics. *Proc Natl Acad Sci USA.* 2008;105(19):7052−7057.

32. Ma Q, et al. Integrated proteomic and metabolomic analysis of an artificial microbial community for two-step production of vitamin C. *PLoS One.* 2011;6(10):e26108.

33. Ren Q, Chen K, Paulsen IT. TransportDB: a comprehensive database resource for cytoplasmic membrane transport systems and outer membrane channels. *Nucleic Acids Res.* 2007;35(Database issue):D274−D279.

34. Dubey GP, Ben-Yehuda S. Intercellular nanotubes mediate bacterial communication. *Cell.* 2011;144 (4):590−600.

35. Shou W, Ram S, Vilar JM. Synthetic cooperation in engineered yeast populations. *Proc Natl Acad Sci USA.* 2007;104(6):1877−1882.

36. Kerner A, et al. A programmable *Escherichia coli* consortium via tunable symbiosis. *PLoS One.* 2012;7(3): e34032.

323

37. Bassler BL, Losick R. Bacterially speaking. *Cell*. 2006;125(2):237−246.

38. Matsuda M, et al. Synthetic signal propagation through direct cell−cell interaction. *Sci Signal*. 2012;5(220):ra31.

39. Hayes CS, Aoki SK, Low DA. Bacterial contact-dependent delivery systems. *Annu Rev Genet*. 2010;44:71−90.

40. Kiers ET, et al. Reciprocal rewards stabilize cooperation in the mycorrhizal symbiosis. *Science*. 2011;333 (6044):880−882.

41. Weber W, Daoud-El Baba M, Fussenegger M. Synthetic ecosystems based on airborne inter- and intrakingdom communication. *Proc Natl Acad Sci USA*. 2007;104(25):10435−10440.

42. Hooshangi S, Bentley WE. From unicellular properties to multicellular behavior: bacteria quorum sensing circuitry and applications. *Curr Opin Biotechnol*. 2008;19(6):550−555.

43. Tabor JJ, et al. A synthetic genetic edge detection program. *Cell*. 2009;137(7):1272−1281.

44. Flemming HC, Wingender J. The biofilm matrix. *Nat Rev Microbiol*. 2010;8(9):623−633.

45. Elias S, Banin E. Multi-species biofilms: living with friendly neighbors. *FEMS Microbiol Rev*. 2012;doi: 10.1111/j.1574-6976.2012.00325.x.

46. Song H, et al. Spatiotemporal modulation of biodiversity in a synthetic chemical-mediated ecosystem. *Nat Chem Biol*. 2009;5(12):929−935.

47. Kim HJ, et al. Defined spatial structure stabilizes a synthetic multispecies bacterial community. *Proc Natl Acad Sci USA*. 2008;105(47):18188−18193.

48. Brenner K, et al. Engineered bidirectional communication mediates a consensus in a microbial biofilm consortium. *Proc Natl Acad Sci USA*. 2007;104(44):17300−17304.

49. Moffitt JR, Lee JB, Cluzel P. The single-cell chemostat: an agarose-based, microfluidic device for high-throughput, single-cell studies of bacteria and bacterial communities. *Lab Chip*. 2012;12(8):1487−1494.

50. Park J, et al. Microdroplet-enabled highly parallel co-cultivation of microbial communities. *PLoS One*. 2011;6 (2):e17019.

51. Stubblefield BA, et al. Constructing multispecies biofilms with defined compositions by sequential deposition of bacteria. *Appl Microbiol Biotechnol*. 2010;86(6):1941−1946.

52. Brenner K, Arnold FH. Self-organization, layered structure, and aggregation enhance persistence of a synthetic biofilm consortium. *PLoS One*. 2011;6(2):e16791.

53. Duan F, March JC. Engineered bacterial communication prevents *Vibrio cholerae* virulence in an infant mouse model. *Proc Natl Acad Sci USA*. 2010;107(25):11260−11264.

54. Saeidi N, et al. Engineering microbes to sense and eradicate *Pseudomonas aeruginosa*, a human pathogen. *Mol Syst Biol*. 2011;7:521.

55. Hu B, et al. An environment-sensitive synthetic microbial ecosystem. *PLoS One*. 2010;5(5):e10619.

56. Hong SH, et al. Synthetic quorum-sensing circuit to control consortial biofilm formation and dispersal in a microfluidic device. *Nat Commun*. 2012;3:613.

57. Prindle A, et al. A sensing array of radically coupled genetic 'biopixels.' *Nature*. 2012;481(7379):39−44.

58. Li L, et al. Removal of methyl parathion from artificial off-gas using a bioreactor containing a constructed microbial consortium. *Environ Sci Technol*. 2008;42(6):2136−2141.

59. Tang X, et al. Construction of an artificial microalgal−bacterial consortium that efficiently degrades crude oil. *J Hazard Mater*. 2010;181(1-3):1158−1162.

60. Ley RE, et al. Obesity alters gut microbial ecology. *Proc Natl Acad Sci USA*. 2005;102(31):11070−11075.

61. Faith JJ, et al. Predicting a human gut microbiota's response to diet in gnotobiotic mice. *Science*. 2011;333 (6038):101−104.

62. Samuel BS, Gordon JI. A humanized gnotobiotic mouse model of host−archaeal−bacterial mutualism. *Proc Natl Acad Sci USA*. 2006;103(26):10011−10016.

63. Zhang Q, et al. Enhancement of methane production from cassava residues by biological pretreatment using a constructed microbial consortium. *Bioresour Technol*. 2011;102(19):8899−8906.

64. Shin HD, et al. *Escherichia coli* binary culture engineered for direct fermentation of hemicellulose to a biofuel. *Appl Environ Microbiol*. 2010;76(24):8150−8159.

65. Eiteman MA, Lee SA, Altman E. A co-fermentation strategy to consume sugar mixtures effectively. *J Biol Eng*. 2008;2:3.

66. Wells JM, Mercenier A. Mucosal delivery of therapeutic and prophylactic molecules using lactic acid bacteria. *Nat Rev Microbiol*. 2008;6(5):349−362.

67. Steidler L, Rottiers P, Coulie B. Actobiotics as a novel method for cytokine delivery. *Ann NY Acad Sci*. 2009;1182:135−145.

68. Kato S, et al. Stable coexistence of five bacterial strains as a cellulose-degrading community. *Appl Environ Microbiol*. 2005;71(11):7099−7106.

69. Xue F, et al. A new strategy for lipid production by mix cultivation of *Spirulina platensis* and *Rhodotorula glutinis*. *Appl Biochem Biotechnol.* 2010;160(2):498–503.

70. Bizukojc M, et al. Metabolic modelling of syntrophic-like growth of a 1,3-propanediol producer, *Clostridium butyricum*, and a methanogenic archeon, *Methanosarcina mazei*, under anaerobic conditions. *Bioprocess Biosyst Eng.* 2010;33(4):507–523.

71. Bayer TS, et al. Synthesis of methyl halides from biomass using engineered microbes. *J Am Chem Soc.* 2009;131(18):6508–6515.

72. Waks Z, Silver PA. Engineering a synthetic dual-organism system for hydrogen production. *Appl Environ Microbiol.* 2009;75(7):1867–1875.

73. Tsai SL, Goyal G, Chen W. Surface display of a functional minicellulosome by intracellular complementation using a synthetic yeast consortium and its application to cellulose hydrolysis and ethanol production. *Appl Environ Microbiol.* 2010;76(22):7514–7520.

74. Goyal G, et al. Simultaneous cell growth and ethanol production from cellulose by an engineered yeast consortium displaying a functional mini-cellulosome. *Microb Cell Fact.* 2011;10:89.

CHAPTER 27

Note: Page numbers followed by "*f*", "*t*" and "*b*" refer to figures, tables and boxes, respectively.

333